"To say that [Alexander] Rose's new book, *Empires of the Sky*, is about the *Hindenburg* is to diminish the genius of the narrative Rose has crafted here. . . . It is a love letter to the airship. And even though we know how the story ends—with the airplane winning our business, and the *Hindenburg* going down in flames and Eckener disappearing from our collective imagination—I read with great urgency all the way to the final page, captivated by what might have been and marveling at what humans can accomplish with the help of engineering, physics, facts. . . . Important history and a true narrative—a definitive tale of an incredible time when mere mortals learned to fly."

—Keith O'Brien, *The New York Times*

"Delightful and informative . . . A great story, and Mr. Rose tells it well. . . . Mr. Rose's account of [the *Hindenburg's*] fate, and that of the survivors, vividly explains why the catastrophe seared itself into American memory and why no Zeppelin ever again visited the United States."

—*The Wall Street Journal*

"An obsessive, decades-long struggle between two equally matched people is always fascinating, and especially when the prize they are fighting for is nothing less than the future of flight. We take the airplane's defeat of the Zeppelin for granted, but in the Roaring Twenties and Dark Thirties it was anything but—and now, in a world aiming for carbon neutrality, we might even regret who won. Alexander Rose is a historian with a scintillating prose style and an eye for the insightful, and often amusing, detail. Whereas dirigibles were heavy, ponderous, and full of gas, this book is the precise opposite."

—Andrew Roberts, author of *Leadership in War*

"An exhilarating journey spanning centuries as humanity slowly took to the air, through the decadence of the Roaring Twenties as [Hugo]

ssly sought to out-maneuver one
pectacular final moments of the
of the Zeppelin era."

"*Empires of the Sky* is a fascinating history of the rise and demise of the Zeppelin."

"[An] exhilarating history of the dawn of modern air travel."

"In this comprehensive history, Rose does justice to the engineering of airships and airplanes, and gives a deep sense of the creativity and vision of the men behind the revolutionary concept of human flight."

EMPIRES OF THE SKY

EMPIRES OF THE SKY

ZEPPELINS, AIRPLANES,

AND TWO MEN'S EPIC DUEL

TO RULE THE WORLD

ALEXANDER ROSE

RANDOM HOUSE • NEW YORK

2021 Random House Trade Paperback Edition

Published in the United States by Random House, an imprint
and division of Penguin Random House LLC, New York.

RANDOM HOUSE and the HOUSE colophon are registered
trademarks of Penguin Random House LLC.

Originally published in hardcover in the United States
by Random House, an imprint and division of
Penguin Random House LLC, in 2020.

LIBRARY OF CONGRESS CATALOGING-IN-PUBLICATION DATA

Names: Rose, Alexander, author.
Title: Empires of the sky: zeppelins, airplanes, and two men's epic
duel to rule the world / Alexander Rose.
Description: First edition. | New York: Random House, [2020]
Identifiers: LCCN 2019024115 (print) | LCCN 2019024116 (ebook) |
ISBN 9780812989984 (trade paperback) | ISBN 9780812989991 (ebook)
Subjects: LCSH: Aeronautics—Biography. | Zeppelin, Ferdinand,
Graf von, 1838–1917. | Eckener, Hugo, 1868–1954. | Trippe, J. T.
(Juan Terry), 1899–1981. | Airships—History—20th century. |
Airplanes—History—20th century. | Aeronautics, Commercial—
History—20th century.
Classification: LCC TL539 .R635 2020 (print) |
LCC TL539 (ebook) | DDC 387.7092/2—dc23
LC record available at lccn.loc.gov/2019024115
LC ebook record available at lccn.loc.gov/2019024116

Printed in the United States of America on acid-free paper

randomhousebooks.com

9 8 7 6 5 4 3 2 1

Book design by Simon M. Sullivan

To Rebecca and Edmund, as always, and forever

The students of the problem are divided into two camps or schools, each of which expects flight to be compassed by somewhat different apparatus. These are:

1: AERONAUTS, who believe that success is to come through some sort of balloon, and that the apparatus must be lighter than the air which it displaces.

2: AVIATORS, who point to the birds, believe that the apparatus must be heavier than the air, and hope for success by purely mechanical means.

Curiously enough, there seems to be very little concert of study between these two schools. Each believes the other so wrong as to have no chance of ultimate success.

OCTAVE CHANUTE, *Aerial Navigation* (1891)

Contents

A PENCIL BALANCES UPRIGHT on its unsharpened end. A glass of water, its contents motionless, waits nearby. A tower of playing cards looms.

A steward bumps the table as he shimmers past. The pencil falls. The water ripples. The cards tumble.

He offers his sincere apologies and hurriedly tidies up.

There are scores of guests aboard today, most dressed immaculately in dark suits and sober ties, the uniform of the prewar American elite class. According to the newspapers, their cumulative net worth is more than a billion dollars—that's in 1936 dollars, when a billion was real money. With so much cash in human form walking around, their hosts have gone to great lengths to ensure a perfect day.

The list of the Great and the Good seems endless on this, the so-called Millionaires' Flight. As John B. Kennedy, the NBC radio announcer broadcasting live from this midday summit, quips for his listeners: "We've got enough notables . . . to make the Who's Who say what's what."

Among the grandees present is Winthrop Aldrich, chairman of Chase National Bank, the mightiest bank in the world. His nephew Nelson Rockefeller works, perhaps not entirely coincidentally, for Chase, and he's also here today.

On the business side, among many others there are Paul Litchfield (president of Goodyear Tire & Rubber), Medley Whelpley (president of the American Express Bank), and John Hertz (owner of Hertz Drive-Ur-Self System).

The airline industry, in particular, has shown up in force. Colorful Jack

Frye is president of TWA, and Eddie Rickenbacker, legendary World War I fighter ace, captains Eastern Air. Lucius Manning is there, representing the secretive Errett Lobban Cord, the transport tycoon who owns American Airlines.

Federal officials are there, too, nearly all from the Department of Commerce's aeronautics branch, a body dedicated to overseeing America's development of civil aviation. Complementing them is a contingent from the navy, including Admiral William Standley, the chief of naval operations, and Rear Admiral Arthur Cook, who heads its Bureau of Aeronautics.

Luncheon is served. In this dining room, furnished in hyper-modern style, barely a corner can be seen; all is graceful curves and vibrant colors. The sleek tables and chairs are made of chromed aluminum tubing, the fittings of futuristic plastic. There is none of the heavy wood and garish brass, the old-fashioned bric-a-brac and busy chintz so characteristic of the passé Edwardian era.

Indian Swallow Nest soup, cold Rhine salmon, and potato salad precede the main course of tenderloin steak in goose liver sauce, Chateau potatoes, and Beans *à la Princesse* (accompanied by a cheeky 1934 Piesporter Goldtröpfchen—a Riesling), followed by a Carmen salad and iced California melon, washed down with a sparkling Feist Brut (1928). Strong Turkish coffee, light Austrian pastries, and fine French liqueurs finish off the meal.

As the staff clear the tables, the guests walk over to a slanted bank of picture windows running the length of the room.

About six hundred feet below spreads a delightful panorama of coastal New England in the fall. To one side, a cobalt sea peppered with yachts; to the other, an emerald coastline edged by endless reddening forest, interrupted only by the occasional town.

It is a grand view even for such exalted company.

But there is hardly time to fix one's gaze. The passing cavalcade is changing every moment, after all.

Soon they're over Boston. In the streets, darkened by a great shadow, the cars stop and the thronging crowds pause in awe and astonishment at the strange object passing directly overhead. Then they wave and cheer.

The men sailing above return the compliment, casting off blue-blooded reserve to exult in the adulation and envy of their audience.

Watching over everyone is Herr Hitler, as the newspapers politely refer to him, whose portrait presides sternly over the room. Few find it overbearing: He did an admirable job, after all, hosting the Berlin Olympics a few months ago. Apart from the brief unpleasantness of 1917–18, Germany and America are friends, always will be, and there is every reason to believe, especially once this clubby, boozy afternoon is over, that Berlin and Washington will deepen their relationship.

Playing gracious hosts are the Germans. They have sent no less a personage than Hans Luther, currently their ambassador to Washington and the former chancellor of Germany and president of the Reichsbank. He's accompanied by a couple of army and navy attachés, who chat about military matters with the American admirals.

But the exclusive focus of attention, the real reason why all are gathered here today, is one German in particular—Dr. Hugo Eckener. In his late sixties, he is buzzcutted and goateed, and as German as one can be. He also happens to be one of the most famous men in the world.

Eckener lacks the privilege, wealth, or station of the others assembled, but his ambition and audacity are legendary.

He is the greatest airshipman of all time. With more flying hours under his belt than anyone else alive or dead, he is master of the aery realm. He is the anointed heir of the father of the airship, Count von Zeppelin, and just as it was Joshua, not Moses, who led the Israelites into the Promised Land, it is Eckener who has surpassed his late mentor's achievements by conceiving the greatest and grandest airship of them all, the most marvelous technological and aeronautical wonder of the age—the very vessel, in fact, in which the titans are soaring.

Hindenburg.

The ultimate transoceanic cruiser. Painted a metallic silver to better glint in the sun, the *Hindenburg* can fly at 84 miles per hour and is capable, or so it is said, of traveling ten thousand miles—enough to go from New York to Berlin and back with a couple of thousand to spare—in a single hop.

It is beyond safe. Since 1912, Eckener's civilian airships have flown 48,778 passengers 1,193,501 miles, over 20,877 hours of flight time, with neither a fatality nor even a serious injury. Considering that in this same year of 1936 alone, 36,126 Americans will die in car crashes and 305 airplane passengers in 1,739 accidents, the Zeppelins' record is second to none.

From a passenger standpoint, the *Hindenburg* is wondrous. It typically makes the transatlantic run between Germany and New York in just over two days, whereas the world's fastest cruise ships need five, and the slower ones up to ten. The *Hindenburg* is so light that it can be docked with two ropes, so nimble that it can revolve on its own axis, and so stable that the passengers can entertain one another with little party tricks involving pencils, water, and cards.

Unlike on aircraft, where stomach-churning yaws and pitches are common, no one ever feels nauseous aboard the *Hindenburg*. During liftoff, there is simply no sense of acceleration, motion, or vibration, and while on airplanes the noise level from the engines can be deafening, on an airship all that can be heard is a dull hum. They're quieter than an unbusy office.

The sheer immensity of the *Hindenburg* astounds. At 805 feet, it is significantly longer than the Golden Gate Bridge's towers are tall, and one could stand a thirteen-story building within its cathedral of elvish latticework delicately lacing together fourteen miles of girders with eighty of steel wire.

Eckener knows his guests are impressed. He needs them to be. Every single one of them has been invited for a reason. Each man represents a potential interest that needs placating, persuading, seducing, so that Eckener can at last fulfill the grand dream behind today's congress: forging an international partnership that will create an airship armada covering Europe, the Americas, and Asia.

Without their participation, the *Hindenburg* may be the last of its kind to roam the skies. Eckener would never admit it, of course, but the airship is falling ever further behind.

The airplane may be a beast that guzzles fuel and carries a relative handful of passengers, and it is far from achieving the Holy Grail of transport—a nonstop transatlantic trip—but sheer quantity has its own quality. In the United States, just in 1936, the airlines transported 931,683 passengers, and its factories produced 3,010 civilian aircraft. There is only one *Hindenburg*, built at colossal expense, and it carries around fifty passengers.

Success today will nevertheless turn the tide against the airplane.

To win, Eckener needs to beat one man. For years they have been bitter rivals, but today is the first time they have met. So far they've said little to each other, but each is watching the other carefully, scanning for weakness.

This other man, young, pudgy, jowly, is the head of the only American airline with major international routes—and lethal when it comes to getting his own way.

Juan Terry Trippe controls Pan American Airways, the greatest airline in the world. While his airplanes can't get across the moat of the Atlantic, they soon will, and Trippe is determined to consign the *Hindenburg* and its like to the dustbin of history. He shall not rest until Pan American's silvery squadrons radiate outward to the very ends of the earth.

Such is the setting on this day of the Millionaires' Flight, as the *Hindenburg* has long since rounded Boston and is moving like a speeding cloud toward its final destination, the naval air base at Lakehurst, New Jersey.

At 5:17 P.M., the *Hindenburg* settles into its quiet, motionless hover as its four mighty Daimler engines cut thrust. From the control car, Captain Ernst Lehmann dumps water ballast and valves hydrogen to ease the airship gently down. Crewmen release two ropes, each four hundred feet long and two inches thick, which snake down through the air.

Below, the ground crew pick them up and heave the *Hindenburg* toward the mooring mast. Ever so slowly, the airship is winched down to the ground. Its giant body subsides and the groundsmen bind the airship down with ropes, hooks, and stakes, as the Lilliputians did Gulliver.

Inside, the tipsy guests lament the finale of their worthy expedition, what Kennedy calls "the end of a perfect day in the air."

And then, just like that, it's over.

The gangway is lowered, and the millionaires, grasping their hats in the wind, descend the steps. Yet another flawless performance by the world's most majestic flying machine.

Trippe, for his part, remains tight-lipped. Eckener's coup is his humiliation. He suspects the deal is done: America will soon be in the airship business and Zeppelin will duel with Pan American for mastery of the coming air empire.

This evening, Dr. Hugo Eckener will soar home in triumph aboard the *Hindenburg*, its future secured, amid the parting fog and clearing skies.[1]

His airship has exactly seven months left to live, but it was born some seventy-three years earlier, when a young German count visiting America asked for a room for the night.

EMPIRES OF THE SKY

1. *The Aeronaut*

ON AUGUST 17, 1863, America was engulfed in civil war. The battles of Gettysburg and Vicksburg had been fought just six weeks earlier, but Mr. Belote, the manager of the International Hotel, the finest in the city of Saint Paul, Minnesota, didn't care about the Blue and the Gray. That night, he was more concerned about the brown—the brown mud, that is—being tracked into his establishment by the hollow-cheeked, rough-whiskered frontiersman claiming to be a "Graf von Zeppelin."

He certainly didn't look like one of the fancy European aristocrats Mr. Belote had read about. Yet he sounded courtly, even if he spoke English, haltingly, with a strong German accent. Upon closer inspection, his clothes, too, were tailored, though torn and ragged and not altogether suited to the backwoods; he was evidently a man who purchased rather than shot what he wore. Still, at the International Hotel, they didn't let rooms to riff-raff or charlatans.[1]

The man, sensing the manager's reluctance, explained that he had spent the last three weeks roaming the wilderness.[2] Fueled by the romantic fantasies of deerslayers exploring primeval American forests he had picked up from reading too many James Fenimore Cooper novels, he had elected to travel along an abandoned fur-trade trail. It was a wonder he hadn't died. Having quickly run out of food and ammunition and beset by mosquitoes and heat, he had been saved by some Chippewa Indians who showed him how to hunt ducks, build a shelter, and gather eggs.[3]

It had been quite an adventure, but he was ready for a comfortable bed, a bath, and a hearty dinner—and had the money at hand. Once he saw the dollars, Mr. Belote relented: He'd be only too pleased to offer such a *distinguished gentleman* his best accommodations. Due to return east on the next day's train, the man paid for a single night's stay.

The following morning, August 18, woken by a commotion outside, Zeppelin drew the curtains and surveyed the open lot across the street. And there he spied a large silken balloon, gaily painted and patchworked, and fitted with a small wicker basket. He'd heard of these legendary, magical things, of course—everyone knew of them—but never had he encountered one.

Right there and then, Zeppelin decided to postpone his trip back home.

ZEPPELIN WAS INDEED a fancy European aristocrat and not a charlatan, but how he ended up in Saint Paul, Minnesota, is something of a roundabout story.

He could trace his ancestry back to a minor thirteenth-century baron named Heynrikus de Zepelin from Mecklenburg in northern Germany whose kinsmen served as mercenaries in the Swedish, Danish, and Prussian armies that occupied their time ravaging and ravishing their way across Europe. For the next five hundred years, successive Zepelins did little other than demonstrate a prodigious talent for drunkenly gambling away the family's estates, ultimately obliging an impecunious, teenaged Ferdinand Ludwig to roam far south and enter the military service of Duke Frederick of Württemberg in the late eighteenth century.

When all-conquering Napoleon upgraded the duchy of Württemberg into a kingdom in 1806, Ferdinand was promoted to count and changed his name to "Zeppelin" (the Württembergers preferred a double *p*). In 1834, his son Friedrich did very well, marrying Amélie Macaire d'Hogguer, the daughter of a wealthy Franco-Swiss cotton manufacturer, and Ferdinand—our count—came along four years later.

He was born into a world of international nobility, where a title served as passport to the elites in Saint Petersburg, Vienna, London, and Paris—or even, in a pinch, Berlin, a backwater. Following the family's martial tradition, Zeppelin entered the Royal Army College at Ludwigsburg in October 1855 and emerged as a lieutenant with one of Württemberg's most swagger regiments, the 8th, based in Stuttgart, the kingdom's capital, in September 1858. During the Franco-Austrian War of 1859 (Württemberg was an Austrian ally), he saw no action while serving on the staff of the quartermaster-general as a specialist in topography and logistics.

That Zeppelin, a curious mix of the unconventional and the traditional, was even in the quartermaster-general's office rather than serving on the higher-status front lines marked him as quirky. Since boyhood, Zeppelin had been fascinated by mechanics, by making machinery work, by practical invention. Before being admitted to the Royal Army College, Zeppelin had attended the prestigious polytechnic school in Stuttgart. Such institutions were in the vanguard of imparting a technical, scientific, and engineering education to smart middle-class boys and ambitious working-class lads. Rich young nobles like Zeppelin were few and far between. Still stranger, during his time with the quartermaster-general, Zeppelin took temporary leave to enroll at the University of Tübingen to study (though he did not take a degree) mechanical sciences—again, a field rather déclassé for a man of his pedigree.

It was a fashion of the era for young officers to tour the armies of foreign nations and report on their armaments and tactics; for those of Zeppelin's breeding, of course, these semi-official visits also allowed them to forge connections with their upper-class counterparts. In 1861–62, the young count visited Vienna, where he was introduced to the Habsburg emperor Franz Joseph I and watched army exercises. Then he was off to Trieste, to visit the fleet, and then the well-known fleshpots of Genoa, Marseille, and Paris, to visit the girls (as he explained to his morally upright, purse-strings-holding father, "In order to know the different people better, I have had to devote some of my time to women"). At Compiègne in northern France he was a guest of Emperor Napoleon III, whose mother had, small world, once been the Zeppelins' neighbor. Later he traveled to Belgium and Denmark before going to England, where he hobnobbed at the Army and Navy Club and the Athenaeum before being invited to watch the Grenadier Guards go through their paces.[4]

America, then enduring its Civil War, beckoned. How could one miss the clash of those gargantuan armies clanking through the Virginia hinterland? Needing permission from his king for yet another furlough, Zeppelin explained that "the Americans are especially inventive in the adaptation of technical developments for military purposes" and pledged to seek information useful for the Württemberg army.[5]

That was pro forma, of course. His real hope, as he confided to his sister, was that, as he had missed all the fun during the Franco-Austrian War, combat "might be revealed to me in its bloody truth and that the phantom [of experiencing real fighting], before which I had hitherto quailed, might become a living reality."

To his father, who was unenthusiastic about the idea, he laid out a rather more elevated motive. He wished to discover the extraordinary vibrancy of American democracy, he said, but Zeppelin senior nevertheless forbade him, saying that the existence of slavery and the fact that commoners could vote—he was unclear as to which was worse—"exclude[d] them from playing a worthy part in civilization." His son persisted, and in the end the paterfamilias gave way, as Zeppelin knew he would. In April 1863, Zeppelin boarded the Cunard ship *Australasia* for the long voyage to America.[6]

After docking in New York on May 6, Zeppelin traveled to Washington, D.C., checking in to the posh Willard Hotel near the White House. His title, as usual, opened doors—even in the great republic. (Zeppelin noticed that "America is definitely a land of contrasts. Everything aristocratic is in opposition to its fundamental ideas, yet nowhere is so much fuss made about a simple traveling count.") The Prussian ambassador, Baron von Gerolt, introduced him to Secretary of the Treasury Salmon P. Chase, who in turn arranged an audience with President Abraham Lincoln, who took time out of a busy day running a war to meet with an obscure junior officer from a small faraway kingdom.

Lincoln was unlike anyone else Zeppelin had ever met in his limited social circle. When the count turned up, dressed to the nines in the traditional frock coat and top hat, he was surprised by the president's utter absence of pretense. When Zeppelin entered the room, "a very tall spare figure with a large head and long untidy hair and beard, exceptionally prominent cheek-bones, but wise and kindly eyes" rose like a specter from behind the desk. When Zeppelin asked for a pass allowing him to travel freely among the Northern armies as an observer, pompously adding that his military credentials included being descended from half a millennium's worth of knights and counts, Lincoln, a commoner born penniless and landless, remarked that he certainly wouldn't hold that against him. A puzzled Zeppelin got his pass.[7]

On May 28, Zeppelin attached himself to the headquarters of the Army of the Potomac, where a friendly German-speaking officer named Captain Frederick Rosencrantz steered him through the inevitable culture clashes. After being mistaken for a major general several times, for instance, Zeppelin cut off his lieutenant's heavy epaulettes and gold-trimmed velvet collar, which so distinguished him from the tousled, down-home look affected by American officers in the field. One night he attended the mess of General Carl Schurz, commanding the XI Corps, and he couldn't believe his ears when Schurz outlined a plan of attack and a mere *captain* declared, "Surely, General, you won't do anything as stupid as that." While everyone laughed, including Schurz, Zeppelin was incredulous that such insolence was tolerated, even encouraged, in any army.

He did get to see the "phantom" of combat, though. Zeppelin volunteered to accompany a detachment to map General Robert E. Lee's positions, only for some Confederate cavalry to hotly pursue them for a short time. Zeppelin found the experience thrilling, but ultimately he wasn't impressed with American soldiering. As he reported, there was "no systematic cooperation, no local patrol work, no enemy intelligence, no general staff, no maps, no corps combining all the different arms, no tactics adapted to local topography—all these shortcomings continue even after several years of war." There was nothing, as far as he could see, that the Yankees could teach Germans about fighting.

And that was why he missed Gettysburg, instead embarking on an arduous trip west to see the famously vast frontier for himself. Like many foreign visitors, he would be amazed by the sheer size of his host country. "After traveling so long [three days] at the speed of an American locomotive over a level countryside," he wrote to his father, "one cannot but feel that this can go on and on and that there is no end to it."[8]

His airships would later sail placidly across such colossal distances, but for the time being he satisfied himself with walking across the street to see the balloonist.

PROFESSOR JOHN STEINER (he sometimes preferred "Captain," an honorary military rank, as his scholarly one was invented) was having another hard day. His first flight had been scheduled the day before for 4 P.M., but at the last minute local gas company officials had informed him that they were rationing his household coal gas because customers were complaining about his monopolizing their supply. Unable to fill the balloon bag, he'd had to cancel the scheduled flights, and today he faced the prospect of disappointing yet more thrill-seekers, if not for lack of gas then for the winds that were picking up. Worse, tomorrow's forecast called for heavy rain in the morning. Saint Paul was looking worryingly like an expensive write-off, a disaster in the making for itinerant aeronauts like Steiner who subsisted hand to mouth.[9]

Steiner's balloon, heroically christened *Hercules*—though the only Herculean thing about it was the effort it took to get it aloft—was a forlorn thing. Normally, by now he'd be astounding crowds eager to accompany him on rides several hundred feet up. On these safe "captive" or "Army" ascensions, as they were known, the balloon would be tied to the ground by rope so as to prevent a "free flight"—the type of untethered flying reserved for professionals like Steiner.

Free flights were inherently risky. It was impossible to judge when, how, and where you would come down. Six years earlier, for instance, Steiner had made his name as a daredevil by attempting to fly across Lake Erie into Canada. Upon encountering a fierce storm, his balloon hit the water and bounced along like a skipping stone. Steiner survived only by jumping out. His balloon, the third he'd wrecked, wasn't so fortunate.[10]

So, when Zeppelin strode toward Steiner on that morning of August 18, the despondent aeronaut only perked up when he realized that here was a fellow German, though at the time "Germany" was merely a convenient geographical expression for a place, jigsawed into dozens of minor kingdoms, grand duchies, princely states, principalities, and free cities, that wouldn't exist until united in 1871.

Steiner, it transpired, was from Bavaria, adjacent to Zeppelin's Württemberg. Aside from this coincidence, Zeppelin and Steiner had little else in common. Steiner had emigrated to America in 1853 at the age of seventeen, and his background was shadowy. He preferred it that way. Even his name was studiedly innocuous. He was a man with no past, but one who had forged, in a manner of speaking, his own future in the New World. A single clear photograph exists of him, taken sometime during the Civil War: Steiner, with the bushy mustache and manly muttonchop whiskers popular at the time, wears a brand-new officer's coat nattily outfitted with store-bought gold epaulettes and brass buttons.

He had volunteered for a civilian position in the Union's nascent Balloon Corps and, experienced aeronauts being few and far between, was given command of the *Eagle* in Cairo, Illinois. Later, he was sent west and in April 1862 was using his miniature tethered balloon to observe Confederate positions and to direct artillery fire. That December, he resigned from the Balloon Corps following a dispute over pay and headed out to make his fortune.[11]

Steiner's time with the Balloon Corps piqued Zeppelin's curiosity. The ostensible reason he was in America, after all, was to report back on military affairs. As he told his father, embroidering Steiner's credentials slightly, "I have made the acquaintance of the famous aeronaut Prof. Steiner, who has invented a new kind of balloon suitable for military reconnaissance."[12]

If there was anything Steiner had right then, it was spare time to chat. The two men talked for some time about the problems of flight. Perhaps Zeppelin related the amusing story of his grandfather, the first count, who in 1811 had allowed a tailor from Ulm named Albrecht Berblinger to fly his rudimentary hang glider across the Danube to impress the king of Württemberg. It ended badly when Berblinger crashed, not fatally (aside from his aeronautical career), into the river.[13]

Steiner, for his part, was accustomed to dealing with customers who treated a balloon ascent as a thrill ride. They asked whether it was dangerous, nervously entered the basket, went up several hundred feet, enjoyed the sensation of seeming weightlessness, gasped at the extraordinary vista laid bare below, and descended a quarter of an hour later. Zeppelin was uncommon in wanting to understand the *technicalities* of the experience.

Steiner was only too delighted to discuss the secrets of flying and its military applications. He pointed out that the reason the Balloon Corps kept its balloons "captive" was that otherwise they would be the sport of the wind. If the currents began to blow the wrong way or too hard, the hapless aeronaut would find himself scudding over hostile territory, crashing into a forest, plummeting helplessly groundward, or soaring high enough to pass out from cold and lack of oxygen.

There was no way, in short, to control free-flying balloons in terms of speed, direction, or altitude, rendering them useless as a reliable means of transport, either for passengers, mail, or cargo. Indeed, it was difficult to keep even a *tethered* balloon stable in the slightest breeze.[14] Balloons held the tantalizing possibility of revolutionizing travel, but their impracticality remained insuperable. It was as if humans had discovered fire but lacked any way to regulate the flame.

Thus was Zeppelin's introduction to the seemingly unattainable fantasia of what had become known as "aerial navigation"—the quest to master the sky and traverse vast distances of the globe by steering a powered, controlled air vehicle.[15]

Steiner believed he had found the miracle solution. In Zeppelin's words, he intended to build a ship of the air that would dispense with the traditional spherical or lightbulb-shaped gasbag and have instead "a very long, thin shape," like a cigar. "Furthermore," explained Zeppelin to his father, "he has added a strong rudder and in that way the balloon is hindered less by wind and it will reach its destination more smoothly and more surely." Steiner boasted that his plan was to return to the army "and make test flights with his newly improved balloon. If those tests bring good results he will go to Europe (Paris first) in two years."[16]

Zeppelin, daring and heedless of risk, instantly suggested that he and Steiner undertake a free flight. Today was out of the question, unfortunately, but why not tomorrow?

THE NEXT MORNING, August 19, 1863, it was raining, but the weather eventually improved. Zeppelin was eager to set off, but Steiner slotted him in for the final flight of the day, when they could sail unhampered by a line of customers waiting their turn.

From the get-go, Steiner was bedeviled by his meager supply of gas. Usually, the *Hercules* could lift around five people, but when the former governor of Minnesota, Alexander Ramsay, and his ten-year-old daughter, Marion, got in for their ride, the politician—discreetly described by an eyewitness as possessing a "specific gravity"—was obliged to hop out of the basket after the balloon sagged beneath the combined weight. Steiner saved Ramsay further embarrassment by taking the adorable Marion up by herself—to much good-natured applause from the local press.

Then, finally, it was Zeppelin's turn. As he recalled in a later newspaper interview, while he waited to go up he "bought all of the spare gas that the Saint Paul gas works would let me have." (When someone was willing to pay so much cash upfront, the company's customers evidently got short shrift.)[17]

He clambered into the basket alongside Steiner. They rose, still tethered to the ground, to an altitude of six or seven hundred feet. Unlike so many other visitors to the aerial realm, Zeppelin was not awestruck by the revelatory panorama surrounding him. Neither then nor at any other time in the course of his long life did he declaim upon a feeling of liberation from the surly bonds of earth or exult in a sensation of kinship with the Lord of Nature. There was not an ounce of heady romance in his stoutly technical mind. He was concerned only with how balloons worked, their utility, and what objectives they might attain. Flight, to Zeppelin, was not proof of the wondrous capacity of mankind but a problem to be solved.

So he approached his first flight with the crabbed perspective of a staff officer of the Topographical and Logistical Department of the Württemberg quartermaster-general. As he dutifully informed his father later that same day, he at once noticed that "the ground is exceptionally fitted for demonstrating the importance of the balloon in military reconnaissance." Saint Paul is situated in a valley; Zeppelin judged that a nearby ridge of hills "form[ed] a very good defensive position against an aggressor." Should the attacker deploy his forces on the other side of that ridge, "there [was] no tower, no elevation high enough" to observe them unless one had a balloon.[18]

This was all very important intelligence if Württemberg ever decided to declare war on Minnesota, but to Zeppelin's bitter disappointment there would be no free flight that day, no cutting of the rope. The gas he had bought turned out to be of such poor quality that Steiner "could not get the bag filled sufficiently to essay a long flight."[19] The duo stayed aloft for a time but eventually had to be winched down.

Zeppelin would not fly again for nearly forty years, and the two men soon lost touch with each other.

The next day, Zeppelin boarded a train and departed, leaving Steiner to attempt a few captive ascents using the last of the gas. The professor-captain was obliged to end his Saint Paul visit early, having lost about $400, and left for Grand Rapids, Michigan, and La Crosse, Wisconsin, doomed forever to lead a peripatetic existence.[20] He would later, perhaps prompted by his travails with the Saint Paul gas works, invent a portable hydrogen-gas generator, but he never built the futuristic ship of the air that would assure him a fortune.[21]

As ZEPPELIN WHILED away the weeks traveling by train through Milwaukee and Chicago, then Baltimore and Philadelphia, until he reached New York, sailing for home on November 19, 1863, his mind nibbled at the problem of aerial navigation, but finding no solution, he put it aside.[22]

Fatherland, and father, called. His country needed him, and Zeppelin senior wanted his son to cease his gallivanting. Zeppelin dutifully joined the king of Württemberg's staff on April 10, 1865, a day after General Robert E. Lee surrendered at Appomattox and brought to an end the ruinous Civil War.[23]

Germany, meanwhile, faced its own civil war. The kingdom of Prussia, by far the mightiest of Württemberg's neighbors, was threatening to assimilate the patchwork of lesser German entities into a *Kleindeutschland*: a "Small Germany" managed by a Prussian hegemon that excluded Habsburg Austria, the ancient protector of the southern states. Zeppelin and most of the others in his circle preferred, however, a Large Germany arrangement in which independent states like Württemberg and Bavaria would continue to enjoy their Austrian alliance. Prussia's intention, he felt, was to provoke a war with its weaker rival in Vienna.[24]

Zeppelin, promoted to captain and named an aide-de-camp to the king in March 1866, observed the calamitous battle of Tauberbischofsheim four months later during the subsequent Austro-Prussian War.[25] He witnessed at first hand the collapse of the Württemberg forces at the hands of the better-trained and better-equipped Prussians. The terms of Austria's surrender were surprisingly lenient: The Habsburgs were to withdraw from the south and to acknowledge Prussian paramountcy in a new "North German Confederation" of twenty-two formerly independent states.

While Württemberg escaped formal annexation into the confederation, the treaty demanded the "Prussianization" of its army.[26] In 1868, Zeppelin, sent by his king to Berlin as part of a stipulated military-exchange program, was attached to the 1st Guards Regiment of the Dragoons, an elite cavalry unit.[27] He found the Prussians rude and arrogant; bored by their interminable conversations about horses, drink, and women (in that order of precedence) in the officers' mess, Zeppelin was relieved to come home in the spring of 1869.

He had other reasons, as well. That May, Zeppelin was introduced to Baroness Isabella von Wolff during a visit to her family's palace (in what is now Latvia) on the occasion of his brother's wedding to her cousin. Zeppelin fell instantly for her. To his father he described Isabella as an "extremely simple but not at all narrow minded" lass who was "clear-headed, with courage and grit[;] a gay, kindly creature, interested and experienced in house-keeping, a pretty doe-like appearance."[28]

The attraction was mutual, and in August they married. As Isabella wrote to her brother, she and Zeppelin were each but half of two: We "live in quiet privacy. No duties and invitations rob my husband from me, and our house is not besieged by strangers. Oh, such a life is indescribable."[29]

Isabella was not as "extremely simple" as Zeppelin had at first imagined. She proved to be a shrewd, clever observer of human nature and advised Zeppelin—who was not—about whom to trust, how to handle critics, and what needed to be done. Her "courage and grit" were undeniable. Over the coming years, Zeppelin relied on her to back him up and to sustain him in his darkest moments.

In their lighter moments, noticed a friend, the pair adopted a "good-natured, teasing tone" in their chats in which they would gently point out the other's small mistakes. He found her quirks amusing (she believed she could cure all sickness with homeopathy), and she tolerated his more exasperating faults (he was terrible with finances).[30]

The outbreak of the Franco-Prussian War in 1870, the last in a series of struggles to unify Germany, allowed Zeppelin to make his mark on military history. He instigated a daring raid on French positions on July 23 in which everyone else in his unit was killed or captured, with only the wounded Zeppelin managing to escape back to his own lines. The exploit gained him the Württemberg Royal Cross, First Class, and got his name in the papers—the *New York Evening Post* commented that it proved "the Germans have got the right stuff in them"—but his Prussian superiors were not as enthused by his unauthorized hotheadedness. This young officer, they noted, was certainly one to watch—but also one to *be* watched.[31]

Their caution was justified. Zeppelin possessed a swollen confidence in his own capabilities, but his true talent was an extraordinary ability to alienate and annoy those above him—or rather, those above him he thought should be below him. In the various spheres of his life, Zeppelin habitually raised the stakes with a frontal assault when withdrawal, taking cover, or a flanking maneuver might have reaped richer dividends. Often, his sheer charisma and indefatigability—and a modicum of luck—forced success, but failure would more than once bring him to the brink of bankruptcy and humiliation.

His raid was a case in point: By conventional military standards, it was unnecessary, foolhardy, and left good men needlessly dead; by Zeppelin's, he had seen an opportunity, made a decision, and acted upon it. Lacking as he did any capacity for second thoughts, *of course* he was right, and anyone who gainsaid his infallibility he regarded as not merely an honest critic but a traitor.

What was always important to Zeppelin was devotion and loyalty, to him and to what would eventually become his life's work: the airship. In later years, he would gather around himself a band of followers equally obsessed with bringing his vision to reality. Along the way, some wavered in their dedication or expressed a mild reluctance to sacrifice their all to the cause, and these quickly found themselves exiled from his tribe, like biblical unfortunates forever condemned to wander the wilderness. He would accept advice only from a very trusted few, and even they, to survive, knew when to give up any hope of changing his mind. Perhaps he could have learned something from that American captain who criticized General Schurz, but he refused to.

In mid-September, Zeppelin was present when the Germans triumphantly began besieging Paris. With the French army in disarray, only the capital still stood defiant. Paris was completely cut off from the world, and the world assumed, as did the besiegers, that its millions of trapped inhabitants would soon submit—or else starve. Parisians were accordingly rationed to small amounts of milk, coffee, bread, and sugar, but this being France, wine was bounteous and the zoo provided much exotic meat, the streets everything else. One upscale menu offered elephant soup, kangaroo stew, roast camel, antelope terrine, wolf in deer sauce, and baked cat with rat garnish.

Even if the crowded restaurants remained brightly lit, the city was otherwise dark and silent. Not a single piece of news, not a letter, not a journal could pass through the blockade. At least not until September 23, 1870, when a small balloon named the *Neptune* shot skyward and sailed over the enemy lines. Its pilot, Jules Duruof, said he could hear the crackle of the soldiers' muskets firing at him from below. The *Neptune* carried a sack containing three thousand letters, one of which Duruof brought to the offices of the *Times* of London after he landed twenty miles outside Paris.

Over the following week, three more balloons would be sent, together bringing another twenty-five thousand letters to the outside world. Many more followed. The plucky Parisians took to christening their air force with names redolent of French genius—*Lavoisier, La Liberté, Lafayette*—to remind the stolid Prussians of their lack of it.[32]

Even so, it was just a matter of time before the city fell, as it did at the end of January 1871. The French defeat resulted in a united Germany, with the king of Prussia, Wilhelm I, anointed German emperor and *Oberster Kriegsherr* (Supreme War Lord).

As for Zeppelin, what he witnessed of the balloons only affirmed everything that Steiner had warned him about: There was no disciplining, no controlling, no navigating them. Of the sixty-five launched, nearly all had landed in friendly territory, but they had been scattered all over the place and obviously could never return. Others, like one that wandered over the Irish Sea and another briefly spotted over Bavaria, were never seen again.

To Zeppelin's mind, the French balloons provided food for thought, perhaps, but in the continued absence of a way to navigate and drive them, they amounted to nothing but empty calories. He spent little time mulling the possibilities of lighter-than-air flight in the aftermath of the war.

Instead, he embarked on a rather dull army career in the new Germany. In January 1872, he was appointed to command a squadron of the 15th Schleswig-Holstein Uhlans (light cavalry), and was promoted to the rank of major in November of the following year.[33] All signs pointed to his making general in another twenty-odd years and retiring, fat and sleek, to die on his estates shortly thereafter.

But then his horse stumbled during an exercise and changed his life.

2. The Fever Dream

THE WORRIED COUNTESS spent her days placing ice wraps on Zeppelin's forehead and her nights watching doctors applying leeches to draw out the bad blood. For a week since March 18, 1874, when he'd landed badly after falling from his horse, her husband had been bedridden with shivering fits and a high temperature. He was on the mend now, but Zeppelin had once fallen into delirium and scared her by raving about "flying ships" and "passengers who [flew] through the sky" faster than a train.[1]

The mystery behind these ravings was solved when it turned out that a staff officer had thoughtfully brought him a stack of reading material to pass the time. He'd included a recently published pamphlet written by one of Zeppelin's distant relatives, Heinrich von Stephan, then serving as postmaster of the newly created Deutsches Reichspost (Imperial German Post Office).

In A World Postal System and Airship Travel, Stephan conjured up an entrancing vision of an era of global communication and trade based on a German-built network of airships. "Providence," he inspiringly concluded, "has surrounded the entire earth with navigable air. This vast ocean of air still lies empty today and wasted, and is not yet used for human transportation."

The electrifying impact the pamphlet had on Zeppelin can be seen in the four dense pages he feverishly scribbled on the night of 25–26 March, headlined "Thoughts about an Airship."[2]

• • • • •

ZEPPELIN'S VISION, BASED on what he'd read in Stephan's pamphlet, was ambitious in both scope and physical size. His dream airship would be gigantic: some 706,200 cubic feet of hydrogen to fill a gasbag 196 feet long and 40 wide (Steiner's *Hercules* had been about 40,000 cubic feet). He didn't bother speculating how he would power such a colossus, vaguely alluding instead to the "forward motion of the machine" produced by a "suitable prime mover" and leaving it at that.

The factor that solely concerned Zeppelin was that of maintaining control over a stable, steerable airship—the key to aerial navigation. To this end, he spoke of adding "planes" or "wings"—large horizontal rudders that could swivel upward or downward, like a Venetian blind—on the sides either to allow the airship to gain height as air flowed past them or to help keep it on an even keel.

As for the structure of the airship, he dismissed the idea of a single huge gasbag. In a cigar-shaped object, if the airship's nose tilted upward the hydrogen would naturally pool in the tip, leading to loss of control. To hold the gas in place, he proposed including eighteen independent "gas cells," or sealed bags of hydrogen, within the outer envelope, or skin. In this manner, even if a few of the gas cells were punctured, the airship should be able to stay aloft. A dangerous flaw with one-big-bag balloons was that if the outer envelope tore or leaked, it crumpled and sagged as gas escaped, leading to a sudden, violent descent.[3]

Over the next several years, Zeppelin pondered the practical problems of his bold and imaginative vision, but the army remained his more prosaic concern. A move to Ulm to join the 2nd Württemberg Dragoons, where the staff duties were not overly onerous, gave him time to sketch out a few more ideas. In his diary for April 4, 1875, Zeppelin wrote that the airship would carry large quantities of premium-priced mail and cargo to fund its voyages. For additional income, Zeppelin planned to build cabins for twenty passengers, who would pay substantial amounts to travel aboard this wonder of the world above mountains and lakes, across trainless tundras, and to distant continents.[4]

On November 29, 1877, he wondered whether ascent and descent could be regulated "by two [propellers] on a vertical axis? The wings [planes] could then be dispensed with." Ultimately, he dispensed with the whole idea of this hybrid airship-helicopter. Seven months later, on June 9, 1878, Zeppelin had moved on to thinking about the fabric of the balloon envelope, which he felt should be "of Chinese silk, very light and, if varnished, almost entirely gasproof."[5]

And then this burst of activity stopped. His promotion to lieutenant colonel and the birth of his only child, Hella, the following year had brought increasing army and family responsibilities.

Notwithstanding the break, what emerges from these jottings is that Zeppelin, thinking as an engineer, conceived of his airship as a package composed of autonomous but interconnected parts. Each piece—the planes, the gas cells, the power plant, the envelope—had to function harmoniously to make the whole thing work in the face of potentially destructive natural forces, like wind or gravity. Such systematic thinking was also a metaphor for his own views on how to ensure a successful future for Germany.

Germany post-1871 was a new country composed, like an airship, of divergent pieces. To unite them, the once-independent states needed what he called a "revitalizing idea," a kind of nationalist magnet, to unite them, or they would be spun apart by—and here Zeppelin exemplified the views of a robustly old-line noble—the centrifugal forces of "liberal despotism," "capitalist industry," "the unthinking mob," and atheism.[6] To Zeppelin, in other words, his airship would be a *political statement*.

It was only in May 1887 that Zeppelin definitively identified the airship with German nationalism when he submitted a lengthy memorandum to the king of Württemberg titled "The Necessity of Dirigible Balloons." In it, the count passed briefly over the possibility that the airship might have "general commerce" applications—his original idea, of transporting wealthy tourists to foreign climes, now seemed childishly naive—and instead he proposed that the very raison d'être of the airship was war against a France keen to avenge her humiliation, which he considered the greatest threat to German unity. A large airship, such as he conceived, would be used to transport "[army] personnel, [military] cargo, and explosive shells" for aerial bombing and to perform reconnaissance over long distances.[7]

His memorandum was read, digested, and ignored. The king of Württemberg, Karl I, had more immediate need for Zeppelin in a rather different capacity.

·····

As THE PROCESS of national unification continued apace, the various constituent kingdoms of Germany retained symbolic "diplomatic" relations with Prussia. King Karl requested that Zeppelin, now a colonel, serve as Württemberg's ambassador to Berlin.

Ambassadors holding army rank like Zeppelin were being groomed for high command: They attended the German emperor's court in the expectation that they would learn how to think politically rather than in a strictly military manner. It was Zeppelin's misfortune to walk straight into the combination of beartrap, snake pit, and lion's den that was Berlin in the late 1880s.

The most pressing issue at hand was completing the integration of the Württemberg army into the Prussian, or rather the German, one. At the heart of the problem was that Zeppelin was wearing two hats, or more specifically a hat and a helmet—those of Württemberg's ambassador and of a German officer—and serving two masters—his king at home and the new German emperor, Wilhelm II.

The latter, aged twenty-nine, succeeded to the throne on June 15, 1888. Wilhelm's was a touchy, prickly, panicky personality consumed by insecurity and prone to hysterics when informed that, occasionally, he could not have his way. Kaiser Wilhelm desired to turn Germany into a great global power, and for that he needed a united, unswervingly loyal army. The vast majority of officers did the sensible thing and happily fell into line. As a soldier, Zeppelin was sympathetic to the idea, but as an ambassador his first obligation was the defense of Württemberg's interests.

Zeppelin had cautiously kept out of the internecine political battles raging around him, biding his time until his posting was due to end in early 1890 and he could return to his airship research. Then Zeppelin submitted to the Prussian Foreign Ministry what he considered to be a helpful memorandum on the need for the Württemberg army to retain some of its autonomy and for King Karl not to become a "mere rubber stamp" (admittedly, not the subtlest choice of words).

The memorandum was passed upstairs to the emperor, who read it with mounting disapproval. Unwilling to overlook any perceived slight, Wilhelm smelled treachery in Zeppelin, who had in the meantime returned to Württemberg blissfully ignorant of the bomb he'd detonated back in Berlin.

Livid, the kaiser scrawled abuse in the margins. He was egged on by General Ludolf von Alvensleben, a Prussian martinet, who advised Wilhelm that "it is the most sacred duty of commanders to monitor the loyalty of their officers, to stamp out [provincialist or anti-German] ideas, and to eliminate any disloyal elements." Zeppelin was a marked man, and his downfall was planned for that fall's *Kaisermanöver*—war games—which Wilhelm had instituted to weed out underperforming officers and to talent-spot the more promising.

For the exercises, Zeppelin was temporarily allotted command of a cavalry division, the preliminary step to promotion to general. On the surface, all seemed so placid that even when his friend General von Heuduck mentioned cryptically to Zeppelin in a conversation that "you must have an enemy in high places in Berlin," the count thought nothing of it.

After three days in the field, Zeppelin was highly recommended by Lieutenant General von Götze as an officer who "expounded correct cavalry principles. . . . His conduct on and off duty is faultless. I consider him in every respect qualified for a divisional command."

In any other situation, such praise would have been more than sufficient for promotion, but instead General von Kleist, the (Prussian) inspector of cavalry, summoned Zeppelin and humiliated him with a caustic dressing-down before his fellow officers. He would never, he was told, be given a division.

It was a horrifyingly embarrassing and shabby way to end a thirty-year army career; even worse, perhaps, was that when an officer was purged in such a manner he was rendered unfit for military, political, or diplomatic service anywhere else but in his home state—and even then he would be the subject of innuendo and smears. The king of Württemberg accepted Zeppelin's resignation with a heavy heart and, to ease the shame, bestowed the rank of general upon him, but only in an honorary capacity. At least it came with a general's pension.

Initially, Zeppelin naively believed that Kleist had mistaken "my pliancy for weakness and my quiet way with subordinates for lack of firmness." It was not until February 23, 1891, that he discovered the truth. Gustav von Steinheil, Württemberg's war minister, confessed that he had been told by Berlin *before* the war games that Zeppelin was finished. Kleist's theatrics had served only as a pretext to force a dishonorable departure. At last the shocked Zeppelin understood the damage he'd inflicted on himself with his memorandum; he could only "take comfort in the thought that I am the victim of convictions openly expressed for the good of the empire."

Zeppelin was fifty-two years old when he left the army in November 1890. As he licked his wounds in forced retirement, Zeppelin might at that moment have shriveled up and vanished from history. But he didn't.

His besmirched reputation prompted neither self-pity nor self-recrimination but instead an inexorable drive to restore his honor and prove to his foes that he was worthy of their respect and admiration. If anything, he became *more* patriotic, *more* devoted to Germany, *more* worshipful of his emperor.[8]

Within a month of realizing the war games had been fixed, Zeppelin embarked on the great project that would occupy him for the rest of his life.

3. The Government of the Air

BLESSED, IF NOT voluntarily, with a surfeit of time, Zeppelin could at last undertake the serious study of aeronautics. No more idle fantasias scribbled in a journal; he had to learn how to fly. He compiled a bibliography, then collected a library, of the texts and manuscripts containing the secret knowledge of the aeronauts, and he studied them with beetle-browed intensity.

At the very beginning, he discovered, nobody had cared that balloons were ungovernable. In the fall of 1783, when the Montgolfier brothers in France sent aloft a bemused sheep, duck, and rooster (King Louis XVI had suggested two convicted criminals instead) on an eight-minute journey in a paper-and-sackcloth balloon filled with heated air, the sheer wonder of flying was more than enough to amaze the world.

Of his own first ascent, Professor Jacques Charles excitedly recalled, "Nothing will ever quite equal that moment of total hilarity that filled my whole body at the moment of take-off. I felt we were flying away from the Earth and all its troubles and persecutions forever. It was not mere delight. It was a sort of physical rapture. . . . I exclaimed to my companion Monsieur Robert—'I'm finished with the Earth. From now on our place is in the sky! . . . Such utter calm. Such immensity! Such an astonishing view. . . . Seeing all these wonders, what fool could wish to hold back the progress of science!'"[1]

The "progress of science" was spectacular enough, yet for some it paled beside the glories fashioned by the Great God of Nature. Prince Hermann von Pückler-Muskau, upon being taken up, was enraptured by the experience of reaching some low-lying clouds, a sight just a handful of humans had ever witnessed. He saw them as "chains of snow-white mountains wrought into fantastic forms, [which] seemed as if they were tumbling headlong upon us. One colossal mass pressed upon another, encompassing us on every side, until we began to ascend more rapidly and soared high above them, where they now lay beneath us, rolling over each other like the billows of the sea when agitated by the violence of the storm, obscuring the earth entirely from our view."[2]

What was dubbed "balloonamania" erupted across Europe at the news that men could fly. One could soon purchase bonnets, walking sticks, clocks, jigsaw puzzles, bed warmers, fans, jewelry, garters, snuffboxes, chinaware, commemorative medals, and even chamber pots festooned with aerial imagery. Almanacs, plays, jest books, novels, poems, and penny ballads were devoted to the ballooning craze, while journals kept readers informed with new sections dedicated to "balloon intelligence."[3]

Fashion, too, took notice. Balloon-shaped straw hats of monstrous size were seen everywhere, and the wealthy could purchase balloon-themed coaches and furniture. In France, a yellow ribbon came to be regarded as the identifying badge of aerial enthusiasm and was quickly adopted by the smart set in England. When the Prince of Wales attended a society event wearing "an air-balloon satin [sash] embroidered down the seams with silver," a newspaper reported, "the seat of majesty was forgot, and all eyes (particularly the ladies) directed towards him."[4]

During 1783 and 1784 alone, all manner of balloons were launched (not always successfully) in London, The Hague, Madrid, Hamburg, Moscow, Saint Petersburg, Berlin, Copenhagen, Vienna, Warsaw, Leipzig, Saxony, and Brunswick. *The Scots Magazine* counted at least twenty-five flights made in the course of a single year.[5]

Americans were particularly taken with ballooning. Since the Treaty of Paris—which ended the War of Independence and acknowledged the United States as a sovereign nation—was signed on September 3, 1783, just weeks after the first Montgolfier and Robert ascensions, it seemed as if the Creator Himself had celebrated the birth of flight with the gift of America, a land conceived in the same spirit of liberty as the desire to soar unimpeded to the clouds or to travel wheresoever one wished.

Even the normally unflappable George Washington succumbed to the excitement, telling a correspondent that "the tales related of [balloons] are marvelous, and lead us to expect that our friends at Paris . . . will come

flying thro' the air, instead of ploughing the ocean to get to America."[6]
Lovers, also, might be reunited thanks to the balloon: John Adams, for too
long separated from Abigail, promised that he would leave France and
"will fly [to you] in one of them at the rate of 30 Knots an hour."[7] The
somewhat less lovelorn Thomas Jefferson cannily noted that since "the
French may now run over their laces, wines, &c. to England duty free [by
air]," so too might his airborne countrymen circumvent British taxes on
trade.[8]

Edward Warren, a thirteen-year-old from Baltimore, became the first
American to fly on June 24, 1784, when Peter Carnes, a Maryland tavern
owner and Methodist preacher who'd enterprisingly built a small hydrogen balloon, asked for someone, anyone, to take his place in its basket.
Carnes, who weighed 234 pounds—colossal for the time—could not in-
duce the balloon to take off, and impatient spectators were clamoring for
their money back. As scarred Revolutionary War veterans and hardy fron-
tiersmen shrank from the challenge, young and clean-limbed Warren had
stepped boldly forward. The teenager went aloft a few hundred feet and
"politely acknowledged" (said a newspaper) the cheers of the crowd with
"a significant wave of his hat." A couple of minutes later he touched down
and everyone took up a collection to reward his bravery. Warren took the
money and vanished from history.[9]

Now that the sky was given to man's dominion, there was no end to the
excited predictions about the fabulous future that awaited. Balloons were
a tabula rasa, a blank slate upon which anyone could inscribe their
dreams—and did.

Explorers would soon cross pathless mountain ranges and soar over
uncharted seas and wander above the most tangled jungles and search for
the fabled North Pole. Neither geographical features—rivers, mountains,
seas, and deserts—nor long-established borders would continue to sunder
one state from another. "The whole order of our existence, our habita-
tions, our architecture . . . will be turned upside down," the Russian diplomat Count Morkov told friends in Saint Petersburg, for balloons freely
floated above natural and man-made obstacles alike. "Like gods they will
descend and ascend," heedless of ancient custom and laws.[10] Aeronauts,
one correspondent suggested to John Jay, might even transform into astronauts by flying to the moon.[11]

Closer to home (and reality), balloon-borne scientists would perfect the
developing field of meteorology by measuring temperature, air pressure,

and humidity at various altitudes. Physicists could investigate the speed of sound, electrical phenomena, and the mysterious effects of gravity. Astronomers could at last peer into space clearly without clouds interfering. Topographers could compile precise maps and please kings with the expanse of their realms. The literary critic Friedrich Melchior, who kept in close contact with the Parisian scientific elite, reported that "among all our circle of friends . . . as in the academic schools, all one hears is talk of experiments, atmospheric air, inflammable gas, flying cars, journeys in the sky."[12]

From out of thin air, an entirely new branch of science was invented to further the new art of flying balloons. Dubbed "aerostation," it amounted to the rudimentary study of what would become aerodynamics—how things fly—and the pursuit of research and technical development to improve performance.

More interesting to some people was the prospect of making fortunes. Merchants would use balloons to transport cargo, mail, and passengers between cities along new air routes. Commercial empires would flourish as businessmen forged international deals in less than a day: One French journalist predicted that soon a director of the English or Dutch East Indies Company could "have breakfast at the Cape of Good Hope, dine and make an expedition to Canton [China], and return to have supper with his family in London or Amsterdam."[13]

Indeed, exclaimed some, the entire System of the World would be overthrown when every man became brother to man as national prejudices dissolved through familiarity and understanding. Neither king's writ nor pope's edict held warrant in the air. For the first time in history, revolutionaries proclaimed, men would be free of ancient authority. Among the clouds, their spirits, their minds, and their flesh would be liberated from earthly shackles. Harmony and universal understanding would pass between once-warring nations.

Well, *perhaps*, said others, more cynical of human nature. More likely, generals would embark their armies aboard balloons and attack the enemy from behind his own lines or reduce the mightiest of fortresses (the Montgolfiers patriotically suggested British-held Gibraltar) to rubble with their air artillery. From proto–aircraft carriers, admirals would launch swarms of bomb-bearing balloons toward rival fleets and sink them at no risk to their own precious men-of-war. Or maybe there would be vast sky battles between what the poet Tennyson later called "the nations' airy navies grappling in the central blue" to decide the fate of empires: A British cartoon of the time depicted four cannon-laden balloons, two flying the Union Jack and two the French fleur-de-lis, unleashing broadsides at each other like Nelsonic ships of the line.[14]

But the future never happened. Owing to balloons' instability and uncontrollability, few scientific experiments were conducted, no explorers discovered any Dark Continents, merchants foolish enough to invest in them lost their money, generals left "airy navies" to poets' imaginations, and budding radicals realized that the most effective method of overthrowing the status quo was to chop off kings' heads (in the French case, that of poor Louis XVI).

Owing to fears of rogue balloons setting fire to wooden buildings and razing entire cities, Russia, the Ottoman Empire, Prussia, the Austrian Netherlands, Genoa, and Rome went so far as to ban aviation.[15] Even in France, birthplace of flight, ballooning's fortunes collapsed: Between 1783 and 1790, no fewer than seventy-six civilian ascents were made, but in the decade following, just six.[16] For nearly a decade, not a single manned balloon was launched in America after young Edward Warren's ride in 1784.[17]

The "aerial phrenzy" quickly waned, and once-awed journalists made running gags about overpromising aeronauts resembling their conveyances in being filled with nothing but hot air.[18] One cynic joked that never "has a soap bubble occupied more seriously a troupe of children" than a balloon, while Horace Walpole, that most waspish of critics, lethally stung aero-enthusiasts by remarking that he expected their "new mechanic meteors will prove only playthings for the learned and the idle."[19]

He was right about the "idle." Ballooning eventually declined into showmanship, the kind of sideshow diversion in which impresarios took passengers up to a couple of hundred feet or tossed cats equipped with makeshift parachutes from the basket. A pursuit that had so recently been exalted as the savior, the redeemer, and the benefactor of humanity degenerated into a vulgar carnival act akin to dancing bears and performing monkeys.[20]

As for the "learned," Walpole was wrong. When the scientist Thomas Martyn pointedly asked, "Of what use are balloons?" the Royal Society, home to some of the finest minds in Britain, quickly distanced itself from ballooning. Sir Joseph Banks, the brilliant botanist who served as its president, warned his great friend Benjamin Franklin that "the more respectable part" of the membership wished to postpone further research into aviation "until some experiment likely to prove beneficial either to Society or Science is proposed."[21]

Advocates retorted that only time would tell. Balloons were so ingenious, and their implications to human existence potentially so staggering, that one day they would prove their worth in ways nobody could foresee. As Franklin, who watched several flights in Paris, remarked to a skeptic scoffing that balloons had no purpose, "What is the use of a newborn babe?" And to Banks he replied, "It does not seem to me a good reason to decline prosecuting a new experiment which apparently increases the power of man over matter, till we can see to what use the power may be applied. When we have learnt to manage it, we may hope some time or other to find uses for it, as men have done for magnetism and electricity, of which the first experiments were [also] mere matters of amusement."[22]

In that letter, the always perceptive Franklin put his finger precisely on the problem. Progress would have to wait until "*we have learnt to manage*" the balloon.[23] In other words, he identified aerial navigation—or, as another supporter put it, "the government of these machines in the atmosphere"—as *the* critical, maybe even the sole, factor in determining whether lighter-than-air flight had a future.[24]

Yet, try as they might, not even their most zealous enthusiasts could work out how to regulate balloons in the air, even if at first glance the problem seemed simple to solve. The Montgolfiers theorized that all one had to do to create propulsion was to rend a hole in one side of the balloon and release hot air or hydrogen through it.[25] In the event, all that happened was that the spherical balloon sagged through loss of pressure and rapidly lost altitude. The mere act of wastefully valving gas to thrust the vessel forward could not overcome the counterforce of wind resistance. Some kind of supplementary motive *power* was needed. But what form should it take?

An enterprising fellow named Mr. Uncles proposed building a balloon propelled by four eagles, which, he advertised, are "perfectly subservient to his pleasure." He actually succeeded in making a fifty-foot-wide balloon out of Persian silk and on July 18, 1786, he took his seat with the seemingly compliant eagles harnessed. After the balloon rose eight feet, Uncles tried to coax the birds to pull forward in unison but he and his contraption crashed to the ground, to the great amusement and little consternation of the ten thousand spectators present. No more was ever heard of Mr. Uncles.[26]

A more promising avenue lay in likening the balloon to a ship. As boats were obviously capable of sailing against or across the wind, wouldn't it make sense to outfit a balloon with a sail? Unfortunately, it didn't, mostly owing to the fact that sailboats can be steered because they interact with two competing environments (they float on water and are propelled by wind), whereas balloons rely on just one. If a boat has its sail removed, all you end up with is the nautical equivalent of a balloon: a drifting vessel at the mercy of the current. So the addition of sails to a balloon was hardly an addition at all: They merely enabled it to head in the same direction and speed as the prevailing wind—which had been the problem in the first place.[27]

A better idea might be to liken apples to apples: Another object that moved exclusively in a single environment was a fish. Why not mimic a sea creature's means of propulsion by adding a rudder and oars as substitutes for its tail and fins?

It was quickly discovered, however, that the density of water—around eight hundred times that of air—was critical to allowing a fish's fins to "push" against something for propulsion, while oars lacked any similar purchase in the relatively thin air. True, it might have been possible to build oars with sails at their ends, but to have any chance of exceeding the current windspeed so many rowers would be needed that the balloon would be too heavy to lift off. And in the absence of sufficient muscle power to generate significant airflow past it, no rudder—even assuming one light enough not to sink the balloon could be made—stood the slightest chance of creating a turning movement, as it does in water.

In France, the Académie Royale des Sciences assigned Jean-Baptiste Meusnier, a military engineer and mathematician, to conceive a solution. Despite his best efforts, Meusnier could never overcome the fundamental dilemma of power versus weight, but he did succeed in illuminating a path forward.

One of his ideas was to throw out the conventional spherical shape of the balloon and replace it with an elongated one. A long rather than round balloon would, he thought, streamline the vehicle to reduce resistance (thus emulating a fish—or a boat, for that matter). He was on the right track, but alas, created another problem. Whereas the Montgolfiers and their ilk could simply inflate their balloons and benefit from a sphere's inherent stability, cigar-shaped balloons were subject to crumpling, rolling, and lurching when gas was added or vented owing to differential pressures at various points and uneven weight distribution. To solve that unexpected issue, Meusnier proposed inserting into the balloon envelope smaller bags that could be filled or emptied during flight to compensate for structural stress. Meusnier's concept of independent "gas cells" within an airship's greater body would later become standard in the Zeppelin industry, but his death in 1793 checked any more development along such innovative lines, and the evolution of the balloon spluttered to a standstill.[28]

Not until the 1840s would balloons, for so long seen as a pitiful folly of wasted potential, be resurrected by enthusiasts as a viable technology, ripe for improvement. It was the SS *Great Western*, a ship designed by Isambard Kingdom Brunel, the greatest and most ingenious engineer of the era, that unexpectedly ignited the imaginations of the beleaguered aeronauts with inspiring visions of adventure and discovery.

The *Great Western* was built with a single purpose in mind: to make transatlantic crossings from England to America in record-breaking time. The most ambitious part of Brunel's vision was that, unlike the other vessels plowing the Atlantic, this ship would be primarily driven by steam power, not sails. For his temerity, Brunel was ridiculed as suffering from a "chimera of the brain": No ship could possibly carry enough coal to feed the colossal steam engine that would be needed to propel it nonstop through 3,500 miles of ocean.[29]

But Brunel had one trick up his sleeve. Historically, naval architects had assumed that a vessel's resistance to motion was in squared proportion to its displacement. That is, if you doubled the dimensions of a ship it became four times harder to propel it through the water.[30] Brunel, however, calculated that while it was true that water resistance rose by a square, the ship's carrying capacity was actually *cubed*. Volume, in other words, increased faster than area. And if that was the case, then not only could you fit more high-paying passengers and valuable cargo on board, but you would do so relatively more economically, in terms of the fuel consumed to drive the ship, than anyone had ever suspected. Put very simply, you could expand a square into a cube but make the cube cost about the same as a square.

On April 8, 1838, the *Great Western* embarked on its maiden voyage to

New York. A paddle-wheeled, wooden steamship 236 feet long—making it the longest ship in the world—it was rated for a speed of 8.5 knots, or nearly 10 mph, thanks to its 750-horsepower steam engine, and had a theoretical capacity of 128 passengers (plus twenty servants). When the *Great Western* entered New York harbor after a voyage of fifteen days and five hours, it had a third of its fuel left over, thereby confirming Brunel's sums and rendering a transatlantic steamship route economically feasible. Just two years earlier, commented the *Quarterly Review*, even the very "notion of navigating the Atlantic by steam, as a permanent, practical, profitable thing . . . had never entered the public mind," but Brunel had rendered an impossibility a reality.[31]

Amid the excitement, aeronauts quickly recognized that Brunel's breakthrough presented them with the opportunity to turn balloons into a rival of the train and even the *Great Western* itself by converting them into ships of the air, or what they now took to calling "airships." Benjamin Franklin's question "What is the use of a new-born babe?" had finally been answered: transporting passengers and cargo over long distances at moderate cost, if they were built large enough to exploit Brunel's cube law.

The great disadvantage of balloons had always been the lack of air density relative to that of water, rendering oars and sails useless appurtenances. But now aeronauts saw that the air could be their invaluable ally precisely *because* the sky was far easier to sail through than the sea. Even better, because hydrogen and silk weighed virtually nothing compared to the wood and iron of Brunel's huge steamship churning through the waves, one could use a relatively small engine while benefiting from greatly increased carrying capacity. In fact, thanks to the cube law and the associated economies of scale, as an airship's gas volume rose, it could carry more cargo and passengers relatively more cheaply than a smaller one. "The bigger the better" soon became, now and forever, the mantra of the airshipmen.

Here was the chance to resurrect the quest for aerial navigation amid the greater purpose of seizing the most glittering prize of all: crossing the Atlantic twixt the shores of worlds new and old. For the victor—and society—the rewards would be magnificent; the benefits, incalculable. Once the passage east from New York to London was established, a route west across the endless Plains, over the impassable Rockies, and ultimately to the fabled land of California across the Great Desert would be forged.

And thence onward to Asia, home to the mysterious empire of Nippon and the Celestial Kingdom of China. America's Manifest Destiny would be fulfilled, not by land or by sea but in the air.

Filled with Yankee pluck, Rufus Porter patriotically took up the challenge.[32] Entirely self-taught, Porter was the founder of *Scientific American*, a tinkerer of renown, and the country's foremost popularizer of mechanics and invention, but it was the prospect of flying that truly animated him.

In 1849, he released plans for an 838,000-cubic-foot hydrogen behemoth — for comparison's sake, a Boeing 747-400 has an interior volume of 31,285 cubic feet, slightly larger than the Montgolfiers' original balloon. It was to be 800 feet long and 50 feet wide and streamlined into a Meusnier-style "revoloidal spindle" (i.e., cylindrical with tapered ends), attached to which would be a 200-passenger saloon and cargo hold 180 feet long. The craft's twin steam engines would apply their 12 horsepower to two 20-foot propellers, prompting Porter to boast that it would hare along at 100 miles per hour, ten times faster than Brunel's laggardly tortoise, and put London or Gold Rush–crazed California within three days' reach of the eastern seaboard.[33]

Had Porter even the most minuscule chance of actually constructing such a marvelous machine, the dominance of the steamship might have come to an end. But, of course, he hadn't, and there was an entirely foreseeable series of mechanical failures, cost overruns, and manufacturing delays. By mid-1852 Porter was reduced to offering his tired investors shares in a rather less impressive scheme: a mini-airship projected to carry five passengers on brief sightseeing trips.

But then an obscure Frenchman surprised the daylights out of everybody by making world history.

Porter dismissed the fantastic reports as mere rumor, but, sadly, they were all too true. Henri Giffard, a twenty-seven-year-old engineer and balloon hobbyist, had built a light steam engine — "light" meaning that the power plant and its *empty* boiler weighed 330 pounds — that could produce 3 horsepower and drive a large rear-facing propeller. This he placed on a small platform suspended beneath a spindle-shaped 144-foot-long balloon that held some 88,300 cubic feet of coal gas. Giffard dubbed it a *dirigeable* (French for "steerable," soon Anglicized to "dirigible").

On September 24, 1852, at 5:15 P.M., Giffard took off from the Hippodrome in Paris and chugged his way to Élancourt, seventeen miles away, on a three-hour, one-way voyage averaging nearly 6 mph. Once he reached his destination, Giffard, dressed for the occasion in a frock coat and top hat, demonstrated his ability to control his craft by making several turning circles in calm air before landing safely at his destination.

Giffard was never able to repeat his success. He had made a great leap forward, certainly, but had collided with a wall. His engine was simply not sufficiently powerful to drive his dirigible in anything but still air. At the merest hint of a breeze, he found himself at its mercy.

In terms of hard numbers, his steam engine had a dry weight-to-power ratio of 110 to 1 (the 330 pounds of the boiler/engine combination divided by the 3 horsepower it indicated). In other words, it required 110 pounds of weight to produce 1 horsepower. (A rule of thumb at the time was that an actual horse weighed a thousand pounds and produced 1 horsepower—itself.) Giffard's ratio beat anyone else's at the time, but it was nowhere near enough to sustain powered flight or to allow controlled aerial navigation in anything but the most favorable conditions.

There were two possible solutions to the problem. Realistically, you could increase the power output *or* reduce the weight of the engine to boost the ratio. Ideally, you would do both, as the Wright brothers would a half-century later, when their first engine (180 pounds) produced 12 hp, yielding a weight-to-power ratio of 15—just enough to allow true flight. Unfortunately, Giffard could do neither. Coaxing more horsepower out of the already taxed engine was out of the question, and paring its weight to the barest possible minimum made little difference because the iron engine remained so heavy that it still ate up a large proportion of his dirigible's "useful lift."

Useful lift was one of two related metrics that were becoming of greater importance. Every airship is defined by its gross lifting power and its useful lift. The former is derived from its gas volume. The more gas there is, the more weight the airship can bear as it floats. But from that figure you need to subtract the total fixed weight the airship is expected to carry, which includes the airship itself, the aeronaut(s), the balloon material, the gas, the engine, propellers, payload, ballast, and fuel. The net result is the useful lift.

Because he had used cheap and dirty household coal gas rather than expensive purified hydrogen, Giffard's gross lifting power was modest to begin with. But after taking away the various fixed weights, Giffard had so little useful lift left that he had been pushing his luck in merely getting off the ground.[1]

Alas, no one could improve on Giffard, and by the Civil War interest had petered out. Instead, Thaddeus Lowe, an impossibly glamorous, imposingly mustached New Hampshireman, successfully built balloons like the 725,000-cubic-foot *City of New York* and the smaller *Enterprise*, but they lacked engines. Not until "a propelling power can be discovered the weight of which shall be but one third of that we now employ," he sadly judged, could "aerial navigation . . . become a practical science."[2]

Lowe would soon found the Union's Balloon Corps, which would perform only occasional reconnaissance duties in small, tethered balloons at an altitude of between 500 and 1,000 feet until its official dissolution in

August 1863—the same month that Zeppelin met Steiner in Saint Paul. While the Civil War would bring to the fore many modern innovations—railroads, land mines, ironclads, rapid-firing guns, submarines—in terms of lighter-than-air flight it was as if time had stopped sometime in the Montgolfier era.

In the absence of a lightweight engine, the quest for aerial navigation was a fool's errand.

THOSE TAKING AN evening stroll down Paris's Champs-Élysées in the late summer or fall of 1881 could not help but notice the yellowish aura eerily emanating from the Palais de l'Industrie, a huge Gothic building on the bank of the Seine. Attracted to the light and intrigued by the growl and whirl of machinery, they entered a spectacular cathedral of science where congregants prostrated themselves not before God but before the high priests of Electricity, a far more mysterious power now finally revealed to mankind.

To demonstrate the new technology's marvels, a model of a lighthouse greeted visitors as they entered. Powered by an electrical generator, its beam revolved and even changed color. Surrounding it was a small lake in which toy electric boats whirred contentedly around. Illuminating the central pavilion were statues bearing scores of electric lamps; the chandeliers were adorned with still more.

Passing through the various halls they saw displays of batteries, telephones, telegraphs, and electro-locomotives. Hall 20 contained a collection of "electro-pneumatic clocks," while Halls 3 and 4 resembled a World of Tomorrow exhibit: A mock-up of a fashionable apartment featured a "tastefully furnished" dining room, kitchen, and bathroom illuminated by battery-powered lamps. Dominating the American pavilion were Thomas Edison's new incandescent bulbs, which promised long-lasting light suitable for homes, all controllable with the simple flick of a switch.[3]

But one visitor, Captain Charles Renard, ignored the domestic novelties and paid close attention instead to an unheralded piece of equipment that served to power many of the smaller displays. Invented by Zénobe Gramme, a Belgian engineer, the "Gramme dynamo" was an adaptable, lightweight generator that could be employed as a general-purpose workhorse for any number of applications.[4]

It was just what he was looking for.

•••••

AFTER ITS CALAMITOUS performance during the Franco-Prussian War, the humiliated French army had undergone a number of reforms, not least of which was the creation of a Military Balloon Establishment. Captains Charles Renard and Arthur Krebs had been recruited to conduct research into powered, lighter-than-air flight, and by 1883 they had adapted a small Gramme dynamo and a bank of batteries (for power storage) to fit into an airship they patriotically christened La France.

Steam, they argued, was obsolete in this new modern age of electricity, and La France was going to be a radical step forward. Their craft was 160 feet long with a 23-foot-diameter propeller placed at the front of a silk-and-bamboo control car that was suspended underneath the balloon. Within the car, Renard and Krebs operated the dynamo and battery array, which produced around 8.5 horsepower and weighed 1,232 pounds, owing to the weight of the batteries.

On August 9, 1884, as officials from the War Ministry watched, they took off from Meudon, a suburb in southwestern Paris. Unfortunately, they didn't have enough hydrogen on hand, so Renard had had to remove half the batteries to lift off, but that helpfully improved the weight-to-power ratio. Ultimately, La France would triumph in an extraordinary feat that marked the very first time in history an aircraft made a sustained, controlled, powered round trip.

Against a light breeze of 3.3 mph, the airship embarked on a twenty-three-minute voyage of 4.25 miles and returned safely to its origin. Though speed dropped drastically during the turning maneuver at the halfway point, its average speed was 12 mph. A triumph!

The only problem was that meteorologists had lately decided that 15 mph was the "average of the rate of air currents in which even in a calm day a balloon floats." They had recently found that wind speeds at ground level differ significantly from those at higher altitudes. Records taken for a 101-day period between June and October 1889 at the top of the Eiffel Tower (984 feet) found that the average daily velocity was 15.75 mph, whereas on the ground in the same location, it averaged a mere 4.9 mph. Until this discovery, any number of aeronauts had unwittingly made the error of building their craft to cope with ground-level winds.

Had the wind blown harder on that atypically calm August day, in other words, La France would have been pushed backward. Even on subsequent flights, with the full complement of batteries, *La France* never traveled faster than 14.5 mph—lower than the wind-speed benchmark—and merely holding the dirigible steady taxed the electric motor so heavily that the batteries drained in less than an hour.

The result was that, as one journal summarized, a genuine "solution of the problem of aerial navigation" would only be within sight when a balloon could "be propelled for some time at a rate of at least fifteen miles per hour." The bar had been raised against an already faltering contender.

Renard pointed out that electric power was in its infancy and would surely mature over coming years, just as steam technology had. "We are now masters of the air balloon," he declared, and "the entire matter [aerial navigation] is now only a question of time and money." Sadly, he was granted neither by the army, and the airship program was killed, cited as a waste of both.[5]

BEFORE HIS FORCED retirement, Zeppelin had read of *La France* with great interest—and mounting alarm at the resurrection of France's military ambitions—and it was during that period that he was prompted to write his 1887 memorandum "The Necessity of Dirigible Balloons."

In 1890, back in his study in forced retirement, Zeppelin reread that memorandum and turned over in his mind what he had learned of the previous century's attempts at aerial navigation. First, based on Meusnier's experiments, it was clear that any successful airship would have to be cigar-shaped and filled with gas cells. Second, as Brunel had proven, it would have to be massive to benefit from the "cubed carrying capacity" principle and the associated economies of scale. Third, Giffard had shown that hydrogen, owing to its impressive lifting ability, must be the gas, and that structural lightness was critical. And fourth, as Renard and Krebs—and everyone else—had unpleasantly discovered, steam engines and electric motors alike were both underpowered and overweight for their assigned task of propelling an air vehicle into the wind.

He had the clues, he had the evidence, and now it was up to him and him alone to solve the mystery of aerial navigation.

O NCE A HOBBY, Zeppelin's interest in airships now turned into a single-minded obsession. He'd once seen the airship as a useful tool, of empire and of war, but now it became a symbol of something much greater: his craving for the restoration of his honor. Languishing under the suspicion of disloyalty for having, as he saw it, told the truth and performed his duty, Zeppelin wanted rehabilitation, both at the emperor's court and within his beloved army. By achieving the impossible he would demonstrate his fidelity to the cause of national unity and transform Germany into a Great Power through air power.

In the Middle Ages, bishops had competed against one another to endow the grandest cathedral; in the eighteenth century, kings strove to build the most magnificent palace. By the late Victorian Age, no Great Power could legitimately call itself one unless it had commissioned a major civil-engineering project to symbolize its national pride and imperial superiority. Immensity was key to boggling the imaginations and stunning the senses of rivals.

America, the up-and-comer among the global players, completed the Brooklyn Bridge in 1883 and cooperated with France, the grande dame lately fallen on hard times, in dedicating the Statue of Liberty three years later. The French, reinvigorated, doubled down by beginning construction of the Panama Canal in 1881 and then scoring an impressive triumph in 1889 with the inauguration of the Eiffel Tower. Meanwhile, Britain built the Tower Bridge in 1887 to keep its hand in the game, but all knew that with its mighty navy, world-spanning empire, and tireless economy, London was the reigning champion anyway.

Wilhelm II's *Kaiserreich* was destined to rule the skies. The airship would dazzle the emperor's enemies in Europe, ferry his legions to the distant plains of Africa, traverse the Atlantic to awe the Americans, wrest the Holy Land from the grip of the Turks, and force the Celestial Kingdom of China to bend its knee. So let France and America and Britain have their Ozymandian steel and stone monoliths, their iron and copper monuments—Germany would outdo them all in silk and gas.

FOR MONTHS AFTER his resignation, Zeppelin confined himself to his study to toil at designing his airship. A monster of conceit, Zeppelin believed himself to be the lone genius who would birth the future. As he wrote at the time, men of grand vision (such as himself) should ignore the naysayers and the petty whiners, "among them highly-skilled technicians, [who] have failed [at making airships]." Rather, he *alone* had "solved" the problem of the airship and aerial navigation. "I have solved it, not because I knew any more than my rivals, but by the simple, sober thinking of a serious man, whom nature has endowed with common-sense."[1] He would always insist that "my system is the best, the only conceivable one for military purposes, and, if airships are ever possible at all, then they will be mine."[2] Later, he would purr with self-satisfied pleasure when they were commonly referred to as "Zeppelins," in recognition of his intense devotion to, his unquenchable enthusiasm for, and his unceasing belief in what he would always call "*my* airships."

Yet, as he soon found, the proverbial "lone genius" will rarely, if ever, be capable of conceiving, designing, testing, manufacturing, financing, marketing, distributing, publicizing, and popularizing a successful product. The prolonged process, rather than a single act, of invention relies instead on connecting networks and leveraging technologies to form a *system*. In Zeppelin's case, he quickly realized that he could not hope to develop a means of propulsion without outside help. Thanks to his (truncated) engineering education, he may have been technically knowledgeable, but Zeppelin was no Giffard, a man who could build his own engine.

Technologically speaking, in these months Zeppelin settled on the optimal source of airship propulsion. The internal combustion engine was only just beginning to emerge as the motor of choice for the coming century. Its rate of adoption would become unstoppable—in 1876, for instance, world-leading Britain nationally generated a measly 2,000 horsepower by means of internal combustion; by 1907, that figure had exploded to 1,528,962—but that triumph was by no means predictable when Zeppelin first began researching the subject.[3]

By being in the right place at the right time, Zeppelin was fortunate in being able to call upon a most useful network: Württemberg happened to be home to the Daimler Motor Works (Daimler-Motoren-Gesellschaft, or DMG), the most advanced automobile-manufacturing company in the world. And where there were Daimler automobiles there were internal combustion engines, and more important, Gottlieb Daimler.

INTERNAL COMBUSTION ENGINES (ICE) worked by combusting a fuel (gasoline, for instance) with an oxidizer such as air to transform chemical energy into mechanical motion. They were simple to run, could stop and start at will, produced a stable supply of power, and needed only occasional refueling. These characteristics gave them some advantages over battery-operated electric motors, which were relatively light and nippy but needed recharging after only moderate use, and steam engines, which could handle heavy loads but required a long time to heat up and a constant fire burning whether or not they were being used.

Like their electrical counterparts, internal combustion engines were, at least in their practical form, a fairly new energy technology. The first ICE to operate on the modern four-stroke principle—intake, compression, power, and exhaust—was made by Nicolaus August Otto in 1876. It was a metal behemoth that stood twelve feet high and produced half a horsepower. Obviously, it was unsuitable for airships.

That's where Gottlieb Daimler came in. The son of a Württemberg baker, Daimler was born in 1834 and attended the same polytechnic academy as the count—they probably even had the same professors and took the same courses. Zeppelin, of course, entered the army while Daimler joined the Brotherhood of Reutlingen, an industrial enterprise located twenty miles south of the capital of Stuttgart that had been established by a charismatic theologian named Gustav Werner on Christian Socialist principles.

Hoping to dull industrial capitalism's sharper edges, Werner had set up a paper mill, a machine factory, and a carpentry shop to give the marginalized—orphans, disabled workers, and the deserving destitute—a hand up by training and employing them. One of the orphans, a fifteen-year-old fellow Württemberger named Wilhelm Maybach, caught Daimler's attention. The teen, Daimler noticed, was pedantically precise, exceedingly perceptive, and creatively brilliant and would focus for days and weeks, even months, on a single niggling problem until he cracked it. Daimler took him under his wing as a machinist, and the two men would be business partners until the older of the two died thirty-seven years later.

The duo made an odd couple. Daimler, an engineer with a talent for self-promotion, had a volcanic temper, while Maybach, shy and retiring, seems never to have raised his voice. The bespectacled Maybach was also content to stay hidden in the background and never minded that Daimler garnered all the glory and most of the money. Daimler always looked after Maybach and made sure he was compensated properly, but it would be Daimler's name that would go down in history as the father of the automobile.

Daimler took Maybach with him to join Nicolaus Otto's internal combustion concern, Gasmotorenfabrik Deutz. Daimler became the factory manager while Maybach served as chief designer. Together they modified the existing Otto Engine to output 3 horsepower. After paying Daimler and Maybach a modest lump sum, Otto patented their improvements, took the credit, and pocketed the windfall of orders that came in. In 1880, Daimler and Otto fell out so bitterly that Otto fired Daimler. Maybach soon followed, and the pair decided to set up their own company to specialize in making small, lightweight engines for vehicles of all types. (It was Daimler who conceived the famous three-pointed-star logo—the spokes representing land, sea, and air—that would be adopted by Mercedes-Benz in 1926 after a company merger.)

In 1882, they moved into a cottage in Cannstatt (now part of Stuttgart) and worked secretly in the greenhouse. By 1885 they had developed a 1-horsepower engine that weighed just 132 pounds. As an experiment they mounted a scaled-down version on a wooden-framed bicycle and Daimler's son rode it—the world's first motorcycle—on a two-mile stretch of road at speeds of up to 7.5 mph.

After proving that a combustion engine could power a human-controlled land vehicle, in the following year Daimler and Maybach created the first four-wheeled "motor car" by installing a 1.1-hp engine weighing 88 pounds into a modified horse-drawn carriage. Locals were amazed to see the contraption zipping along the road at a 10-mph clip. They soon adapted the engine to fit into a fifteen-foot boat that propelled itself along the river at a rate of 6 knots (7 mph). With land and sea conquered, only the air remained.

In August 1888, they lent a balloonist, Karl Woelfert, a small 2-horsepower model that he installed into a hydrogen-filled gasbag. Woelfert diced with death by keeping an open-flame burner close to the gas, but succeeded in making a very short flight. (He would die, after a still shorter one, in an explosion in 1897.) Daimler and Maybach weren't impressed by this performance and decided to focus instead on their car and boat business. In 1889, they built their first custom-made automobile, managing to coax 1.65 hp and 15 mph out of it.[4]

Despite numerous orders for their engines, Daimler and Maybach needed capital for expansion. Max von Duttenhofer, a munitions maker, answered the call and in November 1890—the same month Zeppelin left the army—became chairman of the new business, Daimler-Motoren-Gesellschaft. Within three months relations with Duttenhofer had become so fraught that Maybach resigned and worked from home, where he busied himself on a clandestine project dubbed Phoenix, an engine that he predicted would initially output 2 horsepower but would be capable, he claimed, of exponentially improved performance. Daimler, sensing that his time too would soon be up with Duttenhofer, quietly financed his protégé.

On April 13, 1891, Zeppelin sent Daimler a letter containing an exhaustive list of questions regarding the technical capabilities of internal combustion engines. Daimler politely replied with a summary, but did not mention the secret Phoenix program.[5]

Notwithstanding the limited power and heavy weight of the current generation of engines—which could not hope to propel a craft of the size Zeppelin was projecting—the count had let his imagination soar far, far beyond what anyone had ever before conceived. In one sketch, for instance, the count depicted a long, thin cylinder with semi-spherical ends. The machinery, fuel, and control gondola (where the pilot would sit) were suspended below, amidships. Tiny propellers were located near the rear. The "engine," however, was conveniently left vague, as were the composition and shape of the internal skeleton.[6]

Happily unrestrained by tedious reality, the count added "transport vehicles" for passengers and cargo. Borrowing from the idea of a train, the engine-bearing airship would serve as an "air locomotive" for two unpow-ered balloon-carriages that would trail along in its wake. Zeppelin calcu-lated the useful lift of each of these as 1,100 pounds, sufficient to carry, say, seven soldiers or a small artillery piece. He recklessly predicted that it would be able to stay in the air for twelve hours, which in itself would have been an extraordinary feat, but soon amended that figure to no less than 7.5 days.[7]

In coming up with these incredible numbers, Zeppelin had badly erred in his calculations of air resistance—not entirely his fault, given the prim-itive knowledge of aerodynamics at the time. Zeppelin had assumed that the width or diameter of a vehicle's head is the primary cause of air resis-tance, while the length of its body plays a minor role. Hence he gave his locomotive-airship a narrow head and believed that increasing its length in the form of a train would have virtually no effect.[8] In fact, as the latest research was beginning to show, the total drag imposed by the body is much greater than that of the head, owing to its increased area.[9]

He wrote on June 29, 1891, to General Alfred von Schlieffen, the new chief of the Imperial German General Staff (the powerful body in charge of planning and operations), and urged him to send an officer of the Prus-sian Aeronautical Battalion (PAB) to examine the blueprints and a small model he had built. The general, who had once been friendly with Zep-pelin, replied on July 4 that he was "readily willing" to do so and that Captain Rudolf von Tschudi, the commander of the PAB, would be only too pleased to arrange a visit.[10]

The PAB had been founded in 1884, two months before Renard and Krebs's first *La France* flight, to investigate the military potential of the air. Befitting its origins, the PAB specialized in conducting meteorological experiments, honing reconnaissance skills, and practicing artillery spot-ting from small tethered balloons.

Zeppelin recognized that he needed the PAB imprimatur if he were to stand any chance of interesting the army in his airship, but to his mind, its personnel, too concerned with small-scale experiments and limited uses, lacked the imagination to see the big picture: His airship would be a Won-der of the World, its grandiosity, gigantism, and ambition its greatest attri-butes. He even named his planned ship of the skies the *Deutschland*, all the better to symbolize its connection to a glorious German future.[11]

Captain von Tschudi was agreeably polite when he and Zeppelin met on July 24. Tschudi warned that some of his colleagues were jealous that

Zeppelin was intruding on their bailiwick but he nevertheless encouraged him to build a full-scale prototype to show the PAB. Depressingly, that meant Zeppelin would have to find the funds himself. Since the count himself estimated that cost to be between 250,000 and 600,000 marks, a very large sum, it was clear that Tschudi expected to hear no more from Zeppelin. He would not be the last to underestimate Zeppelin's determination.[12]

A week later, a small article in a newspaper caught Zeppelin's eye. It stated that Renard, he of *La France*, had on July 27 conducted an experiment at Chalais-Meudon—the French army's testing ground—with an improved dirigible with the minister of war present. Nothing concrete resulted from the trial, the chief feature of which, said the newspaper, was "a new engine developing high power with very little weight."[13]

Zeppelin was alarmed at what he believed to be the rapid progress of the French. The next day he wrote that "it is urgently necessary that Herr Daimler should have more faith in the possibilities of a dirigible airship, as soon as there is available a sufficiently light, yet reliable engine."[14]

In the meantime, there was still the matter of the skeleton. This being the small world of Württemberg, Zeppelin introduced himself to Professor Carl von Bach of his old haunt, the Technical University of Stuttgart.[15] The count believed that Bach, an expert on the mechanics of materials, could help him calculate how best to construct the airship infrastructure. Following a prolonged exchange of letters discussing technical details, they finally met on November 5, 1891.[16] Bach, disappointingly, lacked the time to work alongside the count and in any case preferred a background role as a consultant, but he nevertheless soon proved useful.

His first contribution was to suggest aluminum alloy, a wonder metal created by modern science, as the appropriate material for the airship skeleton. Not only was it amazingly light, but it possessed impressive tensile strength. The downside was that refining aluminum was a complex, and predictably expensive, process for much of the nineteenth century, in some instances making it more valuable than gold (the French emperor Napoleon III reserved aluminum plates for formal state dinners). But that was changing. Thanks to the introduction of new electrolytic and chemical processes in the late 1880s to produce it in greater quantities at lower prices (in 1888, aluminum cost 59 French francs per kilogram to make; by 1899, it would plummet to less than 3), the metal was potentially feasible for a dirigible infrastructure.[17]

Bach's second contribution was to recommend Theodor Kober, a twenty-six-year-old former student of his, as Zeppelin's technical designer. Kober had recently passed his diploma examination with a grade of IIb, classifying him only as "quite good to good," but he worked cheap and, Bach added, was an "energetic, farsighted, determined character eager to work." In May 1892, Zeppelin welcomed Kober aboard, saying that "I hope to God that by our joint efforts we will succeed in doing something useful for our German *Vaterland.*"[18]

Kober proved an admirable fit: He was young, mild-mannered, pliable, not so skilled as to question Zeppelin's decisions with any great authority, and easily overawed by the count's gale-force personality. Kober willingly, for instance, signed a contract stipulating that, like the slaves entombed in the pyramids with their pharaohs to serve them in the afterlife, he must "dedicate his whole energy to the execution and testing of the airships planned by me [Zeppelin], and . . . bring this task to its end in case of my death."[19]

The count, as a man who habitually dismissed the opinions of so-called experts who claimed certain technical problems as insurmountable, believed—even when he was wrong—that impossibilities were merely inconveniences that had not yet been fixed. A determined willingness was all that was needed to solve problems, and to that end Zeppelin and Kober immediately set to work making "many hundreds of enquiries at the Royal Materials Testing Office" in Stuttgart regarding the suitability of various metals and fabrics.[20]

In late 1892, Zeppelin had the break he'd been hoping for. That October, Maybach, who had been working quietly on his own since his acrimonious departure from Duttenhofer's DMG, set up a workshop in the ballroom of an abandoned hotel and hired seventeen mechanics and apprentices to help him finish the clandestine Phoenix project, which featured two upright cylinders cast in a single block, his patented spray-nozzle carburetor, and a camshaft for controlling the exhaust valves. The lightweight Phoenix would initially output 2 horsepower, but the design was as adaptable as Maybach had promised: By 1896, the engine would produce 4 horsepower, then 6 in 1897, 8 by 1898, and 16 by about 1900.

Daimler, as has been mentioned, had been secretly funneling money to Maybach even as he remained associated with DMG. The two planned to launch the Phoenix in a new enterprise that would destroy the already ailing DMG—which built just eleven automobiles in 1892–93—and humiliate the hated Duttenhofer.[21]

It was at this point that Maybach, at Daimler's instigation, initiated Zeppelin into the secret of the Phoenix.[22] Henceforth Zeppelin began referring to "the Daimler engine"—as the Phoenix would be dubbed—in his papers and plans and also began calling upon Maybach for advice.[23]

Maybach had gained a client for his and Daimler's covert start-up firm, but Zeppelin gained far more: He now had access to the world's most efficient and advanced engine.

ON SEPTEMBER 14, 1893, Zeppelin emerged from his lair. In a letter to General von Schlieffen, who'd long since assumed that the count had given up the ghost, he confidently boasted that his "safe and fast-flying air trains will confer on the Army many advantages: for example, reconnaissance over hundreds of miles of enemy territory in a few hours; assured supplies for all troops from the nearest depots every day without the delays of road transport; transfer of important officers and important information from one army to the other; bombardment of enemy fortresses or troop concentrations with projectiles."

He proposed that Schlieffen appoint a committee of experts to examine his design; if they favored it, which of course they would since he was Count von Zeppelin, he requested that the War Ministry budget 650,000 marks to build the full-size prototype Captain von Tschudi had earlier suggested.[24]

The blueprints he supplied bore the typical Zeppelin hallmarks—of genius and of ego. Designed according to the count's idiosyncratic dictates, the airship retained his sky-train concept, with a powered front dirigible—the locomotive—385 feet long, the trailing second unit 52.5 feet long, and the third 131 feet long. More than 75,000 square feet of Chinese silk soaked in 660 pounds of varnish for airtightness would cover the front dirigible, with its 336,000 cubic feet of hydrogen capacity. Attached beneath its hull were two open gondolas, one near the bow, and the other the stern. A Daimler engine was installed in each, with Zeppelin claiming they would drive the ship at 20 mph for twenty hours—rather unlikely.

Based on his conversations with Professor Bach, Zeppelin had concluded that the gas envelope needed a rigid aluminum skeleton to maintain its shape and provide adequate support for the weight of the equipment and machinery. Thus he planned to install a series of six-inch-wide tubular rings or hoops, each 36 feet in diameter and spaced about 26 feet apart along the length of the airship. Each was strengthened by a solid vertical strut. The rings were connected and held in place by longitudinal girders of the same material and diameter. There were four of these: One ran along the top of the hull, another the bottom, and the other two along the sides. The rings were braced radially with twenty-two wires each. Within, the gas would be contained, as Zeppelin had always promised, in independent cells.[25]

So eager was Zeppelin for his masterpiece to be evaluated that he soon gained a reputation as a pest, though the count preferred to call it persistence. He sent letters to all and sundry within the government and the military, all of which seemed to vanish into the ether.[26] Finally, on November 4, more than six weeks after his first approaching Schlieffen, the Ministry of War, mostly to make him go away, agreed to form a committee to judge his design. Far from being appeased, Zeppelin pressed harder. He wrote: "May I venture to assume that, should the future committee find no errors in my calculations, the Royal Ministry of War will be willing and able to provide the funds necessary" to build the airship?[27]

No, he shouldn't assume anything of the sort, the ministry annoyedly replied: This was merely the first step in a very, very long process. Then, for more than three months, Zeppelin heard nothing back, despite sending several more letters of inquiry.[28] Just as he began to suspect that yet another conspiracy against him was being hatched in Berlin, the once-laggardly War Ministry transformed into a fireball of energy. In mid-February 1894, it told a surprised Zeppelin that the committee would convene on March 10—lightning-fast by government standards.[29] Berlin had been spooked by something, but what?

It turned out that the Russians were building an airship.

6. The Pivot

A T THE TIME, diplomatic relations with Russia had chilled after Kaiser Wilhelm II had elected not to renew their alliance, and German bankers had cut off lending to Saint Petersburg. The vacuum had quickly been filled by French promises of battleships and financing, heightening the kaiser's fears that Germany might be "encircled" from the west and the east. Those fears became a reality in December 1893, when the Russians concluded a treaty with France pledging mutual military support in the event of a German attack. Hoping to counter the Germans' fearsome army in that unhappy event, the Russians wanted an equally fearsome air weapon and had (mis)placed their faith in an itinerant Austro-Hungarian named David Schwarz, a former timber merchant.[1] At some point, he had happened to read some newspaper articles about *La France*'s success in 1884 that inspired him to dream of building his own powered airship. But his would be different from anyone else's in one major respect: It was going to be made of *metal*, the whole thing, including the gas balloon. And that metal was aluminum.

His countrymen may not have been interested, but the War Ministry in Saint Petersburg certainly was. After much delay, mounting costs, and a torrent of excuses, Schwarz managed to have a finished prototype ready on March 22, 1894, but it was, predictably, a piece of junk.

The airship was small, containing just 115,560 cubic feet of hydrogen within a cylindrical aluminum body, its skin a mere 0.2 mm thick. Unfortunately, lacking as he did technical knowledge, Schwarz had not realized that hydrogen expands and contracts as altitude and temperature change.

In the absence of independent, adjustable gas cells, intolerable pressure or vacuum stresses would be placed on the rigid aluminum shell if a pilot attempted to rise beyond 650 feet—about the same height as the tethered ascensions of the old traveling aeronauts and army observers in their rather less expensive and complex balloons.

Schwarz didn't even get that far. As he filled it with gas in a Russian winter, the sheeting bulged inward and collapsed in on itself owing to a sudden drop in temperature. His employers, furious at the failure and suspecting Schwarz of misappropriating funds, abandoned the project, forcing him to flee the country just ahead of the secret police. Schwarz died a few years later of cardiac arrest, but he had nonetheless given the count the kiss of life.

Before the ignominious demise of Schwarz's airship, the German military attaché in Saint Petersburg had sent reports home regarding its progress, and it was these, combined with the French alliance, that had so alarmed the War Ministry and prompted Zeppelin's out-of-the-blue invitation to present his design. If the Russians had a powered airship, then Germany needed one, too—and Zeppelin's was their only option.

THE LONG-DELAYED MEETING took place in the vice chancellor's office at the well-regarded Royal Technical University in Charlottenburg, just outside Berlin. Zeppelin, at least, could not complain that he wasn't being taken seriously: The most eminent physicist in Germany, for instance, Hermann von Helmholtz, served as chairman. Also sitting on the committee was Professor Heinrich Müller-Breslau, a leading structure analyst and bridge designer; Professor Adolf Slaby, an electrical engineer and an expert on the internal combustion engine; and a certain Richard Assmann, an eminent meteorologist and specialist in measuring atmospheric humidity and temperature in a balloon. If any group of individuals could authoritatively appraise the virtues of Zeppelin's rigidly structured, powered airship, it was this one.

Less happy for Zeppelin, however, was the presence of Captain Hans Gross and Major Stephan von Nieber. The latter had been raised to command of the PAB the year before, and Gross was a rising star in the unit.[2] Aged thirty-four and equipped with a dashing handlebar mustache, Gross was the youngest member of the committee, but he had already made a name for himself by breaking the world record for altitude by ascending to thirty thousand feet.

Neither Nieber nor Gross was inclined to back a project that ran against the grain of the PAB doctrine that airpower was best restricted to short-range reconnaissance and scientific observation conducted by small balloons. With Nieber being new to the job, it fell to Gross to serve as the in-house lighter-than-air expert.

Despite Zeppelin's confidence that the committee would fall in love with his perfect airship, the hearings were a debacle. No one believed it would fly as far and as fast as the count was claiming, and Gross pointed out that it had no discernible military applicability. But the most damning indictment was delivered by the engineer, Müller-Breslau. According to his calculations of how the infrastructure would compress, bend, stretch, and twist under stress and strains, the airship was fatally weak.

In this respect, the inexperience of Zeppelin's Grade IIb technical designer, Kober, had weighed disastrously against him, for he had accounted only for *vertical* rigidity; that is, the airship could probably handle the bending stresses of ascents and descents for a limited time thanks to its struts and rings, but he (and Zeppelin) had ignored the need for horizontal strength. In short, the flimsy longitudinal aluminum girders would buckle at the slightest breath of a crosswind. It was as if Zeppelin wanted to build a house with a roof and a floor but no walls.[3]

Müller-Breslau's embarrassing revelation convinced Zeppelin that Kober had to go. His skills simply weren't up to the task, but he had done the best he could and had been, more important still, loyal to a fault— always a crucial consideration with Zeppelin. The two parted on good terms, the count providing a gracious reference letter noting the young man's sterling character and perseverance.[4]

The committee unanimously and predictably concluded on July 10, 1894, that it must "refrain from realizing the project."[5] Zeppelin rejected its rejection out of hand. At the end of August, he submitted a detailed rebuttal of the committee's findings, including with it an angry denunciation of Müller-Breslau as biased and ignorant. At the War Ministry, the letter was widely ridiculed, its author still more.[6]

Yet Zeppelin was not entirely wrong to complain about Müller-Breslau's conclusions. The science of stress analysis was in its infancy, and Müller-Breslau was not an aeronaut; he was accustomed to studying large immobile structures, like bridges, not lightweight, airborne vehicles.[7] In light of this, Zeppelin demanded in a fiery rebuttal that the committee be reconvened. Normally, such a demand would have been ignored: Zeppelin had had his chance and had flubbed it. But, almost unprecedentedly, he was given a second one.

The reason was that Professor Müller-Breslau, now chairman after Helmholtz's death, was, notwithstanding Zeppelin's derision, a genuine scholar in that he kept an open mind and was committed to arriving at an honest and unbiased conclusion. The professor, having perused Zeppelin's arguments, conceded several points where he might possibly have erred and recommended that the committee take another look at the count's proposal.

Just before Christmas 1894, the committee reconvened to discuss Zeppelin's amended scheme. For this second round, Zeppelin had added a metal truss that ran the entire length of the hull, like a splint to secure a broken limb. It was a crude solution to the problem of lateral bending, but Müller-Breslau conceded that the truss was probably "sufficient for working the ship through wind currents of moderate strength" as long as it traveled at low speed.

Even so, Captain Gross insisted that if the military had to have a powered airship for transporting troops and matériel, it must be capable of a minimum cruising speed of 28 mph to overcome high winds—and the most Zeppelin's (theoretically) could do was a puny 13.5 mph, less than the once-standard benchmark of 15 mph.[8]

On March 2, 1895, the committee released its final verdict: "The highlighted flaws of the project presented have to be considered so essential that we cannot recommend its continued progress" to the War Ministry.[9]

It was over for Zeppelin. In May, he tried one more time to interest Wilhelm II, who asked General Walther Bronsart von Schellendorff, the war minister, whether he might throw Zeppelin a little cash to salve his amour propre? The minister replied that perhaps he could borrow it from Jules Verne, Zeppelin's airship being as fantastical a creation as Captain Nemo's submarine.[10]

Zeppelin's anger only strengthened his resolve. If the military, in yet another snub after his humiliating dismissal from the army, was determined to be fools, and the ivory-tower experts snobbishly refused to acknowledge his genius, then he would show them the error of their ways. He would take his idea to the businessmen, the rising power in a rapidly industrializing nation as the twentieth century approached. *They*, men who possessed the drive to solve problems to make profits, would bring the airship to fruition.

Zeppelin's decision to circumvent the traditional avenues of the military, the court, the government, and the elite universities was a distinct turn in his character. As a noble of ancient pedigree, dealing with vulgar businessmen—people in "trade"—tended to offend his sensibilities, but if he wanted to build his airship, he needed them.

Zeppelin's pivot had been inspired by a single, seemingly throwaway line buried deep inside the committee report's dense pages: "Only if the machines were already in service and had proved their value in civil transport could any attempt be made to use them for military purposes."[11] Zeppelin suddenly realized that he'd been putting the cart before the horse in trying so obsessively to gain army support and funding. So immersed in the military world was he that he had never considered building an airship for civilian use.

In the summer of 1895, he wrote to Duttenhofer at DMG to ask his opinion of the committee's conclusions and was gratified when Duttenhofer agreed that they were pedantic and unfair. Duttenhofer proposed inviting Wilhelm Gross (no relation to Captain Hans Gross), the chief design director of the huge Krupp armaments concern, "to come and examine carefully to see who is right on the points in dispute," as Zeppelin put it.[12]

Duttenhofer had his own reasons for being so helpful. The troubled company's car business had vaporized after Daimler and Maybach departed, but Duttenhofer had patched things up and the prodigal sons were coming back—and were bringing with them the precious Phoenix engine. Perhaps it could power an airship as well as it could a car, thought Duttenhofer, who urgently needed to show his long-suffering shareholders that DMG could make money.

Herr Gross of Krupp soon earned Zeppelin's eternal gratitude by dismissing Müller-Breslau and the other experts as dolts. Jubilantly, Zeppelin wrote that Gross reckoned the ship would have a speed of 28 mph, thereby satisfying the *other* Gross's demand.[13]

Regrettably, Herr Gross's estimates were as hopelessly optimistic as Zeppelin's. Gross was a former artilleryman and, despite Zeppelin's bragging that he was "Germany's leading expert on the resistance of air to moving bodies," he had conducted his experiments with Krupp's artillery shells—which behave very differently as they hurtle through the air than a large, self-propelled vehicle. Zeppelin should have known this, and indeed probably suspected that that was the case, but Gross had nonetheless provided him with a major psychological boost at a critical moment.[14]

In December 1895, Zeppelin published a manifesto based on Gross's calculations intended to raise 800,000 marks to fund a working airship. The manifesto marks a key shift in Zeppelin's concept of an airship. He had finally dropped the cumbersome sky-train plan in favor of a much simpler all-in-one vehicle.

He sent the pamphlet to a large number of industrialists and bankers, but the effort was a fiasco: Zeppelin managed to scrounge together just 100,000 marks from some old friends and a few members of the Württemberg royal family. As he walked down the streets of Stuttgart, Zeppelin was ridiculed and taunted as a lunatic.[15]

Attempting to drum up support, Zeppelin tried a different approach in January 1896. With Professor Bach's backing, he successfully applied for membership in the prestigious Association of German Engineers. A month later Zeppelin found himself presenting his project before its Württemberg chapter of 435 members. Afterward, several professors praised the design as a "very important, expert piece of work" and a "laudable manifestation of [the] deepest German thoroughness."[16]

Determined to bury Müller-Breslau, Gross, and their accomplices once and for all, Zeppelin followed up this success by appealing to the association to form its own committee of engineering experts to evaluate his proposal. If they reported positively, he would then have the ammunition he needed to go out and find willing moneymen.[17]

On October 25, 1896, the engineers' committee favorably presented its case for Zeppelin to the association board, which published the report on December 21. It was better than even Zeppelin could have hoped. Zeppelin's project was important in its own right, the association concluded, in order to explore the world-changing potential of conquering the air. Though many people were "skeptical about everything pertaining to aviation," the association believed that similar objections had been thrown at the inventors of the great steamships and continent-crossing trains. The first step on the path of progress, then, must be "the construction of an airship based on the Zeppelin model."[18]

Another important ally appeared in the form of Hermann Moedebeck, a former PAB man who was counted as the Brahmin of ballooning—much to Captain Gross's envy. He now edited the enthusiasts' journal *Illustrated Aeronautical News* and regarded his job as being to publicize the glorious future of German airships to a patriotic nation. It was Moedebeck who realized that not only did one have to set the German people's imagination aflame with the tantalizing possibilities of flight, one also had to make it *relevant* to them so that the public's excitement would sweep away the naysayers.

Moedebeck railed against the scientists tediously (as he thought) conducting meteorological experiments and writing about them in desiccated academic journals. Instead, he promoted *more* spectacle, *more* showmanship to get the blood pumping and the pulses throbbing. In the April 1897 issue of *Illustrated Aeronautical News*, he influentially weighed in on the Zeppelin airship: "We have complete confidence that the realization of this, the best of all projects so far, will enrich aviation with great experience."[19]

The Moedebeck article caught the eye of Carl Berg, who hailed from a long line of Westphalian ironworks owners and had taken over the family firm in the early 1870s. In 1889 he had visited the Paris Exposition and seen the new, amazing "silver made from clay" metal—aluminum—and brought some samples home. Three years later, he entered the aluminum business full-time and acquired the license from the War Ministry to manufacture the army's canteens and mess kits. Now he offered Zeppelin enough aluminum at a bargain-basement price to build the airship skeleton.

Incensed at Zeppelin's success, Captain Gross of the PAB foolishly accused the count of copying design elements of Schwarz's airship—for which insult Zeppelin challenged him to a duel. A showdown with heavy cavalry sabers was averted only when the kaiser forbade the confrontation on the grounds that "in the battle to conquer the air, both gentlemen were 'Officers before the enemy' and as is well known, there can be no duels during war times."

With the road ahead now clear, Berg, the count, Bach, Duttenhofer, and Moedebeck founded the Society for the Promotion of Aviation (SPA), a company dedicated to building a Zeppelin airship.[20]

THE SPA GOT right to work. In January 1898, the organization sent another appeal for 800,000 marks to members of the Association of German

Engineers and major industrialists. Moedebeck did good work advertising it in the press without burdening readers with an acknowledgment of his own financial involvement. By June, some sixty-five shareholders had purchased bonds totaling 369,000 marks, with Berg alone contributing 100,000 (plus donating another 60,000 marks' worth of aluminum). The rest was purchased by Zeppelin himself, giving him majority control.[21]

It proved surprisingly easy to recruit young talent attracted by the thought of working on such a futuristic project. Professor Bach recommended another of his former students, Hugo Kübler, aged twenty-six, a machinery expert, to serve as project manager and to oversee construction. Zeppelin insisted that he move into his house in Stuttgart, as "our close companionship . . . will be of service for promoting the work."[22]

Soon joining them was the twenty-two-year-old Ludwig Dürr, another Bach pupil from Württemberg. Zeppelin assigned him the job of assembling the finished components of the airship under Kübler's supervision.[23]

Dürr was a very odd duck, it soon became apparent. Long, lean, and missing his right middle finger (engineering accident), with prominent black eyebrows, a precisely cultivated goatee, and piercing eyes that focused on a page of impenetrable calculations with hawklike intensity, Dürr was an utterly humorless loner. Lacking any interest in culture — airships would be his art — he rarely emerged from his office, and when he did, it was only to participate in such solitary pursuits as hiking and bicycling.

At first, he was mocked by the workmen as just another, in the words of one observer, "unimpressive and retiring young man" with an airy-fairy university pedigree and no practical knowledge, but he soon proved them wrong.

There came a day when Dürr expressed disappointment in the quality of a factory-made turnbuckle (a device for adjusting the tension and length of cables), and Preiss, the locksmith in charge of certifying and testing them, decided to embarrass Dürr by challenging him to make one for himself. Dürr went over to Preiss's machine and forged a new one from scratch. The locksmith recognized that no one could have made a better one, but where was the fun in that? So he claimed to have found a minute crack in it. Wordlessly, Dürr disappeared for a few hours and returned with a turnbuckle so flawless that even Preiss was taken aback.

From then on, nobody questioned Dürr's skills. He soon became the only man alive bold enough to disagree publicly with Zeppelin on matters of design and construction: Because Dürr was always right and Dürr *knew* he was always right—even when he was wrong. For his entire life, virtually everyone—apart from Zeppelin, oddly enough—found him "a very difficult man to deal with."[24]

In June 1898, the SPA headquartered itself at Friedrichshafen, a beautiful little town described in a guidebook as a "much-visited health resort and tourist destination" on the shore of Lake Constance. A castle—the summer residence of the Württemberg royal family—lent Friedrichshafen a touch of class, and there were several villas built for retired courtiers. Tourists came, though not in great numbers, to enjoy the spas and to taste the local delicacy, goats' whey—touted as a medicinal tonic. In the evening, there was little to do, so visitors were advised to head to an inn where they could—the guidebook tried to sound enthusiastic—participate in conversations "with a large number of officials employed by the post office, railway, and customs" as well as "several local characters" whose talk "revolves around materialistic and business-related subjects."[25]

Zeppelin built his construction facilities at Manzell (just west of Friedrichshafen) on the lakefront, the king of Württemberg helpfully providing the land rent-free for ninety-nine years. It was an idyllic place, a utopia for those seeking isolation and quiet between the snowcapped peaks of Switzerland, looming just across freshwater Lake Constance, and the emerald fields, vineyards, medieval castles, and forests of Württemberg. The water was green-hued, and the sky, on a clear day, icy blue.

More prosaically, Manzell was as flat as a pancake, the winds there were more constant than anywhere else in the vicinity, and the nearby water would provide a soft, calm landing area—factors critical for airship flight-testing.

Zeppelin's arrival was just one of several signs the modern age was intruding on the area's rustic charms. Electric street lighting was imminent, and lakeside steamship piers and railway stations were being built—all the better to import the industrial products needed by the count to achieve his dreams. The chemical company Griesheim-Elektron sent hydrogen, DMG its engines, Gradenwitz the gas valves and gauges, and the Rhine Metal Works the gas cylinders.

As the airship began to take shape, Zeppelin, now sixty, hired a colorful Swiss balloonist named Eduard Spelterini to give him flying lessons. Spelterini had once toured the world giving acrobatic performances in the company of an exotic trapeze artist named "Leona Dare"—born Susan

Stuart in Mattoon, Illinois—whose star turn was to swing by her teeth from a dangling rope. He and Zeppelin got along surprisingly well. Maybe he reminded the count of old John Steiner from Saint Paul.[26]

The influx of private money allowed Zeppelin to move much faster than would have been possible had he stayed on the military track—so fast, in fact, that he announced that his first *Luftschiff* (airship) *Zeppelin*, LZ-1, would be launched on July 1, 1899. In the event, things went according to plan but not to schedule, and the count was forced to push back the over-optimistic launch date by a year.

Zeppelin himself remained serene, but several of his backers grew worried that every day's delay increased the risk that LZ-1 would be obsolete before it even flew.

Because a new threat was on the horizon: the airplane.

7. A Wonderfully Ingenious Toy

ZEPPELIN PAID LITTLE attention to airplanes. If the history of the airship had seemed unpromising until he came along, then airplanes, if anything, suffered from a reputation still worse.

On the face of it, flying is easy. Or at least birds make it look so. They flap their wings to pick up speed and attain height and outstretch them to soar; they swoop and bank and dive with casual abandon. In nature, they seemingly demonstrated the principle that the Lord designed His creatures so that nothing is either wanting or superfluous. For centuries, then, it had been only reasonable to expect that man, His finest if haughtiest creation, could emulate the birds. If only a man donned artificial wings of fabric or feathers, leapt off a height, and furiously flapped his arms, he too could be a bird.

Many of these deluded souls died horribly at an alarming velocity. Part of the mystery why none of these "ornithopters" (aerial contraptions that replicated the flapping-wing action of birds) worked was eventually solved by a scientist named Giovanni Borelli in 1680.

In his tract *The Movement of Animals*, he demonstrated that birds were physiologically different from humans in critical respects. Ours is a puny species, aviationally speaking. In average healthy humans, muscles are fairly widely distributed and together comprise around 40 percent of body mass, but a third or more of a bird's weight is concentrated in its breast muscles alone: namely, the *pectoralis* (which pulls the wing down) and the *supracoracoideus* (which raises the wing). Given also that our skeletons are relatively much heavier than those of birds, we're simply too cumbrous and weak to conquer the air.

Add to that the disappointing revelations that our feathered friends produce exponentially more power per pound than we can, *and* that they can sustain elevated energy output for immensely long periods (think of those long migratory flights) and it was only natural for Borelli to conclude, sadly, that "it is impossible that men should be able to fly artificially, by their very own strength."[1]

Borelli's proof persuaded virtually everyone to give up on the quest for winged flight as a fool's dream. In 1780, the august mathematician Monsieur Charles-Augustin de Coulomb added to the skepticism by calculating that for a man to fly he would require wings half a mile long beating three times per second. Since no man could ever achieve that tempo, let alone bear the wings' enormous weight, he concluded that "no one but an ignoramus would make an attempt of this kind."[2] Three years later, budding aviators like the Montgolfiers settled instead on the easier alternative path of floating in a balloon.

By 1800, there was just one man left who believed that he could fly: a mild, studious Englishman named Sir George Cayley. He turned himself into the world's first and only aerodynamics engineer. As such he was obliged to invent a fresh branch of science—by himself, from scratch.

Cayley's first great insight was to break the ability to fly down into components, each of which represented a problem that needed solving. Cayley laid out the four factors of flight that, if harmonized into a system, would produce success.

First, aerodynamics. Movement produces a flow of air that exerts two forces on a vehicle. There's perpendicular *Lift*, which raises the craft and is generated almost magically by the stream of air rushing over, around, and under the wings. And there's *Drag*, which slows the aircraft owing to oncoming air resistance.

Second, propulsion. The aircraft must be pulled or pushed through the air by an engine or some other motive force (like muscle) to create *Thrust*. This works in the opposite direction to drag and must be greater than the latter to overcome it.

Third, structure and materials. The aircraft must be built in a way to avoid crumpling, buckling, or snapping as aerodynamic forces stress the structure. The aircraft's total *Weight* (which includes pilot, passengers, fuel, and payload) works in the opposite direction of lift and so gravitationally draws the vehicle down. For steady, level flight, weight must equal lift.

Fourth, flight control and stability. There must be a means of regulating motion to enable banking, climbing, and diving. These include control surfaces such as horizontal hinged elevators on the tail, a vertical rudder, and ailerons, or flaps, on the wings' trailing edge.[3]

Earlier aviators had copied birds by constructing wings that played the dual role of providing thrust *and* lift—and had failed by trying to do too much at the same time. And worse, since they had naturally assumed that the larger the wing the more work it could do, they had contrived ever-heavier, draggier, more uncontrollable craft.

So the first thing Cayley did was dispense with the ornithopter concept: no more all-in-one flapping wings. His wings would be fixed in place and exclusively assigned to provide lift. After that came structure and control. Get those three elements right, and all he had to do was add power to fly. He left that until later. By 1804 Cayley had built a working, yard-long model of a glider—an airplane without an engine. Today it seems like a child's toy, but its radical new form amounted to a revolution.

Cayley had singlehandedly invented the shape that would be the basis of every airplane ever since. Instead of taking one part of a bird—its wings—and grafting them to a man to form a hybrid resembling some beast from a Greek myth, as so many others since the days of Daedalus and Icarus had done, Cayley copied the entire bird and removed the man. A thin wooden rod formed the glider's nose and body, which he dubbed the "fuselage" (French for "spindle-shaped"), with a pair of kitelike wings mounted on it, while at the rear was a cruciform tail. When Cayley hand-launched his glider from atop a local hill that year, with his arm serving as engine, it was the first airplane in history to fly.

In Cayley's triumphant words, "It was beautiful to see this noble white bird sail majestically from the top of a hill to any given point of the plain below it with perfect steadiness and safety."[4]

But he never could find a means of sustained propulsion. In 1807, he conceived a mini-engine that burned gunpowder to generate a single horsepower—for all of fifty-seven seconds. Then he investigated larger steam and hot-air engines, but they too could not defeat the weight-to-power ratio that would similarly afflict Giffard, Renard, and other aeronauts for nearly a century to come: The output was too low to offset their extra weight. No airplane could take off, let alone fly. Depressingly, Cayley could not foresee a day when an engine's weight-to-power ratio would fall below 163 pounds per horsepower.[5]

In 1809–10, he surrendered and published an extraordinary three-part scientific paper, entitled "On Aerial Navigation," that detailed his decade's worth of experiments and thoughts. Nobody had ever read anything like it, this aeronautical equivalent of Darwin's *Origin of Species* or Newton's *Principia*, but anyone who tried to follow in Cayley's footsteps ran into the same wall: lack of power.[6]

Cayley himself may have given up the quest but not the dream. As he put it, wistfully, "an uninterrupted navigable ocean, that comes to the threshold of every man's door, ought not to be neglected as a source of human gratification and advantage." He felt "perfectly confident" that the "noble art [of flying] will soon be brought home to man's general convenience, and that we shall be able to transport ourselves and families, and their goods and chattels, more securely by air than by water, and with a velocity of from 20 to 100 miles per hour." All that was still required was "a first mover, which will generate more power in a given time, in proportion to its weight, than the animal system of muscles."[7]

The aged Cayley derived some satisfaction from Giffard's 1852 dirigible briefly managing to achieve a weight-to-power ratio of 110 — unimaginable in his prime working years — but he always retained his faith that the future of flying would come in the form of the airplane, not the airship.

IN 1866, NEARLY a decade after Cayley's death, a rump of true believers in, or apostles of, his vision of heavier-than-air flight congregated to form the Aeronautical Society of Great Britain to advance the cause of "Aerial Navigation and for Observations in Aerology connected therewith." According to the Duke of Argyll, its president, "the absence of the lighter motive power [of an engine] ought not to stop us from investigating the principle" of flight in order to prepare for that blessed day when a satisfactory one would pentecostally descend upon the world.

Until then, he continued, there was still plenty to explore, so much to know. Not only "are we still ignorant of the rudimentary principles" of airplane construction, but knowledge of bird flight was paltry, and "we are equally ignorant of the force of the wind exerted on surfaces of various sizes, forms, and degrees of inclinations." Thanks to the sainted Cayley, they had been led to the mysterious black box of aerodynamics, but the golden key to unlock it lay hidden still.

In their quest, the society's members enthusiastically conducted their own studies of flight. For instance, Francis Wenham groundbreakingly observed that the swiftest "birds possess extremely long and narrow wings, and the slow, heavy flyers short and wide ones." He speculated that a successful airplane should copy the former in having "high-aspect-ratio wings" (defined as wingspan divided by chord, or the breadth of the wing from leading to trailing edge) to generate additional lift. A couple of years later, Wenham conceived the world's first wind tunnel, allowing him to conduct multiple experiments in a laboratory. By rigging a steam-powered fan to blow air at a top speed of 40 mph through a rectangular wooden duct, Wenham was finally able to begin measuring and quantifying the effects of wind on wings.[8]

A twenty-seven-year-old Londoner named Horatio Phillips heard Wenham's lectures and thought he could do better. Wenham's wind experiments had been conducted with flat plates set at different angles; in 1884, after building a more efficient wind tunnel of his own, Phillips used cambered, or curved, airfoils that dramatically increased lift. Cayley had long before speculated that cambering was a secret ingredient to flight but had done little with it. Phillips, on the other hand, proved that when air flowed quickly over the curved upper surface of a wing its pressure decreased and hence lift was actually generated by the relatively slow-moving, *high*-pressure air underneath pushing upward.

Phillips had at last extinguished the ancient assumption that a bird's wings' flapping was the source of the mysterious power of lift. From then on, fixed cambered wings were the rule in airplane design.[9] Yet these wonderful advances could not, despite myriad attempts, be translated into practical success. Still lacking a suitable engine, their airplanes simply could not fly. It was as if one had learned the rules of chess but lacked the board and pieces to play it.

HIRAM MAXIM INTENDED to become aviation's grandmaster, and he could well afford as many boards and pieces as he wanted. Born the son of a Maine farmer in 1840, he had limited schooling but an unquenchable fascination with machines. In 1881, he emigrated to Britain and two years later marketed the Maxim machine gun, the rapid-firing weapon that earned him fame (and infamy) during several imperial wars in Africa, as well as a very, very large fortune. Having conquered the world of armaments and looking for new ones, in 1889 Maxim set his sights on designing airplanes.

By throwing the enormous sum of £100,000 at the problem of flight, Maxim believed, he could solve it within five years. He constructed a large hangar, hired two American mechanics, purchased a lot of experimental equipment, and knuckled down to build yet another world-shattering invention.[10]

Five years. Maxim was as good as his word. In the summer of 1894, his airplane debuted. A devotee of the principle that bigger is better, Maxim had birthed a colossus. Nothing else like it had ever existed. The appositely named *Leviathan* had a wingspan of 104 feet and a total surface area of 4,000 square feet spread over its four stacked wings (making it a quadraplane). He installed two 180-horsepower steam engines, together weighing 620 pounds, whose boiler alone added another thousand. Their sheer brute force would drive two propellers, each with a diameter of nearly 18 feet, designed to push the 8,000-pound airplane through the air. For once, it seems, engine power was not lacking, though the tremendous weight of the craft—about that of a hippopotamus—militated, to put it mildly, against success.

The machine's four-wheeled undercarriage ran along an 1,800-foot-long railway track with a guardrail attached to prevent the airplane from rising more than two feet above the ground. Maxim's objective at this stage was not to fly a great distance with any altitude, only to demonstrate that his invention was capable of generating sufficient lift to leave the ground under its own power. On his third test run on July 31, the machine's wheels lifted off the track near the end of the rails. Maxim and his two-man crew struggled to contain the beast, but it broke through the guardrail and shot upward for a brief time. "I found myself floating in the air," said Maxim, "with the feeling of being in a boat." He immediately shut off the steam and the plane gently landed.[11]

Maxim held two more public demonstrations over the course of the following year, but after so much labor and so much money, even he had to call it quits, and the mighty *Leviathan* was quickly relegated to the status of a historical curio. Perhaps Maxim was hurt by the eminent British scientist Lord Kelvin's dismissive, if not inaccurate, comment that it resembled "a kind of child's perambulator with a sunshade magnified eight times."[12]

Maxim nevertheless does deserve credit for trying, but what he could never do is achieve true powered or sustained flight, defined as the ability of an aircraft to demonstrate a continual horizontal or rising path without loss of airspeed. Maxim's airplane had in fact taken off by dint of the momentum it acquired as it barreled down the rails, making its "flight" more of a hop than anything.

The oddest thing about it was that Maxim had achieved an astoundingly low weight-to-power ratio of just 4.5—lower than anyone had even thought possible. Put differently, according to the tenets of existing aerodynamics, he *should* have been able not merely to hop but to actually fly, guardrail or not. After all, had his engines not made manifest Cayley's Holy Grail of "a first mover, which will generate more power in a given time, in proportion to its weight, than the animal system of muscles"?

There was still something missing, but what was it?

AT THIS TIME, the 1890s, there was virtually no contact between the heavier-than-air "aeroplanists" and the lighter-than-air aeronauts. The cloud, as aeronauts like Zeppelin saw it, was distinct from the bird, and each faction regarded its rival as foolishly misguided.

From the count's perspective, Maxim's absurdity seemed only to point to the continuing futility of airplanes. It was obvious, he thought, that the only viable aviation technology was the airship.

Lest we think Zeppelin was exceptional in thinking this, his was in fact accepted opinion. Upon being asked to join the Aeronautical Society of Great Britain in 1896, for instance, Lord Kelvin, the mathematical physicist who calculated the temperature of absolute zero, rejected the invitation by replying dismissively, "I have not the smallest molecule of faith in aerial navigation other than ballooning."[13] Three years later, in December 1899, *Scientific American* judged that heavier-than-air flight will either "remain a very hazardous and fatal form of recreation" or at best "a wonderfully ingenious toy."

The magazine believed that the seemingly insuperable obstacle to greater success was that by their very nature airplanes were based "upon the principles which govern the flight of birds." This meant that any pilot must also possess "that God-given faculty by which the bird is able to pre-serve its equilibrium, adjusting the position of its weight and the inclina-tion of its wings to the ever-changing velocity and direction of the wind, and the varying speed and direction of its own flight" to have any hope of staying aloft. Until budding aviators could actually *think like a bird* and react instantaneously to an ever-changing environment, the airplane could never be "a machine of commercial or military utility."

So *that* was where Cayley, Maxim, and all the others had gone wrong. Flying an airplane was not merely a matter of getting the technology, ra-tios, and design right to *copy* the body of a bird; one had to have a bird's brain to control its motion. Without that "God-given faculty," an airplane was merely a lifeless simulacrum of an animal, a replica with no spirit.

The airship, however, was different. Since it was essentially a motorized cloud, "the efforts of the operator[s] may be devoted entirely to steering and propulsion"—which were jobs any half-decent helmsman and a good mechanic could do without much trouble.[14] Ultimately, *Scientific American* voted for the airship as the likeliest to succeed.

Within a few years the conventional wisdom would be proven wrong, not only because Zeppelin would discover that piloting an airship was a *lot* harder than *Scientific American* had so blithely imagined, but because of two birdbrained brothers from the American Midwest.

On May 30, 1899, a month before Zeppelin had first proposed launch-ing LZ-1, one of them, a certain Mr. Wilbur Wright, wrote to the secretary of the Smithsonian Institution in Washington, D.C. In his letter, he as-sured the secretary, who was evidently accustomed to oddballs pestering him, that "I am an enthusiast, but not a crank in the sense that I have some pet theories as to the proper construction of a flying machine," and politely asked for any information he might supply about heavier-than-air flight.[15]

8. The Folly

JUST BEFORE LUNCH on the most important day of his life, now nearly sixty-two years long, Count von Zeppelin signed his will and left it propped up on his desk in his modest, cluttered office above a former fish market in Manzell.[1] He'd been born nearby, and maybe he would die nearby. An attached note to his wife, Isabella, read: "If it is God's will that I meet with an accident during the intended air journey, I will take with me the happy certainty of your conviction that I did not undertake this journey recklessly."[2]

A taste for recklessness had once been part of Zeppelin's nature, but age and bitter experience had long since tempered him. In the dangerous world of the aeronauts, one could be old or one could be bold, but nobody survived being both. Zeppelin was old. The risks were great, but he had thought them through, which was quite a different thing than pitching heedlessly headlong into hazard, as he had done during his ill-fated raid during the Franco-Prussian War. Accordingly, today he betrayed no hint of anxiety as to his prospects. His life was in the hands of the Lord, and the Lord would protect him, as He always did with Zeppelin.

The postponed date scheduled for this, the maiden voyage of LZ-1, was July 2, 1900. He had been working ten-hour days since 1898 (including Sundays and holidays) to build his craft. Every piece of the infrastructure, down to the very rivets, had been fabricated precisely according to his specifications; each segment of the airship had been assembled at the contractors' factories under his beady eye, then dismantled and each part painstakingly labeled and boxed. After arriving in Manzell by train, his workmen had rebuilt it from scratch.[3]

Later that afternoon, when Zeppelin arrived at the shore of Lake Constance, he greeted the quintet of military observers dispatched from Berlin to report on that evening's voyage. There was Captain von Hülsen, a General Staff officer; Major Klussmann, the new head of the Prussian Aeronautical Battalion; two helpful acquaintances of Zeppelin's, Captains von Tschudi and Bartsch von Sigsfeld; and lastly—but by no means welcomely—his old enemy Captain Hans Gross, also of the PAB.[4]

To counterbalance the possible hostility of the PAB, Zeppelin had made sure to make the great event public. It would be difficult for Gross to torpedo the project if the press and the people witnessed it with their own eyes. He was gratified to see the swelling crowds who had come out for the occasion.

Local vendors had set up stalls to hawk snacks as children skittered and ducked around their friends and nannies. The Quality—dressed in their Wilhelmian finest and stiffly polite to one another—promenaded along the boardwalk or picked their parasoled way over the pebbly beach. To watch the day's big event, some of the sightseers had hired boats—ranging from modest fishermen's rigs to flashy private steam yachts—which bobbed gaily upon the placid water.[5]

The workingmen, many employed by Zeppelin, had brought their families and humbly kept their place away from their middle-class superiors. They perched on the dockside walls as their wives fetched beer and sausages and the kids clambered up to the rooftops hoping to catch a sight of the mysterious thing their fathers had been working on.

They were to be disappointed. The "thing" was entirely hidden in a huge barnlike hangar—the largest wooden structure in the world—which, weirdly, was floating in the middle of the lake. Weirder still, it slowly revolved on its central axis as the tide and wind shifted. Little did they know, but the 142-yard-long, 23-yard-high building had a draft of only about 2.5 feet and was anchored to the lake bed 20 yards below by means of a chain attached to a concrete block weighing 41 tons.

Zeppelin took a launch out to the hangar and vanished inside.[6] For the rest of the afternoon, he supervised the final checks of the hydrogen levels and water ballast and made sure that the engines were running smoothly and that the controls were balanced. At around 6:00 P.M., he reappeared on a small floating platform and the crowd stopped chattering and laughing.

The count was not a young man, but he'd retained a youthful energy. Forty years earlier, when he'd visited America during the Civil War, he'd affected a luxuriant blond beard, long since shaved off. He'd kept the (now white) bushy walrus mustache. He'd lost his hair, apart from a snowy halo up top, and his once-trim cavalryman's figure had been replaced by a retired general's paunch, but the twinkling blue eyes hadn't changed and neither had the air of indomitable perseverance. He no longer wore a uniform unless attending an official ceremony, and for the maiden voyage of LZ-1, a ship of the air ocean, he'd decided upon a natty combination of double-breasted navy blazer and a white yachtsman's cap.

Joining Zeppelin on the platform were his friend Baron von Bassus (who had some experience flying balloons), the well-known African explorer and journalist Eugen Wolf; engineer Fritz Burr; and a mechanic named Gross.[7]

The count led them in a short prayer before signaling for the great doors of the hangar to be opened to reveal the marvel hidden within.

THE VAST HANGAR emptied its womb. Gasps of astonishment greeted the birth of the first Zeppelin airship as it hesitantly emerged from the darkness.

At 420 feet, nothing larger had ever been built in the history of aviation. Cylindrical with bullet-shaped ends, the Brobdingnagian baby had skin that was pale, smooth, and taut, made of Pegamoid, a specially impregnated waterproof cotton designed to reduce air resistance more efficiently than (expensive) silk. At the front, for turning, were two vertical rudders made of stretched cotton; at the stern, two of the same. A hand-operated crank could move them simultaneously, but in opposite directions, so if the front pair turned to the left, the rear ones moved to the right.

There were twin twenty-foot-long, boatlike aluminum gondolas connected to the airframe undercarriage by tubular struts and wires. Each gondola was located 105 feet from either end and nine feet below the hull. Cleverly, they could float on water in case of emergency, and in each there was one of Daimler's newest products—a four-cylinder, 850-pound water-cooled engine developing 14.7 horsepower. Personally installed by

Wilhelm Maybach, the engines transmitted power to two propellers four feet in diameter turning at 1,200 rpm.

Zeppelin had connected the gondolas with a wire that suspended a sliding, 220-pound lead weight. To adjust the pitch (nudging the airship's nose up or down for climbing or descending), the crew would winch the weight forward or backward. There was also a rudimentary, and rather precarious, gangway between the gondolas so that passengers could transfer between them, if necessary.

What the crowds couldn't see, or appreciate, was the skeleton within. Alone weighing 4.5 tons — the outer skin added another two — it contained nearly six and a half miles of Carl Berg's finest aluminum, in addition to several more of wire. Its rib cage consisted of sixteen transverse (cross-sectional) rings, each a twenty-four-sided polygon braced with a weave-work of chord and radial wires. Connecting the rings were longitudinal girders between thirteen and twenty-six feet long. Within this rib cage were the organs: seventeen drum-shaped gasbags containing a total of 399,000 cubic feet of hydrogen.[8]

The bare-bones controls in the lead gondola were designed solely for function: some pulleys and toggles to valve gas and drop ballast and a small map table with a spirit level, a barometer, a barograph, and a homemade pendulum to mark the ship's banking angle. Communication between the two gondolas was effected by a rudimentary telegraph, megaphones, and two bells.

Zeppelin beamed, basking in the congratulations of those around him. Handshakes done, he got down to the real work of the day.

At 6:50 P.M., Zeppelin and his passengers gingerly stepped aboard LZ-1 and took their assigned positions. The count, his friend Bassus, and Burr the engineer entered the front gondola; explorer Wolf and mechanic Gross, the rear.

There was supposed to have been one more: the project manager, Hugo Kübler. But at the last minute he had refused to go after Zeppelin told him that he had not purchased life insurance — which the count alleged would have been a weakling's admission of defeat — for the intrepid aeronauts. At this unforgivable act of treachery, not only to him but to the airship, Zeppelin had lost his grandfatherly demeanor and stared at the hapless Kübler with icy disdain. Kübler soon emigrated to Argentina.[9]

Kübler's reluctance to trust his life to LZ-1 was understandable, for no one had any idea whether the airship would actually take off. LZ-1 was the same size as an ocean liner of the time, and suspending a body so huge was no simple matter. For all anyone knew, LZ-1 might immediately spiral uncontrollably upward, explode, crash-land, or crumple when a breath of air hit it. Ludwig Dürr, Zeppelin's chief engineer and designer, had loyally assured him that LZ-1 was airworthy, though he based those assurances on nothing more than theoretical calculations worked out laboriously on paper, making the airship's maiden voyage an open-air, full-size experiment in extreme aerodynamics.

Around 7:30 P.M., the raft securing the airship with ropes was towed out deeper into the lake by a motorboat, the *Württemberg*, captained by Max von Gemmingen, Zeppelin's thirty-eight-year-old nephew.[10] Windspeed was a moderate 17 miles per hour, but the lake surface remained mirror smooth. The airship was turned in the exact direction of the wind. At 7:56, Zeppelin ordered the ground crew—volunteers from the Friedrichshafen fire brigade and a local gymnastics club—to ease their grip on the mooring ropes, colorfully striped to mark each meter, and let the airship rise slowly.[11] At a height of seventeen yards above sea level, LZ-1 stayed steady. So far, so good.

At 8:03 P.M., Zeppelin barked, "Cast off!" At this signal, the handlers were supposed to let go of the ropes at exactly the same time, but the rear team held on a fraction of a second longer and LZ-1's nose lurched sharply upward. Zeppelin urgently worked the sliding weight and the craft soon righted to an even keel. Not many people noticed the slip since "thousands of hoorays sounded from the shore," recalled Wolf, and "all eyes followed the flight of the balloon."[12]

But the military observers stayed silent, their lips pursed. Notwithstanding Zeppelin's rosy prediction that the airship would hare along at 20 mph, it was clear that it was moving just a touch faster than the wind. Their disapproval only deepened as they watched Zeppelin struggle with the sliding weight. Its winch had broken, stuck in the forward position. LZ-1's nose was now headed down and came fatally close to dipping into the water moments before the count dumped hundreds of pounds of ballast to bring it up. Thankfully, the airship regained its trim, but the close call had been noted.

The flight itself lasted nearly eighteen minutes, with Zeppelin and the crew trying to learn on the job how to keep an airship stable. Finally, Zeppelin decided to settle on the water some distance away to let the towboats bring it back to the hangar. He hoisted a large blue flag to warn the sightseeing boats to steer clear. "The air ship sank slowly and rested on the water as smoothly as a sea-gull, with no bump, no crash, no rise, no jumping, no sensation whatever," recalled Wolf.[13]

It had been a moderate, qualified success. On the one hand, LZ-1 hadn't crashed or blown up, Dürr's design was evidently sound, and it had ascended to 1,300 feet and traveled in total some 3.5 miles. On the other, the airship had not made a round trip back to the hangar under its own power despite favorable winds, its top speed was unspectacular, and flight control had been lackluster.

Zeppelin had enough money left over to fund two more trial flights. He and Dürr got to work fixing and improving LZ-1. Among other things, it was clear that the airship's rudders were too small to be effective and were placed in the wrong position. Dürr made them bigger and moved the rear pair from the sides to beneath the hull. Henceforth, steering would be a little easier.[14]

The second flight took place on October 17 at 4:45 P.M. To assist him in managing the ballast and gas during ascents and descents, Zeppelin hired Captain von Krogh, a skilled balloonist. Partly thanks to Krogh's expert hand, this time LZ-1 rose smoothly to a height of nine hundred feet and stayed airborne for eighty minutes. The flight would have gone on longer had twenty-five thousand cubic feet of hydrogen not escaped from one of the gas cells, a rudder not become stuck, and a harried mechanic not accidentally poured distilled water into the confused engines.[15]

Dürr made some more refinements, and on LZ-1's third trial a week later, it was kept fully under control by the increasingly experienced crew and demonstrated an admirable maneuverability against the wind. By now, though, the airship's airframe was dangerously weak and hydrogen was running low, so Zeppelin unsentimentally decided to break it apart and begin work on the craft that was to become LZ-2.

Building LZ-2 would have to be postponed longer than Zeppelin expected. A few weeks later, his company, the Society for the Promotion of Aviation, went out of business. A banking crisis loomed—Berlin's stock market was about to suffer a 61 percent collapse—and no one was willing to extend a loan to a firm with no cash reserves. There wasn't even enough money left to replenish the gas, but Zeppelin managed to rustle up just enough from his own rapidly dwindling bank account to buy up the remaining assets and keep Dürr and two night watchmen on the payroll.[16] He was on his own. No one seemed to care that he'd flown, actually flown.

• • • • •

PRESS REACTION TO LZ-1's first flight had indeed been disappointingly sparse, partly because Kaiser Wilhelm II had that very day monopolized the front pages by delivering a fiery, impromptu speech to an expeditionary force sailing for China, where the Boxer Rebellion raged.[17]

While some local newspapers cheered the hero who had defied gravity, national and regional ones either ignored his achievement or were cutting about it. The *Frankfurter Zeitung,* for instance, had sent a reporter, who observed that the "entire countryside was ceremoniously invited to attend a performance to which not even the overture could be played successfully." In short, the only thing LZ-1 had proven was that "a dirigible balloon is of practically no value."[18] *Scientific American,* one of the few foreign outlets to comment on the flight, was more charitable only in concluding that the airship might have some use "for exploring expeditions that are not of too extended a character."[19]

The subsequent report by the military observers was guarded. The airship was "neither suitable for military nor for non-military purposes," which sounded like a typical contribution from Captain Gross, but the observers added that this had been a first flight. LZ-1, they felt, was "an experimental vehicle" and left it at that.[20] It wasn't exactly a ringing endorsement.

The two follow-up flights, however, barely rated a mention. Hugo Eckener, an occasional correspondent for the *Frankfurter Zeitung,* was surprised when the editor asked him whether he'd be willing to cover the second trial. As a freelance writer on economics, philosophy, and society, as well as occasional arts critic, Eckener had no special interest in either aeronautics or airships and hadn't paid much, if any, attention to the happenings at Lake Constance. *Why him?* he asked, only to be told that since he'd recently moved from Hamburg in the north to the warmer climes of Friedrichshafen for his health (rheumatic knees, weak lungs), it wouldn't cost anything to send him there. Also, well . . . no one else wanted to go.[21]

Eckener, finding himself at loose ends and low on cash, accepted the commission but "adopted a rather cool and critical tone in my story," as he later admitted. Condescending is probably a better description:

> [The airship] hovered purposefully and nicely in the air, made little twists on its vertical axis, perhaps even small turns. It also executed small

turns on its horizontal axis but pretty much stayed happily in the same place. There was no evidence of real movement back and forth or of ascents and descents to higher or lower altitude. I had the sense that the airship was delighted to balance so nicely up in the air [because the] balancing act was the only successful part of the whole affair.

Having so casually dismissed LZ-1 as a folly, Eckener didn't bother going to see the third and last flight. He assumed he'd hear nothing more of this crazy old man, his ridiculous contraption, or his fantasies of flight.[22] But he'd gravely underestimated Count von Zeppelin.

ZEPPELIN HAD SUCCEEDED in getting as far as he had not merely owing to his drive or his temerity, his pluck or his luck, but because he thought differently from his predecessors and his peers. Like Sir George Cayley before him, he had approached the design of his airship *modularly*: From the beginning, he had envisaged LZ-1 as a unified system composed of modules (the engines, the envelope, the infrastructure, and so forth), each of which required a specific solution. Piece by piece, as the problems were resolved, they were slotted into the greater whole to form a high-level system.

Not everything was perfect, of course; far from it. LZ-1, as we've already seen from its flight tests, suffered a host of bugs and issues. But the *concept* was sound. Whereas others had sought to bring the One Big Thing—getting off the ground and moving—to realization, they had overlooked or ignored the Many Small Things in the expectation that these would magically be figured out at some later date or patched together on the fly, or even that they didn't matter. What invariably happened was that seemingly minor issues cascaded into major, usually fatal, flaws.

David Schwarz, for instance, had simply wanted to get his metal airship built and thus hadn't bothered to work out the technicalities of gas management. As early as 1893, Zeppelin, conversely, was laboriously thinking through the means of regulating the hydrogen supply, controlling its expansion and compression, and minimizing leakage by adding, for example, automatic relief valves at the bottom and manual valves at the top of each flexible, sealed gas cell.

It was partly owing to his conviction, then, that LZ-1 was fundamentally "right"—despite the flights being greeted by a collective yawn and his company, the SPA, going bust—that Zeppelin never came close to quitting. In this he was fortunate indeed that he could call upon the support of his dear wife, Isabella, "whom I love to the bottom of my heart, [who] admonished me to not drop my courage and my head."[1]

He was given plenty of opportunities to bow out gracefully but refused to accept them. The kaiser, for instance, congratulated the count for his "epochal" achievement and conferred upon him the Order of the Red Eagle, the second-highest order of chivalry in Prussia.[2] For a man like Zeppelin, still smarting from the humiliation of 1890's *Kaisermanöver*, this was a major step forward in his psychological quest for rehabilitation, and no one would have thought less of him for now retiring, honor restored.

Likewise, his old ally Professor Bach and his friends at the Association of German Engineers urged him to quit while he was ahead. He had achieved so much: Zeppelin had *flown* as no one else ever had. He had come within a whisker of achieving true aerial navigation as no one else ever had. And he had proven every damn fool critic embarrassingly wrong about the practicality of the airship as no one ever had.

Instead, the contrarian old count became ever more vehement that the airship would, must, rise again. He held almost daily meetings with Dürr and continued his correspondence—sometimes pleading, sometimes blustering—with Berg (about aluminum), Maybach (about engines), and Moedebeck (about everything else).[3]

The two primary flaws of LZ-1 emerged from these discussions. First, it had lacked structural strength, and second, flight control had generally been poor.[4] Zeppelin immediately set about rectifying them.

Dürr locked himself away and worked in secret to improve LZ-1's successor's airframe. Berg's people were already developing an enhanced aluminum-zinc-copper alloy that would increase the tensile strength of the material, so that was promising.[5] Dürr also realized that LZ-1's structure had been at once too complicated and too simple.

It was too complicated in the sense that LZ-1 had been based on the inexperienced Theodor Kober's original design, whose transverse rings— the ribs—were twenty-four-sided polygons. Each of the twenty-four intersections had been braced with a cable stretching directly to the one opposite, so forming a kind of wheel-and-spoke arrangement. Dürr reduced the number of sides to sixteen, with three cables now fanning out from each intersection at various angles. The new arrangement would help avoid deformation because it spread pressure more evenly around the circumference.[6]

And it was too simple in that all of LZ-1's longitudinal girders had been flat pieces of aluminum held together by two thin metal strips. Easy to manufacture, they were inherently too weak for the critical strengthening and support work they were expected to perform. Dürr impressed Zeppelin by demonstrating that latticing three beams together to form a triangular girder was exponentially better at preventing buckling, twisting, and bending. Admittedly, it used more aluminum and was somewhat heavier, but the price of hydrogen was dropping, and Daimler's improved engines were lighter and more fuel-efficient than ever, so the marked performance gains would more than offset the extra weight and cost.[7]

For flight control, Dürr was helped by Professor Hugo Hergesell, a friend of Zeppelin's and an eminent meteorologist. Hergesell, dapperly dressed, pencil-mustached, and outfitted in French-style pince-nez glasses, looked like the habitué of a Parisian café, but his expertise in studying the wind was unrivaled.[8] Hergesell advised Dürr to throw out the cumbersome and glitchy sliding-weight contraption (used to adjust pitch, or moving the nose up and down) and to redesign the rudders (for turning left and right, or yaw).

In Dürr's next model, these were all gone, replaced by two sets of three horizontal elevator planes attached fore and aft that resembled Venetian blinds. These could be manipulated from the control gondola using pulleys and ropes. Ahead of the front set and behind their rear counterparts was a trio of movable rudders, arranged vertically. Theoretically, if the pilot turned the horizontal elevators diagonally up or down, the wind flowing through the rudders would adjust pitch, and if the rudders were oriented left or right (and the elevators were kept in neutral) they would help turn the airship.[9]

If LZ-2 were ever built, that is. By early 1903, after a year of development, it wasn't looking likely. Zeppelin traveled to Berlin to ask some "extremely rich patrons" for money but came away empty-handed. "Now, within a few days, I must [sell] everything that made the resumption of work possible," he lamented.[10]

To save Zeppelin, Berg pledged to again donate the necessary aluminum. The Daimler company also proved willing to lend a pair of its newest 85-horsepower engines, which lowered the weight-to-power ratio from

LZ-1's 58 pounds per horsepower to just 11 and allowed Dürr to install propellers double the diameter of LZ-1's.[11]

What Zeppelin was short of, then, was ready cash to pay for everything else, such as hydrogen and wages. He composed an appeal for 60,000 marks to cover his latest research costs. The count, being a count, was still rather uncomfortable with the idea of bowing to the masses and so sent the fundraising letters only to men of the "educated classes." Some ten thousand letters of appeal reaped a mere 8,000 marks.

Part of the failure was owed to Zeppelin's tone. His appeal had been an eye-glazingly dull catalogue of technical descriptions, and he came across as a haughty old noble expecting his social inferiors to tug their financial forelocks. Even when there was a friendly reporter present, Zeppelin could not bring himself to play along. Asked by one to consent to an interview, the count had, with inimitable aristocratic hauteur, brusquely replied, "I am not a circus rider performing for the public; I am completing a serious task in service of the *Vaterland*," and turned his back to him.

Luckily, his friends came to the rescue. Hergesell helped out by writing an article on the benefits of airship travel, while Moedebeck ghosted a second appeal in Zeppelin's name to *Die Woche*, Germany's bestselling popular weekly, and urged subscribers to his *Illustrated Aeronautical Reports* to send in any amount they could afford.

This time, the appeal was considerably more successful: About 375,000 marks arrived.[12] Even so, the bills kept rolling in. After SPA's bankruptcy in 1900, Zeppelin had dismantled LZ-1's floating hangar and sold it for scrap to pay off creditors. Now he needed a new one. Its replacement would at least be cheaper to maintain, since it began on land and protruded into the water. Extending from there, one-hundred-yard-long pontoon-mounted metal rails would aid to roll the airship deeper into the lake, where a boat would haul it into open water.[13]

And this was *before* construction work began on LZ-2. To stave off financial collapse, Zeppelin mortgaged the inherited estates of his remarkably patient and empathetic wife for 130,000 marks to pay for equipment and materials, but he was still short on funds and his credit was tapped out.

As he had once been by the Russians, Zeppelin this time was saved by the French, who were beginning to develop a type of small dirigible known as a Lebaudy for army reconnaissance. Though these ships had nowhere near the same scale, power, or ambition of Zeppelin's machines, their sinister implications seemed clear to the paranoid Wilhelm II. They were intended, or so he believed, as an aerial navy that would bomb his fortresses and ships.[14]

The kaiser was frightened into keeping Zeppelin solvent—just in case he too needed an aerial navy in the future. He approved a grant of 50,000 marks and instructed the Ministry of War to rent the count gas cylinders far below cost.[15] It was just enough to let Zeppelin go ahead with LZ-2.

In April 1905, work began on the airship's construction. In just seven months, it was completed. LZ-2's maiden voyage was scheduled for November 30.

LZ-2 was almost identical to its forerunner in terms of size and capacity, but given Dürr's radical changes in internal structure, the new engines, and improved flight controls, Zeppelin was understandably bullish on its prospects. He promised the press that something special was in the offing.

It certainly was.

DRAWN BY ZEPPELIN's promises of a surprise, a few members of the press, mostly reporters from the provincial papers but also Hugo Eckener from the *Frankfurter Zeitung*, traveled down to the lake for the great unveiling.

The count's entourage was present, of course. One of them was Ludwig Marx, captaining the motorboat *Württemberg*. He was somewhat older than most of the workmen, who played pranks on him. One of their japes involved drunkenly stealing his prized vessel and joyriding it for a midnight run to Switzerland, to indulge in yet more drinks. Though the good old count had thought it amusing, Marx punitively insisted that the men have their wages docked.[16]

Now Marx was towing the LZ-2 cautiously out of the hangar and down the rails. Alongside him in the *Württemberg* stood Professor Hergesell, the meteorologist. Zeppelin and Ludwig Dürr were in the front gondola with two mechanics, with two more and the explorer Eugen Wolf and Captain von Krogh in the rear. Krogh had originally been slotted to work the ballast-and-gas controls, but had graciously surrendered them to placate Dürr, who insisted on doing it.

All was proceeding smoothly—until it wasn't.

According to Marx, a sudden puff of wind at the stern lifted the airship over the motorboat and the towrope became entangled in the elevators at the bow. Marx leaped to cut it as the huge bulk of the airship menacingly bore down on him. As he tussled with the rope, the propellers started to turn. LZ-2's nose hit the water beside the *Württemberg* just as Zeppelin pushed past the frozen Dürr to pull the aft release valve for the gas to right the ship. The count's quick thinking saved the airship from major damage, but the front elevator and rudder assembly were crushed. There would be no flight that day.

Hergesell blamed Marx for the accident, and the two began to swear at each other, prompting Zeppelin to shout down good-naturedly, "Of course, if Hergesell and Marx quarrel, that will make everything all right again," shaming them both into cooperation.[17]

In spite of the setback, Zeppelin was, as usual, indefatigable and announced that the reborn LZ-2 would make its maiden flight a couple of months later, in January 1906. Zeppelin was taking a risk here: No rational person would launch an airship in the middle of winter owing to the likelihood of storms and high winds, but he couldn't chance delaying the flight until the spring or summer. For one thing, he couldn't afford to: The wave of social unrest that had erupted across Russia in early 1905 had destroyed his wife's inherited estates—the ones so recently mortgaged to finance LZ-2—in what would become Estonia and Latvia.[18]

Zeppelin was as good as his word. On January 17, 1906, LZ-2 was ready to go. When the early-morning call came, it was a surprise to the crew: Hans Gassau, one of the mechanics, turned up for work wearing slippers. Ludwig Dürr, who had realized he was not cut out for airship flying, stayed ashore, but there was another notable absentee: Wolf the explorer, who had mildly criticized airships in a newspaper after the November accident. As punishment for his lack of faith, the count had banished him permanently.

This time the launch went well, but LZ-2 was too low on ballast and immediately rocketed straight up to 1,500 feet. A southwesterly wind took hold of the airship, but the engines did their job and held it steady against the current. Zeppelin and Captain von Krogh manned the controls, but they were so stiff that the transmission broke and the rudder jammed. LZ-2 began to drift into a lethal crosswind. Its pitching violently up and down flooded, then stalled, the Daimler engines. As the mechanics urgently tried to fix them, a ventilator belt snapped, making a restart impossible without overheating. If a fire were to break out, they were standing directly below nearly 400,000 cubic feet of flammable hydrogen.

LZ-2 drifted as rudderless as a free balloon of old as Gassau began singing an ironic tune from Verdi's *Rigoletto*: "Oh, do not speak to your wretched father of his lost love." (It was a good thing Zeppelin didn't hear it.)

Minutes later LZ-2 had soared far from Lake Constance as the wind swept it toward the Allgäu, the Bavarian mountain range. Preiss the locksmith, the one who had once challenged Dürr to make a perfect turnbuckle, discreetly remarked to his friends in the rear gondola that "if we did not wish to crash in the mountains, it was time for us to think of landing." Nobody had ever before grounded such a large and rigid airship on hard earth—soft water was more forgiving—but now the luckless occupants had no choice.

Zeppelin cried, "Lower the drag anchor!" and slid the message along the communications wire to Preiss. As Gauss and another mechanic struggled with the anchor chain and threw it overboard, Zeppelin and Krogh pulled every valve they could to eject gas to bring the bolting LZ-2 to heel at a lower altitude. The anchor plummeted to the frozen ground, bit into it, and dragged LZ-2 down with it. The airship hit the ground hard and bounced up. It was now at a height of 325 feet and heading fast toward two desolate farms. On one of them, a girl was putting out some laundry, and, unfazed by the sight of a ten-ton monster bearing down on her, she angrily shouted, "Let me hang my washing!"

The airship rapidly descended and scraped against two trees, which dented the hull, before settling peacefully into swampy pasture. Some local farmers helped anchor the ship fore and aft with cables weighted down at the ends with boulders. As the crew congratulated one another and thanked the Lord for their preservation, "we considered ourselves very clever, not realizing that we were sealing the fate of our airship," Preiss recalled.

An airship that is anchored at both ends is vulnerable to crosswinds—later Zeppelins would be secured at the nose alone to turn with the prevailing wind, not struggle against it. That didn't seem important at the time, just as it didn't seem relevant when the farmers informed Zeppelin that their tiny hamlet was informally known as "Allwind." There was a lot Zeppelin still had to learn about the recondite details of handling airships.

The count returned to Friedrichshafen that evening to arrange repair crews, with most of the crew staying at Farmer Mohr's place (the indomitable laundry lass was his daughter). They were dead tired and slept until 7 A.M.

The next morning they found Zeppelin at the site despondently watching a Friedrichshafen team demolishing the airship, a nightmare of twisted bones and torn skin, with axes and saws. During the night the wind, vengeful at being denied its prey, had repeatedly battered the ship from the side and smashed it to pieces.[19]

A N INTRIGUED HUGO Eckener took a train to the crash site the following morning. There he saw the count "quietly and calmly standing
beside the badly damaged hull and giving orders for it to be dismantled."
Zeppelin was heard to lament, "An airshipman without an airship is like a
cavalry officer without horses. And I am both."[1]

Eckener's report for his paper resisted the impulse to ridicule the count
and adopted instead a studiously neutral tone: "Like the skeleton of a giant
whale, extended over 130 [yards], it lies there before us on the ground,
[and] the speedy destruction is undertaken mercilessly. The balloon
sleeves lie scattered in great heaps. It is as if somebody has given the order
to erase any trace of this airship as quickly as possible."[2]

By the end of that ghastly day, LZ-2 had vanished from existence. Even
the valuable gondolas and engines were hacked apart, the pieces thrown
onto wagons already piled high with horribly deformed aluminum girders,
to be carted away for scrap. This must be, surely, the end of the Zeppelin
enterprise, thought Eckener.

But a couple of weeks later, as he tended his garden, Eckener's house-
maid rushed out and excitedly told him that a Count von Zeppelin was
ringing the doorbell. He found the count in his drawing room dressed in
"meticulously correct morning clothes, with silk hat and yellow gloves,"
the very picture of a traditional aristocrat making a formal call.

Drawing a copy of the *Frankfurter Zeitung* from his pocket, Zeppelin
inquired whether Eckener was the "Dr. E." who'd written the articles
about LZ-1 and LZ-2. Eckener cautiously (since a crusty noble like Zep-

pelin might well challenge him to a duel, as he had done with Captain Gross) said yes, whereupon the count thanked him for the courteous tone toward him personally in the piece. It had made a nice change.

At the time, Zeppelin was more accustomed to being mercilessly lampooned. Even the friendlier locals mocked him as *der Luftikus am Bodensee* ("the aerial dreamer down at the lake"), who tilted, like Don Quixote, at windmills. More uncharitably, a widely circulated article, by some coward pseudonymously writing as "Rudolph von Elphberg" — almost the same name as the ruler of Ruritania in Anthony Hope's *The Prisoner of Zenda* (1894) — had described the count's "pathetic career" trying "to solve the problem of flying" that had led him to squander "his estates and his fortune, after reducing his wife to destitution and his only daughter to penury." Given the failures of LZ-1 and LZ-2, surely it was only a matter of time before "he will be compelled to confess that his life of labor has been spent in vain and that his gray hairs will sink into the grave in sorrow over [his] futile sacrifice."

Eckener, conversely, had restricted himself to observations regarding the airship's technical issues, and it was these that Zeppelin now wished to discuss. The two had a long conversation about what had gone wrong with LZ-2. Much to Eckener's surprise, Zeppelin bluntly denounced his own lack of experience, which he felt had contributed to the disaster. To date, he had flown for four trips totaling around three hours. Granted, that made him the most veteran airship captain in history, but three hours' flying time was nowhere near sufficient to classify him as anything else but a beginner. He was getting better at lifting off and landing, but flying was more than that, just as driving a car amounts to more than knowing how to turn the engine on and off. Zeppelin had realized that successful flight control — that is, maintaining command over the machine as it moved — required the right philosophy, the correct frame of mind, to happen.

An airship, the former cavalryman had concluded, could not simply be inflated and steered in the desired direction (as so many others had blithely assumed). It was in fact not a cloud but a winged horse, and its pilot had to be a skilled equestrian. Without a steady hand at the reins, a horse will take you where *it* wants to go, but a mighty new breed of man, a true *airshipman*, would tame a wild mustang into a Lipizzaner stallion obedient to *his* will.

Successfully maneuvering these colossal vehicles through the air entailed a thorough mastery over dozens of factors, ranging from hydrogen maintenance and wind measurement to precise elevator, engine, and rudder management. Technical proficiency was as important as natural aptitude. The finest airshipmen would exploit the weather, not be at its mercy, and their seemingly effortless skill at piloting would lend their actions a singular beauty and elegance.

Having heard out the count, Eckener accepted his invitation to dinner a few nights later. He "went there with a vague and not completely comfortable feeling that the visit could lead to a relationship which would greatly influence my future life."[3]

LITTLE DID ZEPPELIN know it, but the Wright brothers had reached the same conclusion—the importance of flight control and the right mental outlook. This intellectual breakthrough was what distinguished the Wrights and the count from their many failed competitors. Despite their differences in class and background, Zeppelin and the brothers were very similar in mental outlook.

Born the sons of a stolid midwestern bishop, Wilbur and his younger brother Orville never, like Zeppelin, completed any formal technical training; they were self-taught and driven by an implacable work ethic.

They, too, thought modularly and understood that each component of an airplane, like those of an airship, was important in its own right. Every piece of the system had to work harmoniously with the others in order to build a machine capable of flight, and each part presented a unique problem requiring an individual solution.

As Zeppelin was learning, by far the hardest nut to crack was control: In order to fly they had first to learn *how* to fly. The Wrights studied the birds for inspiration. Whereas aeronautical theorists had investigated the physical means by which birds flew, the Wrights were more interested in their behavior. As Wilbur Wright explained, birds' wings were "undoubtedly very well designed indeed, but it is not any extraordinary efficiency that strikes with astonishment but rather the marvelous skill with which they are used." He and his brother had set about acquiring that "marvelous skill" for themselves.

For years, before adding an engine, they practiced keeping their full-size gliders under control in strong winds, sudden gusts, and rising currents. They designed mechanisms, such as movable tail rudders and wings that could twist slightly to increase lift on one side while reducing it on the other, to cope with and adapt to unexpected weather changes. Aviators, concluded Wilbur, required "skill, experience, and sound judgment" to keep an otherwise unstable machine stable, for "before trying to rise to any dangerous height a man ought to know that in an emergency his mind and muscles will work by instinct rather than conscious effort. There is no time to think."

Only once they had accounted themselves adepts of the air had the Wright brothers placed a homemade engine into the glider—now airplane—that they dubbed the *Flyer*.[4]

DECEMBER 17, 1903, 10:35 A.M.: Another freezing day at Kitty Hawk, North Carolina. Ice-covered ponds. Bitter wind gusts at 20 to 27 miles per hour rubbing raw the cheeks of everyone present.

Wilbur Wright stood by the *Flyer*, a hand steadying its lower wing, as his brother Orville lay on his stomach manning the controls. The engine chugged to life and slowly warmed up. Orville cast off the rope anchoring the *Flyer*, and it started forward. Wilbur ran alongside for a time until the machine rose off the ground, dipped precariously, thrust upward, descended, bounced, and came to a rest after a wingtip scraped the ground.[5]

Four years of unrelenting work had just resulted in a voyage of 120 feet in twelve seconds.

The Wright brothers were the first to make a powered, sustained, piloted, and controlled flight in an airplane. They had achieved the seemingly impossible. Just a year earlier, the science-fiction novelist and essayist H. G. Wells, no slouch when it came to predicting the future, had confidently forecast that it would be decades (though "very probably before 1950" and certainly by the year 2000) before an airplane "will have soared and come home safe and sound."[6]

But few, least of all Zeppelin, cared. Even if he had heard of these Wright characters, he would have regarded their contraption as nothing of consequence.[7] A twelve-second flight? A single occupant? A craft that traveled just a third of the length of LZ-1? For heaven's sake, in July 1900, three and a half years before, and on its first attempt, his own airship had carried five passengers and made a trip of 3.5 miles in eighteen minutes.

The Wrights' follow-up, *Flyer II*, first took off from Huffman Prairie (about eight miles east of Dayton, Ohio) in May 1904. Between May 23 and December 9, the brothers made about eighty flights, their time airborne totaling forty-five minutes. Their longest flight, which took place on November 9, lasted five minutes and four seconds.

But the most important had occurred on September 20, when Wilbur completed the first full circle, a milestone in flight control. This one would later go down in history because Amos Root, the wealthy and rather eccentric publisher of a small trade journal, *Gleanings in Bee Culture*, happened to be on hand that day. His would be the first eyewitness account of a Wright brothers flight.

Root's article in the January 1, 1905, issue of *Gleanings* expressed the awe, shock, dread, and sheer wonder of manned flight that would become the hallmark of the Golden Age of Aviation. An evangelical Christian, Root believed explicitly in the winged gospel—the view that modern technology, just as much a gift from God as the natural world, would forge social betterment. Watching the *Flyer II* circle above him was as astounding, as sublime, and as transcendent as standing atop the Grand Canyon or Niagara Falls. For Root, the overall effect was akin to that of the first Christians being suffused with the silent power of the Cross.

As he described the scene for his readers, "When it first turned that circle, and came near the starting-point, I was right in front [of] it; and I said then, and I believe still, it was one of the grandest sights, if not the grandest sight, of my life. Imagine a locomotive that has left its track, and is climbing up in the air right toward you—a locomotive without any wheels we will say, but with white wings instead. . . . Well, now imagine this white locomotive, with wings that spread 20 feet each way, coming right toward you with a tremendous flap of its propellers, and you will have something like what I saw. The younger brother bade me move to one side for fear it might come down suddenly; but I tell you, friends, the sensation that one feels in such a crisis is something hard to describe."[8]

The Wrights' *Flyer III* would be their first practical airplane. On October 5, 1905, as Zeppelin had been putting the finishing touches to LZ-2, it set a record flight time of 38 minutes, 3 seconds, but more important, it demonstrated a succession of impressive feats, such as banking, circling, and figure eights. In a little under two years, after spending a paltry $1,000 (about $28,000 today) to build their first airplane, the Wright brothers had come an alarmingly long way.[9]

Yet Zeppelin ignored these sharp, punctuated technological leaps. His lack of concern was understandable. First, the brothers' exploits were scarcely reported, and when they were, coverage was inaccurate and exaggerated (usually both). Many others had made similar claims before being exposed as charlatans or fabulists.[10]

And second, as of October 16, 1905—after their astounding successes and just three months before Zeppelin had launched LZ-2 on its doomed voyage—the brothers ceased flying to concentrate on research and experimentation.[11] Fearful of competitors stealing their ideas, the Wrights would vanish until May 6, 1908—two and a half years later. When Zeppelin met Eckener in early 1906, then, it seemed clear that these Americans could be dismissed as failed also-rans, not that the count regarded airplanes as any competition in the first place.

WHEN ECKENER CAME to Zeppelin's house for dinner, it was only because he was more intrigued by the count's personality than he was interested in his ideas on airship construction. He had experienced no Saul-to-Paul conversion to the cause of airships during their initial meeting—if anything, he later said, he was hoping never to talk about them again—but as they ate he found himself strangely bewitched by Zeppelin's energetic monologue on his enemies at the PAB and the newspaper conspiracy against him.

As the night wore on and bottles of wine were emptied, Eckener felt "a surge of admiration and respect" for the count and "without forethought" offered to help him in any way he could. When Zeppelin asked how, Eckener replied that "I am a writer. . . . I know how to write to convince people. As I see it, your problem at the moment is not so much to convince the experts about the practical value of your airship, but to convince the public, the man in the street. Once you win over public opinion, the experts and the authorities will have to give way." If the elites ridiculed and shunned Zeppelin, then the patriotic people of Germany would rush to his defense once they realized the importance of airships to national greatness.

This task had once fallen to Moedebeck, but he was very ill (and would die soon afterward), and Zeppelin needed a replacement. Eckener was perfect for the job. From then on, his reports, Eckener recalled, "were primarily aimed at arousing in the public the necessary understanding of, and interest in, the Zeppelin idea."[12] To that end, in February 1906, about a month after the rotten end of LZ-2, he helped Zeppelin write a pamphlet ("The Truth About My Airship") aimed at countering his critics.[13]

It marked the beginning of a lifelong involvement that would witness the rise and fall of the airship.

IN HIS OWN way, Eckener was as idiosyncratic as the count. With a shock of blond hair, a manicured goatee, and blowtorch-blue eyes, he was the son of a Lutheran tobacco merchant in Flensburg, a small harbor town on the very edge of northern Germany, just four miles from the Danish border and a world away from Zeppelin's Württemberg. His mother, a master shoemaker's daughter, gave birth to Hugo in August 1868, the second of what would be five children. At school, Eckener seemed mysteriously unapproachable and had few friends, and he soon acquired the nickname "Zeus," after the god who atop Mount Olympus remained aloof from the affairs of mere humans. His daughter would later say that "in general he was probably more respected than loved."

He was an indifferent student, preferring instead to read and think alone.[14] Yet one of his teachers, Herr Maass, noticed that the young man (he was twenty when he graduated, having had to repeat a year or two) was an independent thinker with an interest in the arts. Through his influence, Eckener was able to enter the University of Leipzig as a doctoral candidate under the tutelage of the renowned Wilhelm Wundt, who had established a laboratory to conduct psychological research. Eckener's thesis on the "fluctuations of perception of minimal sensory stimuli" was passed in 1892, leaving him at a crossroads. Should he become a professor or work as a private tutor?

In the event, he did neither and joined a local regiment, the 86th Fusiliers, for a two-year stint to give him time to make up his mind. He spent his off-duty hours indulging his love of sailing. Eckener became expert at observing and judging the wind and the weather. During even the roughest squalls, he would man the tiller with, as a fellow sailor put it, "marvelous calmness and circumspection," issuing quiet rebukes to the crew when they were too slow or clumsy.[15]

After leaving the army as a sergeant, Eckener still could not decide what to do and so drifted into journalism as an employee of the *Flensburg News*, a local paper. He turned out all manner of arts criticism, music and theater reviews, as well as social commentary, much of it distinguished by a clever but unformed young man's archness. He wrote on everything under the sun, but what he was truly searching for, as he wrote to a friend, was "a fixed pole [for my] thoughts." Most people "simply go whichever way the wind blows instead of thinking for themselves." He wanted to set a steady course for a known destination, but he had not, so far, found the right vehicle to take him.[16]

He at least took the first step toward settling down by getting married—to the newspaper owner's daughter, Johanna, in October 1897. If the twenty-nine-year-old Hugo, tireless and restless, was a tornado, then Johanna, smart and quiet, was a rock. She preferred to stay at home and was rarely seen in public. For Eckener, she provided a welcome refuge; their marriage was described as one of "unvarying harmony."[17] In private he wrote her excited love letters: "My Johanna, my love and my life, my longing and aspiration; you, my virtue and my vice, my prettiest idea and my craziest thought, my thought to go crazy and my inspiration to feel divine, I'm fond of you! Fond? No, you're my love, my desire, my loving and rejoicing and luck, you're the air that I breathe, the prerequisite of my existence!"[18]

Their first daughter, Hannelise, was born two years later, signaling to Eckener that he urgently needed to find a more reliable source of income. In 1900, the family moved to Friedrichshafen, partly for the sake of Eckener's health and partly because the *Frankfurter Zeitung*, a big step up from the *Flensburg News*, had a spot open for a correspondent. Perform well there, Eckener thought, and he would have a shot at joining the staff in Frankfurt. He enjoyed Friedrichshafen in the meantime, having set up a pleasant little house near the lake, where he sailed, and not so far from the mountains, where he hiked.

It was a delightful middle-class existence, even if money was tight, but with Eckener still waters ran deeper than most suspected. Eckener may have appeared the very representation of bourgeois rectitude, but within him there burned the spirit of a natural rebel, or perhaps a "fighting rooster," as Johanna used to say.

"I'll hardly ever lift a finger to participate in today's so-called politics," claimed Eckener, "to the point that I probably won't ever make use of my right to vote—unless it's to choose a candidate to be overthrown." His "own political system" was a belief in scientific, rational solutions to social problems.[19] He lacked any faith in God (his wedding was the first time he'd been in a church since childhood), argued that the state should run such institutions as the railways and the post office in order to regulate the effect of competition, and felt that government should be in the hands of sensible men (he was not an advocate of female suffrage) of higher learning, such as himself, as normal people were too concerned "for their grossly materialistic interests" to be capable of making wise decisions for themselves.[20]

As he matured, Eckener dispensed with much of this snobbish juvenilia, but there would always remain a core conception of himself, the solutions-oriented supreme rationalist, as He Who Knows Best—especially when it came to airships.

When he met Zeppelin, he'd finally found the "fixed pole" he'd sought for so long.

AFTER THE MODEST performance of LZ-1 and the destruction of LZ-2, it was obvious to critics that Zeppelin's airships were not fit for any purpose. The old man's reach had exceeded his grasp, and while everyone agreed his ambition was laudable, his achievements were laughable.

Zeppelin himself seemed to realize that it might be too late to save his airships. A few days after the demise of LZ-2, he discovered that he had but mere pfennigs in the bank. Tired and downcast, he divulged to Eckener one night that "I shall not go on building any more. The world shall never know how good my airship is!"[21] But as usual his despondency did not last long. The ineffable Zeppelin luck soon rushed to his aid: The army, long after Zeppelin had given up ever trying to interest it in his airships, again was back in the picture.

On January 1, 1906, little more than a fortnight before the LZ-2 debacle, the kaiser had appointed one of his favorite former aides-de-camp, Helmuth von Moltke, as chief of the General Staff. The arrival of the new broom, in combination with the looming threat of a Franco-Russian-British axis, led to a broad reappraisal of German military needs. One of the first changes occurred on January 29, when a committee concluded that "it is time to approach the construction of motorized airships."[22]

The trouble was, which type of airship should Berlin back? The Zeppelin was a rigid model, meaning that the shape of its gas envelope was maintained by an internal structure. As Zeppelin was painfully aware, rigid airships had not, as yet, distinguished themselves in service.

More promising, at least to military observers, was a new generation of semi-rigid and non-rigid types. The French army's Lebaudy airship program that had so alarmed the kaiser, for instance, was based on semirigids, in which a stiff keel ran under a soft envelope. There was also the simple non-rigid version, which we would call a blimp and which was directly descended from the good old balloon.

Inspired by the French, Zeppelin's old antagonist Hans Gross, now a major and at last slated to head the PAB, had recently combined with the balloonist Nikolaus Basenach to work on a 214-foot-long, 177,000-cubic-foot, semi-rigid copy of the Lebaudy they called a "Gross-Basenach." To shorten the odds that the army would rally behind it, Major Gross made sure to put the boot into his rival Zeppelin whenever he could. Upon being asked by the kaiser to "speak the plain truth" about what he thought of Zeppelin's airship, Gross replied, "Majesty, [it's] a load of rubbish!"[23]

The trouble was, the Gross-Basenach existed only on paper for the time being (mind you, so did Zeppelins). To speed up development, Gross tried to woo Ludwig Dürr into working for him—yet another swipe at Zeppelin—but the ever-loyal engineer refused him out of hand.[24] Nevertheless, Gross still benefited from a heavy thumb on the scale: His airship, intended exclusively for the army, was being developed by one of the army's own, and Gross could count on strong support from within the General Staff owing to its in-house pedigree.[25]

For the time being, though, the most immediately attractive prospect to fulfill the army's needs was the basic Parseval blimp, built by August von Parseval. Like the Gross-Basenach, the cigar-shaped balloon was designed expressly for military reconnaissance operations, but it was much smaller (88,000-cubic-foot capacity and 157 feet long). Unlike Zeppelins, the single-engine Parseval could be quickly inflated and deflated and was easily packed and transported so that it could travel with the army. It was cheap to run, used proven technology, and required just a few men and several canisters of gas to operate. The blimp's great advantage was that it was virtually ready; Parseval needed only a few more months before he could begin trials.[26]

On the other hand, the Zeppelin had already attracted the kaiser's eye, Eckener was stoking favorable public opinion, and the count was a force of nature. The choice was obvious, thought Zeppelin with his habitual confidence. The minutes taken of a meeting he had with Moltke give a flavor of his attitude: "He [Zeppelin] must still declare that he was the only expert in Germany who was also an inventor. Moreover, he was not just an ordinary inventor, but a *German general*, and as such he must stress that it was the height of official irresponsibility not to utilize his airship."

The Gross-Basenach and the Parseval, Zeppelin grandly pointed out in the meeting, were no doubt satisfactory in their limited uses, but the former was merely an unambitious knockoff of the Lebaudy ship, and the latter just an updated balloon. Only the Zeppelin was built on the most technologically advanced lines using entirely homegrown German know-how; only the Zeppelin truly represented German greatness; and only the Zeppelin could ever be a Wonder of the World.[27]

Confronted by this dilemma, the Ministry of War came up with an admirably Solomonic solution: a competition. There was no set deadline, but the best airship would win a lavish contract and the eternal thanks of a grateful nation. Zeppelin was at both an advantage and a disadvantage in this regard. His airships took a long time to build, but Gross's semi-rigid was still on the drawing board, which gave him some breathing space. More worryingly, Parseval had already announced that he was planning an experimental flight that May.

Zeppelin needed to build an airship fast and cheap to stay in the game. He lacked the time and resources to make it exponentially better than the LZ-2; it only had to be better than Parseval's blimp to knock him out of contention. A victory in that round would so solidify his lead that by the time Gross's airship was ready, it would be too late for his hated rival to catch up.

It was Dürr who came to the rescue. Why not, he suggested, simply rebuild the LZ-2 with just a few incremental changes to improve control and stability? By not having to design everything from scratch, Zeppelin would save many valuable months and a huge amount of money.[28]

In April 1906, having secured a special dispensation from the kaiser (100,000 marks), a loan of 125,000 marks from Carl Berg, and a promise from the king of Württemberg to hold a lottery, Zeppelin began construction of LZ-3.[29]

By the end of September, LZ-3 was finished. And not a moment too soon, either: On May 26, Parseval had held his first trial as promised, and his blimp had stayed aloft for one and a half hours and traveled at a top speed of 30 mph—more than fulfilling the army's needs for a mobile observatory.[30] But the Parseval tests were scheduled to continue until October 27, meaning that Zeppelin could still steal the show if he could get LZ-3 up in the air before that month was out.

On October 10, launch day, just after Ludwig Marx had towed LZ-3 into the open, an unexpected crosswind caught the airship and threatened to bash it against the side of the lakeside hangar. Zeppelin, increasingly confident at flight control, ordered all ballast dropped, and LZ-3 rose quickly to two hundred feet and freely sailed over the roof. Once the engines chugged to life, the airship soon obeyed his will and embarked on a sixty-nine-mile-long voyage of two hours and seventeen minutes at a top speed of 32 mph.[31]

Without deigning to mention his connection to Zeppelin, Eckener wrote a long article extolling the "victory" of the "rigid system" (Zeppelin's) over those of the Parseval and the Gross-Basenach (which was favored by "the [dismissive] gentlemen who belong to our airship department in Berlin"). Granted, it was not as cheap as its rivals, but then again, it cost just 1 percent of the price of a battleship—and nobody was complaining about building lots of those. (Germany, then in a naval arms race against Britain, commissioned no fewer than fifteen battleships between 1899 and 1908.)[32]

There were still problems to iron out, but the LZ-3 had done what it was intended to do: show up the Parseval as unambitious and backward-looking to knock it out of the running. Interest in the Parseval quickly faded, but the army then claimed that the Zeppelin was *too* ambitious, *too* forward-looking.[33]

The Gross-Basenach remained the favorite, and no decision would be made at least until after its preliminary launch, scheduled for the following March. If that was successful, then there would be yet more trials and so on and on until the army got the result it really wanted: a Gross-Basenach fleet.

As it was, the latest round of flights had brought Zeppelin again to the end of his financial tether, and he was anxious that he wouldn't be able to afford to send LZ-3 up again without additional support.[34]

As always with Zeppelin, he was lucky. This time, the kaiser and the imperial chancellor, Prince Bernhard von Bülow, saw in his airship a possible salvation from the slow-motion collapse of the government. In the closing months of 1906, Wilhelm II was rocked by a succession of scandals caused by nudge, nudge, wink, wink hints in the press that Count Philipp zu Eulenburg, his closest friend and adviser, was an active homosexual who had repeatedly been blackmailed by several disreputable young blades.

Worse still, reporters were now delving into the secretive activities of what was known as the "Liebenberg Round Table," rumored to be a gay coterie of high-ranking and high-born officers surrounding the kaiser. It surely wouldn't be too long before they uncovered Bülow's own intimate relationship with Eulenberg many years before or noticed that he had recently appointed his current lover, Max Scheef, as his private secretary in the Reich Chancellery.[35]

The kaiser and his chancellor urgently needed to divert attention away from this mess, and what could be better than highlighting Germany's astonishing technological and military prowess in the air? General von Moltke was discreetly informed that the emperor would find it most useful for the army to give Zeppelin a boost.

Thus encouraged, on December 19, 1906, the Ministry of War proposed a budget of 500,000 marks to build a new, state-owned hangar for Zeppelin airships called the Reichshalle. It would still be left to Zeppelin to find the money to build an airship to fill it, but the warm aura of official imperial favor was unmistakable.[36]

In early March 1907, Major Gross conducted a successful small-scale test of his prototype semi-rigid as scheduled, but the limelight shifted back to the count when in April the government announced its conditions for further funding of the Zeppelin model. If the count could make a nonstop twenty-four-hour flight of at least 435 miles (say, from Friedrichshafen to Mainz and back), rise to an altitude of 3,900 feet, and return to base safely, the Gross-Basenbach would be relegated to the back burner.[37]

It was a very tall order, indeed perhaps an impossible one, but it was the golden opportunity Zeppelin had been fighting for.

11. Up into the Empyrean

BY SOME STRANGE and inexplicable miracle of state planning, the new Reichshalle hangar at Manzell was completed precisely on schedule and under budget on September 23, 1907.

While the Reichshalle was being built, Dürr had been busy in the existing hangar modifying LZ-3 for superior handling. The major changes centered on the Venetian-blind-style elevators on each side, fore and aft; what had been three louvers now became four. Between the rear horizontal stabilizing fins, he also added three vertical, rotatable rudders.[1] These small changes would pay off significantly in terms of flight control.

On September 24, LZ-3 embarked on its first post-modification trip. In a flawless voyage of four hours and seventeen minutes—the longest a Zeppelin had ever flown—the airship "executed the most diverse evolutions with the greatest ease. It paraded back and forth before the windows of the royal castle of Friedrichshafen, making the turns on a very small radius, and on the homeward stretch it ascended and descended at the will of the pilot, at times almost dipping into the water and then rising to a height of 600 feet," according to a reporter.[2] Two days later, Zeppelin was confident enough of LZ-3's safety that he took his only child, the twenty-eight-year-old Hella, on a flight, incidentally making her the first woman to ride in a powered airship. (His wife, Isabella, owing to a heart condition that required her to "avoid any type of excitement," was never able to fly in one of her husband's creations.)[3]

If Zeppelin were to have any hope of winning the twenty-four-hour endurance competition, though, LZ-3 would need to demonstrate its airworthiness over land, where mountains, unpredictable winds, and valleys were constant dangers. Critics liked to point out that aside from the unfortunate incident with the late LZ-2, Zeppelin had never taken off or landed anywhere but from open water; neither had he ever flown for a prolonged period too far away from the safety of Lake Constance.

Zeppelin was determined that the flight of September 30 would put paid to these long-standing concerns. During this record-setting voyage of seven hours and fifty-four minutes, LZ-3 was intentionally sailed over land toward Ravensburg, twelve miles away. The airship maintained a steady course at an altitude of five hundred feet, and Dürr, entrusted with manning the elevators, was surprised to discover that the air over fields was hotter than that over forests, which meant contending with unexpected up- and downdrafts.[4] There was always something new to learn about airship driving, and the information was added to the ever-growing manuals of instruction that would one day be used to teach Zeppelin captains the art and science of flying.

On that day's trip, LZ-3 flew for 220 miles and reached a speed of 33.5 mph. It was a remarkable achievement, but Zeppelin had realized that LZ-3 was incapable of embarking on an endurance flight. It was too jerry-rigged together, its bones and organs too weakened by the strains of its own efforts, to be risked on seizing the glittering prize. It was clear that the count would need more political support—from which flowed monetary support—to build LZ-4, his supership.

To gain political support, one first needed social respectability, and there was nothing more effective at making one's cause popular or fashionable than the merest nod of royal approval. Once gained, politicians fell into line and the papers sung your praises. To that end, Zeppelin arranged a special flight for October 8.

Under Zeppelin's beady eye, the rickety, rattling LZ-3 was checked, then double-checked, then triple-checked for loose fittings, leaking valves, worn rudders, frayed ropes, and ripped gas cells; its dulled outer skin was given a good spit-and-polish cleaning to bring out its shine, as were the aluminum gondolas, and the cabin controls were tightened up and oiled down for optimal response.

Finally, it was ready. Aboard a steamship, the king of Württemberg and the Austrian archduke Leopold Salvator flanked the emperor's eldest son and heir, the soccer-mad, tennis-playing, women-loving (in the *very* plural) Crown Prince Wilhelm as they watched LZ-3 lift off, rise to five hundred feet, describe a circle, and return to Manzell. As it passed Friedrichshafen, the town's guns were fired, and thousands of onlookers vociferously cheered Zeppelin's name.[5] The flight itself was not a long one—just two hours—but it achieved its intention. The crown prince and his father had a troubled relationship (partly owing to the soccer and the tennis, though mostly to the women), but the younger Wilhelm exerted a powerful influence in Berlin, and his support made Zeppelin difficult to sideline.

Within weeks, the financial floodgates opened. The government agreed to provide 400,000 marks to build LZ-4 and sweetened the deal by offering no less than 2,150,000 marks to purchase both LZ-3 and LZ-4 for the army. Better still, that sum included a 500,000-mark bonus to reimburse Zeppelin for some of the costs he had personally incurred over the years. There remained one catch: In order to trigger the release of the purchase money, LZ-4 had to complete the nonstop, 435-mile, twenty-four-hour flight.[6]

While LZ-3 had been a rush job, Zeppelin could now take time to redesign LZ-4 from scratch, but that didn't mean he had time to waste. In November 1907, he began work on what would become the greatest pre–World War I airship of them all. LZ-4 wouldn't be completed until the middle of June the following year, but in the meantime his LZ-3 successes made news around the world.

The coverage amounted to a complete reversal of Zeppelin's fortunes. Henceforth, there were few of the sneers and jibes that had greeted his earlier efforts; no more "crazy old count by the lake" stories as foreign newspapers became interested in Zeppelin's background. Britain's *Daily Mail* delivered a scoop, supplied by Eckener, when it discovered that the count—a "wonderful old man who has sacrificed fortunes for airships"—had been "engaged as a volunteer in the American war, having a very narrow escape in the battle of Fredericksburg. His interest in the military airship dates from that time, when he made his first ascent in a captive balloon belonging to the Southern Army."[7] (The reporter at least had the spirit, if not the letter, of the facts correct.)

Meanwhile, *The Washington Post* promoted his noble pedigree to prove that the unquestionably Teutonic Zeppelin was actually a quasi-American. The writer made a hopeless hash of the genealogy, yet he was not completely off the mark: Zeppelin *did* have an American connection, though even by the standards of convoluted aristocratic lineages, it was a tenuous one. Put as simply as possible, decades earlier Zeppelin's late aunt had married a widowered French noble whose first wife had been Ellen Sears—daughter of the antebellum Massachusetts senator David Sears. It wasn't quite a direct connection to the *Mayflower*, but it was good enough for the *Post*, which took to calling the count a patriotic "Union Veteran."[8]

Other papers focused more on Zeppelin's achievements than on his past. A *Cincinnati Enquirer* article announced that "Germany is jubilant. Great Britain may be 'monarch of the sea,' but the Fatherland has conquered the air. Count Zeppelin is the hero of the hour. He is the archwizard of the air. His monster airship answers his helm like an ocean greyhound. It rises and sinks perpendicularly like an elevator in a skyscraper . . . and turns this way and that, and cavorts like a well-trained circus horse put through its paces by the hand of its proficient trainer." LZ-3, the paper claimed, "is so far in advance of all [air craft] as the *Lusitania* outclasses the ancient side wheeler of Robert Fulton."[9]

With this heady comparison to the world's newest, largest, fastest ocean liner, Zeppelin was inaugurated into the pantheon of twentieth-century masters of technology. Just before Christmas, the *Chicago Daily Tribune* catalogued the "Seven Great Wonders of Science and Industry Perfected in 1907": The *Lusitania* was there, of course, and so was the world's tallest skyscraper, the Singer Building in New York, as well as railroad electrification, transatlantic wireless communications, "Edison's concrete house for the working man," and the transmission of images of checks, signatures, and photographs by telegraphy, but heading the list was Zeppelin's marvelous airship.[10]

In Germany, a burst of Zeppelin enthusiasm erupted. Amateur aeronauts in Berlin announced that they were forming a club that would "afford its members opportunity to enjoy pleasure trips through the clouds in steerable craft." They were convinced they would make airshipping as popular as motoring, bicycling, or sailing.[11]

For the upper crust, there were balls galore dedicated to an aerial theme. In Berlin ("where even the women take to ballooning," gasped the *St. Louis Post-Dispatch*), there was talk of "a grand ball" to be held "a thousand feet in the air," where guests would be ferried "up into the empyrean" by balloon. The highlight of the evening would be the count arriving in his airship to hover overhead and toast the lords and ladies dancing gaily below. Sadly, this elaborate *Mischianza* never came to fruition.[12]

If Zeppelin was having the time of his life, the same could not be said of the Wright brothers, who finally emerged from their self-imposed exile in May 1908. Neither hide nor hair had been seen of them since October 1905, but they had not been hibernating; in fact, they had been beavering away in North Carolina. On May 14, they took to the air again, Orville for the first time taking up a passenger (Charley Furnas, a mechanic). The day did not end well when Wilbur, unaccustomed to the new controls, mistakenly moved the elevator the wrong way and (in the somewhat over-zealous words of *The New York Times*) was "hurled to the ground with terrific force . . . with the speed of a lightning express." He escaped with cuts and scratches, but the *Flyer* was wrecked.[13]

Considering how far and how quickly Zeppelin had advanced in the same period, the fact that the Wrights had, it seemed to the count, wasted nearly three years building a crash-prone toy only confirmed his suspicion that the airplane was not the future; the airship was.

THE WONDER OF Zeppelins was not just their sheer size. It was that owing to their immensity it required only small changes to make them exponentially larger. Referring to LZ-3, the count (or rather, his amanuensis, Eckener) explained in *Scientific American* that all he needed to do was to enlarge the diameter of the framework by about one yard and increase its length correspondingly, and voilà, thanks to higher gas volume and greater lift, an instant gain of 6,600 pounds in potential payload would abracadabracally appear.[14]

He did much more than that for LZ-4. Whereas LZ-3's length was 420 feet and its diameter 38 feet, LZ-4's dimensions were, respectively, 446 feet and 43 feet. Not so much of a difference to a casual observer, but together the increases raised LZ-4's gas volume to 530,000 cubic feet, up fully a *third* from LZ-3's 399,000 cubic feet, with only a relatively small increase in weight. (Theoretically, you could make an airship any length and breadth you wanted, but in the real world you were limited by hangar size, wages, materials costs, gas supply, and construction time.)

Thanks to the added buoyancy, Zeppelin could at last add a comfortable, windowed cabin located midway between the two gondolas. The bare-bones crew of seven that had manned LZ-3 could now be expanded to eleven crewmen and fourteen passengers.

LZ-4's first test flight was on June 20, 1908, five weeks after the Wrights had most assuredly dropped out for good. Zeppelin flew for around one and a half hours, but maneuvering proved unexpectedly treacherous. The thirty minutes of cavorting Zeppelin executed impressed the crowds, but while LZ-4 made 31 mph on the straights Zeppelin struggled hard to keep the airship under control during turns.[15]

Back on the ground, Zeppelin ordered major changes to the steering controls. Dürr, who had been in the rear gondola with three mechanics, understood immediately what had gone wrong. It was the rudders, bow and stern: They were now too small for the big airship. After larger ones were fitted, the airship's directional stability vastly improved.[16]

LZ-4's second flight, on June 23, and its third, on June 29, were short but flawless. Finally, Zeppelin had his dream airship. But he needed one final run to prepare for the endurance flight that would determine his, and his creation's, future. On July 1, Zeppelin would fly to Switzerland—whose vertiginous peaks, winding passes, and unpredictable winds would test and tax his airship to the utmost.

Major Gross in the meantime made a last-minute effort to steal Zeppelin's thunder by scheduling his own first non-trial flight of the Gross-Basenach semi-rigid (renamed M1, the M standing for "Military") for June 30, a day earlier. The M1 was tiny compared to the Zeppelin, but Gross was no fool. He had copied LZ-3's Venetian-blind-style elevation planes as well as some other Dürr innovations, which made the M1 just the kind of dark horse that might snatch Zeppelin's promised army contract right out from under him.

In Zeppelin's favor was that because the M1, like the Parseval balloon, was an army project, it was necessarily kept secret for fear of giving anything away to the French and the British. The June 30 flight, during which the M1 crash-landed shortly after taking off, was barely reported in the press, though in September Major Gross succeeded in staying aloft for thirteen hours—not that anyone noticed. The dark horse had turned into a damp squib, much to Zeppelin's glee.[17]

By then it was too late to matter, in any case: Zeppelin was already the most famous man in the world.

12. Conquerors of the Celestial Ocean

ZEPPELIN LATER DESCRIBED the *Schweizerfahrt*—the Swiss Voyage—as "the nicest trip I ever made in my life."[1]

His day began just after breakfast on July 1, 1908. Zeppelin, Professor Hergesell, Baron von Bassus, a Dr. Stalberg, the count's longtime lawyer Ernst Uhland, Dürr, and a well-known science-fiction author named Emil Sandt—his novel *Cavete!*, which predicted an era of peace delivered (and enforced) by German airships, had recently come out—arrived at the Reichshalle, where two steersmen and six mechanics were waiting.

Eckener hovered in the background. The previous year, beset by financial problems and with two more children, the Eckeners had been forced to move to a cramped apartment in Hamburg so that Hugo could take up a low-paid position as economics editor at the modest *Hamburg Foreigners' Paper*. Any spare time he had there was devoted to writing a book that he expected would catapult him to intellectual stardom and a lucrative career as a culture critic.

But when his interminable tome, ponderously titled *Lack of Workers or Scarcity of Money? A New Answer to the Old Question and Viewpoints About a Stabilization of Conditions in the Economy and Financial Market*, was published, few bought it and fewer still read it.[2]

It may not have gained him the high-status job at the *Frankfurter Zeitung* he once had craved, but it didn't matter. No longer did he want to be a journalist commenting on great events; he wanted to be part of them. Excited by the ambition and grandeur of the great Zeppelin project, Eckener spent increasing amounts of time moonlighting as the count's publicist.

But that still wasn't enough. Here he was, aged forty, and *missing out*. So in that balmy summer of 1908, he later said, he finally made the connection that had been eluding him when he linked "his moral-political ideals with the purely technical ones."[3]

He had been wandering and lost, but the airship was his North Star, the object whose technical perfection and grand purpose symbolized his own long-held desire to make anew a world striving to overcome its problems rationally and scientifically. He quit his job at the newspaper, moved his family back to Friedrichshafen, and devoted himself uncompromisingly to rendering the Zeppelin airship the salvation of mankind.

While now Eckener chatted with the crew, Zeppelin, as usual, claimed the front gondola, accompanied by Dürr, Hergesell, the steersmen, and three of the mechanics. The others clambered into the rear. Only Sandt, who had been invited as part of Eckener's media campaign, took a place in the newly installed cabin, which he described in a picturesque account published in major papers worldwide as a "room flooded with the yellow light that filters through the translucent balloon fabric of which the walls, the floor, and the ceiling are constituted. Comfortable seats . . . provide a seating capacity for a dozen passengers. For a greater portion of their length the walls are provided with celluloid panes. The floor is also transparent wherever it is not used as a footway."

At 8:30 A.M., Zeppelin gave the order to up ship and Ludwig Marx quickly towed LZ-4 from the hangar. It took just twenty minutes to cross the Bodensee and hover above the ancient city of Konstanz, on the German side of the Swiss border, which also happily happened to be Zeppelin's birthplace. Sandt looked down and saw, to his amazement, "the green earth, water, people, cities, and castles far below. I could also see birds circling around and fluttering anxiously, evidently frightened by the strange giant of the air."

The count didn't tarry long. He had to keep to a schedule, and Hergesell was eager to conduct his meteorological experiments as the airship wound through a series of mountain valleys. "As we swung round one mountain," wrote the professor, "we had our first experience of a vertical ascending current, which pushed the airship strongly upwards, and would necessarily bring our journey to an end if its force could not be resisted.

By means of our . . . rudders we were able, in spite of the disturbing force, to keep the ship at its right level."

LZ-4 bore west and made for Schaffhausen, a Swiss medieval town. From their high gabled houses, said Hergesell, "men and women rush out to look up at us, gradually massing together in crowds. The roofs of the houses become black with people. Handkerchiefs and flags appear everywhere, and shouts reach our ears even through the rattle of motors."

After that, LZ-4 headed for the spectacular Rhine Falls, Europe's largest waterfall. Zeppelin took the airship down as low as he could safely go to ascertain whether the wind eddies would have any effect on the airship. The crew was pleased to discover that, in Hergesell's words, this otherwise "impassable and unchangeable obstacle to navigation present[ed] no difficulty to us. Problems of transit on the surface of the earth have ceased to exist for us in the air."

Meanwhile, Sandt decided to leave the cabin and take his chances walking along the narrow gangway—which lacked handrails—to the rear gondola. Below him writhed the Rhine, and to the north he saw the Hohentwiel, an extinct volcano on the summit of which was a ruined castle, while on the southeastern horizon rose the Säntis mountain, broad and jagged and capped with snow and ice. On a clear day, like this one, an alpinist could see Germany, Austria, Liechtenstein, France, and Italy from its peak. Imagine what the airshipmen could see.

Zeppelin himself, congenitally unromantic, had little inclination for gawping. Instead, he ordered full speed ahead. The airship trembled as its engines strained to drive the propellers, which to Sandt "seemed like disks, revolving with furious speed and yet as transparent as a locust's wings. They gave out a note like that of a deep organ, so loud that the human voice, even when lifted to a shriek, could hardly be heard." The writer was unsettled at first by the sensation but was relieved when "the giant ship obediently sank and rose [and] moved to the right or to the left, slavishly following the slightest pressure of the human hand. . . . At times the forward car lay below us; at times we had to look up at it."

LZ-4 turned south for Lucerne, "a jewel among cities. The lake [Lake Lucerne] itself shimmered brightly where it was struck by the sun; its darker portions lay like an emerald, held in a setting of heliotrope. It was like a melody in colors." Once over Lucerne, continued Sandt, "there was a hubbub and a great jubilation [as] Zeppelin guided his airship down and allowed it to glide at full speed over the city at the height of the church steeples." According to a reporter on the ground, "thousands of astonished tourists from all parts of the world, including hundreds of Americans, greeted her with loud cheering as she sailed quickly over the waters of the lake." Over open water, Zeppelin embarked upon an impressive series of evolutions, including "complicated figures, circles, the figure 8, sharp turns, descents and ascents."

Afterward, LZ-4 headed northeast across the Zugersee toward the town of Zug, where its most difficult task awaited. The airship would have to pass through, according to Sandt, "a narrow gorge where it would be caught in a veritable cyclone." Once inside, the "motors groaned and rattled. The propellers howled a deep droning song. The airship did all that it could. The wind was dead against us, traveling with a velocity of nearly 31 miles an hour. The Count could easily have arisen and escaped the fury of the blast, but it was his purpose not to avoid obstacles, but to court them. Whenever the great airship showed signs of swerving, it was brought back into its course. Far below us in the valley the sharply marked shadow of the airship, crawling slowly from tree to tree, showed us how hard it was struggling. There were minutes when it seemed as if we stood stock still despite the infernal music of the propellers. Gradually the nose of the craft was thrust forward; once more the airship mastered the wind. We had forced our way through the pass, and were dashing on at full speed. The vast shadow below us traveled with the velocity of a bird over mountains, valleys, cliffs, and rocky points, over railway embankments and roads, over water and land."

LZ-4 had passed the test with flying colors. After arriving over Zurich at 2 P.M. — outside of which was an artillery range where the airship "was cheered by the troops engaged in field and firing exercises" — it was an easy shot home by way of Winterthur. Along the way, Zeppelin staged an impromptu race against a train, which LZ-4 won handily.

At 5:30 P.M., the crew caught sight of Lake Constance and saw the Reichshalle "inviting us homeward," but they resisted temptation and instead turned east "in order to keep our promise to run to another point of the Rhine Valley." Zeppelin was intent on making the *Schweizerfahrt* a twelve-hour trip, so they flew to Bregenz, on the eastern corner of the Bodensee, in Austria.

There was perhaps another reason for the scenic detour: As Hergesell would mention, they had "crossed the borders of different nations, always masters of our ship, always champions in the surging sea of air, true conquerors of the celestial ocean," a sentiment that could be read in two ways. The Zeppelin had certainly demonstrated that national borders were invisible to an airship and thus fostered an ethos of international brotherhood. Alternatively, and more sinisterly, the count had shown that a German airship could cross borders with impunity and without permission. One day it might be carrying bombs or troops.

For the time being, the Zeppelin was a peaceful emissary. When it at last arrived over Friedrichshafen, LZ-4 "descended to within 100 feet of the roofs of the houses and was greeted by the firing of a salute from a battery of small mortars." Zeppelin and his exhausted crew waved and cried out to the people below before heading to Manzell, where at precisely 8:26 P.M., twelve hours and 236 miles after her departure, LZ-4 landed.

In terms of time and distance, Zeppelin's airship had come exactly halfway to meeting the government's requirements for the endurance run without mishap. As "the evening sun shone on [Zeppelin's] noble features," rhapsodized the besotted Hergesell, "and kissed them with the breath of immortality," the count knew that all the world lay within his grasp.[4]

THE SWISS VOYAGE sparked a global sensation, partly thanks to Eckener, who helped place Hergesell's resulting article in the popular *Die Woche*, and Sandt's in its rival, Berlin's colorful *Illustrirte Zeitung*. They were quickly picked up and reprinted in France, Britain, and America, allowing Eckener to cover the political spectrum, from the right to the left, from the upscale to the downmarket, the national and the international.

Still better, during the trip Baron von Bassus had served as photographer by employing one of his inventions. He had mounted a downward-facing camera to the muzzle of a hunting rifle held horizontally; by pulling the trigger, he could snap a panoramic shot of the spectacular scenery below. Spelterini, the Swiss balloonist who had taught Zeppelin how to fly a decade earlier, had recently pioneered the use of a camera to capture scenery from a high altitude, but it was Bassus who stunned readers with a revelatory bird's-eye view of the glories of flight in true "as it happened" style.[5]

If having a square in Berlin and a park in Frankfurt newly named after him weren't reward enough, in the days following the trip to Switzerland Zeppelin was gratified to receive congratulatory telegrams from the kaiser, the crown prince, and King Wilhelm II of Württemberg. The latter, having stood by Zeppelin in his darkest years, commanded that he and Queen Charlotte be taken aloft. On July 3, for the first time in history, a monarch flew in an airship. Just in case anything went wrong, the king went up alone for half an hour, followed by his queen. Both times, Zeppelin took them on a voyage around the castle and park at 150 feet. The king responded to the waving of handkerchiefs by his subjects below by waving his own, and after a safe landing, he warmly shook the count's hand, a signal mark of favor.[6]

On July 8, a week after the *Schweizerfahrt*, Zeppelin celebrated his seventieth birthday. Originally, he'd intended to stay quietly at home with his family, but the public had determined otherwise. On that day alone, he received over a thousand telegrams, while Konstanz, Friedrichshafen, Stuttgart, and Lindau bestowed the freedom of their cities on him, the king of Württemberg awarded him the Gold Medal for Arts and Sciences, and his old comrades at the Association of German Engineers presented him with one of the most exalted scientific prizes in Germany. To cap off a wonderful day, the University of Tübingen, from which Zeppelin had never graduated, awarded him with an honorary doctorate of engineering, followed by a fireworks celebration. Zeppelin, blushing, joked that he'd finally taken his degree ninety-nine semesters after going to college.[7]

He also announced that on July 14, God willing, he would captain LZ-4 on its twenty-four-hour endurance flight.

GOD WAS NOT willing. Instead, after just a few minutes aloft, the Lord wrathfully caused a fan blade to snap on the front engine, and Zeppelin had to terminate the flight. The next day, the divine countenance further tested his faith by compelling Ludwig Marx to misjudge his speed, leading LZ-4 to painfully scrape the side of the hangar as he towed it out. The airship needed two weeks of repairs.[1]

Frustrated by the delays, Zeppelin caught everyone off guard early on August 4. That had not been his intention—for such a momentous occasion, the count wanted as many officers and officials as possible present— but it was only after midnight that Hergesell had been able to predict that the coming morning would be a fine one for flying (that is, windless and sultry), as indeed it would be.

Zeppelin and his eight passengers hurriedly gathered at the hangar at 4 A.M. Among them was his old pal Bassus, as were Dürr (to operate the elevators), an engineer named Stahl, Captain Georg Hacker and an assistant named Lau (who were to man the rudders), plus the veteran mechanics Karl Schwarz, Wilhelm Kast, and Kamil Eduard Laburda, a Czech. Workers, woken from their sleep, were already busy preparing the craft for the rigors that lay ahead. Eckener stayed on the ground to help Marx, somewhat embarrassed by his earlier error, direct the airship from its bay. At 6:26 A.M. precisely, LZ-4 set off on its journey.[2]

Among locals, word spread that the Zeppelin had unexpectedly lifted off. Many, along with early-rising summer visitors to the lakeside resort, congregated to wish it well, but their cheering was strangely muted. Given the absence of government and military representatives, most assumed that Zeppelin was simply taking LZ-4 out for a quick trial run.[3]

It wouldn't be long before they realized their mistake. Usually, Zeppelin performed some playful aerial maneuvers to test the steering and impress VIPs, but this time he rapidly accelerated LZ-4 to full speed and aimed it straight at Konstanz across the lake.

By now, the news was rapidly spreading that something amazing was happening. Over the course of the day, reports that the *"Zeppelin kommt!"* were excitedly telegraphed ahead, causing increasingly larger crowds to gather to greet it. In every town and village along LZ-4's route, postmasters and newspaper editors pinned up bulletins recording its progress, each accompanied by cheering and hopeful glances at the sky.[4] As the flight continued, newspapers would publish noon editions, then extra editions, then special editions, all selling out within minutes. Those wealthy enough to own a telephone pestered journalists by calling them directly at the office for updates.[5]

Spectators had different reactions to the wondrous sight. "Some laughed wildly and apparently without control; others raised their voices in ineffectual cries of encouragement; others wept and still others gazed mutely at the apparition in the air." As the airship turned a bend in the river beyond Laufen on its way to Basel, a reporter was touched to see "an old man, feeble, white of beard, and wrinkled of face, . . . crawling painfully to the top of a hill. He reached the crest all out of breath just in time to see it disappear. 'Ah, I have seen it,' he exclaimed."[6]

A sudden emotional outpouring at the sight of a Zeppelin was becoming a common phenomenon. For the faithful, like the old man, witnessing one of the count's celestial creations was like being present at the Ascension of Christ or, as the stoutly atheistic and classically educated Eckener put it, like seeing a "fairy-like apparition . . . coming from another world." It was "an emissary from the [Elysium] in which so many humans still believe in the inmost recesses of the souls."[7]

LZ-4 proceeded on its way through a succession of cities, but mechanical problems plagued the airship. Owing to troublesome engines, at 5:24 P.M.—eleven hours and two minutes after takeoff—LZ-4 gently touched down upon a quiet tributary off the Rhine at Oppenheim, a small town just fourteen miles south of Mainz. They were temporarily fixed, but at 1:27 A.M. the front engine finally quit for good, the rear engine was sputtering and smoking, and the count had almost run out of fuel and oil.[8]

Zeppelin made a tough call, though Stuttgart was temptingly near, after which home was just eighty miles away: He would set LZ-4 down in a field outside Echterdingen, a town 6.5 miles southwest of Stuttgart. There, the Daimler people (from nearby Untertürkheim) could repair the engines properly while LZ-4 was replenished with gas and the men could take a rest. The celebrations and congratulations in Friedrichshafen could surely wait a little.

In the meantime, tens of thousands of curious people headed for Echterdingen to catch a glimpse of, and perhaps even touch, the famous airship. "You don't have to ask where the airship is stranded," said a local reporter who'd endured a packed train to the station. "A stream of humanity clearly shows the way. It is difficult to move ahead in this unbelievable confusion of people and cars. At the church square, the masses make for the country road. We march along a lane, across fields and meadows, always in the same direction."[9] Others traveled by bicycle, car, carriage, and wagon, causing chaos and traffic jams—never before seen in quiet Echterdingen.

Now that LZ-4 had landed, said Karl Schwarz, the mechanic, the first thing that urgently needed doing was to secure the airship. He recruited a horde of sightseers to bury a wagon and attach the nose cable to it. Soon afterward, a contingent of soldiers and police officers came to keep onlookers, now numbering fifty thousand, at a safe distance and prevent them from clambering aboard the gondolas.

The Daimler mechanics arrived with a portable workshop and removed the front engine. As they worked on it, the exhausted crew were driven to the picturesque eighteenth-century Hotel Hirsch in Echterdingen to refresh themselves.

Schwarz had volunteered to stay behind to safeguard the airship. He went to the rear gondola to have a nap, not realizing that his Czech colleague Laburda and a soldier had also remained, but were in the front gondola. They were refilling the water ballast and keeping a Daimler mechanic company as he checked the engine fittings.

At around 2 P.M., the ominous roar of a thunderstorm approaching jolted Schwarz awake. Thunder meant wind, and wind meant disaster. Within a minute, a forceful gale slammed into LZ-4 and began lifting it into the air as violent gusts tore away the jerry-rigged moorings. The soldiers ordered to hold the ropes, not knowing what to do, let them go for fear of being whipped into the air as they lashed to and fro. As he raced along the catwalk to the front gondola, Schwarz was surprised to see the Daimler man leap from it. He didn't spot Laburda and the soldier, but he kept going until he reached the bow controls and pulled the valve releases. Usually, this was a gentle procedure of releasing just a little at a time, but Schwarz cranked them with all his might to try to deflate the ship and prevent it from being swept away. Gas hissed out but not rapidly enough to stop LZ-4 from drifting wildly.

As Schwarz recalled, "The crowd faded away beneath me, and I found myself about half a mile away from the anchorage." LZ-4 struck the ground and the bow smashed into a clump of trees, tearing open the skin and rending several gas cells, but kept on going. A shaken Schwarz leaned over the rail of the gondola and saw the most terrifying sight in the world: "Within the airship envelope there was a suspiciously bright light which seemed to grow and come closer. And suddenly I knew. Fire. The airship was burning."

He paused for the right moment to jump, but the airship was moving too fast to attempt it even as "fifteen thousand cubic meters of hydrogen gas were burning" and the gas cells were "bursting with loud reports. The rings, supports, and struts of the metal frame were glowing, bending, and breaking; the envelope was being torn apart in blazing shreds; and soon the flames were eating through to the gasoline tanks. The heat was becoming unbearable; it was Hell itself in which I was burning alive."

From a distance, a journalist heard a dull thud, and "flames shot up from the hull, a second, a third detonation. . . . A column of fire rose to the sky, immense, horrible, as if the earth had opened up releasing the flames from Hell. Huge flames ate their way up the balloon, piece by piece. . . . An enormous cloud of black smoke marked the spot where the elements had sacrificed the creation of man."[10]

Schwarz lost track of time. The next thing he knew, he was lying outside flat on his face: "Gas-cells, the envelope, and the whole net of girders crackled in livid red above me. As well as I could, I protected my head, breathing fire and trying to sit up and look around. Just then another mass caved in upon me. At such moments, one has terrific strength—I pushed the burden high, wound myself like an eel through the bent girders, slipped under the net of cloth covering me like a shroud, and I did not even feel the flames tonguing at me from all sides. I came free, stumbled to my feet, and said grimly to myself, 'Now, run like hell!'"

With his lungs filled with smoke Schwarz, gasping for cool air, stumbled on for another hundred feet. "When I looked back," he said later, "the proud giant airship was no more; the terrifying pyre had burned itself out, and only a few weak flames rose from the stern of the smoking ruin." In less than three minutes, LZ-4 had been utterly destroyed.

Then "a man in a singed uniform leaped towards me from the wreck, ran around blindly for a few seconds, and then stopped and stared. I got to my feet. He was a soldier. 'Where did you come from?' I asked. He fought for breath, and said, 'I was in the airship; there's someone else back there.'" At this point Schwarz saw the unconscious Laburda lying near the edge of the ruins.

He and the soldier dragged him out of danger, near "an excited crowd of people gathered to stare with dumb horror at the pitiful jumble of ruins which, only a few minutes before, had been a triumph of human endeavor and the symbol of German aspiration."

Ambulancemen helped Laburda away, but Schwarz refused to go. Still dazed, he walked by himself into Echterdingen and wandered around helplessly. Someone noticed that his head and hands were burned and rushed him to a doctor's surgery. The physician had left for the disaster site, but his wife expertly dressed his injuries.[11]

Meanwhile, an army officer had gone to the Hirsch to inform Zeppelin of the dreadful news. "The Count clasped his hands over his head, and then put on his cap and went downstairs, reeling and staggering. The other guests in the hotel, who were still unaware of the catastrophe, gave him a rousing cheer." Outside, a crowd—equally ignorant—surged toward the hero and threw their straw boaters in the air as children tried to crane over the shoulders of adults. Zeppelin "turned as he entered his motor-car, and, greatly moved, motioned away the crowd which was pressing round him, and then drove at full speed to the scene of the disaster."[12]

In Friedrichshafen the mood turned somber when a reporter attached a telegram to the blackboard in front of the newspaper building. Expecting glad tidings that the Zeppelin had left Echterdingen and was on its way, a crowd gathered to read the latest news. One man began to read the message aloud but suddenly stopped, stunned by its contents. Then he cried out, "The balloon has burned up! The balloon has burned up!" Some around him jeered at the tasteless joke, but the doubters soon themselves realized the magnitude of the disaster.

"Perhaps some rascal set the balloon on fire," someone volunteered, and then, with the worst possible timing, Zeppelin's daughter, Countess Hella, drew up in her carriage. She had not heard the news, wrote a reporter, and "a hush fell over the masses as the countess alighted and in wonderment at their strange attitude towards her she went inside her father's office. Soon afterward, through an open window, she was heard to cry, 'That will kill him.'"

Countess Hella had decorated the Zeppelin office with flowers in the count's colors (blue and white), organized a banquet, and hired her father's favorite regimental band to play. When subsequent bulletins confirmed that LZ-4 was no more, "the musicians silently packed away their instruments, and the villagers immediately began taking down the flags and festoons that had been hung in honor of the expected homecoming of the daring aeronaut."

When Ludwig Marx heard of the disaster he raced to the office on his motorcycle and found it closed. He saw only a morose Eckener, tears streaming down his cheeks. Too upset to investigate further, Marx jumped back on his motorcycle and sped off. He didn't return until nightfall.[13]

At the crash site, Zeppelin gazed sorrowfully at the terrible scene before him. Both gondolas had survived, but only a streak of charred grass marked the spot where LZ-4 had died. Nearby there was a twelve-foot-high section of the torn-off bow and a large tree split down its trunk to the roots. Scattered everywhere were remnants of half-burned cloth and twisted splinters of blackened aluminum, which the police and soldiers were trying to protect from relic hunters, not altogether successfully. Taking photographs was forbidden, but one fellow captured a couple of snaps before being arrested by the commander of the Stuttgart garrison, who demanded the negatives. The rascal handed over two of no value and kept the good plates in his camera. The images were soon published worldwide.[14]

Zeppelin said little. He didn't need to: The crowd was already expressing his inner turmoil. David Lloyd George, the future prime minister of Britain, happened to be visiting nearby and had decided to see the moored airship, only to arrive shortly after its destruction: "Of course we were deeply disappointed, but disappointment was a totally inadequate word for the agony of grief and dismay which swept over the massed Germans who witnessed the catastrophe. There was no loss of life to account for it. Hopes and ambitions far wider than those concerned with a scientific and mechanical success appeared to have shared the wreck of the dirigible.

Then the crowd swung into the chanting of *Deutschland über Alles* with a fantastic fervor of patriotism."[15]

Zeppelin drove back to the Hirsch "amid an indescribable ovation from the crowd which thronged about his car," after which he returned to a darkened Friedrichshafen.[16]

THE NEXT MORNING, Ludwig Marx was disconsolately driving to the hangar and happened to pass the Zeppelin office just as the count emerged on the balcony. He looked for all the world as if nothing had happened. "Good morning, Marx! How are you?" he called down to the perplexed tugboat captain. "Why, Marx, you seem to have been hit hard by yesterday's occurrence. Come on up here a moment."

The count led him to a table groaning under a heap of money, postal orders, and telegrams. "That was sent to me by the German people, and there's more coming!" he crowed. "So, Marx, now we'll really begin to build an airship."[1]

Thus began what the press would dub the Miracle of Echterdingen.

Its genesis lay in an impromptu speech by Manfred Franck, a Stuttgart merchant, who had roused the despairing crowds at the wreck with a plea to send donations to help Zeppelin build his next airship. "Everybody who has a German heart in his body will contribute his bit to bring Zeppelin's cause to a good end," he proclaimed. "The balloon body is dead, but the Zeppelin idea lives and will live for all eternity!"[2]

The appeal struck a chord. After newspapers mounted collection drives, the donations began rolling in. By the end of the first day, said Eckener, Zeppelin had received several hundred thousand marks—more than enough to build "LZ-5"—and the flood showed no signs of abating.

The daily press, conservative and liberal alike, led the way. The *Schwäbischer Merkur* raised 5,359 marks by lunchtime. The *Frankfurter Zeitung* started its own appeal, with the newspaper and publisher contributing 5,000 marks to get the ball rolling. Within four days, 62,000 marks had arrived. In Cologne, the *Kölnische Zeitung* received 33,000 marks the first day and another 50,000 by the third.

The aldermen of the city councils of Berlin and Stuttgart donated tens of thousands of marks, as did the Berlin Stock Exchange. The Mining Association of Essen chipped in no less than 100,000 marks. German expatriate communities as far afield as South Africa, Brazil, Java, Bulgaria, Turkey, and Egypt clubbed together and sent significant amounts.

Small towns, small social organizations, and small children also played their part in the grassroots campaign. The tiny hamlet of Haigen sent 50 marks, while Osnabrück, with its modest population of sixty thousand, gave 20,000. A bowling club in Baden sacrificed its annual summer outing and donated 150 marks instead. The passengers and crew of the Lake Constance steamer *Königin Charlotte* passed the hat around and raised 600 marks. A band in Darmstadt threw a special "Zeppelin-Konzert" and donated the proceeds. A little girl in Mainz who'd seen LZ-4 wrote to the count to say that she had broken open her piggy bank and sent every pfennig in it.

In short order, Zeppelin received—in cash—the astonishing sum of 6,096,555 marks. (The worth of that figure in today's dollars cannot be easily calculated. A rough range would be $25 million to $35 million.)

For those who could not afford to send money, there were other options. Over the coming weeks, Zeppelin was snowed under with sausages, hams, woolen socks, poems and songs, and bottles of wine and liqueur. At one point, there was so much incoming mail, the postman gave up trying to sort it and instead merely dumped sacksful of letters, money orders, and packages at the Zeppelin office.[3]

But the most important message of all came by official telegram from Berlin informing Zeppelin that as a result of LZ-4's (near) completion of the endurance challenge the government was pleased to purchase LZ-3 on behalf of the army and offered to buy the next airship built. LZ-3 would soon be renamed Z-1—a military designation—and Zeppelin deployed the lavish funds at his disposal to commission a copy of LZ-4 he initially named LZ-5, which would be acquired by the army as Z-2 the following August. The count had finally made his first sale.[4]

Better was still to come when the kaiser himself visited the Reichshalle hangar and inspected LZ-3/Z-1. Much to Zeppelin's delight, Wilhelm turned to his archenemy Major Hans Gross and said, "You see? That dirigible is quite practical. And now it will be accepted."[5]

The major could only fume in embarrassment as the kaiser then named Zeppelin "Conqueror of the Airways" and awarded him the Order of the Black Eagle, the very highest Prussian order of chivalry. The count, Wilhelm declared before the assembled press, would be numbered among the greatest Germans of the twentieth century.[6]

To complete the count's rehabilitation, he invited Zeppelin to attend the upcoming *Kaisermanöver*, the war games at which he had been ritually humiliated so long before. Zeppelin would soon be photographed in his finest dress uniform, his accumulated medals and decorations proudly adorning his tunic, helmet plume waving in the wind as he pored over maps alongside the Supreme Warlord and the other paladins of the German military caste.[7]

That was just the beginning. The world soon descended into full-blown "Dementia Zeppelina," as the press liked to call it, similar to the balloonamania that had greeted the early aeronauts.[8]

For the sake of dignity, the count refused to endorse Zeppelin perfume, Zeppelin cookies, and Zeppelin beer—not that it stopped entrepreneurs from producing them—but customers could purchase count-approved Zeppelin cigarettes, cheese, dolls, lamps, chocolate, gingerbread, suspenders, tape measures, boot polish, spoons, cigars, detergent, hats, ties, pocket watches, scarves, and firecrackers with his face on the box (rather tasteless, given the nature of LZ-4's demise) to complement the medallions, stamps, and postcards sold to raise additional funds. There was even an amusement park that built a Zeppelin carousel with airship-shaped rides, and the septuagenarian count became an unlikely fashion icon when upscale stores began selling his customary yacht caps and rubber-soled (to prevent sparks) cloth shoes.[9]

The Thermos company, founded four years earlier, advertised its insulated products with references to Zeppelin (allegedly) sipping coffee from them on his long voyages. Enthusiasts could also buy board games, one of them—*Conquest of the Air*—re-creating the Echterdingen flight and another allowing players to choose from a train, a ship, an automobile, and a Zeppelin to compete in a race around Europe. The hottest toys that Christmas would be, *Harper's Bazaar* announced, "the air-ships, air-ships everywhere, and air-ships that work in spite of the wind or weather."

Among the more expensive was a nine-inch-long "miniature reproduction of the Zeppelin Air Ship" equipped with a spring motor and propeller that when attached to a cord would "fly in a circle." A Berlin department store bested even that by exhibiting a fifteen-yard-long Zeppelin replica to delight and awe shoppers.[10]

Schools performed plays and songs based on Zeppelin's exploits, and almost every day newspapers ran a fond new caricature of the count, his twinkling eyes and huge mustache making him instantly recognizable. Membership of Zeppelin fan clubs exploded from nothing to sixty-five thousand.[11] In America, where the many German immigrants were fervently proud of "their" count, Zeppelin was granted honorary membership of the Aero Club for his "distinguished services in the advancement of Aerial Navigation."[12]

Whereas once the press had to specify "Count Zeppelin's airship" in its reports, after Echterdingen reporters began referring to it as "the Zeppelin," as if no other competitor were conceivable. The word *Zeppelin* itself became an adjective betokening "grand," "superlative," and "reliable." It also became a stock-in-trade of a new and fashionable theory: psychoanalysis. Practitioners noticed that their patients had begun to dream about airships. Sigmund Freud made the obvious connection that "the remarkable characteristic of the male organ which enables it to rise up in defiance of the laws of gravity . . . leads to its being represented symbolically by balloons, flying machines, and most recently by Zeppelin airships."[13]

Amid this convulsion of Zeppelinitis, the aristocratic count metamorphosed into that most modern of phenomena: a bona fide celebrity.[14] Now that he had shed the insulting image of an eccentric in favor of a reputation as a pioneer, his life story was mined for improving moral instruction.[15] Herbert Kaufman in the *Chicago Daily Tribune* uplifted his readers by assuring them that the hoots and jeers that had initially greeted Zeppelin and his fellow geniuses (Columbus, Alexander Graham Bell, Victor Hugo) only spurred them "to [develop] more *resource*, bred greater *power*, and aroused a *grit* that a *too-quick* recognition of their merit could *never* have produced." After all, "if you *know* you're right—you are a big enough *jury* and your *own* verdict is the *only* one that *counts*."

Learning to believe in yourself, he counseled, was the lesson everyone should draw from the example of the count.[16]

ERONAUTICAL EXPERTS, HOWEVER, were drawing different lessons from the destruction of LZ-4. After several years of evenhanded treatment, they began to choose sides when it came to deciding whether the airplane or the airship held greater potential. Opinion tended to divide along national lines.

In Germany, Zeppelin contended that Echterdingen was merely an unfortunate, unrepeatable accident on the long path toward perfection. The only reason, he said, that LZ-4 had been destroyed was that a fire had broken out when fuel spilled onto a hot engine fuse. He claimed, rather spuriously, that had the crew been able to restart the engines, "the airship would have sailed safely away."[1]

In truth, it is probable that the crash caused a hydrogen leak, which was detonated by electrostatic sparks set off by the friction of the outer envelope rubbing against tree branches, but Zeppelin's argument went unquestioned in Germany.[2]

Zeppelin was given a pass by his countrymen because he exemplified the quintessential German spirit of work, perseverance in the face of obstacles, and personal sacrifice for the good of the nation. The disaster at Echterdingen was part of a greater heroic story; indeed, the very fact that LZ-4 had been destroyed *was* the story, for as the writer Hugo von Hofmannsthal explained, "a completely unblemished success could never have crowned the genius of this man [Zeppelin] in the same way as this stranger-than-fiction combination of triumph and catastrophe."[3] It would be this near-universal acceptance of the necessity of sacrifice that would in future compel the German public to maintain its confidence in the airship even in the face of repeated mishaps, crashes, and accidents.

The deep resonance of Echterdingen helps explain the emotionally overwrought reactions to what was essentially a zero-fatality accident involving a significant degree of property damage to a rich man's experimental vehicle. Zeppelin's story hit the nexus of three prototypically German cultural touchstones. Like Goethe's Faust, Zeppelin yearned for seemingly impossible knowledge; his airship was destroyed by the kind of violent forces exalted by *Sturm und Drang* ("storm and stress") writers and poets; and his hard work, virtue, and sense of duty were eventually rewarded, as in a Brothers Grimm fairy tale.[4]

Americans paid little heed to these romantic stirrings in the German breast, preferring to focus on the practicalities of flight. They agreed with Zeppelin, rather strangely in hindsight, that fire was a minor risk to airships, even accounting for hydrogen's known flammability. After all, stray sparks from an engine could, and often had, set an airplane's thin fabric skin aflame, too, so fire just seemed to be part and parcel of flying.

Much more worrying, they thought, was the vulnerability of airships to high winds when anchored to the earth.[5] Zeppelin high-handedly dismissed such concerns, saying that he had always known that airships had to be stored in hangars to protect them. From now on, he assured them, it was *verboten* to land an airship in a field or valley in the absence of a dire emergency.[6]

American experts were nevertheless swinging around to the competing technology of the airplane as the way forward—but for reasons that had nothing to do with Echterdingen.

On August 6, 1908, at the very bottom of a long and detailed article about the Echterdingen crash the day before ("Tragic End for Monarch of Air"), the *Boston Daily Globe* appended a brief notice from France mentioning that Wilbur Wright would imminently be demonstrating his airplane.[7]

Wilbur had been in Europe preparing a new model of the *Flyer* for a series of exhibition flights intended to sell the brothers' airplane to a French syndicate. Given the number of charlatans who similarly had claimed to have conquered the air, the French, naturally, wanted proof not only that the Wright product could fly but that it even existed.

Since the Wrights had not publicly shown off an airplane for two and a half years, there was great skepticism surrounding this rumored *Flyer*. Ernest Archdeacon, a prominent member of the Aéro-Club, declared the *Flyer* to be a "phantom machine," while *L'Illustration*, a popular paper, published a photo of the purported aircraft and sternly judged the brothers to be frauds peddling a "fabrication."[8]

August 8 was a beautiful, cloudless, blue-skied day with a gentle breeze blowing. Hundreds of curious, or cynical, spectators outfitted for a summer picnic had ventured to the racetrack at Le Mans to watch Wilbur Wright's inevitable failure. Archdeacon was there, informing everyone around him that this was going to be an entertaining waste of time.

At 6:30 P.M., Wilbur, wearing his customary gray suit and high starched collar, sniffed the wind and told his assistants, "Gentlemen, I'm going to fly." With little ado, he took off and headed for a line of poplar trees but banked and circled back. Then he made another graceful revolution at a height of about thirty-five feet and landed fifty feet from where he had launched.

Wilbur was airborne for nearly two minutes and had covered a distance of some two miles. He enjoyed full control of the craft, and it was obvious that he could have stayed in the air for as long as he wished. At first, the crowd couldn't believe what had happened, but soon it broke out into wild cheers and rushed the *Flyer* as Wilbur primly tried to avoid Gallic kisses on the cheeks (he preferred manly American handshakes). Archdeacon graciously conceded that "for too long a time, the Wright brothers have been accused in Europe of bluff. . . . They are today hallowed in France, and I feel an intense pleasure in counting myself among the first to make amends for the flagrant injustice."

Over the coming days, Wilbur made more flights to prove the point. On August 10, a quick thirty-two-second ride demonstrated a complete turn in a thirty-yard radius, a feat never before achieved, and followed it later that day with two giant figure eights. From then until the end of 1908, Wilbur continued to amass ever more achievements and amaze ever larger crowds. He broke every airplane record there was—for duration, altitude, speed, number of flights, number of passengers—and won every award, prize, contract, and competition going. And on December 31, for his last spectacular display, Wilbur traveled a distance of seventy-seven miles over the course of no less than 2 hours, 20 minutes, and 23.2 seconds. For France, he was not just man of the year but man of the century.[9]

Meanwhile, planning to divide and conquer, Orville had been working at Fort Myer, just west of Arlington National Cemetery outside Washington, D.C. As Zeppelin had done during the Echterdingen voyage, he was intent on fulfilling an army offer stipulating that it would contract with any manufacturer that could build a plane that could fly 40 mph, carry two passengers and sufficient fuel for a 125-mile trip, travel for at least an hour, be reliably controlled, and land at the spot whence it had taken off.

A tall order by any account. American newspapers predicted that the specifications were so tough as to be impossible. As *The New York Globe* pointed out, "Nothing in any way approaching such a machine has ever been constructed." If someone did succeed, said the *Globe*, then it would mean that person had built "the most epoch-making invention in the history of civilization."[10]

Orville would be that person. On September 9, he circumnavigated the parade ground fifty-seven times in an hour. When rumors spread that the Great Orville would fly again that same day, Washington closed as congressmen, officials, cabinet members, and diplomats rushed to Fort Myer. At 5:15 P.M., Orville took off again and, wrote a *Dayton Journal* reporter, "presented somewhat the appearance of an automobile racing about an imaginary racecourse in the air" by circling fifty-five times in one hour and three minutes. For the next two days, he repeated the performance but added figure eights to his repertoire. *The New York Herald* reported that "he dipped down low to earth. He skimmed it at twice a man's height. He rose steadily and gracefully until 150 feet of space lay between him and the ground. . . . He all but brushed the trees in Arlington Cemetery. He tried every combination of the levers and planes in his run of 58 turns around the field." By September 12, he was attracting five thousand spectators at a time.[11]

The year 1908 was a pivotal one in powered flight. America (and France) fell in love with the Wrights as much as Germany had with Zeppelin. Both the brothers and the count had amply demonstrated the viability, potential, and popularity of their respective technologies, and they were now so closely identified with their creations that they actually resembled them, as owners are said to resemble their dogs—or perhaps it was the other way around. Zeppelin, for instance, was as rotund as his cloud-shaped airship, while Wilbur Wright's head (thought the *Daily Mail*) "suggested that of a bird, and the features, dominated by a long, prominent nose that heightened the birdlike effect, were [as] long and bony" as his airplane's skeletal construction.[12]

Character-wise, they were just as different. Zeppelin liked to come across as ebullient and avuncular, a kind of Santa Claus figure, while Wilbur (and Orville, for that matter) took after Jack Sprat—ascetic and parsimonious.

Yet they were often spoken of in the same breath, as aerial conquistadors. The *New Americanized Encyclopedia*, for instance, trumpeted its inclusion of all the latest flights by the Wright brothers *and* Zeppelin, beginning with its 1909 edition.[13] And that Christmas, reproductions of the Wright *Flyer* began to share shelf space in toy stores alongside the popular Zeppelin airships.[14]

ZEPPELIN AND THE Wrights maintained an outwardly polite relationship, but neither regarded the other as a real competitor. Zeppelin still thought the airplane a faddish contraption; the brothers saw the airship as a clumsy white elephant. They were, instead, rivals jealous for attention, prickly about slights, and eager to subtly one-up the other. Matters would come to a head in Berlin in the late summer of 1909.

On August 29, 1909, the count arrived in the capital in his newest airship, the LZ-6, a copy of the late LZ-4 quickly built with some of the Echterdingen money. After circling the city and dipping its nose at the Brandenburg Gate, LZ-6 headed for the immense Tempelhof Parade Grounds, where the kaiser and the royal family excitedly waited. As the church bells rang their merry peals and a military band played the national anthem, LZ-6 stopped its engines and gently glided to the ground. When Zeppelin stepped out, the kaiser saluted, shook his hand heartily, and called for three cheers for the count from the enormous crowd.[15]

It was then that the kaiser introduced Orville Wright—visiting Germany to perform aerial shows with the *Flyer*—to Zeppelin for the first time. It was a meeting of the two "kings of the sky," said one journalist of this great historical moment.[16] Orville stiffly congratulated the count on his airships, and Zeppelin awkwardly returned the compliment.

Orville might well have recalled a widely published interview Zeppelin had given the year before concerning his brother's flights at Le Mans. The count had commended Wilbur on his feats but slid the knife in by adding that the airship "as a means of conveyance certainly is superior to the airplane" given the latter's inability to "ascend to any great altitude" and its "limited scope of action."[17]

Now it was Orville's turn with the stiletto. Interviewed a few days after the meeting, he congratulated Zeppelin on "the graceful ease and apparent accuracy" with which he steered his "balloon"—a little put-down there—yet, he added, "the airship has nearly reached the limit of its capabilities." More insultingly, Orville then likened the airship to a steam engine and the airplane to an internal combustion one: The airship was the past, the airplane the future.[18]

Zeppelin was too well brought up to do much more than bristle silently at the slights, and he only reluctantly allowed Wright to come on a brief, sixty-six-mile voyage aboard the LZ-6 on September 15, though the two exchanged few words.[19] His generosity was not repaid. Two weeks later, Wright again needled Zeppelin by insinuating himself into the bosom of the crown prince, one of the count's key supporters.

On October 2, Wright brought the prince up with him on a ten-minute flight. At first restricting himself to an altitude of just twenty feet—"I felt a great responsibility in having the future German Emperor as a passenger," said Orville, piously—but rising to sixty when the prince cried, "Higher, higher," Wright was rewarded with a signal mark of royal esteem, a crown-shaped diamond and ruby pin. After the ceremony, the aviator took off alone and broke another world airplane record by reaching an altitude of 1,600 feet. According to Wright, he kept climbing "until the field and adjacent country reminded me of the picture I had from Zeppelin's airship [on September 15], only things seemed smaller."[20]

That offhanded comment, "only things seemed smaller," deeply rankled Zeppelin. That Wright had ignored the fact that an airship could ascend to nearly triple that height without breaking a sweat came across as remarkably ungracious to the courtly count. Once again, he said nothing publicly, but he privately believed that the Wrights' machine was nothing more than a sporting toy. Suitable, certainly, for showing off to gawping spectators, but ultimately a footnote in the history of aviation.

The Wrights, for their part, thought exactly the same about the Zeppelin, with Wilbur writing that the airship "must soon become a thing of the past." All the money the count was spending on developing them, he predicted, would "be practically wasted" once the airplane came into its own.[21]

16. Zeppelin City

Y ET IF THERE was something Zeppelin had that the Wrights didn't in the summer of 1909, it was money to burn. The Wrights may have made a small fortune in 1908 thanks to their successes in America and France, but they were now competing against dozens of rivals as the airplane business exploded. Within three years, in the United States alone, there would be 146 airplane companies and 114 different engines on the market.[1] Monoplanes, biplanes, triplanes—all these ate away at the Wrights' once-commanding lead amid a host of meritless lawsuits and wasteful patent-infringement accusations.

Orville and Wilbur sensed that their time at the top was running out. Following a spectacular series of flights in 1910 to show the world they were still the greatest aviators of them all, they essentially retired. Wilbur would die two years later of typhoid, and Orville, who lived until 1948, ceased flying in 1918.[2]

Unlike the Wrights, Zeppelin had no competitors to speak of, and he enjoyed a global monopoly on technical skill and piloting knowledge. He also benefited immensely from the official backing of the government. In 1908, whereas France spent (in contemporary dollars) $235,000 of public money on aviation, with Austria-Hungary coming in second at $27,000, followed by Britain with $25,000 (the United States probably expended about the same), Germany devoted no less than $660,000, the overwhelming majority of which was directed to airship development. And that stupendous figure did not include Zeppelin's own windfall after Echterdingen.[3]

Zeppelin had dreamed of achieving such success since his forced retirement from the army nearly two decades earlier, but now that he'd finally gotten what he wished for, it seemed more a curse than a blessing. With the Niagara of cash came responsibilities, interference, turf battles, and squabbles as Zeppelin's relaxed little cooperative transformed into a corporate conglomerate.

Just a year earlier, Zeppelin's "company" had essentially comprised just him; Dürr, the designer; Uhland, his lawyer; Eckener, his public-relations man; Hergesell, the meteorologist; and various pals like Baron von Bassus, along with a core group of mechanics and workers (like Schwarz, the hero of LZ-4, and Marx, the tugboat captain). They lived locally, and if not native Württembergers then happy immigrants to Friedrichshafen. All had been with the count for years and had stuck by him through thick and thin. He in turn—as befitted a paternalist noble conservative of the old school—had regarded his faithful retainers, almost feudally, as part of his extended family.

Every year, no matter how paltry his finances, Zeppelin would take everyone on board a hired steamer and travel to a lovely park for a picnic. It was a tradition that one of the men would give a speech, but when a mechanic once became tongue-tied, Zeppelin stepped in, telling those gathered in rustic Swabian dialect, "You workmen and officers do not owe me thanks. On the contrary, I am indebted to you. For although this great work was my idea, it was you, my workers and employees, who completed it. Therefore, I thank you from the bottom of my heart, and drink to your health!" He shared coffee with every man who worked for him, courteously inquired after their wives, and marked each child's birthday with a gift of improving books. On the job, he was a strict disciplinarian, but off-duty he always said "please" and addressed every worker almost as an equal.

Everybody knew who was the seigneur of this particular demesne, of course, but the bonds of comradeship forged in adversity had wrought a battle-tested, tight-knit, unhesitatingly devoted band of brothers. Even when Zeppelin had to shut down operations (as after LZ-1), he made sure his staff found employment in his friends' factories and workshops and took them back as soon as he could afford to. Any man who had demonstrated loyalty in the past was granted his consideration. One time, a lanky, gruff blacksmith showed up, and much to the surprise of the other workers, Zeppelin himself showed him around the hangar and allotted him a job. It turned out the man, lately fallen on hard times, was a former dragoon who had been on the count's famous 1870 patrol ride.[4]

This traditional ethos of noblesse oblige eroded faster than Zeppelin ever expected, or wanted, as a more modern form of aggressive industrial capitalism began to intrude upon Friedrichshafen's green and pleasant land. The agent of these "dark satanic mills" was Alfred Colsman, whom Zeppelin had at first welcomed as someone who could oversee the rivers of cash arriving daily after Echterdingen.[5]

Colsman, in his mid-thirties and "an immense fellow, standing six feet two, with a deep chest and broad thick shoulders [with] the tight lips of a self-reliant man, drawn down at the corners in a grim way," had briefly met Zeppelin in 1899, shortly after he'd married Helene Berg, the daughter of the count's friend Carl Berg, the aluminum magnifico. Colsman himself was the son of another aluminum manufacturer, and so the marriage, though a happy one, was truly a dynastic alliance. The second time Colsman met Zeppelin was in 1906, on the occasion of Carl Berg's funeral.

In the immediate aftermath of Echterdingen, Colsman offered to "serve the greater cause." Zeppelin quickly took him on staff, and Colsman started work full-time in August of 1908.

It was Colsman who suggested establishing a formal company to regularize the count's chaotic financial affairs, and in September Colsman accordingly became the manager of the new Luftschiffbau Zeppelin AG—the Zeppelin Airship Construction Corporation, usually called the Zeppelin Company. Overseeing it was the Zeppelin Foundation, comprising a board of directors, which also funded training programs, research, and charitable work.[6]

At first, the relationship between Colsman and Zeppelin flourished. The count was pleased that the bespectacled younger man immediately set to work acquiring the land needed to build a giant new double hangar that could fit two airships side by side.[7]

But the relationship went downhill quickly from there, and Eckener was often dragged in to serve as emollient arbiter. During discussions it was important to "be conciliatory and flexible," he counseled Colsman, and eventually the grand old man would come to see his point of view. Colsman would have none of it, saying, "I will say what I think, as long as I am general manager, whether you like it or not."[8]

Colsman saw quickly that Zeppelin, an aristocratic relic in a new age of profits, mass production, and industrial might, was flummoxed by modern business practices. For instance, during financial negotiations with Herman Sielken, known as the "Millionaire Coffee King" of New York, the lordly Zeppelin instructed him, "Then you will participate in my airship construction, Mr. Sielken." To which the Coffee King replied, "And what would your part of the bargain be, Your Excellency?" Much to Colsman's embarrassment, Zeppelin seemed puzzled that someone beneath him in rank would expect a quid pro quo and airily answered, "I? Oh, I will drink a great deal of coffee." And that was the end of the negotiations.[9]

Unaccustomed to dealing with accountants and tax officials, the count also had a liberal attitude toward budgeting when it came to his beloved airships. For Zeppelin, money was something other people worried about, and he treated company and foundation revenue as his own, and his own as a business asset.

The entrepreneurial Colsman had a hard time convincing the count that airship construction in and of itself did not generate earnings; if anything, it was a potentially bottomless money pit. He complained that "we are not considering the company as a business, but as a duty" to Germany. Yes, exactly so, Zeppelin furiously retorted, and it would stay that way. Colsman did at least succeed in establishing subsidiaries to the Zeppelin Company that would make parts and materials for airships and pass them on at low cost to the holding company. The only reason the count agreed to them was because he believed they were dedicated exclusively to the great god of Zeppelin, but Colsman's intention was for them to make profits from outside sales to give the company's finances some semblance of stability.

At one point, there were ten of these subsidiaries—Eckener later commented that Colsman had an "obsession with founding corporations"— including Maybach-Motorenbau (Maybach Motor Company), which was run by Wilhelm's son Karl Maybach; Zahnrad-Fabrik (gears and drives); Zeppelin Hallenbau (hangars); Ballon-Hüllen-Gesellschaft (outer skin); and Zeppelin Wasserstoff und Sauerstoff AG (hydrogen and oxygen production).[10]

Most were located in or near Friedrichshafen, turning the small resort town into Zeppelin City. Workshops, factories, and laboratories containing mechanical, electrical, telegraphic, and chemical equipment were joined by an aluminum foundry, a large office building, fuel depots, a meteorological station, and a gasometer. New villas, schools, and hotels sprang up along freshly laid out streets.

Until the formation of the Zeppelin Company, the count had usually employed fewer than ten men, rising to around seventy workers at busy times, not including women, who were hired as secretaries or seamstresses sewing together the cotton airship envelopes. By 1914, the permanent workforce would rise to seven hundred, about 10 percent of the population of Friedrichshafen, and the former backwater became the wealthiest town in Württemberg.[11]

Colsman, worrying about the company's high wages, wanted the burgeoning industrial conglomerate to make spending cuts a priority, but on this point he was forced to back down after a battle with Zeppelin, who insisted on providing his workers with the kinds of lavish benefits—known as *Zeppelin Wohlfahrt*, or Zeppelin Welfare—that ensured loyalty and the maintenance of social order.

It was Zeppelin who overruled Colsman by insisting that every employee could take out a company-sponsored life-insurance policy and offering free board and lodging to those who suddenly found themselves in dire straits. The company built an entire workers' settlement with well-designed, picturesque houses and pretty gardens.

There was a grocery, bakery, butcher, church, kindergarten, subsidized canteen, pub (with its own winery), day-care center, school, dormitories for unmarried men and women, savings bank, library, hospital, and first-class sports facilities. There was even a Zeppelin Hall for readings, performances, and concerts.[12]

Such largesse came at a price, and Colsman sought increasingly exotic ways to raise the money needed to pay for it as the count grew ever more irritated by what he regarded as Colsman's horribly bourgeois thriftiness. The low-level hostility eventually erupted into open warfare over the issue of creating a new subsidiary: a passenger airline, the world's first.

On the face of it, forming the DELAG—an acronym for Deutsche Luftschiffahrts-Aktien-Gesellschaft (German Airship Transportation Company)—was a brilliant, bold, innovative, visionary idea. At a time when airplanes could carry, at most, a single passenger and a few fuel tanks as cargo, an airline was a means of putting airships to profitable good use.

Zeppelin didn't see it that way. He *hated* the idea. In the boardroom, shedding his usual amiability and charm, the count subjected the hapless Colsman to a flash of anger. "I detest any notion of commercializing my invention!" he protested and refused to listen to any more talk of an airline. Explained Colsman, writing many years later, "He saw his conception profaned if the airships were used to earn money through the DELAG. That enterprise thus remained for him, the feudal aristocrat and old soldier, a tradesman's venture. . . . In an aristocratic context such as his, a merchant was just not socially acceptable."[13]

Zeppelin had of course dealt before with many "merchants," though they had tended to be wealthy industrialists, and it had been he who back in the 1890s had first suggested going the civilian route when the army proved indifferent. But there was a distinction in his mind between a civilian enterprise dedicated to enhancing German greatness to force the army's hand and a commercial one based on tawdry profit.

On the issue of the DELAG, the board itself was split between the military and civilian factions. Zeppelin, for instance, could number among his backers his nephew, Freiherr von Gemmingen, a traditional career soldier, while Colsman had a quiet ally in Eckener, a man of the modern age like himself.

Snobbery played a role in the count's bitter opposition to making money from a passenger airline, but the military-civilian divide drove at the very heart of a fundamental dilemma harking back to the days of the Montgolfiers: Now that working Zeppelins existed, *what were they for?*

17. The Wonder Weapons

To UNDERSTAND THE issue facing the board, it helps to turn to a very bad (and now, mercifully, obscure) book by Dr. Rudolf Martin, a minor civil servant in the Imperial Statistical Office. In 1907, his extravagant science fiction novel, *Berlin-Baghdad: The German World Empire in the Age of the Airship, 1910–1931*, had laid out the shape of things to come.[1]

Despite its preposterous and convoluted plot, of which the less said the better, *Berlin-Baghdad* contained dual visions of the airship's future.* The

* For the masochistically inclined, herewith a summary. The action opens in 1910, when the kaiser declares that Zeppelin's airship is on a par with the world-shaking importance of the invention of gunpowder. The Reichstag subsequently appropriates a billion marks to build a fleet of them—for defensive purposes, of course. Three years later, after Japan defeats a tottering Russia, a tyrant named Suwarow seizes power in Moscow and uses a secret flotilla of three airships to bomb his way across Central Asia. In 1916, Suwarow, who, like some megalomaniacal Bond villain, lives aboard a two-mile-long battle-airship, turns his greedy eyes toward Europe, where only Germany, the mightiest air power and defender of civilization, stands between him and world domination. The Germans defeat Suwarow by massacring thousands of his soldier-slaves with airship-borne artillery. The tyrant holes up in a Himalayan fastness to lick his wounds, but not before the Russians have devastated Berlin in a vengeful bombing raid. More than 100,000 casualties are the result. The lesson is that the Germans' pacific ways have led to this cataclysm. What Germany must do is expand its borders to provide security for the future. Through conquest and confederation, Germany creates a new empire absorbing Poland, Austria-Hungary, Ukraine, Turkey, the Caucasus, Holland, Belgium, Switzerland, and Morocco. Eventually the empire expands into the Middle East—Mesopotamia, in particular, where German colonizers resurrect Baghdad to its ancient glories. Poles are used as overseers for the hordes of Persians and Turks working in the

first, which attracted Zeppelin's eye, was the extraordinary military potential of these machines. Martin predicted that Germany's future airships
(all four hundred of them) would travel at 250 mph and be capable of ascending to 29,500 feet, rendering them ideal for bombing distant enemy
cities, slaughtering millions of people, and bringing Germany imperial
mastery over Europe and the Middle East. Martin expounded further on
the subject two years later, in the midst of the debate in the Zeppelin
boardroom, in a follow-up bestseller, *World War in the Air*.

This time, as the German airship fleet arrives over Paris, "the most important task . . . was the destruction of the War Ministry at Boulevard St.
Germain 231. Admiral Graf Zeppelin steered his ship to the ministry
[while] the rest of the 150 airships were deployed over army barracks, the
Bank of France, the Foreign Ministry, the Ministry of Finance, and the
presidential palace, the Palais d'Élysée. . . . A minute later, the immense
block housing the War Ministry, from the Boulevard St. Germain to the
Rue St. Dominique, lay in ruins." Before departing homeward, the flying
Wunderwaffen—wonder weapons—leave "the whole of the inner city of
Paris . . . in flames."[2]

In these fantasies, the count—who appreciated being praised as "Admiral Graf Zeppelin"—found fulfillment and validation. To him, Martin
was a genius who shrewdly perceived the military applications of the airship and its place in Germany's world-leading role to come. It was as if a
kindred soul had *finally* understood him.

In the boardroom, Zeppelin turned ever more belligerent. When he
was younger, more liberal, and more desperate to attract supporters anywhere he could find them, Zeppelin had cited the airship as a potential

---

cotton plantations of the new Fertile Crescent. By 1930, the last holdout from joining this
pan-German Utopia is Britain, which haughtily refuses to acknowledge Berlin's legiti-
macy, let alone its supremacy. Tensions rise when the German chancellor reminds the
British ambassador to Berlin that if he so wished he could transport an army of two million to Britain within three hours by means of his unstoppable Zeppelin fleet. As Parliament stalls for time, Suwarow attacks British India and London is humiliatingly forced
to beg Germany for assistance in destroying the Russian menace. The kaiser magnani-
mously assents but stipulates that the price will be eternal friendship between the Anglo-
Saxon peoples—and the transfer of all British possessions in Africa to Germany as a kind
of voluntary contribution to the war effort. Parliament accepts these generous conditions, and together the two great empires march forward to inevitable victory over the
repellent race-traitor Suwarow and his Asiatic/Slavic minions.

troop transport, a tool of imperial expansion, a mail service, and an instrument to bring peace unto warring nations. But his successes after the Swiss Flight and Echterdingen had brewed into a dangerously heady narcotic.

Rehabilitation in the eyes of his emperor, the satisfying defeat of Major Gross and other critics, a weakness for believing his own adulatory press, an aging man's reactionary crustiness, a growing dislike of Colsman's capitalist vulgarity, and the military's purchase of LZ-3 and LZ-5—all these contributed to his curdling into a militarist rabidly eager to wade through blood to raise Germany to paramountcy among the Great Powers. In his mind, airships were now suited *exclusively* for martial purposes and would, he boasted, "assure [Germany] world military domination."[3]

He dismissed those who thought differently as fools and naifs. When he received a letter from one Ernest Dalle of Bremen pleading that he not turn his wonderful airships into weapons of war, Zeppelin—who habitually replied to all who wrote to him—angrily scribbled in the margin, "No answer for this special saint."[4] Another of his correspondents, who had just returned from a visit to Britain, excitedly speculated upon "what incredible destruction of property and war materiel twenty Zeppelins could wreak in one day on London with its billions of marks worth of structures, goods, banks, etc." To him, Zeppelin replied that "I fully share your patrioti[sm], and all the more am I determined to develop my airships as splendid instruments of war."[5]

If war came, the count announced—clearly taking to heart Martin's description of him as "Admiral"—it would be the culmination of his life's work if the kaiser permitted him "to lead the best one of my available airships into battle" since "all Germany expects me to make the first flight over London."[6]

In Britain too, Martin's works were taken seriously, especially after he was quoted claiming that if war came a Zeppelin fleet would "transport 350,000 men in half an hour during the night [and] we would conclude no peace until a German army had occupied London."[7] In March 1909, firm evidence emerged that the Germans were running practice night raids when a police constable named Kettle from Peterborough claimed to have sighted a Zeppelin over Britain. He heard an engine overhead and saw, he said, "a dark body, oblong and narrow in shape, outlined against the stars," shining a light. After the story ignited a bout of Zeppelin paranoia across the country, it turned out that someone had in fact been flying a large kite with a Chinese lantern, and as for the sound of the engine, said a local, "that was the motor which goes all night in the Co-operative Bakery in Cobden Street."[8]

For their part, Colsman and Eckener rolled their eyes at the elderly Zeppelin's increasingly ludicrous pretensions. They drew a different lesson from Martin's aeromania: a peaceful and commercial one. In *Berlin-Baghdad*, Martin had also depicted a New Berlin of 1930, a wealthy capital made prosperous for all by its air-based economy.

There, every high school student learns how to fly; aerial parks have gardens, skating rinks, and tennis courts suspended high above the ground; factory workers can fly to Switzerland for weekend breaks; hospitals float fifteen thousand feet above cities to mitigate the risk of infection; and there are aerial *Autobahnen* for the commuters who take off from flat roofs doubling as runways. Many homes have mini-hangars to dock the family airship.

Relations with Amerika have never been closer. Zeppelin's airships, twelve-engine luxury behemoths carrying more than a thousand passengers, fly the transatlantic route regularly to New York while smaller ones allow tourists to visit the North and South Poles. Within Germany, smaller airships have replaced trains for traveling between cities, and the entire country is connected by a network of airship-ports for rapid and efficient transfers. Colsman and Eckener could not have wished for a better advertisement for the launch of their airline than Martin's anticipations of the glorious Airship Age that lay ahead.

To that end, Colsman proposed, with Eckener's support, building passenger airships that would travel regularly scheduled routes between, at first, German cities, and then to Paris, London, Rome, Copenhagen, Stockholm, Vienna, and Saint Petersburg.[9] Ultimately, the route to the holiest of long-haul destinations—New York—would be opened. Ticket income—the flights wouldn't be cheap—would help offset the spiraling expense of building airships, excited investors would flock to buy DELAG shares, and the Zeppelin Company would establish a new line of business with no competitors, a potentially huge market, and a dizzyingly high cost of entry.

Over my dead body, retorted the count. Airships were only to be sold to the military, he insisted. Zeppelin's argument might have had more force had the military been more interested. Granted, during the Zeppelinitis craze after Echterdingen, General von Moltke had suggested purchasing no fewer than fifteen airships, but he soon slashed the order to just three.[10] For him, Zeppelins remained too expensive, too short-ranged, and too unreliable to be worth diverting money from the budget when generals and admirals were clamoring for more guns and ships.

Zeppelin was told that the huge military contract he hankered after was not in the cards—at least until his airships had proven themselves.[11] With ill-concealed lack of grace, in November 1909 Zeppelin was forced to go cap in hand back to Colsman and agree to back the creation of the DELAG as a commercial airline. But this was merely a tactical retreat in preparation for his strategic offensive to come. If Zeppelins could be put through their paces in the civil market, the count calculated, eventually the military would have to take notice and issue the purchase orders. With Eckener serving as broker, Zeppelin did manage to wring one key concession out of Colsman in exchange for his blessing: Any commercial airship built by the company had to be readily convertible to military use if war broke out.

It was a deal Eckener would come to rue, but for the moment he had big plans for what was to be "his" airline.

18. The Lucky Ship

EVER SINCE ECKENER had first joined Zeppelin, he had stayed in the background and allowed his superior to bask in the glory justly due to, and demanded by, the World's Greatest Aeronaut. But now he was eager to step into the limelight, or perhaps the lions' den, when Colsman appointed him general manager of the DELAG with a director's seat on the board. It was quite a promotion for the former publicity man, but Eckener had proven his loyalty.

Creating an airline from scratch was no simple matter—especially since no one had done it before. The first priority was financing a network of hangars around the country.[1] The DELAG could not itself afford to fund such a large-scale project, so Colsman and Eckener undertook a national tour to charm a succession of burgomasters into paying for their own municipal hangars. It wasn't a hard sell. At the news that the company was considering establishing a prestigious passenger service in their fair city, local papers printed special editions cheering the prospect. Frankfurt, Düsseldorf, Baden-Baden, Stuttgart, and Cologne quickly agreed to the proposal. In turn, the mayors encouraged the local bigwigs and trade associations to invest in the DELAG. In short order, 2.6 million marks flowed into DELAG coffers.[2]

Flush with cash, Eckener demolished the old floating Reichshalle at Manzell and moved the airship-building operation to Friedrichshafen, where a new hangar was furnished with the most modern equipment.[3] The resort town, however, was too remote to serve as the business headquarters of a major enterprise. Frankfurt, the country's financial center and blessed with excellent facilities left over from an aeronautical exposition earlier that year, would be home to the DELAG's head office and main hub.[4]

The nascent company received a huge boost when Albert Ballin, the father of the modern cruise ship, approached Colsman and Eckener with a proposition. Head of the giant Hamburg-America Line (HAPAG), Ballin offered 100,000 marks a year in advertising for the airline in exchange for HAPAG becoming the exclusive seller of airship tickets.[5] The shipping magnate had offices in every city in the country and, better still, had made inroads into the American market. Tourists there would be able to buy their tickets in New York, Chicago, or Los Angeles, sail to Germany on his planned SS *Imperator*, and thence board a DELAG Zeppelin to whichever destination they wished.

For Ballin, it was a neat way of outplaying such rivals as the White Star Line (owners of the much-anticipated *Titanic* and *Olympic*) and Cunard (the new *Lusitania* and *Mauretania*) in the race to dominate the profitable transatlantic route. For Colsman and Eckener, the partnership, by allowing them to sell domestic flights to a global audience, was an immensely valuable one. At this early stage, their plan was to offer scenic Rhine River journeys and short return flights from Frankfurt to and from other regional cities before establishing longer routes to Munich, Hamburg, Dresden, Leipzig, and Berlin, and afterward to major European capitals. Then, several years down the line, the jewel in the crown: New York.

It was a grand and ambitious vision, but reality soon intruded.

IN THE BEGINNING, the DELAG had only one airship in its fleet: LZ-6A, originally built in the hopes of a military sale but modified (hence the "A" designation) to include a handful of seats. It was intended only as a stop-gap until the first airship built specifically for the DELAG was delivered on June 22, 1910. LZ-7—christened *Deutschland* for marketing purposes and in homage to one of the count's earliest prototypes—took off from Friedrichshafen for its new home base of Düsseldorf, three-hundred-odd miles away.

Arriving in Düsseldorf to grand fanfare, Eckener experienced mobs of people, desperate to secure reservations for upcoming trips, quite literally thrusting money in his face.[6] The airship was placed under the command of Captain Kahlenberg, the DELAG's director of flight operations.

On June 28, *Deutschland* embarked on a three-hour flight carrying twenty journalists as a publicity outing before beginning ticketed, scheduled service. Those reporters fortunate enough to be selected were astounded at the luxury on offer. Nothing like it had ever been seen before. At a time when a (single) passenger on an airplane had to don goggles and overalls to prevent motor oil from the engine from spattering all over him or occasionally her, the *Deutschland* provided a lounge equipped with wicker chairs set next to large, sliding windows that allowed optimal viewing of the countryside passing by below. The walls and ceiling were veneered in dark mahogany, with the pillars and roof beams of the same material but richly inlaid with mother-of-pearl. Every sharp corner was swaddled in soft leather, and the floor was thickly carpeted to absorb the engine noise. A tiny galley provided sandwiches and drinks. All the cutlery, plates, and cups were made of aluminum to reduce weight. Even the lavatory—itself a revelation—had aluminum fittings.[7]

Ward Price, a *Daily Mail* correspondent, expressed surprise that the *Deutschland* ascended "so steadily and gently that if I had not been leaning out of the window of the car I could not have believed that we had left the earth," and told his unbelieving readers that "we lay back in our wicker chairs. A plate of caviar sandwiches stood on the window-ledge beside each of us. The surface of the wine in our glasses was unrippled." There was not, he assured them, "the least sensation of giddiness, sickness, or insecurity."

Price gazed down upon the unrolling surface of the earth and "the quaint little toy world a thousand feet below. It was as if we saw it from another planet. How neat and tidy it looked. The roads were so straight; the gardens so charming, with their geometrical displays of color." From above, people looked like "curious mannikins" as they ran with "curious, awkward, jerky movements" to try to keep up with the airship.[8]

Unfortunately, Kahlenberg, through lack of experience, had omitted to check the weather reports and now a fierce storm was unexpectedly approaching. When it hit, the three-hour pleasure jaunt turned into a nine-hour nightmare ride as *Deutschland* fought the unrelenting, turbulent wind. At one point, the airship was actually traveling backward.

Compounding the problem were the airship's overfilled gas cells, which floated *Deutschland* dangerously high and obliged Kahlenberg to valve hydrogen prodigiously to bring the ship lower. After descending, precious fuel was burned to keep the airship steady as the storm's upward gusts violently tried its strained engines.[9] Worried passengers repeatedly asked what was going to happen, only to be told by the exhausted, exasperated officers that "we do not know what is going to happen."[10]

They found out when the fuel tanks were empty and there wasn't enough gas to keep the airship up.

Deutschland plunged slowly through a thousand feet of air. "We were lost," said the man from the *Berliner Tageblatt*. "We heard a frightful crash; the balloon trembled through its whole length."[11]

By sheer good luck, the passengers were saved by *Deutschland*'s falling into, or rather onto, the tall pines of the Teutoburg Forest directly below. "At the absolute instant of the crash into the tree-tops we became for the moment extremely scared human beings," wrote Price, "and hung on grimly to the uprights of the cabin, bracing ourselves against the end that seemed so near." But, amazingly, no one was hurt, despite the enormous tree trunk that pierced the windows of the lounge. The crew ushered the passengers outside onto the bough and they were able to clamber down the thirty feet to the ground.

Price's faith in the airship was undaunted. It was obvious to him that "the practical era of flight had in fact begun. We were, it seemed to us, in at the making of world history. The generations of the future would look back to this June day . . . to find the beginnings of regular practical commercial air-travel, where passengers paid their money and had their flight just as they have a ride in an omnibus."[12]

The crash caused an internal shake-up at the DELAG, with Colsman admitting to his employees that "our pride, our hopes, and intents for this summer [have] been shattered."[13] The hapless Kahlenberg was fired for incompetence, and Eckener took over the company's day-to-day and flight operations.[14]

His first order of business was learning how to fly one of his own products; Zeppelin himself taught him in LZ-6A, the only airship available. After Eckener completed thirty-four flights without incident, an engine failed and LZ-6A went in for repairs but never came out. During a test, an engine roared to life and ignited a small fire. When one of the new workmen mistakenly grabbed a can of fuel to put out the flames, it became a big fire. No one was injured, miraculously, but within minutes LZ-6A was burned to ashes.[15]

Eckener, nevertheless, had high hopes for the coming summer season. *Deutschland II*, or LZ-8, was delivered to Düsseldorf on April 11, 1911, following a triumphant publicity tour of Stuttgart, Frankfurt, and Baden-Baden. Brochures advertised 90- to 120-minute flights for 200 marks (more than the average worker's monthly wages), while pamphlets emphasized the routine safety of the Zeppelin. No one, after all, had ever been killed in a Zeppelin—all the more remarkable considering that almost every week an airplane pilot died.[16]

A little more than a month later, on May 16, Eckener was reluctant to bring out the *Deutschland II* for its twenty-fifth flight—the wind was picking up—but passengers irritated at the delay convinced him to chance it. After he left the hangar, a strong gust hit the airship crossways and it bucked free of the three-hundred-man ground crew holding the ropes. *Deutschland II* was scooped up and dumped unceremoniously down as divine punishment for Eckener's hubris. Aside from some scrapes and sprained ankles, no one was injured when the fire department rescued the passengers, but with its spine broken the pride of the DELAG was a pathetic sight. It was beyond repair and had to be written off.[17]

As with the LZ-6A fire, *Deutschland II*'s end was an unfortunate accident unrelated to actual flying, but the record for the DELAG was unimpressive, to say the least. After all, *Deutschland* had lasted all of a week; LZ-6A, three; and *Deutschland II*, five. The DELAG was an airline without any airships, and it was becoming difficult to explain away all the mishaps.

A contentious board meeting followed. The count was furious that his legacy was being wasted on extravagant passenger airships that were rapidly eroding military interest in his wonder weapon. Eckener, however, stood his ground. The accidents, he pointed out, were a symptom of ignorance—had not the count said the same to him after the death of LZ-2?—and overly ambitious expansion. The DELAG, put bluntly, had been flying by the seat of its pants. The company needed to rethink its entire strategy and start again from scratch.

By a small majority, the board sided with Eckener and voted to throw the dice one last time. It authorized the construction of a new airship—LZ-10, to be christened *Schwaben* (Swabia) in a patriotic, and placatory, homage to the count's native region—that would be the last for the DELAG if it, too, were destroyed.

WITH A GUILLOTINE'S blade hanging over his neck, Eckener worked fast to right the DELAG.

First, the *Deutschland* and *Deutschland II* accidents had been primarily caused by crew errors and inexperience. Both had been avoidable. Eckener mandated that officers needed more rigorous training in handling ships and henceforth must be tested on their expert knowledge of air temperature, wind patterns, gas behavior, the effects of rain, rudder control, and the like before even going near a gondola. No man could be made a captain until he had at least 150 trips under his belt.[18] With veterans in command, flying would become a predictable science. In the meantime, as he had learned, even if the passengers complained, the rule was Safety First: No more taking risks with the airships if bad weather was a possibility.

Second, for that reason, it was critical to have access to accurate weather forecasting. The airshipmen had all too often been left in the dark as to what awaited them above. There could be no more surprises. The DELAG, Eckener proposed, needed its own private weather information network with observation posts all over Germany. Its primary task was to detect and predict high or changing winds—murderous for any unwitting Zeppelin hitting them head-on or crosswise.

The count's friend Professor Hergesell and his colleague Richard Assmann (who'd sat on the 1894 technical committee evaluating Zeppelin's original airship concept) were dispatched to compile a comprehensive analysis of wind patterns based on twenty years' worth of records. From its copious tables, the DELAG staff could infer the daily wind direction and strength in almost any location in Germany and build their passenger schedules around it.[19]

Piece by piece, the secrets of the air revealed themselves to Eckener and the DELAG. Long-held assumptions that gas-filled vehicles were at the mercy of the winds—still cited by aeroplanists as the proof that airships were unviable—quickly became obsolete. Air currents, Hergesell and Assmann determined, were neither static nor binary. In fact, they changed constantly in terms of force, direction, altitude, and duration, and it was never so simple a matter of there being *either* an unhelpful headwind *or* a helpful tailwind. Upon encountering an unfavorable wind, for instance, if a skilled airshipman ascended, descended, or altered course he could find a favorable one.[20]

Owing to its singular geography, Germany was composed of "wind-weak" and "wind-strong" districts, they found, and the frequency and paths of storms could be tracked. They discovered hitherto unsuspected "wind-rivers" that Eckener took advantage of when planning his routes. He and his captains would tack from counter-current to current or make up for lost time by switching from one wind-river to a faster one. Eckener eventually became so proficient at wind-running that he was nicknamed the "Pontiff"—for his infallibility.[21]

The DELAG began producing new "aeritime" maps of the wind-rivers resembling those of the mariners of old. The strength and direction of winds around a given locality were shaped by the surrounding hills and valleys just as much as tides and currents are by bays and harbors.[22] Eckener also commissioned a series of airship-specific maps to be carried aboard all DELAG craft. He had found that ordinary road maps, scaled at a half-inch to the mile, were useless for flying; airships traveled so quickly and so high that even familiar landmarks and countryside were often unrecognizable. DELAG maps now displayed information representing how airshipmen saw the country from above. They were scaled at three miles to the inch, showed terrain heights in various colors, tinted areas that were to be avoided (electricity and telephone wires, for instance), and had rivers, marshes, and forests clearly marked, not so much for location-finding but for how their wind patterns influenced an airship's behavior. Circles were used to show hydrogen-refilling stations; red silhouettes of an airship, hangars.

Eckener installed wireless-telegraphy sets in the control room so that captains could communicate with airport hubs for constant weather updates. He also set in motion plans to build aerial lighthouses that would beam searchlights upward to warn of danger areas, luminous signal balloons as direction aids, and painted roofs with a prominent series of letters to indicate location. Just as the advent of the automobile was generating a need for rules of the road, Eckener set out to systematize the rules of the air for the DELAG era to come. Airships would bear red and green lights at the bow and stern, with yellow ones to mark top and bottom; foghorns were installed to prevent collisions in clouds and instructions were distributed to determine right-of-way when airship met airship.[23]

Eckener also recognized that the practice of using ground crews to maneuver and restrain airships with nothing more than ropes was a failure, proven many times over since the days of LZ-1. To defend against sudden wind gusts on the ground, when Zeppelins were at their most vulnerable, a new system of docking and launching had to be created. Eckener's solution was a several-hundred-foot-long metal track extending from the hangar along which ran little trolleys whose wheels ran in the hollow underneath the surface of the rail to prevent them being swept away by the pull of the airship. Because they resembled cats scampering along the top of a fence, the trolleys were fondly known as *Laufkatzen* (running cats). When its cables were tied to the *Laufkatzen*, the airship was efficiently restrained from lifting, twist and buck as it might.

And finally, and here Eckener had to tread very carefully, Ludwig Dürr's old-fashioned methods had to be updated to reflect the kind of new engineering practices befitting a major corporation. Lacking knowledge of the higher mathematics needed for, say, stress analysis of the airship structure, Dürr, a mechanical artisan, had always proceeded by "feel" and intuition, and among his quirks was to never use anything but the simplest drawings to lay out his design. Once the airship was built, his favored technique was to order a dozen test flights and then improvise solutions as problems arose—often entirely avoidable ones caused by relying on what were essentially sketches.

Dürr may have been a genius at hand-adjusting a recalcitrant rudder or fine-tuning the controls on the fly, and his eccentricities hadn't mattered in the old days, when there was just one airship, but now that the company wanted to expand its facilities to build several at a time, he needed—though did not want—a team.

Much had changed since the start-up days of LZ-1: Now there was a new generation of engineering statisticians eager to work for the Zeppelin Company. Young specialists like Karl Arnstein, a former bridge engineer, and Paul Jaray, an aerodynamicist, could perform the complex calculations that were beyond Dürr's abilities, helping to make future versions of airships stronger, lighter, larger, and more capable than ever conceived possible. It took some effort to convince the stubborn Dürr that he needed assistance, but even he eventually realized that Jaray and Arnstein were unmatched assets.

Many were surprised that Dürr, a tetchy loner, worked so well with these junior upstarts, but together they formed a formidable team by performing complementary roles. Dürr continued in his role as chief designer—the proverbial "big-picture man"—but it would be Arnstein and Jaray who translated his vision into scores of closely penciled pages filled with equations that proved his airships would fly safely. The younger men were always careful to remain respectful when telling their boss that a given design wouldn't work, and for his part, Dürr was gracious enough to concede occasionally that sometimes even he might be (slightly) wrong.[24]

It was Jaray, for example, who gingerly raised the sensitive subject of revising the shape of the airship to improve performance through stream-lining. Using statesmanlike diplomacy, he definitively showed that ever since the 1890s the count had been wrong to focus on length rather than width to increase speed and payload. Hence his airships had been narrow at the head (in the mistaken if widespread belief that it disproportionately cut air resistance) and cylindrically long—which was thought to have lit-tle effect on drag. Instead of adding yards length-wise, as Dürr had always done, Jaray demonstrated that it was more efficient in terms of reducing air resistance to widen the diameter and keep the body relatively short. Even Dürr admitted the strength of his argument and agreed to adopt a "fatter" profile for subsequent Zeppelins.[25]

So, when the *Schwaben* first emerged from its hangar in mid-1911, just in time for the summer flying season, it was quite a different creature from its predecessors. Eckener had instructed Maybach to have three new 145-hp, six-cylinder engines ready come what may. Specifically designed for airships, they delivered a top speed of 44 mph, about 20 percent faster than the *Deutschlands* could muster. In concordance with Jaray's findings, Dürr's people had also hurriedly removed a twenty-four-foot-long section, making this the first time a Zeppelin airship was shorter and less capacious than its predecessor.

Whether there would ever be a successor was now in the hands of the *Schwaben*.

THE "LUCKY SHIP," as Eckener dubbed it, saved the company. From its very first trip, people were agog at the pleasures of flying. A German-American journalist named Carl Dienstbach described for his envious readers what it was like to fly at the very dawn of passenger aviation.

Thanks to the HAPAG cruise-line partnership, said Dienstbach, he bought a Zeppelin ticket in the United States and eventually turned up at the DELAG office in Frankfurt, which was decorated with pictures of Zeppelins sailing over the secluded valleys of the Black Forest.

Because ground winds tend to be weakest in the morning, flights departed at 6 A.M. As Dienstbach's Frankfurt-Düsseldorf flight was predicted to take six hours, the clerk advised travelers to forward their luggage as soon as possible. They could take suitcases on board if they wished, but they were charged an arm and a leg for the extra weight. It was better to make do with a light carry-on bag and a camera. They certainly wouldn't need heavy overcoats, because the air only turned perceptibly cooler at two thousand feet, higher than the *Schwaben* would go. Instead many had brought wide-brimmed straw hats to keep the sun off their faces.

Dienstbach found the *Schwaben*, aglow with lights, lodged in its hangar, abuzz with activity. The hydrogen storage tanks were kept underneath the concrete floor, from which snaked the long rubber hoses used to fill the gas cells one by one. Other maintenance men pumped water into the ballast bags while a truck brought the gasoline to the rear-engine gondola, then the forward.

A thunderous roar suddenly echoed through the shed as the engines were tested. At that point, the captain and two fashionably dressed ladies, the first of Dienstbach's fellow passengers, arrived. It was going to be a full flight, with twenty-four expected. An "able airman" lowered a slender flight of aluminum steps from the lounge door. At their foot stood a steward deferentially helping passengers aboard.

This unnamed steward was another of Eckener's innovations. He was Heinrich Kubis, and he was the world's first flight attendant. Punctilious and dapper, Kubis was then in his early twenties and had worked as a waiter at the most fashionable hotels in Europe—the Carlton in London and the Ritz in Paris were both on his résumé. He would go on to serve as chief steward on every passenger Zeppelin for the next two and a half decades, including the *Hindenburg*.

A carnival atmosphere prevailed as the passengers waved to their cheering friends standing below and the ground crew took their positions. At 6 A.M. on the nose, a siren wailed and all stood clear. The ground crew distributed themselves around the hull and took hold of the trailing ropes. Another siren blast. The two great portals at the front of the hangar groaned and rumbled as they opened, powered by motors. A third siren blast.

The airship beast began to stir. Those aboard realized they were moving only if they watched the girders of the hangar pass slowly by. The men along the hull took the ropes over their shoulders, dug their heels into the ground, pulled with all their might, and began to haul the twenty-ton mass forward. As the airship glided above the new track, it pulled and strained against the *Laufkatzen* but remained stable.

Outside, the airship had to be calibrated for balance. Gas temperature and weight needed measuring so that ballast could be added or subtracted to keep it trimmed. At that point, a gush of water erupted from the bow so that the front of the *Schwaben* was inclined slightly to aid with lift. The engines were throttled out, a final blast of the siren sounded, and the ground crew released the *Schwaben* from its chains. Very slowly, very gently, it levitated. The propellers began turning, at first hesitantly but soon becoming transparent disks, and the airship rose higher and higher, leaving Frankfurt behind.

In the lounge the sensation was like being on a ship moving away from the shoreline and entering the ocean proper, where the invigorating winds flow freely. Every passenger aboard a Zeppelin found the absolute clarity and purity of the atmosphere at 1,500 feet as startling as moving from New York City air to Rocky Mountain air. Those who had also flown in an airplane—admittedly, a tiny number—said that it couldn't compare to the experience of flying in an airship.

Airplanes buzz and rattle, and in them, explained Waldemar Kaempffert, an American science writer, one was always conscious of a feeling of movement and gravity. An airplane pilot "must meet every little gust of wind, and ride over it like a boat over billows. Now and then a swirl, a big wave in the invisible sea of air, catches us. A movement of the lever, which controls at once the front rudder and the ailerons . . . and we ride over the wave, and glide on again. It is as if we are carried by an intangible hand up and over an obstacle that we cannot see."

In contrast, in a Zeppelin, a sense of weightlessness prevailed. For Dienstbach, flying in an airship felt like the "floating fancies of a dream" as flocks of curious birds flew level with you and passengers yearned to touch the clouds above. For Thomas Baldwin, another journalist who sought to convey the experience of flying in an era when so few people had ever left the ground, a passenger believes he "could step from cloud to cloud as I have stepped from stone to stone in the bed of a shallow rivulet." He had even leaned over the side, and instead of feeling dizziness or a fear of falling, as from a tall building, "I experienced a feeling of buoyancy like floating on the water."

There was a modest breeze in the lounge, thanks to a few open windows, yet as the *Schwaben* passed over the plains below it seemed motionless. As it entered the mountain passes the engines roared periodically as the ship changed course to cope with the shifting gusts. The crew and officers worked most urgently at this stage, because negotiating the Rhine Valley was no easy task.

At the Rhine "knee" near Bingen, where the river narrows and bends north, there was always a treacherous wind over the water, which could change direction in an instant, and the verdant side valleys hid dangerous wind eddies. For the passengers, though, it was glorious. On a small island guarding the sacred entrance to the mystical Rhine, the setting for Wagner's *Götterdämmerung*, was the Mouse Tower, so named because there a cruel archbishop was once deservedly eaten by ravenous rodents. To the right loomed Ehrenfels Castle, a ruined pile of ancient pedigree, and a little farther on, to the left, the grand Rheinstein Castle, a royal favorite with a working portcullis and drawbridge.

After a few hours, the *Schwaben* arrived at Bonn, about two-thirds of the way to Düsseldorf, and the crew could finally relax. From there, the scenery was not as spectacular, and the airship ascended into the open air. No longer having to hew to the twists between the cliffs and bends of the river, the *Schwaben* could fly straight and accelerated past 40 mph.

The passengers ate well aboard the *Schwaben*. Borrowing the idea from luxurious ocean liners and first-class Pullman trains, Eckener had introduced decent food (foie gras, caviar, "French fruits," and Westphalian ham) and an admirable wine list (but no beer, alas, considered by Eckener to be too déclassé for his premium clientele).

The company was good, too, for the sound of the engines was so muffled that conversation was not difficult. A few people played cards on the newly added tables, while others sent wireless-telegraph messages to their families from the small cabin at the rear next to the lavatory. Every airship of the DELAG had its own tiny post office and official stamp. Souvenir postcards were sold, and receiving a DELAG-stamped card was always a thrill.

At Düsseldorf, wrote Dienstbach, there was a brief spasm of fear as hands gripped chairs and scared looks were exchanged when the captain turned off the engines and descended headfirst for the airport. The floor was at a steep angle as the bow glided straight for a knot of ground crew. Just as the ship seemed likely to mow them over, another gush of water erupted from the bow tanks and the propellers chugged to life in reverse. The bow tipped up, and the *Schwaben* came to an obedient halt as the groundsmen took hold of the front ropes and guided the ship onto the docking rails.

The adventure ended with the same rhythmic marching of the men that had begun the trip, this time into the hangar. After exiting, the passengers found themselves surrounded by excited friends, who ushered them to the droves of taxicabs waiting next to the hangar to take them to their hotels.[26]

THANKS TO THE *Schwaben*, which transformed a struggling airline into a viable business, 1911–12 proved the turning point in the DELAG's fortunes. Not even the destruction of the "lucky ship" a year after its introduction fazed the company. Having outraced a storm to get to Düsseldorf on June 28, 1912, the *Schwaben* had unloaded its passengers when "frictional electricity" caused by the chafing of the gas cells' rubberized cotton skins set it aflame.

The fallout would have been worse had it not been for Eckener's ruthless reaction. He urgently told Colsman that he sensed a "noticeable flinching of the public after the catastrophe" and publicly blamed the maintenance crew at the Düsseldorf hangar for the accident. Though it pained him to do so, it was a necessary public-relations move, for "otherwise, we won't get another dog to enter the ship."[1]

Internally, the accident prompted Zeppelin to switch to "goldbeaters' skin" to line gas cells. This was a delicate membrane covering the caecum—a food-storage pouch at the entrance to the large intestine—of cattle. Each caecum yielded a skin measuring at most 39 by 6 inches, and up to fifty thousand were required to cover a single gas cell. Sewing together that number of easily tearable skins required skilled handwork—adding to their already prohibitive price.[2] In the interests of safety, however, its belated adoption was imperative, and in future the fire risk posed by chafing was reduced to almost nothing.

To further restore confidence, the company publicly made much of the fact that *Schwaben* had impressively demonstrated its airworthiness over its year of life. It had flown 480 hours without incident and carried 4,354 passengers (including crew, paying customers, and guests) over 234 flights. The company even innovatively commissioned a five-minute movie shown in theaters that depicted a bird's-eye view of a trip aboard the *Schwaben*; people who couldn't afford an airship ticket lined up to be astounded by the vicarious experience of flying.

Shortly afterward, Eckener confirmed to Colsman that all was well: Demand for flight reservations aboard the DELAG's new airships was exceeding pre-accident levels.[3] By then, Eckener had the LZ-11 *Viktoria Luise* (named after the kaiser's only daughter) as well as the LZ-13 *Hansa* (after the Hanseatic League, the medieval merchant confederation that had dominated North German and Baltic trade) in service.

Viktoria Luise and *Hansa* were larger than the *Schwaben*, possessed upgraded and more powerful engines, and had more aerodynamically shaped bows and sterns. Their top speeds ranged between 47 and 50 mph, they carried more passengers, and they enjoyed improved handling. The performance bumps in addition to the growing experience of the DELAG crews meant that *Viktoria Luise* cut the flight time between Frankfurt and Düsseldorf from the *Schwaben*'s six hours to just three, handily faster than the train (but at much higher prices).[4]

Eckener was confident enough to order an even more advanced airship, the *Sachsen* (Saxony), which was delivered in the early summer of 1913. Its home base would be Leipzig, while *Hansa*'s was Hamburg, and *Viktoria Luise* stayed in Frankfurt.[5]

These three hub cities formed an important geographical triangle central to Eckener's plans to expand. Frankfurt lay in the west of Germany, Hamburg to the north, and Leipzig the east. From Frankfurt, his airships could head for France, Italy, and Britain (and, eventually, the United States); Hamburg was the gateway to Scandinavia; and Leipzig allowed passage to Austria-Hungary, Poland, Turkey, and Russia.

To that end, he dispatched *Hansa* on the DELAG's first international flight, on September 19, 1912, to Copenhagen. The following June, *Sachsen* voyaged to Vienna, arriving in half the time an express train could from almost equidistant Stuttgart.[6]

Eckener's hopes of introducing service to Paris and London were frustrated by international tensions, which reached a fever pitch during the Balkan Wars of 1912–13. The British had not forgotten the novelist Dr. Rudolf Martin's threats to land a Zeppelin-borne invasion force of "350,000 men," and the French had no truck with ostensibly civilian German airships—which they suspected of performing double duty as the kaiser's eyes in the sky—overflying their fortifications and compiling accurate bombing maps of their capital.

Beyond London and Paris, the old dream of transatlantic flights to New York beckoned. With its huge German-immigrant population and friendly relations with Berlin, America was a natural destination for Zeppelins. The sinking of the *Titanic* in April 1912, ten weeks before the *Schwaben*'s demise, prompted an ambitious reconsideration of the scheme. The *Titanic* had, notoriously, been touted as unsinkable, and some of the more zealous airship advocates pointed out that while the iceberg had sheared its steel plates like cardboard, you could rip half a dozen gas cells in a Zeppelin and it would remain aloft.[7]

Still, the current generation of Zeppelins was too small, too short-ranged, and too underpowered to undertake such a voyage. One would need a true *Titanic* of the skies to surmount the challenge. Not surprisingly, this type of super-Zeppelin was not yet in the cards, but it was conceivable that the *next* next-generation airship would enable the great technological and engineering leap forward to make the dream a reality and consign the oceanic cruise liner to the scrap heap of history.

This airship of the future, some predicted, would be a 1,000-foot, 120-mph behemoth festooned with scores of propellers and room for promenades, amusement areas, game rooms, restaurants, glassed observation platforms, and deck upon deck of luxuriously appointed staterooms, as well as holds sufficient to transport hundreds of tons of profitable cargo.[8] By 1925, it was thought, a DELAG airship carrying 150 passengers and 42 crew would circumnavigate the globe in eight days.[9]

First things first, though. Successfully forging an Atlantic passage required a great deal of background data. The diligent Professor Hergesell secreted himself aboard a German training cruiser to survey the winds off Florida and the eastern seaboard. It was immediately clear to him that a direct run from Germany to New York was out of the question, because at that latitude in the North Atlantic the airship would be heading right into the prevailing wind. Instead, as Columbus had found, the trade winds arcing westward toward the Caribbean would help speed a vessel on its way. The return trip could be completed farther north using the westerlies at the ship's back.

Since the trade winds, Hergesell argued, blew strongest in winter and there were few cyclones between November and July, airships should aim to depart Europe in the colder months, make pit stops for hydrogen and fuel in the Azores, the Canaries, or the Madeiras in the Atlantic, and then head for Puerto Rico or Havana. After that, they would work their way north to Jacksonville, Florida, or New Orleans and thence up the east coast to New York. The whole thing could be accomplished in five to six days, assuming an average speed of 40 mph.[10]

One advantage would be that DELAG airships working the summer season in Germany could, instead of being placed in storage each winter, be diverted into revenue-generating Atlantic operations. On the other hand, the route added significantly to distance. For instance, Frankfurt to New York in a straight line was roughly 3,850 miles, but Frankfurt to Puerto Rico alone was 4,600, with another 1,260 miles for Puerto Rico to Jacksonville and 835 from Jacksonville to New York. Since the *Viktoria Luise*'s maximum range was roughly 700 miles, crossing the Atlantic Ocean, at least until the advent of super-Zeppelins, could be regarded only as a fantasy.

Eckener's conservatism—his obsession with careful planning, sound training, precise engineering, and preflight trials—put the brakes on the DELAG's ambitions to forge a transatlantic crossing. It was a disappointing, if wise, decision, for the dangers of making such a premature attempt had been apparent for all to see in the saga, or rather fiasco, of Walter Wellman's voyage in October 1910.

Wellman, a fifty-one-year-old Chicago adventurer with a profound talent for self-deluding self-promotion, almost a caricature of a heroically mustached Victorian Explorer, had built the *America*, a large semi-rigid that he claimed would fly from New Jersey to Britain.

Eckener looked upon the *America* with a mixture of pity, marvel, and befuddlement. None of its crew, with a single exception, had ever been in the air before, and neither had the ship's cat, a sourly unenthusiastic feline named Kiddo, who, Wellman complained, had "no imagination . . . no vanity in pioneering."

The trip ended predictably. Within a couple of days, Wellman and his doughty skyfarers had to be rescued by a passing steamer.[11] Not a man (or Kiddo) was lost, amazingly, but the ill-starred venture proved that the airship business was a quintessentially German specialty, a field in which proficiency, experience, and expertise were to be prized above all.

The Zeppelin men were beginning to believe themselves high priests jealously guarding the divine secrets of aerial navigation from the pretensions of the amateurish and the vulgarity of the ignorant. The belief that "no one else can do it" was true enough, it seemed, though it would curdle, once they convinced themselves that they knew everything that could be known, into a lethal arrogance in the years to come.

MORE PROMISING IN the near term, particularly from the count's point of view, was that Eckener's airline success had prompted a reappraisal by the army of its airship requirements. In early 1912, General von Moltke, who had cut airship orders down to three in 1909, now bumped the total to nine more—a huge score for Zeppelin. German High Command had finally decided they could be used for reconnaissance and bombing purposes in a war with France. The first of these new airships, LZ-12 (Z-3), was delivered in July, with the others scheduled over the next two years.[12]

Even better: The navy was expressing interest. Grand Admiral von Tirpitz, its reigning supremo, had long dismissed airships as a diversion from building more battleships. Because the wind at sea is stronger than over land, and the distances greater, Tirpitz felt that previous generations of Zeppelins would be of little help in the final showdown he planned between his High Seas Fleet and Britain's Home Fleet.

Tirpitz remained skeptical, but at least he was no longer dismissing the idea out of hand. The count, smarting after being sidelined by Colsman and Eckener, made a special effort to convince the admiral by promising him that the company would build him an airship to his own specifications. This put Dürr's nose out of joint, of course, and he was bitterly affronted when naval personnel began appearing at Friedrichshafen and ordering his mechanics around.

LZ-14, dubbed L-1 (a naval designation), would be the result. It had nearly 20 percent more gas capacity than *Viktoria Luise*, was thirty feet longer, and was equipped with specially tuned Maybach engines producing 540 hp rather than the standard 450.

Count von Zeppelin, hearkening to Dr. Martin's description of him as an admiral, insisted on captaining L-1 on its thirty-hour maiden flight on October 7, 1912, but, sadly, L-1 was lost at sea the following September, when almost vertical cold gusts lifted it up and heavy rain brought it down. It was almost a repeat of Captain Kahlenberg's *Deutschland* crash into the Teutoburg Forest, but this time those aboard weren't so fortunate. When the rudders failed, the airship plunged from a height of three hundred feet into the turbulent water off the coast of Heligoland, and while six officers and men were rescued, fourteen drowned, including the head of the new Naval Airship Division. It was the first time anyone had ever been killed on a Zeppelin, yet the disaster was chalked up to an act of God and the company was not blamed.[13]

Tirpitz had considered the L-1 quite impressive, but still inadequate, and contracted for a second airship. Felix Pietzker, a naval designer, was chosen to oversee construction at Friedrichshafen—again, over Dürr's vociferous opposition. Pietzker ignored Dürr's advice and demanded that the engines be placed closer to the hull and the characteristic Zeppelin external keel—the spine from which the gondolas hung—moved inside the body of the craft to enhance streamlining and speed. These changes were dangerously drastic, complained Dürr, but he was dismissed as a fuddy-duddy who worked too slowly. The count backed his own man, agreeing that the redesign was premature, but Pietzker went ahead with it anyway, and Zeppelin, realizing that further opposition would terminate the project, quietly surrendered.

A *four*-engine monster of 840 hp and 20 percent larger than even L-1, L-2 was considered a terrific success until its tenth flight on October 17, 1913.

Shortly after takeoff at 10:05 A.M., witnesses registered a flame issuing from the front gondola, the fatal prelude to L-2's plummeting 450 feet to earth, with several fiery explosions shaking its aching, arching body on the way down. The entire complement of twenty-eight, including Pietzker and five Zeppelin executives, burned to death. The subsequent investigation determined that a hydrogen leak had flowed into the tunnel-like internal keel, which lacked exhaust shafts, then into the forward engine car (set, as Dürr had warned, too close to the hull) and was ignited by a carburetor flame.[14]

Tirpitz this time did blame the count, who in turn accused the late Pietzker of incompetence. At the otherwise solemn military funeral for the lost crewmen, a livid Zeppelin instigated a very public argument with Tirpitz, and the two came almost to blows. Colsman spent the next few months patching things up with Tirpitz, who eventually agreed to not cancel the naval airship program but refused to ever deal with the count again.

These two successive fatal disasters seem to have finally knocked the wind out of the once-indefatigable Zeppelin, now seventy-five. He realized that everything does indeed come to an end. When Colsman and Eckener next convened the board, Zeppelin barely raised a word of complaint when he was removed, respectfully but ruthlessly, from day-to-day involvement with his own company. Relegated to the status of the Grand Old Man of German airships, Zeppelin made appearances, usually for publicity, only a few more times over the following years.[15]

Perhaps surprisingly, German public opinion remained overwhelmingly favorable toward airships after the L-1 and L-2 accidents. As the years and decades passed, advocates like Eckener would cling ever more tightly to the unshakable principle that there was nothing inherently flawed about the airship concept and that any problems that arose could easily be rectified. In the broader sense, they were right: Airplanes (and automobiles and trains) crashed all the time, too, yet no one thought of them as fatally misconceived. Between 1908 and the outbreak of World War I in August 1914, for instance, more than 500 people would be killed in airplanes but only 42 in rigid airships (L-1 and L-2) and none in a civilian DELAG vessel.[16] When you flew in an airplane, you took your life in your hands; not so in a Zeppelin.

A few years earlier, ground-level gusts had been regarded as the greatest menace to airship safety, but now the role hydrogen had played in the L-2 accident was alarmingly noticeable. Eckener calmed these fears by explaining that hydrogen was only potentially hazardous when, because it was expanding from the sun's heat or from lowered air pressure, it was expunged from the gas cells to maintain equilibrium *and* allowed to build up, as in the case of L-2's internal keel, in an unventilated space.

Hydrogen, Eckener liked to remind people, is not inherently explosive; it becomes violently volatile only when mixed with at least 15 percent of air. If hydrogen remained contained and pure at a constant temperature and pressure—as Zeppelin specialized in doing—one could light a candle three inches from an insulated gas cell with no fear of fire. Had L-2, in other words, been outfitted with simple vents to allow the accumulating hydrogen to escape from the "tunnel" into the open air, the disaster would never have happened.[17]

Still, despite the company's sunny claims of hydrogen's safety—when used correctly—there was quiet acknowledgment that a different gas, one that was immune to fire, might in future be used to further mitigate risk. Zeppelin conceded as much in a lecture when he mentioned, rather briefly, that "finding . . . a non-inflammable gas, even if it were not quite so light, is not impossible." Of course, this necessarily "heavier" replacement—hydrogen was incredibly buoyant owing to its unparalleled lightness as an element—would require much larger airships than those currently existing in order to compensate for the loss of natural lift. But if a satisfactory gas could be discovered, then "the very safest vehicle conceivable would be procured."[18]

There were rumors, however, that one called helium might suffice. Unfortunately, scientists had only recently extracted minute samples from the world's only known deposits of natural gas in Texas, Oklahoma, and Kansas, so for the time being it was hardly a realistic proposition.

MEANWHILE, THE DELAG boomed. Between June 22, 1910, and July 31, 1914, its airships made 1,588 flights, flew for 3,176 hours, covered 107,205 miles, and carried 34,028 passengers.[19] Eckener proudly reported that as of April 2, 1914, two airship ports had been opened at Dresden and Liegnitz to complement the existing eight, a second airship-construction facility had been built, and the new airport at Potsdam was attracting hundreds of visitors a day.[20]

Such rapid expansion did not come cheap, so while the DELAG was generating around 200,000 marks in annual revenue, its costs were in the millions.[21] But soon, Eckener forecast, the financial situation would ease.

All the money invested at the outset in infrastructure, he pointed out, was already beginning to pay off: Now that the company had its own hydrogen plant, for example, the cost of inflating a Zeppelin had plummeted by 83 percent.[22] Airships, through longevity, were also earning out their construction costs. Whereas the *Schwaben* had lasted fifty-two weeks, by mid-1914 the other three had all enjoyed two-year operating lives and showed no signs of obsolescence.

Perhaps most important, one key metric was rising fast as airship travel slowly became part of the fabric of everyday life in Germany: the proportion of paying customers to non-paying ones. Initially, Eckener had needed to impress burgomasters, journalists, businessmen, and officers, and so fully two-thirds of the seats on LZ-6A and the *Deutschland* had been provided for free to garner publicity. By 1913, the DELAG having proved itself, three-quarters of passengers aboard *Hansa* and *Sachsen* had paid full price for their flights, and that number was increasing.[23]

The real issue facing the airline, however, was becoming obvious: Eckener could not earn a profit employing large, expensive airships on short hops. It was like trying to make money using an ocean liner for daylong rides along a canal. As Eckener later admitted, only small, cheap airplanes ("of which no one then thought seriously") enjoyed sufficiently low running costs and high enough speeds to make quick domestic trips financially worthwhile.[24]

According to his own careful analyses, Zeppelins were best suited to voyages of 1,800 miles and above, and he accordingly planned to pivot his next-generation airships toward exclusively long-haul routes in the coming years. Soon, scores of Zeppelins would ply the world's wind-rivers and plant the flag in faraway lands once considered exotic.

But it never happened.

In the summer of 1914, just as the airship season was taking off in earnest, the Great Powers of Europe slithered into the cauldron of a war for which they had ached so long.

20. Z-Ships

O**N THE BALMY** evening of July 31, 1914, Captain Ernst Lehmann, commanding the *Sachsen* on an otherwise unremarkable DELAG flight, was passed an odd telegram by his wireless operator. The sender, the War Ministry in Berlin, directed him not to fly *Sachsen* more than thirty miles from its home base until further notice.[1]

He knew what it meant. Earlier that day, Russia had ordered its army to mobilize against Germany's ally Austria-Hungary. A German mobilization in response was inevitable, with the French and probably the British following suit to support the tsar. Overnight, quite literally, the airline Eckener and Colsman had so laboriously conjured into being vanished. According to the terms of their previous agreement with the count, all DELAG airships were to be turned over to the military in the event of war.

Vindicated in his belief that Zeppelins were meant for combat not peace, the elderly count proclaimed to the Prussian State Assembly that because "the most devastating war is ultimately the most merciful," Germany must launch airship-bombing attacks against her enemies as soon as possible.[2] General von Einem recalled that shortly after the outbreak of war, he happened to bump into Zeppelin, who exclaimed, "All England must burn!" as "his eyes sparkled."[3] Zeppelin's fiery vigor—he called for the Reich's immediate annexation of Belgium and France—would later become so embarrassing that even the Chancellery had to request that he refrain from making any more public statements regarding the war.[4]

A week after receiving the fateful telegram, Lehmann piloted the *Sachsen* to the hangar at Potsdam for conversion into a military vehicle.

Viktoria Luise and *Hansa* were already there, and the three airships had their wicker furniture removed while their central gangways were hastily transformed into bomb-storage areas. A few machine guns were mounted in the gondolas and later onto a platform built atop the dirigibles.[5]

Lehmann was joined in Potsdam by Max von Gemmingen, Zeppelin's nephew and a member of the board. At fifty-two, he was too old for conventional military service, but he insisted on serving aboard the *Sachsen*.[6] Zeppelin was also trying to acquire a commission as an airship admiral, but that would prove fruitless, so instead he traveled to visit his old regiment. The coming war was just too good to miss, he told the young officers.[7]

Eckener did his patriotic bit by volunteering for the Naval Airship Division, commanded by his former student Peter Strasser, who had trained aboard the prewar *Sachsen*. Eckener remained a civilian (he wore the blue DELAG uniform with the cap of the Imperial Yacht Club) and was engaged as an instructor for the naval crews then urgently being recruited.[8]

Training began almost immediately. Lehmann, for instance, went aloft at 4 o'clock each morning and had his airshipmen drop dummy bombs. When they ran out, so short was their supply, they then had to land and pick them up. For the time being, they used old artillery shells, which spun wildly as they fell, but soon discovered that to help make them hit nose-first they could tie strips of cut-up blankets to their bases.[9]

If suitable ammunition was scarce, a greater scarcity was airship crews. Each early airship needed a captain, three officers, and around fifteen men.[10] In the DELAG era, a year of intensive schooling was required to weld a green crew into a cohesive unit, while the captain and his first officer, navigation officer, and flight engineer needed at least two before they could even be considered for an active post. It fell to Eckener to accelerate the training program to just a few months, and the cigar-chomping ex-DELAG chief became notorious for his gruff, curt directives to recruits as they struggled with the unfamiliar mechanisms and the idiosyncrasies of airship flying.

Eckener transferred seasoned former DELAG men from regular military service to leaven the new crews, but for the war's first couple of years any number of otherwise avoidable accidents and losses were attributable to inexperienced and undertrained personnel.[11] He was particularly exercised by the presence, which he later managed to remove, of General Staff officers assigned by headquarters to oversee operations on board the army airships; they knew nothing of airship capabilities and demanded cavalry-style turning maneuvers that, had they been attempted, would have torn the ships apart.[12]

If there was an element of improvisation to German preparations, it reflected a wider confusion among European and American military experts as to the role and importance of aviation in the war to come. Both airships and airplanes as working vehicles were little more than a decade old, and no one could predict their future—if, indeed, they even had much of one.

Despite the excited pronouncements of prewar novelists like H. G. Wells and Rudolph Martin that thousands-strong armadas of airships and airplanes would devastate whole civilizations, the great majority of war planners conceived a limited role for these new aerial weapons.[13] They would be used to scout ahead of the main army, it was assumed, and a few radicals daringly suggested that maybe they could drop bombs on enemy troops or on important chokepoints, like railway junctions and bridges.[14]

But the real question facing Germany's enemies was, what would the Germans do? A lack of hard information about capabilities and intentions, tinctured with fear and anxiety, colored predictions of what the kaiser's airshipmen had in mind—and in store.

On the eve of war, the British and the French knew, or at least thought they knew, three things.

First, that the count himself commanded the airship fleet, estimated to be upward of forty giant Zeppelins.[15] It was certain that Germany planned to increase this number precipitously. Economic intelligence reports noted with alarm that the price of aluminum in January 1914 was about $400 per ton, and by June it had almost doubled to $700 thanks to German demand. What other reason could there be but to construct a still vaster Zeppelin armada?[16]

And second, that the Germans possessed military-grade airships of startling might. Their latest naval Zeppelin, for instance, could travel at 94 mph and cruise, undetected and unstoppable, to deliver its payload. Germany had long been secretly practicing bombing and machine-gun target practice and reportedly had already sent scout ships equipped with muffled engines to England at night to confirm targets. In adducing the airships' astonishing bomb-aiming ability, *Scientific American* cited Eckener's "startling revelations" that airship crews could place a bomb right in the center of a fifteen-foot-diameter circle from no less than five thousand feet up.[17]

And last, that against Zeppelins there was no defense. At 94 mph, airships not only were faster than airplanes but also effortlessly outmatched them in climbing to altitudes beyond any airplane's ability. Even if an airplane succeeded in closing in on a Zeppelin, an airship could be "riddled . . . with a thousand bullets [and] still keep afloat." In any case, its defensive machine guns, thanks to the absence of "vibration," were astonishingly accurate and could destroy an enemy aircraft at more than two thousand yards, all the while "plunging through the air at express-train speeds" and dropping two hundred bombs at a time on columns of soldiers heading for the front lines. As one commentator wondered, "Who can describe the frightful panic that would ensue?"[18]

The truth, however, was that Zeppelin had long since been ousted from any position of real influence, and at the beginning of the war the German army and navy together possessed a grand total of nine airships, including the three DELAG dirigibles that had been dragooned into military service.[19] The army's vehicles, moreover, were all based on the civilian models and would soon prove their utter uselessness in combat.

As for the feared L-3, the latest (and currently only) naval airship, it had a top speed of 52 mph — much slower than any airplane. Neither had there been any top-secret missions over England, and the science of "aiming" consisted of tossing handheld, sometimes handmade, bombs over the side and hoping they hit somewhere in the general vicinity of an *area*, let alone the bull's-eye of a small circle. Eckener, last but not least, had never made any sort of "startling revelations" to the contrary.

So it was that Lehmann recalled that the High Command, far from having diligently considered airships' combat role, actually "had no idea what to do" with the few they had. The army's Z-8, for instance, was based in Metz, just a few miles from the French border, but was kept in storage even after the outbreak of hostilities; when the captain petitioned his superiors for permission to inflate it in case enemy cavalry advanced, he was turned down.[20]

When the orders eventually did come to launch airship operations against French and British forces, it was quickly discovered that Zeppelins "could fly neither high, nor fast, nor far enough, nor carry sufficient bombs, to accomplish the feats the General Staff demanded of them."

Captains received absurd directives, such as the time headquarters ambitiously ordered Z-9 to "carry out bombing attacks on Antwerp, Zeebrugge, Dunkirk, and Calais. Return via Lille, also to be bombed." No one in Berlin seems to have realized that Z-9's full complement of bombs amounted to just ten small artillery shells, making for an unimpressive two bombs for each location.[21]

Worse, Zeppelin losses were unsustainable. No fewer than three—a third of the fleet—were lost in the opening weeks of hostilities on the Western Front when troops armed with nothing but rifles turned out to be more dangerous than anyone had anticipated. Badly holed by small-arms fire after a fruitless attack on Belgian forts in Liège, Captain Kleinschmidt's Z-6 limped back to Germany and crashed in a forest outside Bonn. Then, on August 21, both Z-7 and Z-8 were forced down. The former managed to land behind German lines, but the ship was a write-off; the latter also came down hard, but in no-man's-land near Bandonvilliers. As the crew tried to set fire to it—they were under orders not to allow the enemy to capture an airship and its precious technology intact—there was too little gas left for Z-8 to burn and a party of zealous French *chasseurs* chased them away. The crew escaped to a friendly advance post only because the high-spirited cavalrymen were having such a good time hacking away at the airship envelope with their sabers.[22]

Even when airships weren't lost owing to their low altitude and speed, the minor damage they inflicted hardly justified their cost. In late August 1914, for instance, Lehmann's *Sachsen* and Z-9 launched a series of raids on Antwerp. Shortly before dawn, recalled Lehmann, as they approached the city the alerted defenders beamed up the "white fingers of searchlights" and their improvised "anti-aircraft guns" (really, just field pieces pointed as high as they would go) fired blindly, sending explosions several hundred yards behind them but getting closer. The men stood ready, holding their bombs, as Gemmingen spotted the targets (the central railway station, a major wireless station, and the defensive forts).

Some two thousand pounds' worth of bombs tore holes in ten houses, pocked a field, wrecked a diamond-cutting store, destroyed the home of a luckless factory owner named Van Geel, and blew up an inn. All the ordnance fell more than half a mile from the targets—testament to how unexpectedly difficult it was to drop bombs with any accuracy from a moving aerial vehicle. The raids in total seem (estimates vary) to have left twelve civilians dead and ten injured. One journal commented that "in a densely populated city this can hardly be called slaughter," and even Lehmann, who had an interest in promoting the efficacy of airships, conceded that little damage had been done.[23]

After these disastrous episodes, German High Command called a temporary halt to all airship missions in France and the Low Countries. Losses so far had been confined to prewar airships, and it was abundantly clear that they were not up to the task required of them. Commanders would have to wait for the next generation of airships, due for delivery in the coming months. One captain recalled that for a time it seemed as if "the High Command forgot all about us. . . . We waited for weeks and no orders came."[24]

If the army's airships had been a disappointment, the Naval Airship Division promised great things. Peter Strasser, its dapper commander, was a silent tornado of a man who rarely lost an argument—primarily because he refused to listen to what opponents said. After dinner he would finish off a large glass of wine and a potent black cigar before sitting alone in his room to cut himself off from the rest of the officers and draw up his plans. "A strangely close-mouthed oddball" driven "by the demon of his will and his ambition," Eckener once called Strasser, whom he considered an iconoclast more ruthlessly obsessed with airships, and *winning*, than even Zeppelin.

That said, the two always maintained good relations, and Eckener thought highly of his skills. Strasser, he told his wife, was "a great fellow and stands tower-high above the average and above his personal detractors." Even if the cohort of airship captains did not like him much, they still respected Strasser's authority, nicknaming him "God Himself" (Eckener remained the "Pope" or "Pontiff").[25]

Eckener and Strasser agreed that the army had been using its airships incorrectly by risking them over the Western Front, where they were easily shot down. Deploying them at sea was a different matter: "The crews can see everything coming from a distance," wrote Eckener, and they weren't faced with massed artillery or infantry.[26]

Nevertheless, they had to contend with an ingrained navy bias against airships—an understandable one given the still-raw L-1 and L-2 disasters. It was only thanks to Strasser's prodding that Admiral Tirpitz had even (reluctantly) approved the construction of L-3 in early 1914; now they had to prove the importance of the Naval Airship Division by sending their new airships, like L-4 and L-5, on attention-grabbing missions to spot minefields in the open sea.[27]

When a spate of losses temporarily soured naval opinion on the airships—L-3 and L-4 were blown out to sea in February 1915, never to be seen again, and L-8 was lost with all hands the following month—Eckener and Strasser made sure to organize publicity-friendly missions to restore confidence.[28] Gradually, they pulled the navy around to believing "that it can, at least occasionally, obtain great service from the Z-ships," as Eckener informed Colsman in Friedrichshafen.[29]

Eckener and Strasser's campaign was almost *too* successful. As a result of all the skill, all the expertise, developed at prewar Friedrichshafen being leveraged to matchless effect under wartime conditions, when funds, manpower, and resources were virtually unlimited, the pace of airship construction was beginning to exceed demand. Whereas it had once taken the count years to build a single airship, by mid-1915 production time had been slashed to a mere two to three weeks, with several airships being outfitted at the same time in multiple factories (the Friedrichshafen facilities alone would employ twelve thousand people). The era of the painstaking craftsman was over, replaced by industrialized efficiency. Airships were now assembled using mass-produced girders rapidly bolted together by teams following set instructions.[30]

Indeed, so many Zeppelins were coming onstream that Eckener couldn't hope to train sufficient crews for them. Strasser nevertheless refused to take a breather. He clamored for bigger, better Zeppelins, which would integrate the newest advances in structural and streamlining research.

At Friedrichshafen, Karl Arnstein (the stress expert in charge of structure and load distribution) and Paul Jaray (his aerodynamicist colleague investigating drag and performance) urgently worked to design the next Zeppelin model. By late 1915–16, they had again altered the shape of the airship. They had gone some way toward this goal before the outbreak of war with the DELAG airships, but now they gained Eckener's (and Dürr's) approval to go much further. The new design, which greatly enhanced performance, resulted in teardrop-shaped Zeppelins becoming less blunt at the nose, even "fatter" and "rounder" at their midriff than before, and curving gracefully toward a point at the tail.[31]

Airships were changing quickly in other ways, too. In December 1914, the first of a new war-class type of Zeppelin (Z-12) was made of a recently invented aluminum alloy containing copper, manganese, and silicon called duralumin that was both stronger and lighter than its predecessor.[32]

From this perfect metal, Dürr was able to conjure ever larger, ever more potent airships at an astonishing rate. The *pre*-duralumin Z-10 of October 1914, for instance, was 518 feet long and carried just 1,200 pounds of bombs at a maximum altitude of 7,900 feet. The *post*-duralumin L-10, which first flew seven months later, carried 5,400 pounds of bombs at up to 11,600 feet. By the end of the war, three years later, L-70 would be nearly 700 feet long, bear 10,500 pounds of munitions, and be able to rise to 23,000 feet.[33]

Now equipped with proper weaponry, the Germans could begin their most important mission: bombing Britain into surrender.

21. Pirates of the Air

IN THE HUMILIATING aftermath of the Zeppelin losses in 1914–15, the British had laughed.[1] For so many years, they had been afraid of what the count's vaunted airship fleet would do, and now, it seemed, it could do embarrassingly little. Whereas the nurse and writer Vera Brittain had worried at the outbreak of war that "trouble & disasters are menacing us the nature of which we cannot even guess at," another woman now comically likened one of the notorious airships to the "biggest sausage I ever saw in my life."[2]

So relaxed were the British about the prospect of a bombing campaign that the anti-airship defenses of London and other cities barely existed. There were perhaps just 12 guns, whose projectiles could barely reach normal airship altitude, assigned to defend the 700 square miles of metropolitan London. Paris, of some 49 square miles, was crammed with 215.[3]

When the first raid came, it was duly unimpressive, and wasn't even on London. On the night of January 19, 1915, Strasser's Naval Airship Division struck Great Yarmouth, Kings Lynn, and a few other towns in the eastern county of Norfolk with a ton of bombs, causing four deaths and a minor amount of damage.

Shocked British newspapers were quick to dub the surprise attack an "act of barbarism" against peaceful civilians—a deliberate violation of the Hague Conventions forbidding bombardments of undefended towns and attacks on civilian life and property.[4]

In Germany, though, there was ecstasy. Schoolchildren chanted a song, "Fly, Zeppelin! Fly to England! England shall be destroyed with fire!" and souvenir commemorative medals were struck to mark the wonderful event.[5] German papers cheered the brave airshipmen who had taught the arrogant British a lesson in humility. No longer could they remain safe and sound behind the moat of the English Channel while Germans suffered the naval blockade the Royal Navy had imposed.

As for the accusations of barbarism, *Tageblatt*, a Berlin newspaper, claimed that since the airships were (allegedly) fired upon by coastal guns they had naturally "answered by the throwing of bombs" onto the "fortified place" of Yarmouth.[6]

The German point of view was not altogether persuasive. In America, the impression was that Berlin was scuttling behind pedantically legalistic definitions of what constituted a "fortified place"—Yarmouth may have been a naval supply base, yes, but it was also a popular seaside resort—and the murder of civilians led to an outcry.[7] (A few months later, a still louder one would erupt when a U-boat sank the passenger liner *Lusitania*, claiming that it too was a legitimate military target.)

The German authorities prevailed upon the aged Count von Zeppelin to explain the facts of the matter to Karl von Wiegand, a Berlin-based American journalist and a friend of Eckener's. His exclusive two-part interview was published widely in the United States and did much to calm the gathering storm.

While Zeppelin admitted that, tragically, airships had killed civilians, he claimed that British artillery had also killed German civilians. The indignant yelps now emanating from the British had been prompted by their lack of airships, and they feared, he believed, that the war would be won by the Zeppelin.

The count's master class in evasion skipped over the fact that the German casualties caused by British artillery had been an unintentional byproduct of military operations, whereas he himself had been calling for airships to be used as terror weapons against the general British population. Wiegand did not press further, instead letting Zeppelin recount his sentimental tale of ascending with Professor Steiner in Saint Paul all those years ago.

Finally, in a bid for warmer German-American relations, Zeppelin announced that he had "always hoped to be the first to pilot an airship across the Atlantic." When Wiegand gingerly asked about a British-circulated rumor that he was intending to sail a Zeppelin to New York to bomb it, he exclaimed, "Throw bombs on people who have been so kind to me when I was among them? Never! Never! I want to sail on a peaceful mission when I go to America."[8]

Those who had been bombed by Zeppelin's creations were less enchanted by the old man's reminiscences than was Wiegand, who was dubbed a "notorious" propagandist when the interview made its way to Britain.[9]

WARTIME SERVICE ABOARD a Zeppelin was by no means the pleasant summertime jaunt of the old DELAG days, especially once British defenses improved and the airships, despite the count's assurances, found themselves at an increasing disadvantage.

Bombing missions took place at night, timed to arrive over the target just before dawn when the defenders were sleepiest and the attackers could return undetected ahead of the rising sun. The less moonlight the better, as it lethally illuminated the shimmering silver mass of an airship. In the control car, now wholly enclosed behind glass, the only light that shone was the dim bulb of the telegraph machine. "The two helmsmen stood like phantoms beside the wheel," recalled Ernst Lehmann. "In his narrow cubicle the radio operator sat with his headset over his ears, listening to the confusion of signals and voices whispering in the infinity of space."

Wartime control cars had been thoroughly redesigned to make room for more bombs. Now, crammed within a space of 6.5 feet by 9 alongside the helmsmen, were the rudder man, who turned them back and forth using a small spoked wheel and a compass, and the elevator man, who manipulated a similar wheel to regulate direction upward and downward.

Behind them stood the captain with his chart table and special maps, rather like the old DELAG ones, only these "looked like an astronomical chart full of stars. The Thames was the Milky Way; the stars denoted searchlights, [anti-aircraft guns], and everything else which was to be avoided or destroyed." The captain relayed orders via a small telegraph device and speaking tubes to the mechanics in the rear, the bomb room amidships, and the machine-gunners on top and at the rear of the airship. The executive officer worked near the captain; he operated the bombsight during the raid.

In the rear corner of the control room an aluminum ladder extended into the body of the ship and thence to the machine-gun deck on top of the ship, a dangerous post, for "it sometimes happened that a man was overcome by vertigo and slid off," never to be seen again. Behind the control car there was the euphemistically named "officers' lounge," really a space with two large open windows from which protruded a pair of machine guns.

In the belly of the ship was a bay with high-explosive and phosphorus bombs hanging like "rows of pears." There a bombing officer would lie on his stomach "staring impatiently through the open trap-door" waiting for the orders to release the bombs from their womb.

The crew quarters nearby were rudimentary. A few hammocks hung from the girders, and there was a bit of plywood serving as a makeshift table. The amenities were seldom used. No one slept, ate, or drank much, on the principle that "either we returned in twenty hours or we did not return at all." Some captains ran dry ships (Lieutenant Commander Mathy claimed his was "the strictest Sunday school institution"), but others were more lenient and issued the occasional nip of schnapps.

In the engine compartments, which resembled metal nutshells, the clanking din of the Maybachs ruled out even shouting, and there was scarcely room to turn around. Gasoline fumes and exhaust gases filled them, and the mechanics were widely regarded by the rest of the crew as a strange breed who suffered an above-average incidence of hearing loss and mental instability.

In the warmer months, especially in the opening stages of the war, men wore their service uniforms, greatcoats, and rubber-soled boots. Regular hobnailed military footwear was banned, owing to the risk of sparks. Even so, everyone was always cold, a problem that only worsened as summer turned into fall and winter. One February, Captain Lehmann's thermometer dropped to minus 50 degrees Fahrenheit.

In later years, as Zeppelins were forced to remain at ever higher elevations to avoid being shot down, temperatures of minus 27.5 degrees Fahrenheit were the norm, and airshipmen adopted fur-lined coats and thick gloves. Those in the control car had to contend with snow blowing in through chinks in the window frames that formed into small drifts at their feet, but the "lookout men"—those assigned to observation duty on the top platform of the airship—had it the worst. They could be up there alone with the wind howling in their faces for hours at a stretch and masochistically looked forward to when the captain valved hydrogen (which was colder than the air) because, after a few minutes of enduring its icy current, said August Seim, "the air round you felt as warm as a hothouse."

Crews quickly learned new lessons about the weather, too, especially when an airship found itself combating the forked flashes and earsplitting roars of a storm. During one thunderstorm, recalled Hans von Schiller, "hailstones rattled on the dripping outer envelope, and the airship shuddered like an animal whipped by a lash. Sometimes it was tossed a hundred yards high and sank again two or three hundred yards before it steadied itself."

In these instances, those aboard frequently experienced the disconcerting but beautiful phenomenon of Saint Elmo's fire. A man stationed on top of the Zeppelin could spread his hand and see "little flames spurt out from [his] fingertips," one said, while Schiller noticed that wires and cables "glowed with a bluish-violet light."

The most important revelation, however, was that lightning was not dangerous so long as the gas remained contained. After one fierce encounter with Mother Nature, Lehmann inspected his Zeppelin and found only a few melted metal joints and some pea-sized holes where bolts had struck.

But after all the hardships and the toil came that wonderful moment when, in the words of one veteran, "there bobs up in the deep black night below us the weak rays of light of a city or a village," and the crew experienced "the same feeling as a bird of prey when it spies its victims." Once he spotted the dimmed lights, Heinrich Mathy, probably the war's most audacious airship captain but insinuatingly described in the British press as being an "entirely bald, smooth-faced figure, [as] slender and supple as a young woman," always directed his craft toward the darkest spots, assuming that the enemy would make sure to blacken their most important assets, like the Bank of England and the dockyards.[10]

The Zeppelins had arrived.

FOR THOSE BELOW, as the British soon realized, a Zeppelin raid was terrifying. In 1915, roused by the sound of an explosion, Lewis Freeman, an American living in London, went to the window of his apartment and within moments the "awakening searchlights" were slashing the darkness "to ribbons." For several minutes they searched until they focused on two or three "slender slivers of vivid brightness" gliding above the city that "swiftly, undeviatingly, relentlessly" came on. One Royal Engineer manning a searchlight recalled that a Zeppelin locked in a beam looked "just like a goldfish in a bowl."

Freeman could hear the "wooden clackity-clack" of the Zeppelin propellers as the crew dropped bombs a hundred yards away. Then the raid leader flashed a light to signal "rapid fire," and "a number of sputtering fire-trails—not unlike the wakes of meteors" erupted from the Zeppelins. As the bombs fell, "the hiss of cloven air—similar to but not so high-keyed as the shriek of a shell—became audible, and a second or two later, the flash of the explosion and the rolling boom were practically simultaneous."

Freeman was rare in staying inside, where it was safer. Outside, there was pandemonium. During a raid on Walthamstow, Perriton Maxwell saw "men, women, children, all ludicrously clothesless, [as they] swarmed aimlessly like bees in an overturned hive. Stark terror gripped them. . . . The night-gowned and pajamaed throng could not be persuaded that safety lay not in sight of the Zeppelin but away from it. The hypnotism of horror lured them on to where twelve houses lay spread about in smoking chaos." There, "women frankly hysterical or swooning were roughly swept aside. Children shrieking in uncomprehending panic were swept along with the crowd or trodden on. Lumbering men ran and shouted and cursed and shook hairy fists at the long blot in the clouds."[11]

As time went on, however, it became harder for airships to bomb London and other towns with this kind of near impunity. Defenses, once so rudimentary, rapidly improved as British and French commanders frantically raced for solutions to the airship menace.[12]

One effective method was to establish concentric rings of defenses connected by telephone and including more high-angle, high-velocity guns supplied with fragmenting or incendiary shells. Searchlights would fix the airship with their beams and allow the guns time to adjust for range, speed, and height and then concentrate their fire.[13]

By the summer of 1916, noticed Freeman, there had been a vast sea change in the London defenses from a year earlier. The alerted searchlights now came on *before* the airships arrived. "There was no wheeling or reeling of the lights in wide circles, as a year ago, but rather a steady, persistent stabbing at the clouds, each one appearing to keep to an allotted area of its own. . . . There was nothing 'panicky' in the work of the lights this time, but only the suggestion of methodical, ordered, relentless vigilance."

Once a searchlight caught sight of a Zeppelin, the others quickly latched on and created a glare so intense it dazzled and disoriented the airship crews. Then a hundred guns opened up in unison. Within ten seconds a staggering, helpless Zeppelin would be "straddled" (shells exploding above and below before finding their range), and for a minute and a half the air would be "vibrant with the roar of hard-pumped guns and the shriek of speeding shell"—enough to ensure destruction.[14]

FOR ZEPPELINS, OPERATING in enemy airspace was infinitely more dangerous than encountering a storm. Unlike an airplane pilot, who stood a fighting chance of landing safely in a bullet-riddled machine, the survival rate among airship crews who were shot down was essentially zero, especially after the advent of incendiary projectiles in 1916.

There were no parachutes in Zeppelins, as a full complement reduced bomb load, so the men aboard a stricken airship knew they were dead as soon as they saw the first licks of flame.[15] Only four men, as far as can be ascertained, survived a burning Zeppelin during the war.

One of them, Otto Mieth, the bombing officer of L-48, left behind an account of the collapsing world inside a doomed airship. In the early morning of June 17, 1917, L-48 was caught by a searchlight and an airplane swooped in, pumping incendiary bullets. The gunner on the upper platform cried, "Fire in the stern!" Mieth leaned out of the control car and saw a reddish flame creeping slowly up the body. "Dancing, lambent flames licked ravenously at her quickly bared skeleton," remembered Mieth. He turned to Captain Schütze, but it was not necessary to make a report. Standing motionless as he fixed his eyes on the blaze, Schütze calmly said, "It's all over."

The engines were still running, and the mechanics called from the rear to ask whether to turn them off. Schütze replied that it didn't matter now. The fire rolled forward from the stern as the men fled along the walkway to the control car, and the flames folded over it like a "purple canopy" as black smoke enveloped everyone within. One man, lighting up his last cigarette, joked, "No smoking allowed!" L-48 held an even keel for a few seconds and then turned into a flaming ball ten thousand feet above the ground. The airship broke asunder with a frightful jolt, plummeting like a burning newspaper.

In the falling control car, the men tumbled into one another and struggled helplessly, choking on the gas fumes and smoke as they roasted. It struck a tree and the wireless room was torn off, with the radioman and Mieth inside. Mieth only remembered "a thrill of horror as I opened my eyes and saw myself surrounded by a sea of flames and red-hot metal beams" before blacking out. The radioman's neck was snapped, but a broken-legged, burned Mieth woke up in a hospital to hear a friendly British voice asking whether he wanted a cigarette.[16]

Predictably, a gallows humor prevailed aboard the Zeppelins. When once asked if he carried a medic aboard his airship, Captain Mathy quipped that "if we were brought down [in enemy territory] I guess there would be doctors there if we were to need any, which would be unlikely."[17] The joke disguised Mathy's own troubled psychology. He felt that "it is only a question of time before we join the rest. Everyone admits that they feel it. Our nerves are ruined by mistreatment. If anyone should say that he was not haunted by visions of burning airships, then he would be a braggart."[18]

Mathy himself was shot down on the night of October 1, 1916, his final act being to jump from the control car.[19] According to Michael Mac-Donagh, who went to see the crash site, "I saw the imprint of his body clearly defined in the stubby grass. There was a round hole for the head, then deep impressions for the trunk, with outstretched arms, and finally the widely separated legs. Life was in him when he was picked up, but the spark soon went out" of the "first and most ruthless of these Pirates of the Air."[20]

WHEN IT CAME to administering the killing thrust, the British had discovered, the most lethal method was the airplane, matador of the skies. On September 2, 1916, Strasser sent sixteen airships—the largest raid of the war—to deliver a grand knockout blow, only to be frustrated by Lieutenant William Leefe Robinson's shooting one down in an incendiary-armed B.E.2c fighter.[21] The gigantic fireball that erupted so discouraged the other commanders that they turned tail and left. A single man in a single airplane had defeated an airship armada. Between that night and the end of 1916, seven more airships were shot down by airplanes.

In a rare moment of unblinkered acceptance of reality, Strasser, stunned by the losses, called a halt to the raids. They would not resume until October 1917, when a thirteen-strong fleet equipped with the newest models dared again to venture across the English Channel. They inflicted more than eighty casualties, but five Zeppelins were destroyed. Friedrichshafen, where most of their crews had lived, went into mourning.[22]

So awful were the losses that Eckener could scarcely believe that Strasser refused to give up. In fact, he had ordered four new airships. It was madness, thought Eckener; the air war was already lost.

By 1917–18, it was obvious that airplanes had utterly eclipsed the Zeppelins' initial offensive advantage. A typical 1915 fighter had required forty-five minutes to climb to 10,000 feet, for instance, but current models could hit 15,000 feet in less than eight, allowing them plenty of time and fuel to hunt raiders.[23] Zeppelins could still stay out of harm's way by flying high, but not for much longer. By 1917, Zeppelins could operate between 16,000 and 18,000 feet, but the newest British fighter, the Sopwith Camel, could reach 17,300 feet.

In response, Dürr and his colleagues designed new classes of "height-climbers" that raised the ceiling by stripping Zeppelins of every ounce of extra weight. Hull girders were shaved to their thinnest feasible width, the control car was made even smaller, the crew's quarters were eliminated, and most of the machine guns were removed—saving about seven thousand pounds and allowing a maximum height of 20,700 feet.

Unfortunately, human frailty became their weak point. The crews, from 12,000 feet and up for prolonged periods, relied on oxygen masks to avoid hypoxia, or altitude sickness. Impurities in the gas caused intense nausea and vertigo, but anyone who removed the mask, as many did, would feel at first a throbbing in the teeth and blurred vision, followed by an expansion of abdominal gases—the symphony of farting aboard a Zeppelin was something to behold—before exhibiting symptoms similar to those of carbon monoxide poisoning or a severe hangover. Whereas fighter pilots stayed up so high only for short periods, continued exposure by airship crews could fatally result in fluid in the lungs and brain swelling (leading to bladder and bowel dysfunction, loss of coordination, paralysis, and confusion). At some point, the Zeppelins simply could not climb any higher without killing their crews.[24]

Worse, in the spring of 1918, the British devised a rudimentary aircraft carrier that allowed their long-range planes to hit Zeppelin bases. In July of that year, the Tondern sheds were bombed by seven sea-launched Sopwith Camels. Now the Zeppelins were soft targets not just in the air over Britain but on the ground in Germany itself.[25]

So impotent, indeed, had the Zeppelins become that in January 1918, the novelist (and grinding snob) Virginia Woolf complained that the desultory raids were not so much dangerous as mildly annoying—in the sense that when she was obliged to take shelter she had to talk to the servants, which she found "boring."[26]

WHY DID GERMANY continue with the raids for the duration of the war when they were clearly unable to achieve their goal of bringing Britain to its knees? The simple answer is that German airshipmen, with the major exception of Eckener, were convinced that they *were* bringing Britain to its knees. Among the Old Guard it was an article of faith that a "million" soldiers—as Ernst Lehmann was still claiming as late as 1927—had to be kept in Britain to counter the Zeppelin threat. These men, who would otherwise have been sent to France to fight, were a huge drain on the British war economy—all thanks to the mighty Zeppelin.[27]

But these numbers were a delusion. If anything, Germany spent more of its *own* blood and treasure on quixotically pursuing the airship campaign than the British did in withstanding it. In June 1918, there were precisely 6,136 men dedicated to fending off the bomber menace—a drop in the bucket relative to the two million British Empire soldiers on the Golgotha of the Western Front.[28] Put another way, the number of able-bodied men working in the Friedrichshafen facilities alone doubled that assigned to defend Britain against their products.

The reality was that over the course of the war Britain suffered £1.5 million worth of property damage—not an insignificant sum. But on the other hand, in early 1918 one expert estimated that Germany had hitherto spent up to £13.25 million on airship construction, maintenance, personnel, hangars, gas production, and fuel, thus making the Zeppelins nearly nine times costlier to their owners than to the assailed.[29]

The losses were enormous. Of the ninety-one Zeppelins built and operated during the war there were just sixteen left (including an unfinished experimental model, some training ships, and a few obsolete ones) at the Armistice in November 1918. The list of the fates that the others experienced makes for depressing reading: "shot down in flames," "forced down," "dismantled," "destroyed in explosion," "wrecked in landing," "burned accidentally," "bombed in shed," "rammed a mountain," "crashed," "lost in North Sea," and so forth.[30]

What the Germans received in exchange was hardly worth the effort. In terms of loss of life, when in May 1915 the submarine U-20 sank the *Lusitania*, its captain, Walther Schwieger, inflicted more than twice as many fatalities with a single torpedo as did three years of Zeppelin raiding (1,198 to 557). Indeed, deaths among the Zeppelin crews came close to equaling the number of their victims.

Nevertheless, the German bombing campaign continued almost to the very end, motivated by an unshakable conviction that the raids were inflicting enormous damage on the British economy and morale—an impression only strengthened by seemingly authoritative articles in the German press. The *Magdeburgische Zeitung*, for instance, reported that two raids, on August 1 and 3, 1916, killed and wounded some 21,000 people, yet British figures showed that to date the total number of fatalities in *all* raids was 334 civilians and 50 soldiers.[31]

The Germans dismissed any British claims contradicting their doctrine of faith as disinformation, and in this they were not completely wrong.[32] The Zeppelin raids did cause more damage than the British censor let on: A raid near a major gas facility at Greenwich on January 19, 1917, for instance, combusted eight million cubic feet of gas, wrecking it so completely that even two and a half years later it had not been fully repaired.[33]

But this was a rarity. More often, the censor didn't have to bother spinning a story, for the great majority of Zeppelin raids were a complete waste of time, money, and life. In early 1916, for example, when L-19 set out on its first and only mission, its modest contribution to Germany's war effort would end up being the razing of a pub in the quaint town of Tipton and the murder of several farm animals; on its way back, it suffered engine trouble and crashed at sea, and the entire crew died.[34]

Lying at the heart of the matter was that German strategists relied on the reports of their airship captains, but these were often highly inaccurate. Of a January 31, 1916, raid, Germany officially stated that the Zeppelins "dropped large quantities of explosives and incendiary bombs on the docks, harbor, and factories in and near Liverpool and [iron foundries and smelting furnaces in] Birkenhead [and] on Manchester factories. . . . Everywhere marked effects were observed in the gigantic explosions and serious conflagrations." A stunning success, it seemed, except that no airships had gone anywhere near Liverpool, Birkenhead, or Manchester. Admittedly, some breweries, several railway sheds, a couple of factories, several workmen's houses, and a few churches were damaged elsewhere.[35] Such misestimates were not mere propagandistic fodder for public consumption but were swallowed whole at the highest levels.[36]

It was not that airshipmen lied outright; it was more that they believed untruths to be true or perhaps told untruths to serve the greater poetic truth that Zeppelins *must* be war-winning weapons. It was very easy to get confused. At night, over an alien country where landmarks were elusive and unfamiliar, using maps purchased in souvenir shops before the war, airship captains often found themselves not just a few but dozens of miles away from their objective.[37] Hans Gebauer, the observation officer aboard L-40, saw the nine-hundred-foot-wide Thames River below him during a March 1917 raid and, assuming his airship to be directly above London, was gratified to see the bombs raining death and destruction upon the hated enemy. Unfortunately, L-40 was actually hovering over the sixty-foot-wide Royal Military Canal sixty miles southeast of the capital. The bombs fell into a marsh.[38]

Gebauer's was by no means a unique error: Up to a third of all missions throughout the war were abandoned or ineffective simply because the crews got lost.[39] During the course of just one five-Zeppelin raid on August 9, 1915, L-10 claimed to have reached London when it was actually thirty-five miles distant, L-12 ended up over Dover instead of the Harwich naval air base sixty miles to the north, L-9 found itself over the Humber River in the north of England, L-11 turned back at Lowestoft in Suffolk claiming navigational errors, and L-13 didn't even make it past the coastline before heading for home.[40]

Eckener, who knew airships and their flaws, quirks, and capabilities better than anyone else, was one of the very few observers who was skeptical of the claims of immense damage being inflicted on the British war effort. As early as February 1916, he was confidentially telling Colsman that the airship captains "never report anything that diminishes their competence and contributions. Based on the radio notifications that we've received, I'm convinced that most of them didn't know for sure where they were this time. . . . It'll be more residential buildings than factories that are hit, I'm certain of that."[41]

Unfortunately, he was ill-placed to do anything about it. So dispirited was Eckener by his wartime experience that he would devote just a single page in his 184-page book, *My Zeppelins*, to the subject.[42] In 1917, he requested permission to resign his post as head of training but was refused.[43]

The fatal decision to pursue the Zeppelin raids come what may lay, Eckener thought, with Strasser. Through remorseless force of will, he seemed to exert a Rasputinesque hold over the minds of his superiors. When they wanted to call a halt, Strasser would persuade them not only to stay the course but to give him more resources to build bigger and better.[44]

Disastrously, Strasser accepted no weakness, either in himself or in others. When it was discovered that Zeppelin crews were passing out from lack of oxygen at high altitudes, said Eckener, Strasser dismissed them as "listless and cowardly," and he attributed airship losses to airplanes as "accidents."[45] It was only Strasser himself coming to grief in August 1918, just a few months before the Armistice, that startled the Germans out of their trance.

Commanding the behemoth L-70 as his flagship, Strasser embarked on a giant raid against London on August 5. He had boasted that L-70, containing more than two million cubic feet of hydrogen, outfitted with no fewer than seven Maybachs, and carrying six thousand pounds of bombs, was invulnerable.

It wasn't. When L-70 was shot down off the English coast by a fighter, it burst into flame and Strasser was last seen jumping out. He was, commented Eckener sadly to his wife, "simply too confident that he was unassailable."[46]

Within the week, the new head of the Naval Airship Division canceled an order for Strasser's planned swan song: a super-Zeppelin so colossal that he had decided to skip the usual format of sequential numbering and jump directly to "L-100." It was to have been a staggering 3.8-million-cubic-footer, able to climb to 28,600 feet (nearly the height of Mount Everest), propelled by ten engines, and·insulated from incendiary bullets by an innovative fireproof cover.[47]

T HE TRUE END to the era of the wartime Zeppelin was not, however, Strasser's death but Count von Zeppelin's, in March 1917.

Eckener had known that Zeppelin was faltering for some time, though for much of 1916 the count had kept up a busy schedule visiting hangars, workshops, and factories; he even took a trip on the maiden voyage of the first of a new class of Zeppelin and toured army headquarters at Pless, where he was hosted by Field Marshal Paul von Hindenburg, the de facto commander-in-chief of the German armed forces and, like Zeppelin, the subject of a cult of personality.[1] The count had lost weight, recalled a friend, Rudolph Stratz, and had turned into "a very old man . . . with an aloof expression in his eyes—the fanaticism of a discoverer of new worlds. A sacred fanaticism. Otherwise he looked like the typical old, reckless hussar."[2]

The "touching simplicity" to his manner that Hindenburg would later mention in his memoirs as well as, possibly, the aloof, fanatical look in his eyes may have been due to the onset of dementia, and he certainly seemed to suffer from a failing memory. In February 1917, Stratz bumped into him at an aeronautical exhibition in Berlin. The count, looking lost, shook Stratz's hand. "I am sure I know you well," Zeppelin said softly to his old friend. "But I do not recognize you. Who are you?"[3]

At the end of the month, the count was taken violently ill and rushed to the Sanatorium des Westens, where he was diagnosed with an obstruction of the intestines. An operation was successful, but he soon contracted pneumonia. His daughter, Hella, hurried to Berlin, but doctors had advised his wife, the countess, that he would probably recover and she arrived, too late, a couple of days later. He had died on March 8.

When the sad news was announced, all the houses in Friedrichshafen hung out black-bordered flags, and a special train was hired to take hundreds of employees to Stuttgart, where the count was to be buried.

The funeral on March 12 was a stately affair, befitting a Hero of Germany. Ten thousand mourners were present, including his family, the kaiser, the king and queen of Württemberg, a bemedaled gaggle of field marshals, generals, and admirals, and a consort of leading scientists. As the funeral procession, headed by a military band and trailed by a column of his old regiment, marched toward the family vault along roads adorned with black-draped stone columns topped by urns aflame, bells solemnly pealed and twenty-four guns fired in salute. His casket was borne by twelve cavalrymen, while up above, two giant Zeppelins, their hulls draped in black crepe, paid homage to the master of the air. They dropped wreaths as the casket was lowered into the grave.[4]

Most foreign observers, notwithstanding their hatred for what his creations had become, generously remembered Zeppelin for the man he once was. He was ranked, in *The New York Times*, as equal to the Wright brothers as a "very great inventor" and in other places praised "for the unfaltering courage with which he pursued his vision of a practical dirigible through disappointment to ultimate success."[5]

There was, however, no disguising the fact that the airship had been a total loss in war, with some outlets' obituaries alleging that Zeppelin's death had been hastened by his realization that his namesakes had so signally failed to deliver on their initial promise.[6]

The passing of the Old Gentleman, as Eckener and Colsman called him, finally released, or perhaps liberated, them from his deadweight, his heavy hand, his long shadow. Eckener had loved the count but, all too aware of the challenges of dealing with Zeppelin before and during the war, was relieved that he had *finally* departed. Even a year after the count's death, when Eckener was asked to write an official obituary for him, he refused, "as a kind of horror still grips me when I think I'm supposed to sing a hymn in his honor. An objective appraisal—yes, that would be a different matter. But he's still too alive in the people's love to allow that. He was certainly a fine person, but we shouldn't overdo it with the praise."[7]

• • • • •

ZEPPELIN'S DEATH ALLOWED Eckener at last to think about the future—the future of the airship in a postwar world. He wanted to pick up where the DELAG had been so abruptly terminated at the outbreak of war, but on a grander scale.[8] If nothing else, the war had accelerated airship development to an unprecedented degree—the most modern Zeppelins were about as similar to the prewar ones as a new passenger jet is to its 1950s ancestor—and bitter experience had offered valuable lessons about flying in a matter of years as opposed to decades.

Over the course of the war, the Naval Airship Division alone built 68 airships that flew 325 raid missions, 1,205 scouting flights, and 2,984 other flights, for a total of 1,491,600 miles.[9] This trove of experience otherwise unobtainable during the limited runs of peacetime was analyzed carefully at the company. For instance, instead of trying to fly above thunderstorms as per previous guidelines, read one instruction, commanders should ride *through* them and reel in their radio antennae (to avoid electrical charges).[10]

Thanks to Eckener's training before and during the war, Zeppelins were manned by a large number of veterans—some seventy crews of twenty apiece.[11] Granted, several hundred had died in the fighting, but afterward many, especially officers, would form the core of a resurrected airship empire.[12]

Of these, the greatest was Ernst Lehmann. A native of Ludwigshafen, Lehmann, blessed with "unfailing courtesy and [a] charming personality," was the son of a chemist and had studied at the Technical University at Charlottenburg before joining the DELAG. He flew dozens of missions during the war and was a favorite of the count's.

Lehmann and Eckener, though, had a prickly relationship, Lehmann's ego equaling Eckener's own when it came to confidence in, or perhaps arrogance about, his own airship expertise. Their personal brushes later turned into bitter divisions over the direction and future of the Zeppelin Company, though in Machiavellian fashion they occasionally joined forces against a mutual enemy. Eckener, husky, gruff, and bluff, with prominent jowls and eyebags, presented a stark contrast to the wiry, sharp-featured smaller man (Eckener stood a full head taller) whose deviousness and conceit he often found intolerable, even as Lehmann loathed Eckener's grandstanding and envied his prominence. During the war, they avoided each other as much as possible, Lehmann opting to fly army airships simply so as not to fall under Eckener's oversight at the Naval Airship Division. In their respective memoirs each referred to the other as infrequently as they could respectably manage, and when they did, it was invariably in the stiffest manner.

Lehmann was known as the count's most hawkish acolyte, and Eckener's alienation from Zeppelin's bloodthirstiness inevitably put them at odds. Eckener, once a liberal, was, like Colsman, a business-minded centrist of moderate views, while Lehmann moved in ever more extreme right-wing nationalist circles. To Lehmann, Eckener by 1917–18 was a treacherous snake willing to sell out sacred Germany in search of a humiliating peace; to Eckener, Lehmann was a militarist lunatic digging Germany's own grave with his fanatical desire to fight until there was nothing left to save.[13]

But the root cause of their enmity lay, ironically, in their total agreement on the Zeppelin's potential as a long-haul vehicle. For Eckener, the road ahead was lit clear as day even in the darkness cast by war: He would turn the dream of passenger and commercial travel across the Atlantic, and perhaps even around the world, into reality.

Lehmann, however, was keener on creating the mortal threat of a global bomber that could, if push came to shove, punish New York. To that end, he creatively claimed that in 1915 Zeppelins had carried out thirty reconnaissance missions along the U.S. coastline (simply an impossible feat) and would later brag that he would have carried out the first transatlantic bombing mission himself had the war continued into 1919, which he dearly wished it had.[14]

Self-interest brought them together during the war to strike a deal: Eckener and Lehmann would send an airship to Africa to determine whether such distant voyages were even possible.

THE ADVENTURE BEGAN in 1916 when Dr. Max Zupitza, the chief staff surgeon to the German colonial troops fighting in Africa, was captured in Togo. After his release, he suggested to the Colonial Ministry in May 1917 that they send a Zeppelin to support General Paul von Lettow-Vorbeck, known as the "Lion of Africa." The general was then commanding an astounding guerrilla campaign in which fourteen thousand German and

African soldiers were fighting up to three hundred thousand British, Portuguese, and Belgian troops, and he desperately needed emergency supplies and medical aid. The Colonial Ministry thought the idea crazy but passed it on to Eckener, who didn't.

In September, Lehmann returned to Friedrichshafen to work with Eckener to outfit a new airship, L-57, for the mission. Back in July, Lehmann had achieved the world record for flight duration by staying airborne in LZ-120 for 101 hours, so he assumed he would captain the Africa expedition.[15] Eckener had other ideas, not least of which was preventing his rival from hogging the glory, and managed to block the appointment by arguing that Germany needed to keep its best men on the job—at home.

Korvettenkapitän Ludwig Bockholt, a compromise candidate Eckener suggested, was instead offered the command. Bockholt had a decent if not spectacular record, and, similarly, the crew chosen was experienced but not considered the elite of the Naval Airship Division. Eckener had decided that the Africa voyage was to be a one-way trip: The men aboard L-57 would stay and fight alongside Lettow-Vorbeck, which meant they had to be, in the nicest possible sense, disposable. None of them was aware of this, unfortunately, as to maintain secrecy they were told they were being dispatched for service in the Balkans, a cushier posting than the frigid North Sea or the deadly skies above London.

Bockholt did not start well. On October 7, during a trial run, L-57 was wrecked shortly after taking off when a sudden high wind dashed it into a metal fence. No one was hurt, but the airship was unsalvageable.

Eckener arranged for its replacement, L-59, to be significantly modified. Dürr stretched the airship to measure almost 750 feet, but the biggest changes concerned its adaptation for one-time use. Once L-59 landed in Africa, it would be cannibalized, its every part repurposed for military use. The soft gas-cell covering was intended to be cut up and turned into bandages and sleeping bags, the outer fabric into waterproof tents and clothing, and some of its duralumin skeleton into radio masts, with the rest of the longitudinals and girders repurposed as building materials. The internal catwalks were covered in leather, which would be ripped up and used to make belts and shoes.

Once the outfitting was done, L-59 was loaded to maximum capacity. It carried no fewer than 311,100 rounds of rifle ammunition, 230 machine-gun belts containing 57,500 cartridges, 54 machine-gun ammunition boxes with an additional 13,500 cartridges, 30 machine guns, 22 rifles for the crew (including the hardy Dr. Zupitza), 9 spare machine-gun barrels, 61 sacks of medical supplies (including enough antimalarial quinine to last a year), rifle bolts, binoculars, bush knives, spare radio parts, mail from home, sewing kits for new uniforms, and a case of wine for the crew to celebrate their arrival.

On November 3, 1917, L-59 set off for its European departure point at Yambol in Bulgaria, a German ally. It was only then that the crew were informed of their real mission. Eckener, in the meantime, traveled there by train. When dawn arrived on November 21, everyone gathered in the hangar as L-59 waited, bathed in the pale glow of the arc lights. All shook hands, and the crew embarked on their grand adventure.

By 9:45 A.M. L-59 overflew Adrianople in Turkey and that evening left the Turkish coast and headed for Crete (8:30 P.M.), where it adhered to protocol by winding in its trailing radio antennae before combating a storm. At 5:15 A.M., L-59 crossed into Africa over Libya.

L-59 flew in complete radio silence. As it happened, the Colonial Ministry had recently received intelligence indicating that the British had advanced upon Lettow-Vorbeck's position and that he would surely be forced to surrender. The ministry informed Admiral von Holtzendorff, who decided to call off the mission. Eckener and the station staff at Yambol urgently tried to contact L-59, but between their weak radio transmitter and the airship's reeled-in antennae, it was too late. Eckener informed naval headquarters that "L-59 can no longer be reached from here, request recall through Nauen," home of the most powerful transmitter in Germany. Nauen broadcast all night long, with nary a reply from Bockholt.

L-59 continued on through the desert. On the afternoon of November 22 it became the first airship to cross the Tropic of Cancer. Trained in northern Europe, the crew experienced strange and unexpected weather phenomena in the unfamiliar environment. During the day, the sun beating down on the airship super-heated the gas, which had to be valved off gently to maintain a stable cruising altitude. Powerful thermal currents rose from the shimmering sands and threw the ship this way and that. Crewmen experienced nauseating airsickness as L-59 bobbed up and down like a roller coaster. Lacking sunglasses, many of them had to endure violent headaches, hallucinations, and half-blinded eyes.

Still, L-59 dutifully continued, eventually reaching the Dakhla Oasis, where the appearance of this wondrous sky god astounded the local bedouin. (In December 1933, when a pilot had to make a forced landing at Dakhla, he found outlines of a Zeppelin scratched on the wooden huts. The bedouin chief told him that they paid homage to "a mighty sign which, as many years ago as there are fingers on both hands and toes on one foot, appeared in the heavens above the desert.")

Bockholt headed for the Nile, intending to follow its course to Sudan. At 4:20 P.M., an engine gave out, but the other four Maybachs continued working perfectly. The night, however, brought another unpleasant surprise: As the temperature plummeted it super-cooled the hydrogen, forcing Bockholt to dump ballast just to stay airborne. At 3 A.M., after descending from 3,100 feet to 1,300 feet, he came close to hitting the side of a mountain that loomed up from out of nowhere, and nearly the last of the ballast was desperately thrown overboard. L-59, much to the crew's relief, slowly righted itself and rose to a safer altitude.

Bockholt was just 125 miles west of Khartoum and two-thirds of the way to relieving Lettow-Vorbeck when, at 12:45 A.M., he received a curious, crackly, coded message on L-59's prearranged wavelength. Nauen had finally managed to contact him: "Break off operation, return. Enemy has seized greater part of Makonde Highlands, already holds Kitangari. Portuguese are attacking remainder of Protectorate Forces from south." Lettow-Vorbeck was finished, or at least so it seemed.

Or more accurately, that is how the British *wanted* it to seem. At the Admiralty in London, Room 40 was home to naval intelligence, whose cipher experts had cracked the German naval code. Room 40 had also been dimly aware that the Germans were planning *something* for many months before L-59 set out, thanks in part to an obscure British agent (code-named Mortimer) who had parachuted into Austria to meet an anonymous American of Bulgarian ancestry working for the Secret Service.

The latter said that he'd heard rumors of an airship that was due to go to Yambol, for reasons unknown. The operation was called China Show. British intelligence later picked up Naval Airship Division radio chatter mentioning China Show, but the pieces couldn't be fitted together until L-59 was reported to be heading for Lettow-Vorbeck's location in Africa.

To prepare a surprise welcome party, London instructed the East African Royal Flying Squadron to keep its fighters on standby to shoot down L-59 and at the same time sent a coded message, purportedly from Lettow-

Vorbeck, telling of his encirclement. This was, as intended, picked up in Nauen and thence transmitted to Bockholt. Ultimately, the fighters were not needed as the message did its work all too well.

On board L-59, Bockholt received the transmission and paused. What should he do? He was so close to Lettow-Vorbeck, yet this was a clear order from an unimpeachable source to break off the mission. He put the decision to the crew and for the next two hours they debated it, with one faction agreeing with Bockholt that they must turn back and another arguing that to depart now would abandon the Lion of Africa to his dreadful fate. Discipline won. At 2:30 A.M., Bockholt ordered L-59 to turn and go home. At 7:30 in the morning on November 25, L-59 arrived in Europe only to discover that that same day, Lettow-Vorbeck had actually put the enemy to flight.

In a strictly military sense, then, the L-59 flight was a failure, but privately Eckener was jubilant. The airship, because it had never landed in Africa, had completed a nonstop voyage of 4,225 miles—about the same distance as between Friedrichshafen and New York—in 95 hours at an average speed of 45 mph over troublesome terrain and in harsh conditions where the temperature had oscillated between 23 degrees and 95 degrees Fahrenheit. Bockholt still had fuel enough for another sixty-four hours of flight, translating into an extra 3,750 miles.[16] L-59 had also borne cargo and passengers—of a military nature, true, but a valuable payload nonetheless—making it evident to Eckener that once the damned war was over the Zeppelin Company had a clear path to developing, at long last, large-scale intercontinental commercial travel.

The only problem was, would there even be a Zeppelin Company?

23. The Beginning or the End?

WHEN THE ARMISTICE that halted the Great War was signed on November 11, 1918, Eckener was in Friedrichshafen, desperately seeking a way to save the ailing Zeppelin Company.

Even before the Armistice, chaos had engulfed a collapsing Germany, and Friedrichshafen, like many other places, experienced a dangerous wave of popular unrest and outright rebellion as millions of demobilized and deserting soldiers returned home to unemployment and impoverishment. They found their wives and children stricken with tuberculosis and pneumonia—consequences of malnourishment—with the death rate among women a quarter higher than before the war.[1]

Meat, eggs, butter, and milk were scarce, as were clothing, shoes, and furniture. In Friedrichshafen, a military official noted, "bread was of the worst quality and damp, the potatoes were already rotten when they arrived from northern Germany, and there were no vegetables." Zeppelin employees were fortunate in one way, however. They were permitted to purchase up to five yards of the fine Egyptian cotton cloth so recently used to make the airships' outer skins; the material could be turned into shirts and trousers that could be bartered for produce from the local farmers.[2]

Even so, at the Friedrichshafen hangars and workshops, an official report recorded, employees were suffering from "a total collapse of morale, an overall hopeless emotional state, a collapse of nerves, and a deep-rooted bitterness." In an alarming sign of things to come, on October 22, eight hundred workers had left the Maybach factory and marched to the town hall shouting, "Down with the war! Up with the German Republic!" A strike had been called for October 26, but the Zeppelin factories were shut down to prevent it. Angry men, four thousand strong, then marched again to the town hall, where one radical shouted at Colsman, "For you, too, Herr Colsman, we have a bullet."

The new Workers' and Soldiers' Council called another strike for November 5, this time to force an immediate peace and the installation of a Socialist republic. Colsman bravely demanded to speak to the crowd and managed to calm the marchers down, but on November 10 red flags were for the first time raised in Friedrichshafen, and up to ten thousand people crammed into the streets to cheer the coming revolution. Ultimately, police and troops dispersed the workers, but they did not return to work—for there was no work to return to.[3] Thousands of jobs had vanished overnight, and by the end of the year there were just one hundred part-time employees left.[4]

With the factories shuttered, the hangars uncharacteristically quiet, and the once prim and pristine employees' village falling into disrepair, Friedrichshafen had tumbled hard from its glory days. An American reporter who visited the town noted seeing "a barbed wire [fence] and a bored-looking bulldog" guarding the forlorn workshops, and a member of the Workers' Council spoke the once-unspeakable: "We are not proud of these Zeppelins. [They] have aroused the indignation of the world against Germany. They must go."[5]

Eckener, surveying the gloomy scene that greeted him each morning as he passed the noiseless hangars and headed to the emptied offices, promised himself that the Zeppelin Company would rise again from the ashes of defeat. But how?

THE WAY FORWARD, he believed, was "to begin again in 1918 where we had stopped in August 1914."[6] That is, to resurrect the DELAG and commence domestic passenger service while the German government negotiated the final peace terms with the Allies as quickly as possible to stave off anarchy. He was convinced that the British and the French were "damn eager to finish us off" by imposing such harsh terms that Germany would be under their heel forever—which made it all the more important to invigorate German industry while he still could.[7]

Colsman, as a fellow businessman of moderately conservative temperament, agreed that a peace treaty was in the best interests of all but was adamantly opposed to talk of bringing the DELAG back from the dead.[8] He had helped birth it, but the DELAG was in the grave and should stay there. Here was a golden opportunity to forge a new direction for the Zeppelin Company, he argued, pointing out that while Germany was poor and the factories closed, the company was, thanks to his efforts, rich.

During the war, the government had spent lavishly on its airship fleets and paid for hangar upgrades and modernized facilities for Zeppelin. Colsman had wisely—especially in light of the inflation that was then beginning to grip Germany—invested the rivers of cash flowing into Zeppelin coffers in material assets, such as the large stockpiles of duralumin sitting in the company's warehouses.[9]

Now that that gravy train had crashed to a halt, Colsman wanted to get out of the airship business. Airships were a losing proposition, and an airline nobody would want to travel on (given the Zeppelins' dreadful wartime record) would inevitably be a bottomless money pit. He argued that the only means of survival lay in diversifying into a multitude of industries: Maybach would stop making engines and become a luxury automobile manufacturer; Zahnradfabrik Friedrichshafen (ZF), which made the gears and transmissions for the airships, would produce them for budget carmakers; the assembly workshops could repair railway carriages and hammer out cheap furniture. He even hired out an airship hangar as a movie set and turned the factory that had made the airships' outer skin into a textile plant.

But most shockingly, most unforgivably, the Zeppelin Company would be converted into a manufacturer of consumer goods. Colsman had it all planned out. Use the precious duralumin trove to churn out pots, cake pans, and pitchers under the brand name Ellzett, based on the German pronunciation of the initials LZ, for Luftschiffbau Zeppelin.[10]

Colsman's proposal horrified the Zeppelin Foundation, the board of directors overseeing the sprawling operation Colsman had assembled as general manager of the Zeppelin Company, which owned the subsidiaries like the DELAG, Maybach, and ZF. The foundation was headed by Gemmingen, the count's nephew, and it was, in Colsman's eyes, a fusty old boys' club packed with Zeppelin's aristocratic cronies and relatives.

The board members retained their emotional attachment to the Zeppelin airships but lacked the expertise to combat Colsman's financial acumen. Colsman succeeded in selling the idea to the board only after agreeing to Gemmingen's demand that the diversification effort would be only a tactical maneuver to keep afloat until, as Gemmingen put it, "the underlying purpose of the Zeppelin enterprise, *the development and advancement of air-transport,* [comes] to the fore again."[11]

Colsman accepted the terms because he knew it would comfort the old gentlemen in their dotage. Baron von Bassus, who had traveled in the count's very first airship, was frail, and Gemmingen was ailing. Once they were gone, he would inevitably control the foundation and through it the company—the short-term fix would naturally turn permanent. Gemmingen, however, was not quite as doddering and naive as Colsman believed. He quietly recruited Eckener, who was not a director, and set about promoting him to the board to counter Colsman and uphold the Zeppelin Company's original vision.

On the face of it, Gemmingen arranged a system of checks and balances in which Eckener became vice chairman but was nominally subordinate to Colsman, who retained control over financial and business affairs. If a majority of the board voted with Colsman, he could overrule Eckener, but if the other directors unanimously sided with Eckener, it would be Colsman who was overruled. Throwing this otherwise stable arrangement badly off kilter were the unfortunate facts that both Colsman and Eckener were autocrats accustomed to getting their way and were diametrically opposed when it came to deciding the future of the company.

The former allies soon became rivals. Whenever Colsman complained that reestablishing the DELAG would drain money out of the subsidiaries, Eckener retorted that airships were the company's genes, its sinews, its blood, not "pots and pans," as he dismissively said. Zeppelin, for Eckener, was not a normal business concerned with profits and losses; its special mission was to bring the benefits of the German airship to the world and damn the expense.

In any case, Eckener liked to add, it was actually Colsman's diversification push that was bleeding money: Maybach was close to insolvency (nobody was buying expensive cars in post-Armistice Germany), Ellzett was in the red, and ZF's gear business was a flop (carmakers preferred to make their own). If anything, concentrating the combined assets of all the subsidiaries toward one shining goal—airships—and risking everything in the process were what the Zeppelin Company stood for. It would either succeed, or they would all go down honorably together.

Colsman, naturally, objected to Eckener's Sparta-style strategy, accusing him of being a "false prophet" who was misleading the directors into thinking airships could ever be a viable business. They were living in a different world now, and under the new economic conditions "the issues that were accommodated while Count Zeppelin was still alive are not tolerable." Eckener's romanticism, as he saw it, was but a futile gesture that would drag them to perdition.

Eckener and Colsman's relationship, which had once been friendly enough to counterbalance Count von Zeppelin's considerable avoirdupois, quickly deteriorated to the point that they only communicated via stiffly formal interoffice memoranda in between their "fierce" and "numerous" fights, as Eckener put it.

At some point in this epic contest of wills and competing visions, it was clear to both men, one of them was going to have to knife the other.[12]

ECKENER IN THE meantime had to fight on a second front against another rival, Captain Lehmann, who was threatening to destroy the Zeppelin Company by different means. At war's end, Lehmann had moved back to Friedrichshafen to take over the construction division, or what was left of it.[13] Lehmann considered himself the voice of the loyal Zeppelin crews, who had sided, like him, with the conservative nationalists. Deludedly believing that Germany had not *really* lost the war but had been stabbed in the back at the Armistice by cowards, liberals, and Jews at home, Lehmann and the airship crews, refusing to truckle to the British and the French, vowed to fight on if necessary to secure Germany's rights and honor.

As Germany and the Allies negotiated the Treaty of Versailles in the months following the Armistice, the victors' dictates stuck in German craws, especially the ones concerning the country's disarmament. To supervise Germany's adherence to the terms, three Inter-Allied Commissions of Control were formed (one each for army, naval, and aerial matters). The American reaction to the treaty, however, remained a wild card. The United States, under Democratic president Woodrow Wilson, would certainly sign the treaty in Paris, but the Senate still needed to consent to its ratification—and *that* did not seem at all likely, given Republican hostility and the loud objections of the German immigrant population. As a result, in right-wing German circles, there was a growing belief that America was a closet ally in the struggle against what was seen as Franco-British vindictiveness.[14]

In November 1918, Germany had sixteen Zeppelins left, nearly all housed at the northern airbases of Nordholz and Wittmundhafen. At the end of that month, their crews composed a declaration proposing a trans-atlantic voyage to help forge a German-American axis against the French and British. "It may sound dangerous to speak of crossing the ocean, yet we have no fear of it," the airshipmen continued. "We remind you of the Africa flight of the naval airship L-59, and of [Lehmann's] 100-hour flight of the airship LZ-120. . . . With reasonably favorable weather we could make the crossing in a flight of 100 hours and together with 30 passengers, we could transport a quantity of leaflets and other propaganda material."[15]

Lehmann took up their cause. He persuaded Gemmingen that he, Lehmann, should captain L-72, the last and most modern giant Zeppelin commissioned during the war, and undertake the mission. It was what the count would have done, he argued; never would he have allowed the pride of the German airship fleet to be surrendered to the enemy. Leh-mann proposed to fill L-72 to the gills with fuel and hydrogen, recruit a crew of patriots, and then, late one night, take off and secretly—even the government in Berlin would not be informed—head to America.[16]

When he heard about it, Eckener couldn't believe that Gemmingen was backing Lehmann's lunatic scheme. Did he really think the Ameri-cans, who had so recently suffered more than three hundred thousand casualties in the war fighting Germany, would conduct business as usual when a stolen Zeppelin appeared above New York—an airship piloted by the same man rumored to have been the one who would have bombed that city had the war continued?[17]

More dangerous still, Eckener added, had Gemmingen not considered the adverse reaction by the British and the French to an escapade of this nature? At the moment it remained to be seen just how diligently the (Aerial) Inter-Allied Commission would take its job of confiscating Ger-many's remaining airships and razing the infrastructure. With any luck, hoped Eckener, the commission would honor its remit more in the breach than the observance, and he would soon be able to get back to business.

But stirring up the hornet's nest, as Lehmann intended to do with his futile gesture of defiance, meant that "the Allies will jump on us and knock everything to pieces. They will destroy everything so thoroughly that we'll never again be in a position to carry on Graf Zeppelin's work!"[18] A buccaneering bit of bravado by Lehmann would not just bury any hope of ever reestablishing the DELAG but also hand Colsman the right to claim that airships were indeed a dead business.

Much to Lehmann's dismay, Gemmingen changed his mind. To make sure it stayed that way, Eckener discreetly informed the government of Lehmann's scheme. Berlin, not wanting any trouble at this delicate time, contacted the Inter-Allied Commission and sent a telegram to Friedrichshafen forbidding the departure of L-72 from German soil.

Lehmann was mystified as to how the Inter-Allied Commission "had learned of all the particulars of our plan" but soon fingered Eckener as the perpetrator. "Treason was cheap amongst us at the time," he later, darkly, hissed.[19]

ECKENER ALSO HAD other reasons to thwart Lehmann's designs. He was, in fact, *already* secretly in touch with the Americans, and they *did* want to conduct business as usual, but Lehmann's antics threatened Eckener's plan of *selling* them L-72.

Even before the war, the Americans had been interested in acquiring or developing airships but had enjoyed no success.[20] Lady Fortune had finally smiled in October 1917 when L-49 lost its bearings over France after a raid on London. A fighter patrol intercepted it; according to the eyewitness Joseph Ganson, an American officer, "when the Germans saw luminous [incendiary] balls whizzing past they knew they were lost, and hung out a white flag."[21] The fighter leader ordered it to descend at Bourbonne-les-Bains. Upon landing, its captain tried to set it on fire but was prevented by a local man armed with a shotgun. His friends took the rest of the complaisant crew prisoner.

Hearing that a Zeppelin had come down, Colonel William "Billy" Mitchell, one of the leading American airpower advocates, rushed to the site and was awed by his first vision of the downed monster. "I immediately climbed through it and inspected it from end to end," he wrote. "It was one of the latest models and was in a perfect state of preservation."

The "Super-Zeppelin" L-49, just four months old, provided a windfall of technical knowledge thanks to its being captured nearly intact. For the first time analysts saw how the Germans so magically controlled their airships when they discovered an apparatus containing thirty-eight red and white buttons—the top-secret mechanisms that regulated the expansion and compression of the gas cells. The Americans could only gasp at the excellence of Arnstein's and Jaray's handiwork. By their calculations, since 1914 the Germans had *halved* aerodynamic drag resistance, a feat previously thought impossible.[22]

In July 1918, the army-navy Joint Airship Board recommended building four airships based on L-49, but after the Armistice, defense budgets were slashed—by 90 percent in the navy's case. With such a small pot of money available, the Americans were forced to search for bargains.[23]

In the spring of 1919, eager to keep L-72 out of Lehmann's hands and knowing that the Americans were hoping to get their own cheap airship, Eckener had sounded out the U.S. military attaché at The Hague, Colonel Edward Davis. Davis asked Washington for instructions and received a top-secret cable from the War Department informing him that it was sending Colonel William Hensley, a skilled balloonist, to Germany for a clandestine meeting with Eckener.[24]

ECKENER WAS, AS real estate agents like to say, "highly motivated" to sell. Though a first-class airship in perfect condition, the very apex of Zeppelin's technological skill, and the pride of the fleet, in fact L-72 would have been no better than scrap if it remained in Germany.

In the proposed Treaty of Versailles there were two key provisions directly affecting the Zeppelin Company. Clause 198 explicitly stated that "no dirigible shall be kept" by Germany, and Clause 202 added that "all military and naval aeronautical material" (including airships) that "are or have been in use or were designed for warlike purposes" must be handed over to the Allies within three months of the treaty becoming operative.

Eckener read and reread those passages, then read them again. Depending on how one looked at it, they were both clear and unclear, definitive yet ambiguous. First, in Clause 198, what did "kept" mean, exactly? The Allied negotiators assumed that the Germans could not "keep"—in the sense of possessing them—any airships whatsoever. But Eckener could argue that "kept" alternatively meant that Germany could not retain any airships that, like L-72, had been built *during the war*. Clause 202 was equally slippery. Unquestionably, it forbade keeping any army and navy airships left over from the war, but from Eckener's perspective it did not unequivocally declare that *civilian airships built after the Armistice*, airships that were neither used nor designed for "warlike purposes," were outlawed.

In other words, Eckener had just discovered a Zeppelin-sized loophole in a treaty painstakingly sweated over by scores of international jurists. There was nothing, according to his interpretation, legally preventing the Zeppelin Company from building a civilian airship for the new DELAG. That did not, however, make it wise to twist a tiger's tail. If it so wished, the Inter-Allied Commission could, like Humpty Dumpty in *Alice in Wonderland*, make words mean whatever they wanted them to mean.

To avoid rousing the Inter-Allied Commission's ire, in March 1919 Eckener instructed Dürr, Arnstein, and Jaray to design and build an airship from scratch. He had one condition. Under no circumstances could it be seen as being in any way suitable for "warlike purposes." The ship had to be small and humble and patently unable to carry bombs. It must have no connection to the Naval Airship Division, was to be operated by a re-launched DELAG, and would be crewed by civilians. He even chose an appropriately harmless name for the new craft: *Bodensee*, an homage to the lovely prewar *Hansa*, *Viktoria Luise*, and *Schwaben*. Who could ever fear an airship christened so bucolically?

In private, Eckener had bigger things in mind. *Bodensee*, he promised the board, was merely the stepping-stone to the great water jump across the Atlantic. Once the principle was established that Germany could build small civil airships and run a domestic passenger service, there was no logical reason why Zeppelin could not build large civil airships and run an international one.[25]

BUT FIRST HE had to get rid of L-72. The negotiations with Colonel Hensley were proceeding smoothly, or at least they were until the night of June 23, 1919, when a band of former airship crewmen sneaked into the hangars at Nordholz and Wittmundhafen and destroyed seven deflated Zeppelins by tearing away their supports and letting them crash onto the concrete floor.[26] Now there were just nine left (including an unfinished experimental model), and the British and the French were, understandably, furious at being deprived of their loot just five days before the Treaty of Versailles was signed (June 28).

Eckener could not believe the magnitude of the stupidity at work. The timing of the sabotage was particularly suspicious: Could it have been coincidence, or had someone—a vengeful Lehmann came to mind— tipped the crewmen off that L-72 was about to be sold?

It didn't matter now. Eckener was dealing with a major crisis. The Inter-Allied Commission swooped in to confiscate the surviving airships as prizes, allocating them thus: L-72 and LZ-113 to France, L-71 and L-64 to Britain, L-61 and LZ-120 to Italy, L-30 to Belgium, and L-37 to Japan. Within a few years nearly all of them would crash, be dismantled, or fall into decrepitude. In the end, a sullen German crew duly delivered the controversial L-72 to France; upon entering, its new owners discovered the inside daubed with pornographic graffiti as a welcome gift.[27]

As the Romans had sowed salt into the ruins of defeated Carthage, a livid Inter-Allied Commission insisted that all airship facilities were to be immediately leveled or repurposed—the twenty-five-airship base at Ahl-horn, for instance, became a children's home, a tuberculosis hospital, and a convalescent center—to ensure that Germany would never again men-ace their cities from the air.

The victors, aside from the Americans, had their spoils. They had orig-inally been promised two of the now-destroyed Nordholz Zeppelins as war booty; in the revised allotment, the Americans received none because they said they would be satisfied to be compensated in cash by the Ger-man government for the loss of their two airships. With an eye on the fu-ture, they also asked the commission to save the Friedrichshafen hangars and assets as a favor to them.

With L-72 now in French hands, Eckener obviously had no airship to sell. Colonel Hensley suggested using the compensation the United States was owed to commission an all-new giant airship (LZ-125) at Friedrichs-hafen. Eckener was enthused with the idea. Building an airship for a for-eign government as a form of reparations was not *expressly* forbidden by the treaty, it would keep his workers busy, and he'd earn points with the Americans, who would form a useful counterweight to further Franco-British machinations.

A major downside, however, was the enormous financial risk: LZ-125 would essentially be an airship built for free by Zeppelin on the strength of a government IOU to Washington. As it was, inflation was daily eroding the value of any cash reserves the company still had, and none of Cols-man's business ventures were working out, so whether Zeppelin would remain sufficiently solvent to finish the airship was the real issue.

The decision was taken out of his hands when—speaking of machinations—the French and British spies tailing Hensley informed the Inter-Allied Commission of his activities. The commission reminded Washington that because the United States had not yet ratified the Versailles Treaty it was, legally speaking, still at war with Germany. Newton Baker, the U.S. secretary of war, was horrified to discover that Hensley was in effect, then, running a covert arms-smuggling operation in violation of the Trading with the Enemy Act. On December 1, 1919, Newton canceled and covered up the entire project. (The United States and Germany would eventually sign a bilateral peace treaty in August 1921.)[28]

For Eckener, it was a blow, but not a wholly unwelcome one. Building LZ-125 would have brought, aside from economic jeopardy, increased scrutiny from the Inter-Allied Commission just as the DELAG was, against all odds, performing unexpected wonders.

24. Bringing Back the Dead

THE ZEPPELIN MEN had outdone themselves. Drawing upon their wartime experience, Dürr, Jaray, and Arnstein managed to conjure *Bodensee* from nothing in less than five months.

The airship was like none other. Integrating the latest aerodynamic advances and a masterpiece of craftwork, its graceful streamlined curves made it an elongated Art Deco teardrop. A straight line or a right angle could scarcely be found. Eckener's friend Colonel Hensley, who would make sixteen flights on it, said *Bodensee* "is really so far advanced in every way over anything I have seen, that one is led to express the opinion that in airship construction (and operation as well), all other countries are mere Babes in the Woods compared to the Germans."[1]

The combined passenger and control car resembled a railway carriage on the inside, with twenty passengers facing one another across tables. An arched window graced each four-person compartment, and another six people could sit in wicker chairs in the aisle, weight permitting. There was a bathroom and a buffet and overhead luggage racks. For double the price, there was even a first-class private cabin at the front for a VIP traveling alone. A visiting American named Robert Thompson was enraptured by the cleverness of its construction, marveling that the removable "glass" windows were actually made of paper-thin mica, that the comfortable chairs were feather-light, and that the apparently substantial wooden columns decorating the cabin were as "hollow as a paper tube."[2]

Yet at just 396 feet long, *Bodensee* was considerably shorter than any other previous Zeppelin—by some 25 feet when compared to the first of its kind, the count's LZ-1 of 1900. At 706,200 cubic feet, *Bodensee*'s volume was comparable to the prewar *Sachsen*; it was almost as if, just as Eckener had pledged to begin where it all had ended, the war had never happened. But of course it had happened, and the sacrifices of those four years had not been wasted.

Whereas *Sachsen*'s useful lift had been just 16,300 pounds, *Bodensee*'s was 22,000, and the latter's four war-tested 245-horsepower Maybach Mb IVa engines drove it faster than any airship that had ever been built: a stunning 82.3 mph, a record not beaten until the advent of the *Hindenburg* seventeen years later. Henley was amazed when after an airplane drew up alongside the airship, Eckener decided to have some fun ordering *"Allekraft"* (full speed). *Bodensee* jumped forward as the engines roared, and the race was on. For twenty-two minutes airplane and airship were neck and neck before the pilot waved, dived, and headed for home, leaving Eckener the jubilant victor.[3]

With just a single airship, Eckener could set up only one route. Sightseeing tours, a mainstay of the old leisurely DELAG in its grand imperial days, were rejected as out of date and worse, unprofitable. With the kaiser gone, the new DELAG was part of a modern republic and was directed at the time-conscious business traveler—the men and perhaps even women (it's interesting that neither Zeppelin's nor Eckener's wives would ever fly in an airship) working hard to get Germany back on its feet.

Eckener selected Friedrichshafen-Berlin, with a stop in Munich a few times a week to pick up passengers. Before the war, the trip had taken, thanks to the remorselessly efficient German railway system, fourteen hours. In 1919, revolutionary strikes, smashed engines, coal shortages, and cracked rails had turned the once-pleasant journey into a twenty-eighthour descent into Hades—when the trains actually ran. Eckener said he could do it in between five and seven hours.[4]

On August 24, 1919, the *Bodensee* embarked on its first voyage and was an instant success, with tickets on subsequent flights being sold out four weeks in advance. Prices were set relatively low to lure affluent customers away from the railway, but they were not cheap. The DELAG brought in extra income from the mail and packages it carried on each flight, as well as from add-on fees for heavy luggage. Each passenger could take a carry-on bag and had a free allowance of thirty pounds in a suitcase. For passengers unaware of the weight restrictions, an unwelcome surprise awaited: One traveler noticed an "elegantly fur-clad" woman protesting the bill for the more than "half a dozen trunks" she had brought with her. The add-on fee for the extra pounds amounted to more than her ticket.[5]

In its first month of operation, the *Bodensee* brought in 500,000 marks—not bad at all in a country where potatoes were now luxuries. Costs exceeded that figure, but as larger Zeppelins plied ever-longer routes, Eckener predicted, operating costs per mile would fall.[6]

The DELAG operated a daily schedule. On the odd days of the month, *Bodensee* departed Friedrichshafen and sailed north; on the evens, it returned south from Berlin. With a good tailwind, the trip sometimes took just four and a half hours.[7]

The DELAG was so successful that Eckener went to Sweden in October 1919 to discuss opening a route—an international one—to Stockholm with a new airship he'd commissioned: the *Nordstern* (North Star), a slightly larger (thirty-passenger) clone of the *Bodensee*. He received an enthusiastic reception and gave, he told his wife, Johanna, twenty interviews to the press. He even presold 50,000 marks' worth of tickets for an airship that didn't yet exist.[8]

Spain was a strong possibility for the next expansion.[9] Like Sweden, Spain had been neutral during the war, and Eckener was still wary of needlessly antagonizing the French and British with talk of running a route across their territory. But a flight south through Switzerland would require only the permission of the more easygoing Italians before crossing the Mediterranean.

Spain was important to Eckener for three reasons. First, its warm weather meant that Eckener could operate flights year-round rather than having to close during the winter.[10] Second, he could benefit from the coming boom in postwar tourism by selling the spectacular views of mountains and sea to leisure travelers. The views in Germany, in contrast, were unremittingly awful, with its pyres of grim black devastation. The journalist Ernst Klein, while traveling on the *Bodensee*, lamented the smokeless factory chimneys of once-vibrant Nuremberg and the immense new cemetery outside Leipzig where soldiers' graves stretched in endless rows.[11]

Last, and most important, Spain was the gateway to South America, whose commercial and tourism possibilities were only just beginning to be realized. Seville or Cádiz, in particular, provided an ideal jumping-off point for a trip to Rio de Janeiro or Buenos Aires along a southern route, an easier one than the North Atlantic, where fog, gales, and low temperatures often prevailed. That the Canary Islands and Cape Verde were on the way for repairs and emergency respite was an added bonus. From South America, the next natural jump was Miami, then Washington and New York. If Eckener could set up airports in Spain, Brazil, and Argentina, he'd be well ahead of anyone else in establishing a global network. No one could ever claim Eckener lacked ambition.[12]

It was not to be. On December 5, 1919, four days after Secretary Baker canceled Colonel Hensley's "compensation airship" scheme, the DELAG closed for the winter. *Bodensee* went into its hangar for refitting, and *Nordstern* was scheduled to undergo final preparations for the coming season's Stockholm expansion.

Then the Inter-Allied Commission of Control struck its death blow: Starting January 10, 1920, when the Treaty of Versailles was due to come into force, all DELAG operations would be suspended and the *Bodensee* and *Nordstern* confiscated.

ECKENER WAS SURPRISED, and bitterly disappointed, at the guillotining of his airship line. In its ninety-eight days of existence, the DELAG had flown on eighty-eight days and made 103 flights totaling 32,300 miles. *Bodensee* had carried 2,253 passengers (an average of 22 per flight, making for an impressive 85 percent load factor) plus 11,000 pounds of mail and 6,600 of cargo with not a single accident.[13]

The DELAG had also struck a hopeful chord in the German breast. Not only was the *Bodensee* a symbol of a new and proud Germany, it represented youth and optimism, the "returning energy" and "physical reconstruction" of a vanquished nation. As one writer noticed, on his flight there had been two Dutchmen, one of whom remarked as they disembarked in Berlin, "Do you know, a nation that can set up such an institution as this right after its defeat will never go down."[14]

The mastermind behind the shutdown soon revealed himself: Air Commodore Edward Masterman, the head of the Inter-Allied Commission and an airship expert of some note. Masterman just happened to be in charge of Britain's own lackluster airship program, and the DELAG move, to Eckener's mind, was a blatantly protectionist attempt to strangle the competition.

During the war, in late 1916, the British had commissioned R-34 (the *R* stood for "rigid"), a near-exact copy of the German L-33, which had recently crashed. It had not been finished by the Armistice, and the Admiralty managed to get rid of the white elephant by dumping it on the newly formed Air Ministry, which hoped some private investors would take it off its hands. To whip up interest, the ministry organized a long-distance voyage: R-34 would travel to America and back.[15]

Annoyingly, just as R-34 was being readied, the honor and glory of conquering the Atlantic by air fell to the U.S. Navy. Setting out from Newfoundland, three of its flying boats had made the attempt, though only one, the stripped-down NC-4, had made it to Britain on May 31, 1919, after an arduous journey of nearly two weeks, including three prolonged stops.[16]

Still, R-34 could at least salvage something by making a relatively rapid *return* trip. On July 2, R-34 departed Scotland bound for Roosevelt Field on Long Island, near New York City. Owing to constant mechanical problems and high winds, the flight west was very slow, taking 108 hours and 12 minutes—the steamship record for a crossing was held by the *Mauretania* in 1909, better by roughly two hours—but R-34 had beaten the pants off the U.S. Navy aviators. It had been a close thing, though. On arrival, R-34 had just 140 gallons of fuel left—enough for a mere two hours' flying time. The return flight, thanks to a nice tailwind, took only seventy-five hours.[17]

The New York Times may have claimed that "the [NC-4] airplane flights of 1919 were gallant adventures" and that "the voyage of R-34 is the real beginning of the new age" of transatlantic travel, but the plaudits were premature.[18] Disappointingly, still no commercial bidders appeared, probably because R-34 had shown that while a return flight was indeed possible, no one could see how to make a paying business out of it. R-34, old and worn ragged by the flight, had left no room for revenue passengers, mail, or cargo, and more to the point, what madman would want to risk his life in a contraption that had made transoceanic landfall with a mere two-hour margin of safety?

In Germany, Eckener couldn't understand the hubbub over NC-4 and R-34. They were both one-off, go-for-broke stunts undertaken for the sake of being first. Had not his L-59 *two years earlier* traveled farther (to Africa and back) with no stops and no mishaps? Eckener, always thinking about the long term, understood that the key to ultimate success was endurance. Making a single moonshot effort was a very far cry from flying safely, repeatedly, and profitably.

Echoing Eckener's views, the German press dismissed the Atlantic crossings as a mere "sports feat."[19] As indeed it proved (as Eckener had shrewdly suspected all along): R-34 flew domestically a few more times before ignominiously bumping into a hill and being sold for scrap.

That was cold comfort. To save the Zeppelin Company, Eckener had thrown the dice when he'd tried to sell L-72 to the Americans—and lost. He'd thrown them again to build Hensley's LZ-125—and lost. And he'd thrown them once more to set up the DELAG—and lost.

But this was no time to quit.

WHAT FORCE CONTINUED to propel Eckener forward with almost demonic energy in these dire years for the Zeppelin Company? Why was he compelled to fight off rivals and challenges that would have felled a man less driven? It could not have been only a sense of loyalty to the count's memory—though Eckener proved himself adept at exploiting that memory for his own ends—and neither was it due to some lust for power and station, for he could easily have found himself a place in the government or a directorship at a major conglomerate.

It was because, very simply, he knew he was right. He unshakably believed that if he could steer the Zeppelin Company through the storms he was onto a winner. And he was not wrong about that.

The fact of the matter is, the Zeppelin, condemned and mocked as the world's worst weapon by 1918, was by 1919–20 being exalted as the silvery herald of global travel. That such a surprising turnaround, partly engineered by Eckener himself, occurred when memories of the Zeppelin raids were still raw says much about his canny publicity skills.

It was Eckener who assiduously promoted the idea that the jolly old count's marvelous invention had been perverted and brutalized by the German war machine for nefarious purposes, whereas he had remained loyal to Zeppelin's (allegedly) original intent of forging ever closer union between the peoples of the earth. It was a tale undoubtedly true in certain respects—Eckener had always fantasized, and always would, of peace-through-airship—but he took care to obscure his own wartime role in the Naval Airship Division and downplayed the count's less wholesome aspects.

Stage One in the rehabilitation process had been the appearance in March 1919, a few months before the *Bodensee*'s first flight, of a report (leaked by Eckener) in the German aeronautical weekly *Luftpost* on the still-secret L-59 Africa flight of 1917. The amazing story was picked up by the British, and then international, press.[20] The revelation reignited interest in the airship as a viable transatlantic transport—hence the R-34 voyage—with advocates deriving inspiration from President Woodrow Wilson's famous address to Congress unveiling his idealistic "14 Points" that would reform an anarchic, shattered world.

When Wilson promised free trade, freedom of navigation, open borders, and liberty to all, airship enthusiasts resurrected the rhetoric of those happier days when balloon flight promised to unchain mankind from its manacles and proclaimed the imminent arrival of a new age "when giant ships of the air will be carrying mail and passengers to all parts of the globe, ignoring boundary lines and spreading the democracy of the world."[21] When Wilson spoke of establishing a League of Nations to bring world peace through disarmament and negotiation, the airshipmen echoed that an "internationalization [of] the air" was at hand: It would create a Utopia where people never again had to "live permanently in a border-defending state of affairs."[22]

For the nations mourning ten million dead in a war to end all wars that had swept away five emperors, eight kings, and eighteen dynasties like so much detritus, *their* future would be dominated and safeguarded by the airship. And *this* future was the Dream—Eckener's Dream. Eckener was compelled to pursue the airship, even when all seemed lost, as a means to revolutionize the system of the world and attain glory immortal and peace eternal. Building airships was his destiny, and he would let nothing stand in his way.

Eckener might otherwise have been dismissed as a monomaniacal crank but for the fact that the pie-eyed fantasists were actually airplane lovers like Willis Abbott, a well-known journalist who boldly imagined an alternative future where aerobuses, private biplanes staffed with liveried pilots, four-seater family fliers more numerous than Ford Model Ts, and huge triplane passenger expresses plied the crowded sky above America with flights between New York, Chicago, Boston, Pittsburgh, and Washington. They would take off every hour on the hour, Abbott confidently forecast, and "delays are most unusual."[23]

Eckener, at least, had a record of reliable flying undergirding his utopian dreams, but readers of Abbott's article would have been struck by the gulf between his rosy predictions and the parlous reality that was aviation in America at that time.

25. The Visionary

SINCE ITS HEYDAY in the Wright brothers era, the United States had fallen so far behind in aviation that it barely scraped a mention in the world rankings. In 1914, on the very eve of war, the entire U.S. airplane industry employed a grand total of 168 men, and the largest firm, Glenn Curtiss's Aeroplane Company of Hammondsport, New York, could build just one aircraft per week.

Meanwhile, that same year, the French military possessed 162 modern airplanes compared to the United States' all-obsolescent air force of 23. By April 6, 1917, when America entered the war, the situation had worsened: The armed forces now had at their command 109 airplanes (half were non-combat trainers; the rest were so dated they couldn't be used in combat) and just 83 pilots. In comparison, France, Germany, and Britain each had thousands of aircraft.

On May 24, the Allies requested that the United States provide 4,500 airplanes as soon as possible for the war effort. Full of vim, the army and navy went above and beyond and in 1918 alone the United States trained some 10,000 pilots and built 14,020 aircraft.

An extraordinary achievement, but the vast majority of these aircraft were kept at home as bare-bones trainers. A mere 260 U.S. airplanes were operating on the Western Front in November 1918—France alone had 2,820 front-line aircraft—and their technological backwardness was embarrassingly apparent: Not a single American ace flew a U.S.-made fighter in World War I.

At this critical moment, the Armistice cut the legs out from under the industry. Within days, the U.S. government canceled contracts for another 13,000 airplanes and 20,000 engines, bankrupting hundreds of firms and leaving the survivors with crippling financial losses. Within three months, 90 percent of the airplane industry's 175,000 workers had lost their jobs. In the following year, 1919, just 780 airplanes were manufactured, and by 1922 that figure had fallen to 263.[1]

Despite the tremendous activity during the war, at its end no domestic market existed to pick up the slack. Of those 780 aircraft in 1919, for instance, just eight were intended for civilian use.[2] Worse still from the long-term perspective, aeronautical research and development was frozen in time, which meant that as the years passed American airplane technology remained stuck somewhere in the 1914–18 era, with its wood-and-fabric, strut-and-wire biplanes held together by glue and screws.

The only thing America *did* have, in spades, was war-surplus trainer planes that were sold off for a fraction of their value to former military pilots who emulated the old nineteenth-century balloon aeronauts and set themselves up as traveling barnstormers (a term borrowed from itinerant acting troupes that had played in barns during the westward expansion). After returning home from Europe, these pilots scraped together a few hundred dollars for a Curtiss JN-4 ("Jenny"), a tandem-cockpit biplane with a 90-horsepower engine, and wandered the country giving shows at out-of-the-way towns and county fairs.

As Professor Steiner had done generations earlier in his balloon, they took thrill seekers up in the air for their first ride (urinating down onto their own town was peculiarly popular) but also served as death-defying jacks-of-all-trades: One of young Charles Lindbergh's business cards advertised his specialties as "Fair and Carnival Exhibition Work, Offering Plane Exchange in Midair, Wing Walking, Parachute Jumping, Break-aways, Night Fireworks, Smoke Trails, and Deaf Flights," while others dangled from trapeze bars suspended below the aircraft, fought aerial duels, and performed loop-the-loops.[3]

Lindbergh's fellow barnstormers were, as one Slats Rodgers recalled, "a sort of mixture of the cowhand of the Old West, the hot-rod driver of today, and the real gypsy."[4] When the weather was good, the living was easy: Pilots ate steak for breakfast, stayed at the finest hotel in town, and tried to lure impressionable young ladies into their clutches. (Lindbergh remembered one flier whose only, and only occasionally successful, chat-up line ran, "Do you or don't you? That's all I want to know.") But in winter or in foul conditions, they often slept under a wing with nothing but peanuts for dinner.[5] The vast majority were old war veterans, but there was a surprising number of women and minorities admitted to the cadre. Bessie Coleman, for instance, grew up poor in Texas but earned a flying license in France in 1921 and toured extensively until she was killed in a crash four years later. In the black press, she was eulogized as "Miss Bessie Coleman— The Race's First Aviatrix."[6]

The barnstorming era was, for many Americans, their introduction to aviation. Up to a million people may have taken a five-minute flight, including Paul Tibbetts, who would go on to pilot the *Enola Gay* over Hiroshima, and Howard Hughes, who first flew with his father on a five-dollar plane ride. One famous barnstormer, Ivan Gates, estimated that he had put on 1,836 exhibitions in 1,042 towns across forty-one states by 1926.[7]

Less admirable was the barnstormers' safety record. This was a time when there was neither aviation law nor any governmental oversight whatsoever of civilian plane inspection, pilot training and licensing, traffic control, or the use of aircraft. Quite literally, *anybody* could fly any plane he or she wanted, and fly anywhere he or she wanted, joyfully careless about the condition of the aircraft. One youngster named Jack Chapman, for instance, became a pilot at the age of eleven and liked to fly solo.

Barnstormers exploited the lack of regulation by ascending in biplanes with but a quarter gallon of gas in the tank, wings of rotting fabric holding on for dear life, and control sticks jerry-rigged from a convenient piece of wood. The useful life of a military Jenny, which had come with a complement of army mechanics and a hangar, had been estimated as six months at most, but barnstormers patched and wired and glued theirs for *years* after they should have been scrapped. The rate of mechanical and structural failure alone was astounding, and that without even factoring in the risk of running into power lines, loop-the-looping straight into the ground, or plowing into the crowds watching directly below.

Fatalities were, unsurprisingly, common. Between 1921 and 1923, 196 people were killed and 351 more injured in barnstorming accidents; in the latter year alone, what one observer decried as "winged imbeciles" caused 179 crashes, or about one every two days. For some spectators, it was all part of the entertainment to see if the barnstormer would come to grief. Roger Q. Williams, who advertised himself as the "King of All Dare Devils" who "Flirt[s] with Death in Mid-Air," later said he felt as if paying customers were disappointed when he didn't kill himself.[8]

. . . . .

JUST AS CLACKETY and as dangerous, but ultimately of greater value, was the establishment of the Air Mail by the Post Office. In the fall of 1919, seeking to speed up delivery, the Post Office acquired over one hundred surplus De Havilland DH-4 biplanes and succeeded in delivering mail from New York via Cleveland to Chicago. A San Francisco–Sacramento leg opened, followed by a Chicago-Omaha one by May 1920. The hardest route to traverse was across the Rocky Mountains, but by September Post Office airplanes were picking their way through, with numerous stops and plane changes.

Soon after, on the first westbound trip from New York to San Francisco —the first transcontinental airmail run in history—the Post Office pilots averaged 80 mph, carried sixteen thousand letters, and shaved nearly a day off the fastest express trains, which took about a hundred hours, or more than four days, coast to coast. Once the Post Office began flying at day *and* night the following year, delivery time was slashed to around thirty-three hours.[9]

The savings added up quickly. A long-distance phone call cost $4.65 for three minutes, but a letter could be airmailed for 10 cents. Banks discovered that for every $10,000 check they sent from San Francisco for deposit in New York they avoided having to pay two days' interest; by such means, one bank saved $33,000 annually, and on a larger scale, the Federal Reserve in New York was immensely pleased to discover that in 1924 it had reduced interest charges by $809,589 thanks to the Post Office.[10]

Yet, like the barnstormers, the Air Mail lived on a wing and a prayer. As the De Havillands had a maximum range of just 250 miles but carried minimal fuel, pilots could not make transcontinental runs by themselves but instead shuttled between local "airports"—really, a short grassy strip and a hut—picking up and dropping off the mailbags as and when needed. The lack of infrastructure and organization, combined with delays when bad weather hit, meant that costs climbed precipitously: Despite increasing revenue, the Post Office lost $12 million between 1918 and 1927.[11]

The toll in human life rose as well. Being an Air Mail pilot was the most dangerous job on earth. Not for nothing was it known as the "Suicide Club," whose members were all young men with an insatiable appetite for risk.[12] Looking, said one observer, like "Eskimos in their ungainly flying-suits, with helmets and goggles on their heads and great fur-lined moccasins over their shoes," they carried sidearms "strapped to their thighs in approved 'bad man' fashion" to ward off not only mail thieves but bears in the event they had to crash-land.[13]

Crashing was a virtual certainty: Everyone expected at least a few accidents (one pilot, Wesley Smith, pranged no fewer than fifteen airplanes), and if they were lucky they would happen in a nice, flat area; trying to land in mountainous or forested terrain was usually a death sentence, perhaps at the paws of said bears. Dean Smith, however, survived a particularly tricky situation: As he telegrammed his boss to explain, "Dead stick—flying low—only place available, on cow. Killed cow—wrecked plane—scared me."[14]

The pilots led short but eventful lives. One in six died in 1920 alone, and within eight years all but nine of the original contingent of forty pilots were dead.[15] That figure doesn't include the forty-nine other people killed and seriously injured in accidents, in which more than two hundred Air Mail planes were lost.[16]

The Air Mail nevertheless endowed flying with utility as much as the barnstormers associated it with entertainment. What neither was able to do—unsurprisingly, given their dire safety records—was persuade Americans that taking a plane to where they wanted to go could be as natural as catching the train, steaming aboard a ship, or driving a car.

It wasn't for want of trying. As early as 1913, various entrepreneurs had tried to set up passenger airlines, with occasional, if brief, success. In Florida, in December of that year, the Saint Petersburg–Tampa Airboat Line began to provide service—it was the world's first non-airship line—as an alternative to the two-hour ferry, the twelve-hour railroad, or the drive along unpaved roads that took up most of a day. The flights took about twenty minutes and cost five dollars each way. But when the tourist low season came the following April, the passengers vanished, and so did the airline.

After the war, some returning pilots set up airlines, but few ever managed to either get off the ground (literally) or to keep flying long. Their routes were invariably short (Aero Ltd., for instance, ran a New York–Atlantic City service taking less than half an hour—about as long as Aero Ltd. lasted), running the planes on any significant scale was prohibitively expensive, and passengers were hard to attract owing to, first, the wide availability of comfortable, cheap trains; and second, their understandable fear of crashing en route.[17]

Most of these airlines were seat-of-the-pants, fly-by-night outfits run on luck, booze, and debt that didn't stand a chance of overcoming the challenges of operating a successful business. And it is here that the man who would eventually rise to become the Eckener of the Airplane steps into the picture.

JUAN TERRY TRIPPE was an effortlessly superior, enviably wealthy, upper-class Anglo-Saxon Protestant of ancient pedigree—or at least that was what he wanted people to think.

In truth, in 1899 he had been born to the plush if not the purple, and his family background was more colorful than he ever admitted, beginning with his exotic name—Juan—which he loathed.

It certainly didn't come from his father's side, where a Henry Trippe had grown tobacco in Maryland in the late seventeenth century and John Trippe (Juan's great-great-grandfather) had distinguished himself fighting the Barbary pirates in 1805 with the U.S. Navy. Afterward, the family maintained a state of modest gentility by working as stockbrokers, engineers, and pharmaceutical wholesalers.

"Juan," in fact, came from his mother's side. Trippe's grandmother, Kitty Flynn, was born poor in Ireland and emigrated to Liverpool in 1860, where she worked as a barmaid and married an American bank robber on the lam named Charles Bullard (in polite society, he called himself Charles H. Wells, "man of independent fortune," but his underworld sobriquet was Piano Charley). Kitty herself was no mean con woman and was described by William Pinkerton, the detective later dispatched to find the rogue pair, as "a beautiful woman and a brilliant conversationalist who dressed in the height of fashion."

The couple had a daughter named Lucy and moved to Paris (England was becoming too hot, Pinkerton-wise) to open a bar, where they specialized in fleecing patrons at faro. The French police closed the establishment after Bullard returned to his old habits and robbed a diamond merchant. Kitty and Piano Charley fled to London and had another daughter, Katherine, in 1877, Bullard vanishing soon after.

He would eventually turn up in Boston, where he was sentenced to twenty years in prison, but escaped to Canada and thence to Belgium, where the authorities finally caught up with him. During the trial, it emerged that Bullard had bigamously married Kitty, thereby making their daughters illegitimate.

Kitty, now calling herself Kate, had by then moved to Brooklyn, New York, where she worked various minor stock-market frauds, ran a gambling den, and owned a hot-sheet hotel masquerading as a boardinghouse. Then one day in walked blue-eyed, dark-haired Juan Pedro Terry, aged thirty. He was the son of Tomaso Terry, a Venezuelan of Irish ancestry who owned the biggest sugar plantation in Cuba. In 1881 Kate, inexplicably attracted to the multimillionaire, made sure to marry him, her checkered history conveniently vanishing into the mists of dim memory.

The kindly Juan Terry gave her daughters fine educations with sojourns in Europe. In 1886, Tomaso died, leaving behind a gargantuan fortune of $50 million, of which Juan inherited $6 million, but he mysteriously died a few months after his father, bequeathing a fifth of his existing estate to Kate. By the time she died of Bright's disease in 1894, she had spent her Terry money and left just $5,000 to Lucy and Katherine.

Eighteen months after Kate's death, on November 24, 1895, Lucy married Charles White Trippe, a civil engineer, in the Episcopal Church of the Heavenly Rest. The bride gave her father's name as "Charles Wells"— the only time she ever acknowledged his existence, even if she did employ his upscale alias. Lucy was as adept at socially camouflaging herself as her son would be: Through her mother, she was not only a Catholic, but worse, the lowest of the low: an *Irish* Catholic at a time when the recent arrivals from that green and moist isle were regarded as slum dwellers fit for nothing more than mindless fighting and menial labor. If it meant entrée into New York's elite society, it was worth an Episcopalian ceremony.

Lucy did keep a connection to her Terry relatives by insisting that her second son be named after her stepfather, Juan. It was a name, as mentioned earlier, that the boy grew to hate. It made him stick out as much as his darker, olive-skinned looks did, and it was a constant, niggling reminder of his embarrassing and never-to-be-spoken-of Hispanic-Irish ancestry. He called himself John instead, an appropriately anodyne name for an outsider trying to blend in.

No matter how hard Trippe tried to cover up his background, there were whispers. When he went to Yale, his classbook vaguely stated that his mother had lived in London and New York before marrying, thus allowing Trippe to let others assume she had led some kind of Henry James/ Edith Wharton–like existence, but behind his back his friends gossiped. As one recalled, "He was supposed to have Cuban connections, but we never tracked them down."[18]

Trippe made sure nobody could ever "track down" his background. At Yale, he acquired an unwanted nickname, "Mummy"—as in "to keep mum," or silent—when asked to give a speech to introduce himself during his initiation into an undergraduate dining club. Mute terror had gripped him. He revealed nothing, for fear of ridicule if the truth came out.[19]

After college, still presenting himself as a banal cipher, he went to work on Wall Street, if that is the right word for selling bonds to friends in the morning before heading to his club in the afternoon. He paid his golf club dues, smoked stolid cigars rather than devil-may-care cigarettes, wore stiff old-man suits, and associated with *Social Register* types far wealthier than himself so that they, naturally, assumed he was One of Us.

But of course he wasn't, not really. In these years, Trippe may not have been poor, at least as conventionally defined, but the plutocratic cigars cost fifteen cents for two, his budget Lower East Side tailor hadn't learned to cut cloth on Savile Row, he lived in a modest apartment on the distant outskirts of a fashionable neighborhood, his universal tip was a lowly dime, and he drove an ancient, used Pierce-Arrow car.[20]

To wend his way in the world, Trippe adopted a silky sneakiness that would become legendary among those who later had to deal with him. Trippe was so habitually devious that "if the front door was open, he would go in by the side window"—and that odd plaudit came from one of his very few trusted *friends*. Later, Franklin Delano Roosevelt, no stranger to silky sneakiness, half-admiringly called him "the most fascinating Yale gangster I ever met" and "a man of all-yielding suavity who can be depended on to pursue his own ruthless way."[21]

"All-yielding suavity" was a perfect description. Blessed as he was with a seductive politeness, a quiet voice, a modest air, a cherubic smile, and an exaggerated obsequiousness toward his elders and betters, the quiet, pudgy Trippe was often assumed to be weak or soft, yet, inexplicably, he would emerge from tough negotiations as the winner.

What nearly everyone missed was that Trippe had been obsessed by airplanes ever since his father had taken him to see Wilbur Wright circle the Statue of Liberty in 1909. During the war, he had volunteered and trained as a bomber pilot (but saw no combat), and at Yale, he joined a flying club and raced against other Ivy League schools until, citing its dangers, the university banned the practice.

In 1923, he quit his tedious Wall Street job and set up Long Island Airways (LIA).[22] LIA comprised nine old navy biplanes and six pilots and operated from a makeshift hangar on Coney Island. That summer, he started offering tourists rides up and down the beach but couldn't make a profit, partly because his pilots "were show-offs who [got] drunk and [took] the next day off" but also because his biplanes were two-seaters that could accommodate only one passenger.[23] The revenue accruing from that passenger was not enough to cover expenses—a common problem for early airlines.

Hoping to discover a solution, Trippe spent his off-hours at the New York Public Library studying the history of other modes of transport, especially trains and ships.[24] As one who'd worked so hard to ape the manners of the elite, Trippe was a meticulous planner who prided himself on being better prepared than anyone else.

In this case, he noticed that from small, scrappy beginnings, the railroads and shipping firms had forged huge stable oligopolies; could the same be done for airlines? For months, he pored over dense manuals detailing their economics (operating costs, break-even points, logistics management) and regulatory environment, and he concluded that if aviation were ever to succeed as a business, three conditions needed to be fulfilled.

First and foremost, an airline needed capital to invest in new equipment. That was actually the easiest problem to fix, because it wasn't really a problem for Trippe in the first place. His college buddies like William H. Vanderbilt, John Hambleton, and Cornelius ("Sonny") Vanderbilt Whitney were wealthy in their own right and, as fellow members of the Yale racing club, were almost as airplane-obsessed as he was. Without a doubt, they would help finance any new enterprises Trippe asked them to, if only as a bit of fun.[25]

Second, the carrier had to be able to charge a rate that would ensure a profitable return, not only to stay in business but to fund further expansion. On Coney Island, since anyone could take people for thrill rides, Trippe's many other rivals squeezed out any possible profits in a race to the bottom. Market competition was fine and good, of course, but not, as Trippe believed, when it came to establishing an untried and expensive new form of transportation.

Trippe determined that an airline—if it were ever to amount to more than providing joyrides in increasingly geriatric aircraft—had to own the exclusive right to operate between given destinations. Such a right could be granted only by government regulation (as had been the case with the railroads) so as to protect the fragile business against overcompetition.[26] Given the free-for-all of the barnstorming era, this seemed a rather remote possibility, but some politicians were already talking about introducing legislation to improve safety standards and to shore up the Air Mail.

And last, there had to be an *urge* and a *need* to fly among the public. To see if he could pump up demand, Trippe adapted his planes by replacing their engines with larger ones and moving the fuel tanks from the inside to the outside to make room for an extra person. Instead of beach flights, he began renting out airplanes for charter trips to Atlantic City, the Hamptons, and Newport for affluent couples heading to their weekend houses. Income accordingly rose for a brief time, but Trippe was bedeviled by the same underlying issue: Few people relied on airplanes when trains, boats, and automobiles, despite their time disadvantages, were already servicing their needs. Airplane trips were still regarded only as occasional extravagances with no larger purpose. People went aloft, told their friends how exciting it was, and never flew again.

Like so many others, LIA shut down operations several months later, but the information Trippe had picked up about the perils and pitfalls of running an airline was priceless. He would try again.

In a sense, Trippe had seen *too* far into the future to make the present a success. But he had put his finger on the nub of the problem. Airplanes were too small and underpowered to serve as anything more than summertime rides, but anyone who tried to build larger and more expensive ones to accommodate more passengers would get eaten alive by the costs because the demand simply was not there. It was impossible to fill several seats consistently at an acceptable price.

Seen in this light, it's clear why Eckener was so certain that he was right in believing that airplanes would never threaten his airships' supremacy. The *Bodensee* was a magnificently designed aerodynamic marvel expertly captained and crewed, whereas the barnstormers were nothing but stuntmen flying creaky deathtraps. The DELAG carried passengers, cargo, and mail imperturbably, almost profitably and (one day) internationally, whereas the U.S. Air Mail was a maniac-piloted, high-fatality money loser requiring several stops just to get a letter from New York to Chicago. And the Zeppelin Company built machines capable of flying thousands of miles, whereas Long Island Airways and its bankrupt ilk considered it a major accomplishment to take a couple of passengers on a twenty-minute trip.

Compared to airships, airplanes were nothing more, surely, than a gnat is to a lion.

EVEN LEAVING ASIDE the success of the DELAG, the airplane and the airship were hardly competitors. In the days when Zeppelin had competed against the Wrights, the two forms of air transport had been considered as an "either/or" question: *Either* the airship *or* the airplane would survive. The answer, it had turned out, was "and/but": The airship *and* the airplane existed *but* would play separate roles.

As Ladislas d'Orcy, an influential editor at *Aviation and Aeronautical Engineering*, put it, the "airplane is mainly a high-speed short-distance carrier" while "the large, rigid airship is essentially a medium-speed long-distance carrier."

Airships—luxurious, reliable, safe, efficient—would be used to carry heavy payloads and large numbers of passengers on flights between Europe and America or Cairo and Delhi.[27] Airplanes—small in terms of both size and carrying capacity, as well as supremely uncomfortable but fast—would be employed shuttling high-salaried businessmen between regional cities and transporting lightweight, time-sensitive objects (documents, securities, newsreel films, and so forth).

In other words, as one magazine explained, "airplanes and airships have their own peculiar powers and limitations; and because of their differences actual rivalry cannot seriously exist between them."[28] The airship's real competitor, then, was the oceangoing passenger liner, and the airplane's, the intercity railway train.[29]

On this last point, the big question of airships versus ocean liners, it was now a matter of "instead." Eckener was confident that it would be a matter of people gradually converting to traveling quickly by air *instead* of slowly by water. If airships' speed could be pushed to 100 mph—not so much more than their current capability—the travel time between San Francisco and New Zealand (6,500 miles) would be cut from the prevailing twenty-two days by ship to just six.[30] And why restrict oneself to sea, lake, and river when these and mountains and jungles and deserts and plains meant nothing to an airship? *Anywhere* on the face of the earth was within reach.

Eventually, inevitably, the Zeppelin must dominate the world.

BY FOCUSING ON ocean liners, however, Eckener, like so many others, had fundamentally misunderstood the real threat posed by the airplane by looking only at its *present* condition and assuming that that state of affairs would continue, unchanged, into the future. The possibility that one day, not long in coming, the airplane would vanquish the airship (and the ocean liner and the train) as a safe, comfortable, large, high-speed, long-distance passenger carrier was unforeseen by virtually everyone.

The very idea of building an airplane big enough to make the transatlantic jump in a single bound, let alone roomy enough to accommodate numerous passengers and valuable cargo, was almost universally deemed ludicrous.

Juan Trippe was among the very, very few who believed in the airplane's potential. As early as May 1919, while still a student, he had written perceptively of the navy's attempt to bridge the Atlantic by airplane (the one culminating in NC-4's success at the end of that month). A crossing, he predicted, would "demonstrate that a flight across the Atlantic Ocean is a perfectly safe and sane commercial proposition and not a gigantic gamble in which the prospective transatlantic pilot or passenger has big odds against his safe arrival."[31]

Trippe had been influenced by the views of the American writer Alfred Lawson, a frequent contributor to aviation magazines. In 1916, he'd written a piece prophesying that by 1930 passenger airplanes would be crossing the Atlantic nonstop at more than 180 miles an hour. Aside from Trippe and a few other enthusiasts, no one believed him. Then again, Lawson was a hard man to believe in, having also predicted that by 1970 there would be so many people flying that some kind of air traffic control system might be required.

Neither did it help Trippe's case that his fellow zealots were widely perceived to be mad. Lawson, for instance, said that by 10,000 A.D., there would be godlike, ethereal superhumans he called "Alti-Men" living in the upper atmosphere who would control the weather and rule pitilessly over the mere mortals existing on earth.[32]

While the perspicacity of that forecast remains to be seen, Lawson himself was a well-known egomaniac who claimed that his emergence from the womb "was the most momentous occurrence since the birth of mankind," though he at least pretended to have the modesty to say it under a pseudonym. Just as fantastically, in 1920–21 Lawson built a prototype of what he dubbed an "airliner" for between eighteen and twenty-six passengers for a planned thirty-six-hour New York–San Francisco service. Unfortunately, it promptly crashed and "Lawson Airlines" went belly-up. (Undaunted, Lawson later founded the University of Lawsonomy, which banned books not written by himself. It was later investigated for fraud.)[33]

Given the eccentricity of Trippe's bedfellows and their not altogether successful record, it was understandable that most people considered the prospects of a large, long-range passenger airplane to be very dubious.[34] In fact, it simply could not be done, at least not until perhaps 1980, thought one aeronautical expert.[35]

Granted, the airplane had already made astonishing advances in its short life. In 1909, for instance, competitors in the international James Gordon Bennett Aviation Cup race had to complete a 12.4-mile-long course; just four years later, range and endurance had improved so greatly that it was extended tenfold, to 124 miles.[36] During the war, the airplane made additional leaps. A British B.E.2c fighter of 1915, for instance, had a maximum speed of 72 mph and a service ceiling of 10,000 feet, whereas a Sopwith Camel of 1918 could make 113 mph and ascend to 19,000 feet.

As impressive as they were, some pointed out, these improvements were surprisingly modest compared to the Zeppelin's gains. In 1914 a typical airship's engine power had been 630 hp, but by 1918 it had tripled to 1,820 hp. In the same period, maximum range had shot from 2,000 to 7,500 miles, useful lift had more than quadrupled from 9.2 to 44.5 tons, the altitude ceiling had risen from 6,000 to 21,000 feet, cruising endurance at a steady 45 mph had leapt from 20 to 177.5 hours, and speed had increased from 47 mph to 77 mph even as the ships had tripled in gas capacity.[37]

The airplane's best days were already behind it, was the aeronautical world's conventional opinion. Its technical development was slowing and soon would plateau. That of the airship, in contrast, was accelerating.[38] The critical factor, as the nineteenth-century aeronauts had discovered, was believed to lie in the fact of the airship's deriving its lift from the cubic feet of volume whereas the airplane did the same based on its square feet of wing area.

Scientific American, using some questionable figures, explained why the airplane was destined to fall ever further behind: For an airplane to carry as much as a contemporary Zeppelin could, it would need a wing area of nearly half an acre (17,700 square feet). To put that figure into perspective, the then-largest civilian American airplane—the Curtiss F-5L—had a wing area of about 1,400 square feet. Building such a mammoth was an impossibility.

Even if one could, it wouldn't matter because as an airplane's wing area increased so too did the vehicle's gross weight, not only owing to the greater quantity of material required but to the need to make the larger structure stronger. Mathematically speaking, it was thought that if one doubled wing area an airplane's lift would quadruple (good)—but to obtain adequate structural strength weight would have to rise eightfold (bad). There would necessarily be a point beyond which every additional square foot of wing surface weighed more than it could lift (very bad).

To multiply an airship's lift by the same amount, however, it was a matter of lengthening *just* the body by the cube root of four (or by about 1.5 times), as a result of which total weight would only quadruple while maintaining the same proportional structural strength.

Owing to the Zeppelins' insuperable dimensional advantages, all Eckener had to do to make a world-beating airship, it followed, was to add some girders to the framework, pump in several more million cubic feet of cheap hydrogen, and soup up the engines a bit.[39]

It was not quite as easy as that, because when it came to airships, politics would *always* get in the way of physics, as Eckener was beginning to realize after the DELAG shutdown, the confiscation of the *Bodensee* and the *Nordstern,* and the end of Colonel Hensley's scheme to acquire the proposed LZ-125. If he wanted to build a transatlantic airship, he was going to have to learn to play the game.

Juan Trippe was reaching the same conclusion by a different route. He didn't care about scholastic disquisitions on theoretical aerodynamics. For him, the problem was quite simple: In the real world, why go to all the trouble to square wing area when he could just square politicians? Then they would allot him profit-protected routes, and if he had those then he could open service to underserved or difficult-to-reach destinations to stimulate demand in other communities for the same. Once people realized the benefits of fast and convenient flights to places they wanted to go, he would just build a bigger airplane to take them and charge accordingly.

The mysterious workings of politics would somehow make the physics work.

Trippe was almost unique in believing that airplanes could double their capacity—and then double it repeatedly without major penalty. Had he not, with but a modicum of effort, increased his passenger load at Long Island Airways from one to two? All he had done, really, was add a bigger engine and carve out a bit more space behind the pilot.

Technology, for Trippe, served politics. Where he had previously failed was in trying to compete in an unregulated environment (i.e., Coney Island) amid falling prices. Trippe wasn't the type to make the same mistake twice. Next time he would play for *position*.

26. The Stolen Horse

AFTER THE ARMISTICE, Japan, a wartime ally of the United States, Britain, and France, had acquired some German possessions in the Pacific as its spoils. These included the isolated Marshall, Caroline, and Mariana Islands, which happened to lie astride the route from the American naval base at Pearl Harbor to its glittering colonial jewel, the Philippines. If war ever broke out between the two rising Pacific empires, then, Japan could blockade the Philippines and use its island outposts to launch attacks against Guam, Hawaii, and other American assets.

It was up to the U.S. Navy's battleships, the mightiest weapons afloat, to destroy the Japanese fleet before that happened. In the vast open spaces of the Pacific, however, finding the enemy was almost impossible without the ability to scout thousands of square miles around. That's where airships came in handy. With their long ranges, low fuel consumption, and ability to see to the horizon, airships would be invaluable in searching for the Imperial Japanese Navy.[1]

To that end, in mid-1921 the navy was intent on becoming the world's number-one airship power, a minor problem being that it didn't have any airships. Previous American attempts to acquire a good one had been frustrated: L-72 had been allocated to the French after the sabotage at Nordholz and Wittmundhafen, and the scheme to commission LZ-125 from Eckener had been scuttled by Air Commodore Masterman and his Inter-Allied Commission.

An obvious solution was to build its own, so the navy opened a naval air station near Lakehurst, New Jersey, where work was set to begin on ZR-1, the USS *Shenandoah* ("Daughter of the Stars"), an airship based on the downed German L-49 in 1917.

Another was to buy one. Luckily, the British had just the thing: R-38, a remaindered wartime airship in so-so condition that the long-suffering Air Ministry was eager to sell for a song. The budget-minded navy quickly signed the contract—an American admiral crowed that the fire sale price amounted "to practically a gift of the ship"—and the British agreed to deliver R-38 to Lakehurst.[2]

That summer of 1921, excitement built as R-38—rechristened ZR-2 by the Americans, with the name still to be decided—was readied for its flight to Lakehurst. The public was as thrilled as the admirals by their airships. The *Shenandoah* was going to be proudly "Made in the U.S.A." while ZR-2, at 2.7 million cubic feet and 700 feet long, was so big, it must be the best.[3]

Fueling the excitement was the news that the navy's airships were not going to be ordinary airships, they were going to be the world's first *helium* airships.

By any definition, this was an astounding development considering that in early 1917, just four years earlier, the entire world's supply of helium had been kept in three small glass flasks—about sufficient to fill a child's birthday balloon—perched on the top shelf of a chemistry laboratory at the University of Kansas.

Until America's entry into the war that April, nobody had known what to do with it, but the realization that the gas might be used in the Allies' airship program turned helium into a strategic natural resource, jealously guarded. The U.S. Bureau of Mines and the Linde Air Products company financed an experimental helium-processing plant, as a result of which extraction costs fell from an eye-popping $6,000 per cubic foot to $1,700. By the Armistice in 1918 the United States had 750 steel cylinders filled with 147,000 cubic feet of helium sitting on the docks in New Orleans ready for export to Europe. That amount would barely fill a small weather balloon, admittedly, but had the war continued production would soon have risen to 50,000 cubic feet per day.

A year later, Major General George Squier (the first military passenger carried on an airplane, by the Wrights in 1908, and the man who would later bless the world with a system of piping music to businesses he called

"Muzak") informed the American Institute of Electrical Engineers that the cost of helium was predicted to plummet to ten *cents* per cubic foot, or $100 per thousand cubic feet, by 1920.[4]

That was still extremely pricey compared to hydrogen—between $2 and $3 per thousand cubic feet—and for the time being the navy could afford just enough to fill *one* of its new airships. Because it was smaller, the 2.1-million-cubic-foot *Shenandoah* was selected as the lucky recipient, with ZR-2 sticking with hydrogen for the next year or two.

IN FRIEDRICHSHAFEN, ECKENER watched the celebrations enviously. To his mind, R-38/ZR-2 was a dated piece of junk, helium or not; Zeppelin could do so much better, if only given the chance, but that was hardly likely given the toxic political environment in Europe. A few months earlier, the Allied powers had accused Germany of backsliding and foot-dragging on the war reparations payments it had agreed to at Versailles. To crack the whip, the French threatened to send troops to occupy the Ruhr Valley industrial region while the British began interpreting the Versailles terms more stringently.

As part of this effort, Masterman and the Inter-Allied Commission clamped down on the Zeppelin Company. After suspending the DELAG, the air commodore persuaded the Conference of Ambassadors, a new body based in Paris responsible for enforcing the peace treaties, to define *any* German airship as being *inherently* "military" in purpose, intention, and effect—and thus specifically banned—if it exceeded one million cubic feet in size.[5] Eckener's prized "loophole" had been efficiently sealed.

The new ruling vaporized Eckener's hopes of ever crossing the Atlantic. The *Bodensee* and the *Nordstern* had been tiny creatures of seven hundred thousand cubic feet, good only for short-range journeys, and Masterman had confiscated them. For an Atlantic route, Eckener would need an airship of at least two million cubic feet, more realistically three, but if he tried building one of that size it would suffer the same fate as its DELAG predecessors. The worst of it was that even as he had the theoretical proof that giant airships were the future, he was powerless to do anything about it.

There was nothing else for it; Eckener had to take drastic action to circumvent Masterman. On June 13, 1921, he wrote to Major Benjamin Foulois, the assistant to the American military attaché in Berlin. His offer: Zeppelin would build a super-giant, brand-new, top-of-the-line airship of 3.5 million cubic feet for free. Was he interested?

Eckener knew that *that* would get his attention. Foulois immediately informed Secretary of State Charles Hughes, who took it to President Warren Harding. Yes, they were very interested indeed. The navy was particularly enthusiastic, rubbing its hands at the prospect of so unexpectedly acquiring a third airship to join ZR-1 and ZR-2.

Eckener outlined the terms. First, the construction cost of the airship ("LZ-126") would be counted against what was still owed the United States in compensation for the two promised airships destroyed by the sabotage at Nordholz. To be absolutely clear, Eckener said, LZ-126 was a "compensation airship" freely and independently negotiated between two partners, not a government-endorsed "reparations airship," in which case it would be part of the Versailles settlement—and *that* would involve the troublesome Masterman and his hostile Inter-Allied Commission.

Since the compensation owed had been assessed as 3.2 million gold marks (the "gold mark" was a theoretical rate of exchange set at the prewar value of 4.25 marks to the dollar) and Eckener calculated the cost of a new airship at 3.56 million gold marks, its construction would more than wipe out the debt.

So eager were the Americans to get their hands on this airship that they didn't stop to ask how Eckener was performing such a feat of financial magic. The trick, which Eckener did not reveal to the audience, was that since the near-bankrupt Zeppelin lacked the money to build an airship for nothing, he had secretly arranged with Walther Rathenau, the minister of reconstruction, to advance some 3 million gold marks to keep the company solvent. At a time when Germany was claiming penury as the cause of its delayed Versailles payments, this arrangement was, to put it charitably, a sleight of hand.

Second, in exchange for LZ-126, the Americans had to pledge to pressure the Conference of Ambassadors to overturn Masterman's ban on large airships so that Eckener could start building his own for the transatlantic route.[6] Without cover from the Conference of Ambassadors, there was a danger that Masterman would ignore legal niceties like the difference between a "compensation airship" and a "reparations airship" and swoop in to confiscate it.

By anyone's definition, Eckener's was a brilliant, if extraordinarily risky, jujitsu move. At a stroke, if all went according to plan, he would get approval to build big airships, save the company from extinction, frustrate the hated Masterman no end, gain powerful American backing, and buy more time to pursue the dream of a transatlantic crossing.

If all went according to plan.

If it didn't, though, and he failed to produce LZ-126, it was over for Zeppelin.

At first, all did go according to plan. Two weeks after writing to Foulois, Washington informed the Conference of Ambassadors that it was willing to accept a Zeppelin in lieu of cash for the losses at Nordholz.

Then it didn't. Masterman bluntly told Myron Herrick, the U.S. emissary, that he would never lift the million-cubic-feet restriction. On August 17, 1921, the Conference of Ambassadors accordingly voted against approving LZ-126.

Eckener had fallen at the first hurdle.

ON AUGUST 24, a week after the ambassadors' vote, R-38/ZR-2 set off on a final trial flight before departing for America. On board were thirty-two British crewmen and officers and seventeen from the U.S. Navy. Flight Lieutenant Archibald Wann, in command, performed some turning tests at 62 mph at low altitude, which so stressed the flawed framework that it crumpled amidships. As the ship bent in two, its structural failure was soon followed by two violent explosions—caused either by sparks or by a fuel fire igniting a very impure and thus very flammable mix of air and hydrogen in the gas cells. The last message sent from the airship came from Wireless Officer Wicks: "Ship broken; falling."[7]

Quite apart from the careless failure to maintain hydrogen purity to reduce the risk of fire, which stunned the Zeppelin experts at Friedrichshafen, R-38's designer, Commander Charles Campbell, in an omission that boggled Dürr, Arnstein, and Jaray, had not bothered to strengthen the skeleton to compensate for the increased aerodynamic loads at low altitude. They had designed their wartime height-climbing Zeppelins—R-38 was essentially a clone based on captured German technology—to operate in the thin air at high altitude; closer to the ground, the stresses on the airship rose precipitously when the vehicle deviated from a straight line to turn or pitch. The inexperienced Wann compounded the error by under-

taking high-speed, high-risk maneuvers that would never have been tolerated in an Eckener-trained Zeppelin captain.

Campbell died in the disaster, along with all but five of his countrymen
(Wann lived but suffered serious injuries). Only a single American survived. The dead were buried with full honors in Westminster Abbey, leaving behind a livid U.S. Navy and an American public suddenly bereft of
its anticipated treasure.

In the United States, the press almost unanimously pointed to "structural weakness"—a euphemism for "British incompetence"—as the cause
of the accident, perhaps explaining why, despite the loss of life, public
support for the airship remained resolute.[8] *Scientific American* commented that the accident may have been "deplorable" but it would "not
prevent airship travel. Deplorable, also, was the loss of the *Titanic*; but
people still travel in steamships and will continue to do so."[9]

The U.S. Navy had by no means lost its faith in airships, but never
again would the Americans trust British airship engineering or construction. The next acquisition would have to be German, Masterman be
damned.

WHEN R-38 EXPLODED on August 24, so too did the Americans. They returned to Paris accusing Masterman of being obstreperous and self-
interestedly mounting a personal crusade against Zeppelin. The U.S.
ambassador to London delivered a note to 10 Downing Street expressing
Washington's "surprise"—a bitingly harsh word in the diplomatic world—
that Britain was preventing the acquisition of LZ-126 in light of the deaths
of so many American personnel aboard one of its own third-rate products.

Caught in this embarrassing position, His Majesty's Government
opened negotiations with Washington. They eventually agreed to meet in
the middle between Eckener's proposed 3.5-million-cubic-foot airship and
Masterman's one-million-cubic-foot restriction. On December 16, 1921,
the Conference of Ambassadors voted to approve a 2.5-million-cubic-foot
airship, though to save face it demanded two conditions.

First, the airship would be used only to "determine the feasibility of
rigid airships for commercial purposes"; and second, this would be a "special exception" to the rule forbidding Zeppelin construction. There was to
be just this one non-military airship, no more, after which Eckener would

either liquidate the business or start selling pots and pans, as Colsman had once proposed. Three weeks later, on January 6, Masterman was ordered to approve the deal, notwithstanding his fierce objections.[10]

Neither Eckener nor the navy had any real intention of adhering to these conditions. LZ-126 was being built for the American military, and it was hardly likely that Zeppelin, the navy's co-conspirator, would voluntarily close up shop upon finishing it. As the very Hungarian proverb goes, if you trust somebody, then you can steal horses together, or in this case, an airship.

SOON AFTERWARD, THE navy's representatives arrived in Friedrichshafen to establish the "Office of Inspector of Naval Aircraft" to supervise construction of LZ-126—now dubbed ZR-3—which they expected Zeppelin to finish within fifteen months.[11]

Led by Lieutenant Commander Garland Fulton, the Americans quickly ran into trouble. At the welcoming party, Eckener, Gemmingen, Dürr, and Karl Maybach (in charge of developing the engines) were "most courteous and pleasant," Fulton reported home, but "it soon became obvious that they did not propose to give us the slightest bit of information in regard to the proposed airship."

Over the coming weeks, Fulton and Commander Ralph Weyerbacher, his colleague, also kept hearing mysterious rumors of a "Dr. Karl Arnstein," considered, said Weyerbacher, "the brains of the outfit." Yet "the Zep people took pains to hide this," and secreted him away. (Jaray, the aerodynamicist, was sick with tuberculosis and about to leave the company to move to Switzerland.)[12]

The reasons for the Zeppeliners' curious behavior soon became clear. To the Germans, the U.S. Navy had bought an airship—but *not* the arcane know-how that created it and made it work. That information was the exclusive property of the Zeppelin Company and must remain a closely guarded secret—or else it would seem as if Eckener had sold his customer the keys to the kingdom. The Americans, conversely, believed they owned the complete package, including details of German design methods, aerodynamic and structural calculations, and stressing processes.

Legally speaking, the Germans were right. The contract stated specifically that the inspectors must have full access to the manufacturing facilities for the structure, engines, and other equipment, but it said nothing about being given the crucial data.

On the other hand, the contract also stipulated that the airship must be finished within fifteen months and "operational in every respect," including engines, mechanical and signals equipment, the control car, and even the tools commonly carried aboard. If Zeppelin failed "to comply with one of the provisions of this contract," the navy could demand full repayment of more than 3 million gold marks in cash—which, of course, Eckener would no longer have after building the airship.[13]

Confronted by a smokescreen and a stonewall, poor Fulton tried to figure out who was really in charge and what on earth was going on at Zeppelin. In a lengthy report for his boss, a bemused Rear Admiral William Moffett, the chief of the U.S. Navy's Bureau of Aeronautics, Fulton described the existence of three groups: "The 'Colsman faction' and the 'Eckener faction' with possibly a third 'Dürr faction.' The Eckener faction is in the saddle and the Dürr faction pulls very well with it."

With Colsman sidelined by Eckener, Fulton only had the Eckener-Dürr alliance to worry about. The latter was certainly the weaker, but also more troublesome, partner. The hermitlike Dürr "has never traveled and has very narrow views. Believes implicitly in German theory of absolute secrecy about work; also in building just about as he pleases." Explaining his resistance, thought Fulton, was that "he has never quite accepted the ZR-3 contracts Eckener signed and it was rumored he declined to affix his name to a contract that 'sold' the Zeppelin 'birthright.'" The Zeppelin captain and off-and-on Eckener rival, Ernst Lehmann, he added, was siding with Dürr in wanting to give the Americans as little assistance as possible.

Fulton fixed on Eckener as the easiest to deal with. "Dr. Eckener is a very fine type and the most broad-minded and reasonable man in the organization," he judged. "He creates an excellent impression and is very highly regarded throughout Germany. He is an excellent negotiator—sizes up a situation and makes decisions quickly."

Weyerbacher was more hostile toward Eckener, pointing out that he was the same "just plain rotten" man who had so ruthlessly "stabbed Colsman in the back and tossed him over the side" to seize control of Zeppelin. As an airshipman, Eckener was admittedly "a shrewd, cautious operator," and the same could be said for his negotiating skills.

Fulton would soon come to agree that Eckener was more cunning than he at first appeared, but he understood that Eckener had acted against Colsman to save the company at a desperate time. Fulton also began to appreciate the pressure Eckener was under to maintain the facade of mutual support with Dürr and Lehmann—called a "conceited little man" by Eckener—in order to keep the Zeppelin flying.[14]

Once the scales had fallen from Fulton's eyes, he had what he called a "very frank conversation" with Eckener to impress upon him the precariousness of his seemingly strong position. He acknowledged that, yes, the contract excluded technical data, but reminded him that he had full power to cancel it if Zeppelin failed to deliver on each and every one of its other provisions. It probably wouldn't be too difficult to find somewhere Zeppelin had slipped up, and if the Americans walked, by the way, Eckener could look forward to the tender ministrations of Masterman.

In short, Eckener was within his legal rights to withhold information, but it would be . . . *strategically* unwise to choose to insist on it. With a start, Eckener realized that Fulton was not quite so naive as he had originally thought, and promised to persuade his colleagues to become more helpful.

Aside from shuffling the never-seen Dr. Arnstein out from his lair in November to meet Fulton to show goodwill, Eckener initially had little success.[15] Dürr and Lehmann continued to be uncooperative.

Dürr insisted that he was willing to sacrifice the contract if it meant keeping the count's birthright in German hands, somehow not realizing that without a contract there would be no birthright to keep. "I have encountered," wrote one of the exasperated navy observers, "that old bird in the conference room and have noticed that for unadulterated, square-headed resistance, he wears a gold medal, or if he don't, they ought to give him one."[16]

For his part, Lehmann stalled, then repeatedly submitted inaccurate plans while claiming to be protecting "vital business secrets." Fulton, who'd earlier been impressed by Lehmann's seeming helpfulness, revised his earlier judgment and now saw him as two-faced, underhanded, and more than willing to betray Eckener if ever given the chance.[17]

LEHMANN INDEED MAY have been those things, but he had his own reasons to be obstreperous. In this period, Germany's relations with France reached their lowest note over Berlin's continuing defaults on the reparations payments due under the Versailles Treaty. From the French point of view, following negotiations the huge sum had already been nearly halved, and yet still the Germans were loath to pay what they themselves had agreed they owed. In thirty-six months, for instance, Germany had not sent its promised coal delivery thirty-four times.

In mid-January 1923, French troops marched into the Ruhr region and occupied the mainstay of German coal, steel, and iron production. With strikes and civil unrest convulsing Germany in reaction, a coalition of right-wing parties emerged to present a united front against Versailles, and the nationalist-minded Lehmann made common cause in opposing any compromise with the Allies, who he believed were intent on destroying Germany. Like Dürr, but with a different motive, he found it hard to accept selling the Zeppelin Company to the Americans.

Fulton had another one of his straight-shooting talks with Eckener, passing on a remark made by Admiral Moffett that Americans had a prejudice against "slippery contractors" who appeared to be evading their responsibilities.[18] Eckener was vexed at the insult but was conciliatory. At some length, he explained to Fulton how best to deal with his people. All the problems stemmed from a culture clash, Eckener said.

As Fulton reported to Moffett in Washington, first, "it should be realized that German psychology, methods, habits, and language are all different from our own. Secondly, that the Zeppelin Company is a peculiar organization. They always enjoyed very special privileges from the Government, their only previous customer, and have therefore had a free hand in regard to building airships. This independence coupled with secrecy carried to an extraordinary length even for Germany . . . has created a state of mind which requires a process of education and some time to dispel."

Eckener, meanwhile, emphasized the need for trust: Dürr was admittedly somewhat eccentric, but what he and the others feared was that the Americans were going to steal their stolen horse. They would soak the company for its data and then cancel the contract, "leaving them in a hole. They have been afraid of this all along."

Fulton satisfied Eckener that he had no intention of thieving Zeppelin secrets and offered to pacify Dürr by reducing the number of requests for information and visitors to his triple-locked research office. Eckener in turn promised that he would vouch for Fulton's honor to get Dürr on board.

The more politicized Lehmann would be a tougher proposition, but Eckener assured Fulton that within a few weeks it wouldn't matter very much what he thought. Gemmingen was stepping away from his position of chairman (he would die the following year), and the Zeppelin Foundation board was going to be reorganized with Eckener serving as chairman.

In fact, once installed, Eckener reorganized the corporate structure. The Zeppelin Foundation remained the Zeppelin Company's sole shareholder, but now it became also the majority owner of the autonomous subsidiaries run by Colsman. From his chairman's seat, Eckener was able to direct the activities of the entire enterprise toward a single goal: building airships.

As magnifico and prince of the Zeppelin empire, Eckener didn't even have to bother holding board meetings anymore, and there wouldn't be one for years. In one swift move, Eckener had defenestrated both Lehmann and Colsman, neither any longer having a say in company policy-making. Both, predictably, took it badly: Lehmann's rivalry with Eckener only sharpened, whereas Colsman avoided visiting the Zeppelin offices if he knew his new boss would be there.[19]

By mid-April 1923, Zeppelin and the U.S. Navy were happy partners, and work on the airship, which had been delayed due to the tussles over technical data, picked up steam.[20] In August, Fulton could report that, aside from the fins, the hull was complete and the cover had been twice coated in the clear lacquer known as "dope" to render the skin waterproof and airtight. (The third and fourth coats would have aluminum powder added, giving the airship its distinctive silvery glow.)[21]

By November, though, months past the fifteen-month deadline with no announcements from the U.S. Navy or Zeppelin, newspapers speculated as to the mysterious cause of the delays. American officials at Friedrichs-hafen, reported *The New York Times*, remained tight-lipped, and "access to the shed or any discussion outside is taboo." Every entrance was under lock and key, and visitors were required to show identification constantly. A correspondent found the naval inspectors there "so guarded in their remarks within hearing of alien ears and so impervious to quizzes that one might as well ply questions at the Sphinx." The most popular rumor was that workmen at the plant referred to the airship as "our meal bag" and were dragging out construction to stretch their paychecks.[22]

It's certainly possible that there was some procrastination on the job, but the truth of the matter was that ZR-3's structure was finished, but its engines weren't. Rather late in the day, Fulton had discovered that if Dürr had been a handful, Karl Maybach was an armful. Intense and prim, he was so secretive that even Lehmann complained about him. After several more months of delays, Fulton finally succeeded in entering his realm to speak with the reclusive genius, whose attitude was "Leave me alone. I know what I am doing. You don't, and I object to your making suggestions."[23]

To be fair, this man with no master used to say the same thing to Eckener, but now there was nothing anyone could do but wait until Maybach finished the new engine design he was working on.

A s MAYBACH TWEAKED and twiddled in his workshop, Eckener was painfully aware that he had to plan something big to ensure Zeppelin's survival. Uppermost in his mind was that in its vote to approve the construction of ZR-3, the Conference of Ambassadors had insisted that it was a "special exception" to the rule banning German airships. The U.S. Navy was therefore a one-time client, leaving Zeppelin without a future once ZR-3 was delivered.

Possible salvation came in the form of an inquiry from Paul Litchfield, a vice president of the Goodyear Tire and Rubber Company in Akron, Ohio. Litchfield had excellent navy contacts, and the navy, its appetite whetted by the ZR-3 project, wanted more sea-scouting airships—which had to be built in America by an American company to circumvent the "special exception" rule the Conference of Ambassadors demanded.

Litchfield and company chairman Edward Wilmer traveled incognito to Friedrichshafen in September 1923, when ZR-3 had been all but completed (aside from Maybach's engines), to negotiate a contract with Eckener. They proposed that a new U.S.-based company, Goodyear-Zeppelin Corporation, would build airships for the navy in Ohio at Wingfoot Lake Aviation Field with Zeppelin-provided experts on loan. (There was nothing in the Versailles Treaty forbidding German personnel from helping other countries build airships.) Once flush with navy cash, Goodyear-Zeppelin would develop transcontinental routes across the United States.

Warned by Eckener to be sensitive to German concerns, the ones that had so recently put Dürr's and Lehmann's noses out of joint, Litchfield and Wilmer made sure to clarify that while Zeppelin would be allocated one-third ownership in the American business, the Americans would *not* be involved in Germany, where the company would remain independent and free to pursue its own transatlantic ambitions, if ever Eckener managed to squirm free of the Versailles restrictions.

There was nevertheless a minor breach of etiquette during the discussions. Litchfield at first proposed having Dürr, of all people, serve as chief engineer in Akron, but he was soon disabused of the notion when Dürr refused to ever leave homey Württemberg, refused to shake hands, and refused to learn a word of English. The designer Karl Arnstein, his next choice, quickly agreed. Anti-Semitism was on the rise in Germany, and the Land of Plenty promised a bounty of food, silk stockings, a car, and fur-lined boots—all of which had been either unavailable or unaffordable to Germans in these hard years. (Even Eckener, the head of a major conglomerate, couldn't afford an automobile and pedaled around on a bicycle.) Lehmann, who spoke excellent English, was also eager to go, probably to get away from Eckener.[1]

Helping to speed things along was that Eckener was uninterested in working for the U.S. Navy once ZR-3 was handed over. He'd had enough of military airships during the war, and all he wanted was to try to turn his dream of building a transatlantic airship into reality. As far as Eckener was concerned, Goodyear-Zeppelin was important only with respect to its cementing American-German friendship and bringing in hard foreign currency when hyperinflation—soon to reach 4 trillion marks to the dollar—was ravaging the economy.

Time, then, was of the essence, but the longer Maybach dithered on the new engine, the more likely it became that Goodyear would lose interest in the deal, or perhaps just try to build its own airship. In the new year of 1924, Eckener brought the hammer down and ordered Maybach to demonstrate what he had. It was not good. During testing, the engines malfunctioned multiple times and Maybach had to be "persuaded" to bring in two engineering professors to help him rectify the problems. On August 4, 1924, nearly a year after the contracted deadline, Maybach deigned to announce that they were ready.[2]

It was worth the wait, for Maybach demonstrated a masterpiece, the VL-1. Combined, the five VL-1s intended for ZR-3 produced 2,000 horsepower, nearly 300 more than *seven* of the mightiest of their wartime predecessors put together.

By the end of the month, the VL-1s had been installed and Eckener prepared to test-fly Friedrichshafen's finest product. He couldn't afford to waste any time, and the subsequent trials were considerably more abbreviated than was customary. There were no major technical issues, thankfully, and Eckener deftly exercised his publicity skills by releasing "exclusives" to popular magazines and inviting movie cameramen and newspaper correspondents along for the rides.[3]

As ZR-3 sailed over a succession of cities or visited Count von Zeppelin's grave, Germans, for the first time in what seemed like forever, had something to be proud of: their own "swift-winged silver tenant of the air, reflecting the bright sunbeams from her glossy sides, and wending her course proudly and confidently whither she willed," as the onboard correspondent for the *Frankfurter Zeitung* rapturously put it.[4] Tears streamed down spectators' faces, traffic stopped, people climbed atop their roofs for a better look, and there were spontaneous outbreaks of the national anthem: Nothing like it had been witnessed since the storied days of the count's flights to Switzerland and the miracle of Echterdingen.

Eckener was very clever in trying to appeal to all Germans, whatever their class, age, or political stripe, by varying how he emphasized ZR-3's benefits in the media. To appeal to the powerful Socialist movement, he stressed how the company employed thousands of workers and was a shining example of the virtue of government subsidies. For business-oriented newspapers, Zeppelin was a beacon of profitability, innovation, and enterprise that had been brought back from near bankruptcy. For older conservatives, the airship was a living link to the grandeur and majesty of imperial Germany, a return to the good old days before the "yoke" of Versailles was harnessed to their necks. Younger liberals in turn were attracted to Eckener's message that ZR-3 symbolized modernity and its marriage of science and technology in the service of peace and mankind. For Germans as a whole, the Zeppelin relieved their war guilt and represented pride in the country's products.

He was less successful in gaining the support of the more extreme, and noisier, factions of the left and the right. There were sour notes from the Communists, who claimed ZR-3 was part of a sinister capitalist-imperialist conspiracy to steal the labor of the proletariat. Ultranationalists accused Eckener of treachery for handing over German assets to the United States. "What is it with this airship built in Germany by German workers and engineers, paid for with German money, but which belongs to America?" grumbled one curmudgeonly Berlin columnist.

From there it was but a small step to attacking Karl Arnstein, who was charged by the usual anti-Semites with being a member of an Eckener-backed Jewish cabal at Zeppelin apparently intent on selling out Germany. That probably no one else at the company was a Jew—even the Jewish Telegraphic Agency believed he was "the only Jew in Friedrichshafen"— and that Arnstein was a designer, not a board executive, did not seem to faze his detractors. The viciousness of the accusations only affirmed Arnstein's decision to depart Germany to join Goodyear in Akron.

In the fevered charges of Jewish plotting and German backstabbing were shades of the fate that had awaited the Jewish politician-industrialist Walther Rathenau, who had approved the 3-million gold mark advance to build ZR-3.

He had been assassinated by the Organization Consul, a secretive ultra-nationalist murder gang so deranged even the nascent Nazi Party kept its distance. Rathenau was only the most prominent of hundreds of German "traitors" executed by the Organization Consul and its kin, so when Eckener's own life was threatened by these blood-and-thunder zealots, he took it seriously. A sensible decision it proved to be when a rifle-armed student who'd sworn to assassinate him was arrested.[5]

More unexpectedly, Eckener was running into opposition from within the government. One might have assumed that Berlin, having paid for it, would have been excited by the prospect of ZR-3, but the Foreign Ministry was raising objections.

The immediate cause stemmed from Germany's fortunes turning earlier that summer when the ongoing resistance in the Ruhr against the French occupation had persuaded Britain and the United States that the reparations system, as it stood, was unworkable. Paris did not agree, and the franc came under heavy assault from Wall Street and London banks as investors bet against the French government's surviving the reparations crisis (they were right). Soon afterward, Édouard Herriot and the leftish Radical Party came to power and, under Anglo-American pressure, accepted the need to restructure Germany's debt.

An international committee chaired by Charles Dawes, the former head of the United States Bureau of the Budget, convened to calm the roiling economic turmoil. The resulting Dawes Plan was implemented in

September, coinciding with Eckener's test flights. Among other things, the plan guaranteed American and British loans to allow Germany to make reparations payments of just 1 billion of the new, post-inflation marks in the first year, rising to 2.5 billion marks annually by the fifth year. French troops would in the meantime vacate the Ruhr to help German industry get back on its feet and make the higher payments demanded by Paris once the five years were up.

From German foreign minister Gustav Stresemann's perspective, it was critical not to antagonize the French in any way for fear they might not leave the Ruhr. Stresemann's main worry was what would happen if a German-crewed ZR-3 overflew France, where memories of the once-feared Zeppelins were still raw, on its way to America.

As there was no way to stop the airship from taking off—American fury at being denied could scarcely be imagined—Stresemann at this point had a quiet word with Eckener, who had announced that he would captain ZR-3.

Under no circumstances was he to provoke the French by flying over the Ruhr, and neither would any photographs be taken of French fortresses or military bases. He was to keep strictly to a preapproved flight path over French territory and must not fly at night so that he could be observed from the ground at all times.[6]

Lastly, to avoid any appearance of official backing of what might be seen as a political statement at a sensitive moment, the government would not underwrite the flight to America. Eckener was shocked. If Berlin didn't insure ZR-3, then no else would, either. And without insurance against damage or loss, Eckener was at the mercy of the contract he had signed with the U.S. Navy, which, burned by the R-38 disaster and the loss of its deposit, had stipulated that "until accepted delivery at Lakehurst [naval base] the company is to assume full liability for the airship."[7] This meant that if ZR-3 were wrecked for *any* reason—an ill-judged turn, an unexpected gust, a gas leak, a weak girder, a saboteur, even an act of God—then the United States would come calling for its compensation money.

Eckener took the biggest risk of his life and, without consulting anyone, "bet everything on one card," as he said, by pledging the entire Zeppelin Company and its subsidiaries as surety for the endeavor. If he failed, he said, Zeppelin—its patents, its assets, its facilities—would be owned by its creditors, the world would lose its "trust in airship travel," and he would be "out—done with work—forever!"[8]

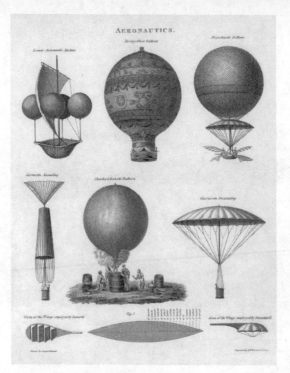

Early balloons' endless forms most wonderful—
yet mostly impracticable
GETTY IMAGES

A young Count von Zeppelin (second from right),
with some Union friends during the Civil War
LIBRARY OF CONGRESS

The aeronaut John Steiner, who first
sparked Zeppelin's interest in airships
and the secrets of aerial navigation
COURTESY OF MILITARY IMAGES AND SCOTT
VEZEAU

The first Zeppelin, LZ-1, on its maiden flight, July 2, 1900

The stern of LZ-4, with its complicated, if much improved, control system of fins, elevators, and rudders

Hugo Eckener (right) on board the *Schwaben* with an elderly Count von Zeppelin

Eckener and his reclusive engineer Ludwig Dürr, the latter looking not altogether happy about being photographed

The wreck of LZ-4, immediately before "the Miracle at Echterdingen"

The DELAG airship *Viktoria Luise* salutes a sailing regatta on the Bodensee.

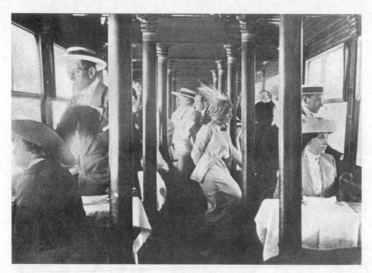

The pleasures of early air travel aboard Eckener's DELAG.

Iit Zeppelin kühn voran!

"With Zeppelin Boldly Forward!"—a fanciful depiction of a Zeppelin leading the troops in the Great War. Any airship attempting to do this would quickly have been shot down.

Eckener's graceful, modest *Bodensee* of 1919

The front section of the U.S. Navy's *Shenandoah.* Because it used helium, it did not burn, yet the crash still killed fourteen crewmen.

The *Los Angeles* (ZR-3), the airship that saved the Zeppelin Company, dwarfed by the U.S. Navy's *Akron* in the Lakehurst hangar

"Wild Bill" Hopson of the U.S. Air Mail's "Suicide Club." He was killed, as many were, in an airplane crash.

IT WAS TIME for ZR-3 to leave the homeland. Everything—the Goodyear alliance, Zeppelin's survival, an Atlantic passenger service, U.S. Navy support—now hinged on one thing and one thing alone: delivering ZR-3 safely.

Eckener selected his best men to ensure nothing went wrong. He tapped the three most experienced of his wartime and DELAG commanders—Lehmann, Hans-Curt Flemming, and Hans von Schiller—to come with him. Lehmann, whatever his other feelings about Eckener, was dedicated to the cause of Zeppelin, and his skills were formidable. They all understood what they were signing up for. If ZR-3 was destroyed in a storm, or blew up, or crashed on landing, they would all die together, but what was there to live for if the dream had no future?[1]

In public, of course, Eckener betrayed no misgivings as to his chances. He, an inveterate smoker, made sure to be quoted as saying, "The idea of leaving my pipe behind is what is bothering me and not how I will get the ship across."[2] Privately, he was less cocksure. It was not the size of ZR-3 that worried him—it was slightly shorter and a little wider than previous Zeppelin giants—but running out of fuel. With 43 tons of useful lift available, Eckener had to keep crew and ballast to a minimum to make room for the 33 tons of fuel and 2.2 tons of oil he calculated he would need to make all five Maybachs last the entire voyage. If he encountered sustained headwinds, the engines would guzzle so much fuel he'd be left drifting aimlessly across a trackless ocean.[3]

Keeping the weight down was no joke. On the eve of departure, Flemming calibrated the ship and found some mysterious excess poundage. "The vessel is tail heavy," he informed Eckener, who ordered a search, which uncovered two stowaway journalists. He eventually consented to take a canary they had bought to give them a self-created scoop: "The First Bird to Fly the Atlantic as a Passenger!"[4]

At dawn on October 12, the date when Columbus had first sighted land in 1492, which Eckener had cannily chosen to appeal to the Americans, the crew of twenty-eight Germans (nattily attired in new uniforms) and four American naval officers gathered to prepare the ship and wait out the fog. At 7:30 A.M., ZR-3 lifted off, and the gathered crowd spontaneously sang the "Deutschlandlied," better known to many as "Deutschland über Alles." They were "the last words we heard before leaving our native land," remembered Lehmann.[5]

ZR-3 floated "above a shimmering ocean of mist lit by the rising sun, while the chain of the Swiss Alps rose through the mist to the south." Soon they saw the stony hills of Burgundy's vineyards and the level fields of Bordeaux rolling beneath them. By midday, as the crewmen were settling down to a sumptuous lunch of turtle soup, Hungarian goulash, pudding, and coffee, Eckener reached the mouth of the Gironde estuary on the Bay of Biscay and saw the surf "stretching out to right and left like a bright, dividing line between land and sea."[6]

In the late afternoon, ZR-3 made for Cape Ortegal on the northwest coast of Spain. As it grew darker, the local fishermen returning with their catches craned their necks to watch the wandering silver star float overhead. Once they were past the cape at midnight there was nothing but four thousand miles of the boundless gray-blue plain that would determine whether Zeppelin had a future.

Some way out into the Atlantic, they flew over the freighter *City of Boston*, whose crew had been at sea so long they hadn't heard the news. Mystified at the apparition above, they wirelessed: "Who are you? Where are you going?"

After thirty-two hours in the air, and with the ocean strewn with patches of seaweed as big as small islands, ZR-3 reached San Miguel in the Azores, a smidgen less than halfway between Friedrichshafen and New York. There were some low-lying clouds but nothing to worry about as Eckener flew past the 7,500-foot peak of Mount Pico. "A strange spectacle," he exclaimed, "this fantastic mountain isle apparently floating in the air in the midst of the ocean!"

The white-capped waves below signaled that the wind was picking up. Eckener's fuel calculations had been based on a 15-mph headwind, but this one was already driving at them at 22 mph. By sunset, it had risen to 31 mph, eating away nearly half their speed. If it continued, Eckener was alarmed to discover, he would need *seventy* hours to reach New York, but he had less than fifty hours of fuel left in the tanks.

After huddling with Lehmann, Schiller, and Flemming, Eckener decided that the only way to beat the wind was to change course and head northwest to Newfoundland. They would burn more fuel than they'd like getting there, but once in position they could exploit the wind's counterclockwise direction as a powerful tailwind propelling them southwest to Boston.

It was quite a thrill ride off Newfoundland. At one point they hit 105 mph, but after encountering fog banks they had to descend from 5,000 feet almost to sea level to take bearings. As the temperature dropped from 77 to 41 degrees they ran into a storm but emerged with no more damage than a single torn wire. "Heartfelt congratulations on the magnificent achievement of [ZR-3]," Eckener messaged Dürr back in Friedrichshafen, "which tonight easily mastered" the heavy weather. He also sent Maybach a signal thanking him for the wonderful engines, "which already have run over sixty hours without any interruption."

At 10 P.M., Eckener spotted the lights of Halifax, Nova Scotia, and the flight was as good as over. Navigating down the coast from Canada to New York was child's play for the crew of ZR-3. They arrived at 7:15 A.M.

"The tremendous city with its stone forest of heaven-raking skyscrapers made a deep impression on us," wrote Lehmann. "We steered for Long Island, crossed the little towns strewn along the shores, and then followed the course of the East River, and circled the Statue of Liberty while all the sirens in the harbor howled their welcome toward us. Then we cruised over the city, flying here and there over Broadway and the Battery. Shrill factory whistles spat out white clouds of steam and from the roof-tops great crowds of people waved a greeting. In the narrow stone canyons of the streets, omnibuses, trolley-cars, and automobiles stopped, the passengers rushing out to stare raptly into the sky. The sun rose and painted a halo around the 'Queen of the Air' as the Americans called the Zeppelin."[7]

After several more leisurely circles over Manhattan and the Statue of Liberty, Eckener headed for Lakehurst. At 9:37 A.M. on October 15, 1924, after eighty-one hours in the air, Eckener landed ZR-3. When he declared,

"A new world's record—5,060 miles of continuous flight," he was handed a most important slip of paper—an official receipt for the airship's safe delivery from the U.S. Navy—and the crew was surrounded by a mob of well-wishers and newsmen, souvenir hunters, and autograph collectors.[8]

Some German mechanics had brought a large brown teddy bear with them and posed with it for the cameras; they charged five dollars for the privilege, as they needed the foreign currency. Every man aboard was given a carton of American cigarettes, whose fine Virginia tobacco was greatly prized over the rough, austerity-imposed roll-ups they were used to.[9]

Just as important to Eckener was the oohing and ahing over ZR-3's accommodations, installed as part of the ruse to claim that the airship was intended only for commercial purposes, as demanded by the Conference of Ambassadors. Eckener had welcomed the opportunity to show off his concept for a transatlantic passenger airship, and the furnishings, which did not disappoint, provided a mountain of free advertising.

There were five compartments resembling first-class railway sleepers, each equipped with two velvet sofas (convertible into beds) and two fold-out berths above. Electrical lights, a curtained window, a vase of fresh flowers, and wood paneling completed the picture of luxury in the air for twenty lucky passengers. Two lavatories were down the hall, and fine cuisine (heated by "the white-hot exhaust pipes") was freely available. Nothing like it had ever been seen.[10]

In a telegram, President Calvin Coolidge congratulated Eckener on a "trail-blazing exploit [that had] proved as never before the ability of lighter-than-air craft to make long-distance flights with considerable quantities of freight and passengers."[11]

THE SCALE AND ambition of his astounding achievement entitled Eckener to the honor of a New York ticker-tape parade, after which he embarked on an exhausting series of celebrations.[12]

His first stop was to give a speech at the Capitol Theatre, where nearly a quarter of the rapturous audience were German-Americans. Before the event, the orchestra played Wagner to accompany a short film extolling ZR-3. Eckener, who disliked the composer's politics and listened instead to the more internationalist music of Beethoven, kept diplomatically silent but later murmured that "the older one gets, the more one forsakes Wagner in favor of Bach, Beethoven, and Mozart."

In his address, Eckener expressed the hope that Germany and America would forever be friends and was greeted by "stormy applause" when the orchestra played the German national anthem. It was the first time it had been publicly performed in the United States since before the war. The audience rose to sing along and, some years later, Eckener patriotically recalled, "my eyes clouded over, and to this day I cannot think of the occasion without tears coming to my eyes. As a German, I sensed how the national anthem once again paid tribute to me and my people. I left the theater as I were in a dream." It was as if the ice had been broken between the two nations.[13]

After a visit to Philadelphia, he took the train to Washington and was escorted to the Department of the Navy to meet its airship experts. The navy secretary, Curtis Wilbur, announced that ZR-3 would be christened *Los Angeles* "because it came to us like an angel of peace."

Afterward, Eckener was whisked to the White House to see Coolidge, then back to New York, where he met the mayor and was given the Freedom of the City. Eckener was then rushed to the Polo Grounds stadium in upper Manhattan to watch a soccer game with seventy-five thousand spectators. "The most excruciating thing I had ever experienced" was when Eckener had to walk around the field as the crowds cheered.[14]

After that, it was all business. Eckener, joined by Lehmann, headed out west. By now, Eckener was genuinely famous and complained that he couldn't enjoy a minute to himself on trains or in restaurants without people shaking his hand and introducing themselves. His celebrity also had the benefit of opening doors, all in the service of Zeppelin. In Detroit, Henry Ford gave the Germans a tour of his workshops. He told Eckener, "The next time you come to Detroit, you should bring your airship with you. I might be impressed into taking a ride with you."[15]

Lehmann and Eckener were getting along, but after too much time together their always tense relationship was beginning to chafe. Lehmann was jealous of Eckener's fame and his own delegated role as number two. He especially disliked how at the unending succession of dinners and toasts, it was always Eckener who, as the main attraction, spoke first and left it to Lehmann, as if he were some kind of flunky, to translate. In his own speeches, Lehmann then had "to improvise my own comments so as not to repeat what Eckener had just said." It quickly became an annoyance.[16]

For Lehmann, Akron couldn't come fast enough. That was the location of their most important meeting, with Zeppelin's Goodyear partner, Paul

Litchfield. From Goodyear's point of view, there could have been no better start to the relationship than the success of ZR-3, and Litchfield looked forward to receiving a cascade of navy orders for airships. When the others left, Lehmann, with a great sigh of relief, stayed behind in Akron to help start the new company. His first task was to prepare to welcome Arnstein and a dozen Zeppelin engineers and designers, instantly dubbed the Apostles.[17]

In Germany, when news came through that ZR-3 had landed at Lakehurst, thousands gathered in public squares to sing the national anthem, students from the various university fraternities marched with flags, and there were band-led processions to nearby war memorials.[18]

In Friedrichshafen, cannons boomed 126 times (LZ-126 being the German designation for ZR-3), and church bells joyously pealed. Residents held a torchlight parade to the statue of Count von Zeppelin and then made their way to the homes of Arnstein and Dürr to give them "tumultuous ovations." Dürr, as curt as ever, dourly murmured a few words of thanks and closed the door, but Arnstein more loquaciously handed reporters the inspirational quotes they wanted: "All of us . . . who saw the work rise from its beginnings, never even for a moment doubted that our Zeppelin would succeed in crossing the ocean. It was a piece of ourselves that we set free into the air."[19]

Politically, though, the situation was much more complicated than in America. The Socialist daily *Vorwärts* congratulated Eckener for using "German engineers and German workers, with German machines and German tools" to build his airship: The more Germans Zeppelin employed, all the better for the labor movement.[20]

But, as Eckener's wife, Johanna, told him, the "completely right-wing-oriented papers aren't joining in the general elation."[21] If older conservatives retained their soft spot for Zeppelins, seeing in them nostalgic echoes of the vanished glory of the kaiser's Germany, their younger and more zealous successors were more inspired by the dynamic, driven Fascism of Mussolini, who had recently led the March on Rome, when his Blackshirts seized power over Italy.

Mussolini, a self-described "aviation fanatic," had learned how to fly and proclaimed that pilots were natural Fascists. To govern a Fascist country, one must be an aviator in practice because "life must be risked and risked daily, continually, demonstrating that one is ready to throw it away when necessary."[22]

Considering that Eckener emphasized his airships' sedateness and safety, one can see how a movement attracted by the speed and violence of airplanes—"we'll cut off their heads with our propellers," threatened Mussolini of those who stood in the way of Fascism's progress—might have dismissed Zeppelins as quaint relics with no place in the coming age of revolutionary might.

To that end, Hermann Göring, a former fighter pilot who had recently participated in an abortive coup (the Beer Hall Putsch), said that the rising National Socialist Party would one day "recapture the German Empire" by building fleets of military airplanes.[23] He made no mention of Eckener's airship or its inconsequential, bourgeois flight to America.

PARTLY EXPLAINING THE nearly universal ecstasy that had greeted his arrival in America was Eckener's auspicious timing. A week before he touched down, the U.S. Navy's American-built *Shenandoah* (ZR-1) had, in a lovely complement to Eckener's transatlantic voyage, made headlines by beginning a transcontinental one.[24]

Under Lieutenant Commander Zachary Lansdowne, *Shenandoah*, proudly emblazoned with "U.S. Navy" painted amidships and a red, white, and blue star-in-a-circle on the rudder, had departed Lakehurst on October 7 and headed for a cruise down Pennsylvania Avenue in Washington, D.C., and thence on a tour of the South before swinging toward Fort Worth, Texas, on its way to the naval air station at San Diego, culminating in a visit to Seattle. *Shenandoah* arrived back in Lakehurst on October 25.[25]

Eckener sent his congratulations to Lansdowne for his excellent, *almost* German, mastery of the arcane secrets of static lift, dynamic lift, superheating, supercooling, weather forecasting, navigation, and loading and weight distribution that had made the voyage a success. Privately, though, he had mixed feelings about *Shenandoah*'s most radical innovation: the use of helium instead of hydrogen.

In the previous five years, the price of helium had fallen to $55 per thousand cubic feet, but it remained dauntingly high compared to that of hydrogen.[26] Maybe the rich Americans could afford to throw money away, but what worried Eckener more was the performance hit airships took from what he regarded, despite its immunity to fire, as an inferior substitute for hydrogen. It was helium's "heaviness" compared to hydrogen that was the main problem: *Shenandoah* contained roughly the same gas capacity as a 1918 Zeppelin but enjoyed less than half its useful lift. The disparity was staggering.

Helium's reduced useful lift had the knock-on effect—a fatal one in Eckener's eyes—of cutting an airship's cruising range by between 30 and 40 percent because, once you included the weight of the crew, machinery, engines, and infrastructure, less fuel could be carried, and less fuel, of course, meant more frequent refueling and an inability to make the giant oceanic leaps he wanted. Indeed, the primary reason that Eckener could bridge the Atlantic in a single bound while Lansdowne had had to stop to refuel in Fort Worth and San Diego was that Eckener used hydrogen rather than helium.

If that weren't already enough of a negative, the logistics of using helium were formidable. To transport millions of cubic feet of fresh helium from the Texas fields to Lakehurst and the West Coast required specially built freight trains bearing giant steel canisters. The charge for subsequently shipping the helium to Germany would have been astronomical.

Leakage and loss, as well, added to the headache of using helium. Helium unavoidably trickled out through valves and loose connections during repurification, transport, storage, and refilling. Even when the airship was docked in the hangar, it leaked: The *Shenandoah* lost 150,000 cubic feet of helium *each month* doing nothing. If you included normal flight operations—Lansdowne had blown off 640,000 cubic feet, more than a third of *Shenandoah*'s capacity, on his trip—it was estimated that every year the navy would have to replace at the very least one and a half times *Shenandoah*'s total ship volume.[27] Hydrogen leaked, too, but it was cheap and easy to replace and was less prone to wastage as it was manufactured and stored locally.

For a time, Eckener weighed whether to switch to helium anyway, but soon decided to rely on hydrogen. His crews were accustomed to handling it, the gas was perfectly safe in experienced hands, it was economical, it yielded high useful lift, and, most important, it bestowed the priceless advantage of immensely long range on his airships.

AS OF OCTOBER 1924, then, Eckener had not one, but two, proofs of the airship's gleaming future. A better advertisement for the coming supremacy of the Zeppelin Company's products could scarcely be imagined than the twin triumphs of ZR-3 and *Shenandoah*.

In a lengthy interview with the American press, Eckener boasted that in a few years customers would be able to travel by airship from Hamburg to New York in the same amount of time as taking the train from New York to Chicago (twenty-five hours) and in the process lop four days off a sea voyage. While the plan was for Zeppelin to reserve the Atlantic route for itself, Goodyear-Zeppelin would handle the transcontinental ones. Its airships, he claimed, would cut the New York–Los Angeles route to a third of the time a train needed.[28]

Eckener's vision seemed so enticing, so futuristic, so *inevitable*, because it appeared precisely when America had reached the very nadir of the airplane business.

In that same year, 1924, the only "airlines," to put it charitably, existing in the entire United States were a small seasonal service to Catalina Island, two tiny mail contractors in Puget Sound and the Mississippi Delta, and an experimental mail run in Alaska, which after a single round-trip to prove that airplanes were faster than dog teams gave up the ghost.[29]

Compared to airships, airplanes were laughably inept when it came to long-range flying. An around-the-world attempt mounted by the U.S. Army Air Service using four grandiosely named "Douglas World Cruisers"— single-engine, two-seater biplanes—required 175 days and 69 stops to complete the circuit, with two of the airplanes crashing on the way.[30] In contrast, Eckener's nonstop eighty-one-hour cruise across the Atlantic seemed effortless.

With good reason, perhaps, Eckener was confident that long-haul passenger airplanes were a dead end—and he wasn't afraid to say it. François Nitschke, a trainee pilot, later recalled that Eckener visited his school at this time to give a lecture, during which he told them the harsh truth that none of them would have a job if they stuck to airplanes. "The future of air transportation, especially transatlantic air transportation," warned Eckener, "rested solely with the airship."[31]

29. Annus Horribilis

IF 1924 HAD been Eckener's annus mirabilis, 1925 would be a year of unrelenting disappointment.

With his stock riding high, Eckener wanted to begin work as soon as possible on a new Zeppelin, one that would dwarf ZR-3 and finally initiate true transatlantic service. The problem was, the company had no money to build one. The twenty-fifth anniversary of the count's first flight, however, was coming up in the summer, and Eckener realized all the attention would provide a perfect opportunity to launch a fund-raising campaign using the Zeppelin as a symbol of German pride.

He immediately ran into trouble with the foreign minister, Gustav Stresemann, who had the temerity, as Eckener saw it, to ask him to postpone any celebration. Stresemann was then in the midst of delicate negotiations to end the French occupation of the Ruhr. The last thing he wanted was for Eckener to be stirring up memories of the war with what might be seen by the Allies as nationalistic calls to arms.[1]

During a stormy meeting with Stresemann, Eckener angrily refused to postpone anything, to which Stresemann threatened that he would not attend the event, followed by Eckener disinviting him. The meeting ended a moment later.[2]

At the celebrations in July, just a few minor functionaries attended, but the big news was Eckener's announcement of what he called the Zeppelin-Eckener Spende des Deutsches Volkes (the Zeppelin-Eckener Fund of the German People), which was quickly abbreviated to the Eckener Spende.[3]

Eckener rousingly reminded Germans of Zeppelin's heritage and prospects for the future, and embarked on a national publicity tour in which he gave hundreds of speeches, sometimes six a day. Schoolchildren went door to door shaking their collection tins, volunteers sold postcards, and foremen made the rounds in factories on Fridays to coincide with payday.[4]

In naming the fund after himself, as if *he* were the main selling point, Eckener had made the mistake of believing his own adulatory press in America. The likening of him to Columbus, the White House handshakes, the talk of airships dominating the world, all of it had gone to his head, and he was about to learn a sharp lesson in humility.

A huge amount of money certainly flowed into Zeppelin's coffers, about 2.5 million marks—but Eckener's announced goal had been 7 million. Several factors contributed to the shortfall, but the primary one was that this wasn't the good old days, when the count more or less jerry-rigged his small airships, his staff worked for a pittance, kings were tapped for gifts, and noble estates could be remortgaged to pay a few bills. In the modern era, Zeppelins were simply too expensive to build, maintain, and operate without major government and corporate financial support—at least not until Eckener had a large fleet in service that could exploit economies of scale and leverage his monopoly power to turn a profit.

It fell to the always annoying Colsman to point out that a money-making airline would need at least 70 *million* marks of start-up capital to pay for six airships and nine hangars.[5] Yet Eckener had been stuck selling postcards and rattling collection cans to render his dream a reality.

WORSE NEWS ARRIVED on September 3 concerning the *Shenandoah*, then beginning a routine publicity tour of midwestern county fairs.

On its way to Ohio, *Shenandoah* had encountered a powerful wind gust and suddenly began rising 200 feet per minute. Frighteningly, the rate accelerated up to 1,000 feet per minute—greater than any experienced by any airship anywhere before. When Lieutenant Commander Lansdowne ordered all engines to be run at full power and the elevators pushed down as far as they would go to try to regain control, *Shenandoah*'s nose steeply declined to a nearly unheard-of 25 degrees.

The airship's climb slowed, but the engines overheated. Lansdowne had run out of options but wind velocity was increasing, and *Shenandoah* was lifted, as if by an invisible hand, to 6,060 feet as the crew heard girders crumpling and wires snapping. Then a blast of icy air blew through the ship, shoving it downward at 1,500 feet per minute. The angry 55-mph gust that then slammed into the ship's underside and rolled it severely to port was its death knell.

The colossal stress on the infrastructure tore *Shenandoah* apart. Crewmen were thrown out (their bodies were found half a mile away) as the control and engine cars were wrenched away from the stricken ship, disintegrating as it fell. Luckily, sections of the bow and stern formed rudimentary balloons and the men in them (including Lieutenant Commander Charles Rosendahl, the second officer) landed in a field. Twenty-nine men survived the *Shenandoah*, and fourteen didn't—but all of them would have died in an explosion had the airship been filled with hydrogen.

That helium had saved some lives was the sole consolation Americans could draw from the disaster. The subsequent Court of Inquiry found no evidence of incompetence or negligence on the part of Lansdowne or the crew, judging instead that "the final destruction of the ship was due primarily to large, unbalanced, external, aerodynamic forces arising from high velocity air currents." In other words, *Shenandoah* had run into a massive flow of cold air from the northwest that had overrun the warm air from the southwest.

The inevitable question, then, was whether the resulting fatal turbulence had been a unique fluke. If it was, then airships could be considered sound in concept and design. The problem, unfortunately, was that these kinds of extreme superadiabatic conditions were determined to be relatively common in the eastern half or two-thirds of the United States, where the solar heating of large inland areas and walls of mountain ranges forced air masses upward to produce thunderstorms characterized by deadly vertical currents.

It was an extreme weather pattern unfamiliar to Eckener, who was accustomed to the milder storms of Germany and the Atlantic, and he had failed to account for it when talking up Goodyear-Zeppelin's proposed transcontinental flights. As Karl Arnstein later said, "No [German] airship was designed for [such] gusts—there weren't supposed to be any," adding that "no European designer could possibly imagine the violence of weather conditions in the American Midwest."[6]

During the Court of Inquiry, Eckener escaped censure for the accident, and justly so, for *Shenandoah* was American-built and American-crewed, and he could hardly be held responsible for the navy's basing its design on a wartime Zeppelin originally intended for the very different geography of Europe.

He also remained diplomatically silent when it came to judging Lansdowne's response to the gust. Based on wartime experience, Eckener's habit in high wind was to cut the engine speed and let the airship function as a kind of free balloon, secure in the knowledge that he could always get back on course in calmer weather. Lansdowne, however, following textbook American practice, fought the gods at full power in order to keep control over the vessel and lost.[7]

In his heart, Eckener knew the loss of the *Shenandoah* was a major setback to ever establishing a New York–California route. No matter how fortified Zeppelins were, they would still be subjected to the Midwest's unpredictable gusts and moving walls of air.

He needed the Midwest, and he could not have it.[8] Crossing the Midwest was the fastest, most direct way to get to Los Angeles or San Francisco, but it was rather unlikely, at least not while the horrible deaths suffered by the crew remained uppermost in their minds, that many passengers would be eager to buy a ticket.

Adding to Eckener's sense that the bloom had come off his rose, American newspapers trumpeted ominous headlines like "No More Zeppelins" and "Why Not Abandon Airships?" in the days following the disaster. There hadn't been such negative reactions toward the airship since the darkest days of the wartime bombing campaign. Thankfully, the papers soon calmed down and decided that "it is not the American way to give up in the face of defeat."[9]

Nevertheless, the military romance with airships experienced a sharp dip. Admiral Moffett at the navy's Bureau of Aeronautics was severely criticized for his penchant for publicity flights—Lansdowne's widow accused him of murdering her husband—and naval airshipmen, so recently the belles of the ball, found themselves marginalized by a new school of strategists arguing that aircraft carriers equipped with scouting aircraft, torpedo planes, fighters, and dive-bombers represented the future of fleet operations. Two years later, the navy's first carriers—*Lexington* and *Saratoga*—put to sea.

The aftershocks of the *Shenandoah* disaster were felt most immediately by Goodyear. The plans to build passenger airships for cross-country travel were put on hold, and many of the Zeppelin personnel (including Ernst

Lehmann) returned to Friedrichshafen. The Goodyear-Zeppelin partnership fell into dormancy.

The year 1925, which Eckener had misguidedly believed to be his year of destiny—of taking the first steps toward a transcontinental route, of raising enough money to build his own airship—had ended in despair. Worse, he was condemned to the sidelines at the precise moment the airplane business began taking off.

30. The Fox

As REVOLUTIONS SO often are, this one was sparked by an otherwise minor and rather banal incident: the big railroad bosses complaining that the Post Office's transcontinental airmail route was eroding their profits. Since instituting day-and-night runs, the Post Office airplanes had been beating their express trains by seventy-two hours, and the intercity air routes were stealing their bread-and-butter mail—all with the apparent connivance of the government. Representative M. Clyde Kelly, Republican of Pennsylvania (a major railroad state), took up the cause and proposed privatizing the Air Mail to put trains and planes on an even footing. No more lavish Washington subsidies for the Post Office.[1]

The Post Office would contract out most of its routes to private firms that would bid, on a per-ounce or per-pound basis, for a license to carry the mail by air. To placate small-government conservatives, who had long disliked the European-style direct subsidies being paid to the perennially money-losing Post Office, the plan stipulated that the Post Office would pay the winning carriers 80 percent of the revenue generated by customers purchasing air-mail stamps.

In this way, the financial burden would fall on the users of air mail rather than on the American taxpayer. If people decided not to use pricey air mail, well, the carriers would see reduced income and eventual bankruptcy. Obviously, since aviation was in such a mess, railroad executives were pleased as punch to be seen posing as great liberalizers of the economy even if the scheme was really a protectionist play to keep airplanes from taking their business.

That the Airmail Act, better known as the Kelly Act, of February 2, 1925, would inadvertently create an immensely powerful airline industry and transform the little putt-putt airplane into a world-changing dynamo was an irony these railroad executives appreciated too late.

That measure, and its follow-up, the Air Commerce Act of May 1926, gave the country's struggling aviation industry a reason to live. Each renewable air-mail contract was to last for four years, in which time the carriers were expected to have invested enough money in route development, training, and equipment to stand on their own two feet, financially speaking.[2]

In order to accumulate that money, the first order of business was to reduce risk, the primary obstacle to growth. Because flying was so dangerous, insurance was exorbitantly expensive—when it was even offered. At this time there was just a single company left in the country willing to write a policy, and its premiums alone ate up nearly a fifth of one fledgling airline's annual expenses.[3] Without insurance to cover losses, few respectable businessmen had been tempted to invest in any aerial venture. And without capital to fund research, development, and expansion, aviation would remain mired in its primeval state.

To make flying safer, air advocates demanded stringent new rules to ensure their own products' safety. At this, Secretary of Commerce Herbert Hoover dryly noted that aviation "is the only industry that favors having itself regulated by government."[4] To that end, William MacCracken, the assistant secretary of commerce, was appointed to serve as the nation's first civil-aviation regulator. He quickly established a committee to investigate accidents, recommend improvements, impose pilot-certification procedures and equipment checks, and accredit flying schools.[5]

As a result, insurance soon became much cheaper and more available— within a few years, a person could buy $5,000 worth of personal coverage for just two dollars—even if a number of MacCracken's innovations ran into ground-level resistance.[6] His inspectors often alienated barnstormers with arbitrary determinations as to airworthiness. One took a pocketknife and slashed some wing fabric, saying, "I don't think this is strong enough," while another inspector judged, "I don't like that plane and I'll tell you why—I just don't like it."[7]

The pilots' complaints went unheeded; they only proved the reforms were working. MacCracken's intention was to impose uniform regulations and to do that he had to, as he said, "ride herd on . . . barnstorming" because "the time had come for a new kind of aviation to emerge in this country, perhaps less colorful but certainly more responsible."

Wing-walking, among the most dangerous of feats, was the first target. MacCracken mandated that henceforth wing-walkers had to wear parachutes, which sucked all the fun out of it. Spectators now also had to be fenced off well away from the aerobatics to avoid accidental deaths when a pilot crashed his plane. Jessie Woods, a barnstormer, grumbled that the government "had so many rules, just choking us down. They got very nasty about us being there. Every time we did something [an inspector] would jump into the cockpit to check and see what we were doing." Within a few years, most of the barnstormers had quit and gone to work as Hollywood stuntmen, where their skills were put to good use in such movie epics as 1927's *Wings* and 1930's *Hell's Angels*.[8]

The reforms came not a moment too soon, for the country's war-surplus airplanes were literally falling to pieces, often in mid-flight: By 1925–26, wartime aircraft were suffering severe structural failure at twice the rate of postwar ones, themselves of uncertain manufacture. Compared to even the older Post Office airplanes, which though unreliable themselves neverthe-less benefited from stricter maintenance, private aircraft were nearly thirty-six times more prone to fatal accidents.[9]

One of MacCracken's key allies was the philanthropist Daniel Guggenheim. In early 1926, he and his son Harry (who'd flown bombers during the war) established the Daniel Guggenheim Fund for the Promotion of Aeronautics with an endowment of $3 million to "realize for humanity the ultimate possibilities of aerial navigation, and to give America the place in the air to which her inventive genius entitles her."

Their money funded a host of new aeronautical schools, as well as technical research and development into engines and airplane design, heralding a technological shift from traditional wood-and-fabric biplanes to streamlined modern monoplanes. Thanks to Eckener, Arnstein, and Jaray, the Zeppelin had already undergone the equivalent updating in the war; now it was the airplane's turn.

Within three years, from just ninety-six students in the entire United States undertaking degrees in aeronautical science and engineering, en-rollment would rocket to 1,400, and American R&D, which had lagged behind the rest of the world by a decade or more, gestated a host of innovations.[10] Giant new wind tunnels allowed pioneering advances in, among other things, retractable landing gear, propellers, flaps, wing design, and high-octane fuel. New radial, air-cooled powerplants like the Pratt & Whitney Wasp and the Wright Whirlwind boasted dramatically better performance than the wartime, water-cooled Liberty engines they replaced.

Overall, the concerted campaign to reduce risk and increase innovation inspired a tsunami of investment into the privatized aviation business. Ford, for instance, introduced the revolutionary three-engine Trimotor plane in late 1925, with Henry and his son Edsel personally loading the mailbags for newly founded Ford Air Transport's first mail route between Dearborn and Cleveland a few months later.[11]

JUAN TRIPPE, TOO, seized his opportunity. With Long Island Airlines disbanded, he set up a new outfit, Eastern Air Transport, intending, like hundreds of other hopefuls, to apply for one of the new airmail routes being spun off by the Post Office—in this case, the Boston–New York line. But he soon realized he didn't have a chance. With no assets, Eastern existed only in name. There was a likelier contender for Contract Air Mail Route No. 1 (CAM-1), and it was called Colonial Air Transport.

Colonial was owned by a group of well-heeled New England worthies, with the current Connecticut governor, John Trumbull, serving as chairman. Like Eastern, it owned no airplanes and was little more than a shell company, but it was still the odds-on favorite to win CAM-1 thanks to its board's connections in Washington. In that event, Colonial would have to find someone to run it, and men with hands-on experience were few and far between.

In mid-1925, Trippe approached Trumbull with an idea to merge the two airlines, with Trippe coming on board as a vice president and operations manager.

From Trumbull's point of view, it was an easy deal to make. Little did he know he was letting an insatiable fox into his comfortable henhouse.

TRIPPE'S FIRST MOVE was to instruct his lawyer to "fix it so that our crowd is in control." By "our crowd" Trippe was referring to his college friends and LIA backers, John Hambleton, William A. Rockefeller, and Sonny Whitney. After the merger, Colonial's board had swelled to no fewer than thirty members, giving their meetings something of an enjoyable country-club flavor, but the arrangement was hardly efficient. Trippe helpfully set up an executive "voting trust" to make the important decisions and allow the board to get back to playing golf.

This trust comprised seven members, divided between four New Englanders under Trumbull and three New Yorkers (Trippe, Hambleton, and Rockefeller) so as to balance the Colonial-Eastern blocs. On the face of it, Trumbull controlled the majority, but Trippe had seen to it that one of the New Englanders was a potential fifth columnist. His inside man was Theodore Weicker, the father of one of his old fraternity friends. Weicker was pliable enough to throw his weight against his own side if necessary; with the turn of his single vote, then, Trippe could take over Colonial any time he wanted.[1]

In booby-trapping the voting trust, Trippe was playing for greater stakes than mere control of a company. He had to have Colonial to win the wife he wanted. Ed Stettinius, a fraternity pal, had recently invited Trippe to stay at the family's Gatsbyesque estate at Locust Valley, Long Island. There he'd met the aged patriarch of House Stettinius, Edward, a partner at J. P. Morgan & Co., and the two men had chatted about the future of aviation. Edward senior would die shortly afterward, but not before advising his delightful daughter, Betty, that this Trippe fellow knew his onions. If anyone could find a way to make money in airplanes, Trippe could.

Betty was quickly taken by this odd young man, so very different from her boring suitors at the local country club. To better fit in, Trippe had turned himself into an excellent golfer; after Betty hit a fine drive on the fourth hole at Piping Rock, he summoned up the courage to inquire, "Do you ever come to town on Saturdays or Sundays?" It was a curious question to ask of a debutante—no decent person went into the city during the summer, especially on a weekend—and in asking it Trippe accidentally let slip the mask he habitually wore. Betty decided to come to town.

At dinner a few days later, Trippe showed a different side. Recalled Betty, he "was so quiet, yet so dynamic, full of interesting, intelligent conversation, yet with a real twinkle to his eye and a great sense of humor." The two were soon courting.

In the years to come, Betty would complement Trippe's otherwise habitual taciturnity with her engaging, cheerful disposition. Whereas Trippe could only, with some effort, be the soul of bonhomie when he needed something from someone, Betty softened his social utilitarianism with a genuine interest in others' well-being. A manager who worked thirty feet away from Trippe for years never heard a "Good morning" when he walked in, but Betty made sure he received a birthday card and asked about his kids.

Because he needed them more than they needed him, Trippe had a harder time persuading Betty's mother and her older brother William of his potential. They were very protective of the lass, and while they considered Trippe a nice young man, he seemed thoroughly impractical. Airy dreams of aviation's future were all very well, but why didn't he have a real job on Wall Street like everyone else? Being in charge of an airline with no airplanes was no way to provide for a family.

Betty indignantly defended Trippe, but the decision was made: No match could be approved until "Juan has a business," and until such time as Trippe made a success of himself at Colonial she would be dispatched to Paris under the care of her formidable Aunt Hazel.[2]

Lost on the Stettinius clan was Trippe's penchant for calculated risk-taking and a habit of presenting fait accomplis to get his way. Within weeks of the Eastern-Colonial merger going through, he committed to buying—without bothering to ask the board—two as-yet unbuilt Fokker F-7 airplanes and two new Fokker Universals as the airline prepared to start its Boston–New York service.

When Trumbull heard about the purchase, he was furious. Colonial was an exclusively air-mail operation, which required small, cheap, single-engine airplanes, but these Fokkers were three-engine beasts designed for a specific purpose: carrying between four and eight passengers. What on earth was Trippe playing at?

He'd have been angrier still if he knew that Trippe had recently returned from a secret visit to Havana, where he had met with General Gerardo Machado, the new president of Cuba. Machado had run on a modernization program for his country, and Trippe seized the opportunity to pitch the idea of operating a Key West–Havana service.

The visit was an experiment in applying the lessons of domestic Air Mail privatization. The Post Office was contracting out its routes on an exclusive basis to eligible bidders, like Colonial. Once a route, such as Boston–New York, was allotted no one else could compete. What Trippe wanted to know was, could you do the same *internationally* for the mail and, potentially, passengers?

What Trippe understood, years before anyone else, was that, no matter the idealistic intentions of those who declaimed upon the freedom of the air, the air is not free.

It had been once, sometime before the war, but no longer by the mid-1920s, when a body of aviation law was taking shape. From the pre-1914 utopia was emerging a system of air sovereignty that differed from the age-old law of the sea, which allowed that no merchant vessel in peacetime could be excluded from calling at any port in the world. But since airplanes, unlike ships, could cross borders at will and penetrate far inland, countries were making sure they controlled all air movement above their territory. This meant that every airline had to acquire legal permission before landing.

Trippe grasped that those landing rights were priceless, for they could only be acquired by international agreements or by private arrangements with the relevant governments. If you owned the exclusive rights, your rivals were locked out of competing against you.[3]

Which is why Trippe treasured a letter from General Machado he kept tucked in his pocket. The letter gave him exclusive landing rights at Campo Colombia (a military training field outside Havana), tax exemptions, and preferential use of customs and immigration facilities. By fixing Machado—some money might have changed hands—Trippe had quietly prevented anyone else from landing in Cuba. He had acquired his first Holy Grail: a government-protected international route—albeit without the means to fly it.

In typical Trippe style, he kept this rather important information to himself and his allies on the voting trust. His ambition was to expand Colonial from its Boston/New York base to Florida and the Caribbean, but he knew full well that he would never get the New England "old fogies" (as he called them) to approve any such scheme.[4]

The reason was that the Kelly and Air Commerce Acts had been intended to liberalize the Air Mail, not to create a passenger industry. No airline in the United States at this time, aside from the odd local sightseeing operation, regularly transported people between destinations. That was what trains, boats, and cars did.

It was not unheard of for one or two people, such as the newlyweds who spent $2,000 for a short honeymoon flight, to take a one-off trip on a tiny, juddering mail plane. But the law was that the mail always came first. It was permissible, even encouraged, to strand a passenger, no matter how much he or she had paid, at some remote airfield in the dead of night if there was a mailbag waiting. One poor fellow was left behind to take in the delights of Omaha for five days before he could be picked up.[5]

You also took your chances with the weather: Since mail didn't feel cold or heat, airplanes were not insulated against the elements, and the brave soul who wangled a "seat"—a rough canvas mailbag—had to put up with searing blasts in the desert or icy winds above the Great Plains and the Rocky Mountains. That is, if you could actually get into the airplane, whose door was designed to accommodate mailbags, not humans, who had to fold their bodies with origami-like precision to enter or exit.

Airlines actually went out of their way to *discourage* passengers, whose revenue per pound was less profitable than air mail. (And neither did mail whine about being uncomfortable.) So unenthusiastic was one company, National Air Transport (NAT), about providing passenger service that it doubled the price of a one-way ticket to $200 to intimidate all but the most determined travelers. The tactic worked: In 1926, NAT was gratified to discover that just 168 passengers had decided to fly with them.[6]

Colonial was of a piece. Aside from Trippe, for whom Colonial was a vehicle to greater things, its board members had many other investment interests, and the airline was for them just another. All they wanted Trippe to do was deliver the mail and hand them government checks.

For the moment, Trippe let the matter rest and focused on making the Post Office deadline to begin airmail operations. All went well, and on July 1, 1926, a Colonial Fokker Universal took off from Boston at 6 P.M., right on time.

Much to Trippe's irritation, though, the board had refused to install seats in the Fokker Universals and had bought a couple of Curtiss Larks (small single-engine biplanes) to round out the fleet. Even using the cheap Larks, however, Colonial could not make money. As summer turned to fall and winter, fog and snow too often delayed flights or forced cancellations. To break even, the airline needed to carry three hundred pounds of mail per month but was currently achieving just thirty. For businessmen, the time saved versus the money spent to send a letter 220 miles was not worth it, not when mail-express trains ran hourly. Trippe, following a common practice (or scam) known in the trade as the "airmail-augmentation program," ordered employees to send one another one-ounce letters in one-pound pouches to inflate the government's weight-dependent subsidies. (Put it this way, there were an awful lot of bricks sent by airmail in those days.)[7]

But such tricks could only postpone a financial reckoning. Trippe agitated to change the board's disposition and pursue passenger revenue to save the company, but he succeeded only in alienating its already skeptical members, who began to believe they had made a mistake in hiring such a young man. With the Fokker F-7s' scheduled delivery date approaching, they recalled him to Boston to answer for his decision to buy them on credit. To deflect criticism, Trippe proposed installing Major General John F. O'Ryan as Colonial's president; placing such an eminent "senior man" in charge would do much to calm worries as to the airline's financial condition.

On September 1, 1926, O'Ryan was duly appointed. Trippe, in one of his few strategic errors, had labored under the misapprehension that O'Ryan, like Trumbull, was just another foolish old duffer he could easily manipulate. But O'Ryan, who had commanded the 27th Division in France during the war (and thought he still did), proved a stickler for hierarchy and intended to keep Trippe in his place.

At first, Trippe and O'Ryan seemed to get along, even if the general irritatingly called Trippe his "assistant" (and Trippe called O'Ryan "*my* assistant" behind his back). Relations quickly deteriorated to the point where O'Ryan downgraded Trippe to "lackey," and the latter told anyone who would listen that he lacked any "confidence whatsoever in [O'Ryan's] ability to successfully manage an air transport company."

Tensions erupted into open warfare in January 1927, when the Post Office announced that it was opening bidding for the prime New York–Chicago line (CAM-17), the most lucrative and prestigious of the mail routes.

O'Ryan was seemingly intent on acquiring the route for Colonial but, in a strange move, pushed for entering an absurdly low bid—so low that Colonial would go bankrupt if it won the contract. As it was, Colonial was losing $8,000 a month, its cash reserves had fallen to $100,000, and the banks were owed $75,000 for the two Fokker F-7s.

Trippe suspected something was up, but he couldn't understand exactly what. He also wanted Colonial to win New York–Chicago, because businesses would surely pay a pretty penny to save days over the train, but Colonial obviously needed to bid higher, even if it meant risking losing the route to a competitor.

In mid-March O'Ryan called a board meeting in Hartford when he knew Trippe was in New York. The members approved the O'Ryan plan, and it was only then that Trippe and his allies belatedly realized its true objective. O'Ryan had no intention of ever delivering the mail. Simply by securing New York–Chicago with a bid lower than anyone else's, asset-rich Colonial would be a prime target for a sale to a larger and richer rival like National Air Transport that could afford to take a temporary loss on the route, thereby allowing the board to get out of the aviation business at a profit.

After "Old Man Weicker" told him that "you can count on me," Trippe triggered his trap: the voting trust, now assured of a 4-3 majority. At 12:01 A.M. on March 24 (the last day of bidding) at the Greenwich train station, Trippe summoned his tame trustees, who obediently voted to approve his own bid. A few minutes later, Trippe and Hambleton caught the early milk train to New York, commandeered a company Fokker, and flew to Washington. At 11:30 A.M., half an hour before the deadline, they ran into the very surprised O'Ryan and Trumbull outside the Post Office building, barged past them, and submitted their bid.

The next morning, a Colonial lawyer called Trippe to inform him that the voting trust was illegal: It had been drawn up for a term of ten years whereas Connecticut law stipulated a maximum of seven for such an entity. The new bid, therefore, was invalid and the New Englanders were going to sell Colonial.

Trippe was purged from the company, along with Hambleton, Whitney, and Rockefeller. As a parting gift, O'Ryan permitted him to take, at cost, the two accursed, and still undelivered, Fokker F-7s that had been at the heart of Trippe's troubles.

Trippe's ousting was so sudden that few outsiders even knew that he was gone until they visited the office. When a manufacturer, accustomed to seeing Trippe there, stopped by shortly afterward, he asked what had happened. "He got too visionary," O'Ryan explained. "He couldn't keep his feet on the ground any more and I had to let him go. Too much youth and enthusiasm." (As for CAM-17 to Chicago, the route went to National Air Transport, and Colonial was merged into what would become American Airlines soon after.)[8]

Trippe was coming up to his twenty-eighth birthday and was unemployed. The prospect of marriage to Betty seemed further away than ever.

But he still had that letter from Cuba's President Machado.

32. The Boom

T RIPPE HAD BEEN exiled from the airline business just as it was about to experience its Big Bang. He mooned about New York for a few months but happened to travel to Roosevelt Field on Long Island on Friday, May 20, 1927. It was a rainy day, but something exciting was in the offing. An obscure former barnstormer and Air Mail pilot, twenty-five-year-old Charles Lindbergh, was attempting the impossible: to fly across the Atlantic to France, by himself, nonstop, in a small Ryan single-engine airplane.

Trippe was acquainted with Lindbergh. The two had met sometime the year before, when Lindbergh was still flying domestic mail. Trippe had considered Lindbergh ripe for recruitment to Colonial, but the young aviator had bigger things in mind. Now at least Trippe knew what those "bigger things" were.

When he took off at 7:52 A.M., Lindbergh was widely ridiculed as the "Flying Fool." After he landed in Paris thirty-three and a half hours later, he arguably became more famous than anyone in history. *The New York Times* devoted five full pages, including its front page, to the flight, and the city's papers would print no fewer than three hundred thousand stories about Lindbergh, his astounding feat, his family, the weather, his navigation techniques, his views on aviation, and his love life over the coming months.

The demand for Lindberghiana was insatiable. Movie cameras recorded 7,430,000 feet of newsreel footage, significant portions of which were shown day in and day out in the nation's thousands of movie theaters. He received a ticker-tape parade attended by some four million New Yorkers (Trippe watched from the windows of the Union Club), the Medal of Honor, the Distinguished Flying Cross, and the "Man of the Year" cover of a new magazine called *Time*. Hundreds of songs were written about him ("Columbus of the Air," "Eagle of Liberty," et cetera), and readers sent the *Times* more than two thousand poems of widely varying quality.

After his return, Lindbergh embarked on an eighty-two-city national flying tour, during which up to thirty million Americans—a quarter of the population—saw the great aviator. Even Henry Ford, who owned an airline but had never flown, agreed to take his first flight on the *Spirit of St. Louis* with Lindbergh at the controls.[1]

Eckener had experienced a similar phenomenon, albeit on a smaller scale, after delivering ZR-3 *Los Angeles*, but he, like most other Germans, wrote off the Lindbergh venture as just another one-off, publicity-driven stunt, like the navy's NC-4 trip and the British R-34 crossing. In France, Lindbergh may have received the laurels once reserved for Roman emperors returning from their latest conquest, but Eckener dismissed him as a dollar-hungry entertainer whose feat could hardly be said to mark the beginning of profitable, sustained transatlantic air travel.[2]

And once again, Eckener was not wholly wrong. The *Spirit of St. Louis* was an exception to the rule that airplanes were not up to the rigors of long-distance flight. Lindbergh's aircraft was a thoroughly customized job attuned to his exacting specifications. To be capable of making the 3,600-mile trip—a range easily within a Zeppelin's reach—the *Spirit of St. Louis* had been stripped down to the barest essentials in order to carry the required fuel. Even Lindbergh admitted that his airplane was essentially "nine barrels of gasoline and oil, wrapped up in fabric"; to further reduce weight he'd brought only a few sandwiches, sat on a compact wicker stool, foregone a radio or life raft, and obsessively cut off unnecessary parts of his maps to save a couple of ounces more.[3]

What Eckener had not accounted for was that the plane didn't matter; it was Lindbergh's clean-shaven, go-it-alone individualism that attracted Americans. Lindbergh, uniquely for a barnstormer or an Air Mail pilot, didn't drink, didn't smoke, didn't womanize, and didn't curse. He was self-taught, modest, stoic, and idealistic, a Minnesota man who as a boy had hunted and fished like Tom Sawyer and resembled a skyfaring cross between Davy Crockett and Daniel Boone. There had been plenty of pilots before, but none had ever transfixed the American imagination like Charles Lindbergh.

It also mattered to many that, in an era when membership of the Ku Klux Klan ran into the millions, Lindbergh, a lanky, six-foot, blond Viking type, had the "right" blood flowing through him. Yet the cult of Lindbergh overcame even racial boundaries. Josephine Baker, then in Paris, had interrupted her cabaret show at the Folies Bergère to greet Lindbergh, forgetting that he "was a white man and that he came from [heavily segregated] St. Louis and might not have liked Negroes," she said in 1952. "I only remembered that he was an American and that he had done something great for the progress of the world."[4]

Lindbergh transformed Americans' perceptions of aviation. Despite the immense dangers involved with flying, Lindbergh demonstrated that it could be made safe if one analyzed the risk factors rationally, compiled checklists, and applied skill and knowledge to the mission, as he had done.

Until Lindbergh, pilots had resembled storm-tossed sailors of old, adrift and helpless amidst the roiling ocean of America's mountains, forests, valleys, and plains; they were hostages to the gods of good fortune to see them to their destinations. After Lindbergh, who advertised the benefits of "blind flying" (using instruments alone for guidance), pilots increasingly began to rely on a host of new navigational aids, such as directional gyroscopes, radio-homing devices, barometric altimeters, and artificial horizons, to avoid dangerous terrain and stay aloft. Initially, there was some resistance from pilots, a hardy breed suckled on the principle that *real* flying meant hands-on control and "feeling my way around the sky like a prowling cat," as one fondly said, but, as they adjusted to the changes, crash rates began to fall.[5]

Thanks to Lindbergh, model-airplane building became the nation's number-one craze almost overnight. Radio stations broadcast hobbyist shows, the biggest of which was the Jimmie Allen Club, whose members, like Mickey Rooney and Shirley Temple, received a monthly newsletter filled with tips to improve the aerodynamics of their little balsa-and-glue planes with rubber-band engines. In Los Angeles, there was a special playground reserved for flying model airplanes, and crowds flocked to the cinemas to see Walt Disney's new creation, a certain mouse, dolled up as an aspiring Lindbergh in *Plane Crazy*, as well as a movie about Great War pilots called *Wings* (released on August 12, three months after the flight) that would go on to win Best Picture at the first Academy Awards.

In junior high schools, courses in aeronautics, the history of aviation, and meteorology were introduced into science curricula; Air Cadet programs taught students to fly, and at college they joined flying clubs and took advanced lessons at reduced rates from one of the 575 Guggenheim-backed aviation schools that had sprung up by 1929. In America's burgeoning aviation industry, the Department of Commerce advised, there would be good jobs available as mechanics, engineers, airplane designers, airport builders, ground crewmen, radio operators, advertising copywriters, and ticket salesmen.

Flying quickly became a part of everyday culture. In 1925, one newspaper, to give an example, had carried eighty columns and three pages of pictures (mostly crashes) about aviation; by 1928, those numbers had swollen to 450 and 36, respectively. Whereas there had been just one or two special-interest journals covering aviation in the early 1920s, in 1928 there were twenty popular magazines dedicated exclusively to the subject.[6]

The urge to fly had finally arrived—the demand Trippe had dreamed of—but the wherewithal was lacking, not only because flying remained prohibitively expensive compared to traveling by train or car but also because there were so few seats.[7]

The airlines were slow to realize that passengers, annoying as they were, might be worth accommodating, and it was not in fact Lindbergh who prompted them to start offering service but a boring-but-important change in air-mail rates in January 1928.

When the Post Office halved the cost of a stamp, it dramatically boosted the number of letters and packages carried, but not by enough to allow airlines to offset the increasing costs of operating the larger airplane fleets, in terms of both size and quantity, that the volume of extra mail necessitated. Unexpectedly, then, they needed to attract paying human customers, otherwise known as "self-loading cargo," to bring in additional revenue. As a result, a few primitive seats were hastily added to the mail planes flying the busier legs.

The humorist Will Rogers, the country's most popular columnist and an aviation aficionado, took advantage of the new pro-passenger atmosphere to thrill his forty million daily readers by relating in a series of articles his jerry-rigged trip from Los Angeles to New York. On one of Boeing Air Transport's (BAT) new two-passenger mail planes (the Boeing 40A) from Salt Lake City to Chicago, he luxuriated in the side-by-side arrangement, which resembled the cheapest coach seats on a local train. After a couple of stops, unfortunately, an executive with the Labor Bank in New York joined him. "Now when I told you this seat was narrow, I didn't just put that in there to make more words," wrote Rogers. "It is either terrible narrow or this old Labor boy keeps spreading out. He is a big husky thing." Rogers believed the seats were designed for "one frail woman accompanied by a male contortionist."[8]

Once airlines grasped that there were people willing to spend $200—a quarter of the price of a new car—for a one-way ticket from Chicago to New York, shuttling mail at five cents a letter quickly became of secondary interest. They turned their attention to selling the benefits and glamour of passenger travel.

A concerted campaign was highly successful in persuading Americans that flying was no more dangerous than driving—in fact, a great deal less dangerous—with advertisements emphasizing that the airlines' only concern was, as one slogan went, "Safety—First, Last, and Always."[9] Now, there was, of course (and here the public-relations people discreetly ahemmed), the *occasional* accident—actually, they were more than "occasional," there being 234 airline crashes and 29 deaths between 1927 and 1929—but these paled beside the number of fatalities caused by cars, which slaughtered 80,619 men, women, and children in those same years.[10]

Even if you did die in a plane crash, an off-message (and anonymous) operations chief of a large midwestern carrier casually remarked, it really wasn't so bad. Some people, he said, will "tell you how terrible it must be to fall through a mile of air. But I don't see that it is any more terrible than to fall through a mile of water—that's what you do when a ship sinks under you. And in the air there aren't any sharks to nibble at you." What was *really* important, the chief went on to say (in yet another unfortunate choice of phrase), was that as air travel grew more popular, "fares will be falling faster than a punctured parachute."[11]

So they did. Within the year, a cross-country airplane trip from Los Angeles to New York would cost $400 compared to between $200 and $350 by train, and the price would continue to decline as passenger numbers exploded.[12] In 1926, U.S. airlines had carried a mere 5,782 passengers; the following year, 8,679; in 1928, 48,312, which more than tripled in 1929 to 161,933, hitting 384,506 a year later—an increase of 6,650 percent in four years.[13]

A significant proportion of that growth came from women. Early attempts to drum up customers had focused exclusively on attracting businessmen, but airline executives soon realized that women formed an excited contingent of potential travelers. Ford's magazine ads depicted them waiting at the terminal and cheering: "Man No Longer Flies Alone!" In 1929, a quarter of passengers were women, rising to a third by the mid-1930s.[14]

The push was not, maybe, an altogether altruistic one. Airlines had cannily understood that *if even a helpless woman* was willing to go aloft, then hitherto reluctant males had no excuse not to, unless they were cowards. One company came up with an innovative ploy to lure husbands into taking their first flight by offering a free ticket to their accompanying wives. The scheme came to a sudden but perhaps all too predictable end when it was discovered that a number of the young, attractive "wives" the men were bringing on their business trips were not, in fact, married to the passengers in question.[15]

In order to keep up with demand, the industry, which had built only 344 airplanes in 1925, doubled and redoubled output, hitting a peak in 1929 of 5,414.[16]

A large number of these were small private planes employed in a variety of new ways. Politicians chartered them to travel around their states; merchants began offering express delivery; the American Tobacco Company hit upon the idea of skywriting "Lucky Strikes" in smoke; companies purchased them to ferry executives around; they were used as crop dusters and seeders; and the Coast Guard used them to spot schools of fish for the benefit of trawlers.

Airplanes were not only being woven into the warp and woof of everyday life; they were also opening new vistas. All twenty-nine thousand square miles of Alaska was mapped for the first time between 1926 and 1929, while farther south airborne archaeologists discovered pre-Columbian ruins in the Mexican jungle and stumbled upon tribes cut off from civilization for millennia.[17]

Technologically, the major development lay in the advent of big, three-engine passenger aircraft. While Will Rogers had been wowed by the Boeing 40A, introduced in 1927, having two seats, the Boeing 80A, which appeared a year later, was capable of accommodating eighteen passengers and three crew; the Fokker F-10 (a larger version of the F-7 that had so infuriated Trippe's Colonial partners), hauling twelve, was introduced to fleets at the same time; and the faithful eleven-seater Ford Trimotor sold in numbers greater than all the others put together.[18]

Thirst for the new airplanes was hard to quench. In 1928 alone, 23 airlines were launched. By the following year, there were 62 passenger airlines, 47 air-mail lines, and 32 air-cargo lines flying in the United States—and all they wanted were airplanes, more airplanes.[19]

To service them on the ground, a new infrastructure came into being. Airports, for instance, could no longer be the grass-and-dirt runways equipped with a shack that had once been used by the Post Office.[20] By January 1928, there were suddenly 503 airports in the United States, and by December 1929, 834, many with paved, lit runways, hotels, sturdy hangars, fueling stations, and terminals with waiting rooms, marbled lavatories, and offices. Washington-Hoover, serving the District of Columbia, even had a swimming pool, and Enterprise, Alabama, living up to its name, built a three-runway airport in the midst of a nine-hole golf course where players had fun hitting long drives over taxiing planes.[21]

To service customers in the sky, airlines rapidly improved their in-flight service. Will Rogers, again, had counted himself lucky to be gruffly handed a chicken sandwich, a Thermos of coffee, and an apple when he'd flown cross-country, but a year later, purred the travel journalist Arthur Phillips, his luncheon "was no uncouth handing out of tomato sandwiches from a paper sack but a most delicious repast, carefully prepared and daintily served."

Chicken remained a staple, but now at least it came in a variety of forms (cold roast chicken, chicken salad, creamed chicken, fried chicken). Potato salad, white and rye bread, string beans with onions, baked ham, chocolate cake, strawberry shortcake, fruit cups, and, for the few vegetarians present, egg salad sandwiches were served. For beverages, passengers could choose from coffee, iced tea, orange juice, tomato juice, and lemonade.[22]

Cutlery was enviably expensive. As the inimitably named columnist Velva Darling noticed as she flew from Los Angeles, the knives, forks, and spoons were gold-plated and heavy enough to stay more or less put in turbulence. Cups were made of metal or enameled; plates, porcelain; napkins, linen. On most airlines, a lap pillow served to balance them, but a few others pioneered the use of detachable, one-legged folding trays fitted into holes in the floor that were secured by two clips to the cabin wall.

Arthur Phillips was, again, in heaven when he found his tray covered "with a linen spread of fine quality . . . of a beautiful lavender shade [that] formed an effective and delightful color combination with the modish dirigold tableware."[23]

An early innovation had been the introduction of "stewards," as on ocean liners, often the sons of major investors in the airline. One smitten passenger described her steward as "a college boy, tall, clean-cut, intelligent . . . and very eager to make each passenger as comfortable and happy as possible."[24] They adjusted vents, hung coats, distributed cotton earplugs (a Ford Trimotor could generate 115 decibels of noise), handed *Vogue* and *House Beautiful* to the ladies (gentlemen read *Fortune*), and showed new fliers maps of the terrain and landmarks they should look for.[25]

Training was rudimentary for these fine young men, the natural assumption being that their good breeding ensured decorum and discretion; the most useful advice they received was to gargle with Listerine before serving food and to kneel in the aisle, rather than lean over passengers, when pouring drinks to avoid spillage.[26]

They were eventually replaced by "stewardesses"—"sky girls," "hostesses," "courierettes," "skipperettes," and the unfortunate "escorts" were all considered as job titles—who were selected on the basis of their sturdiness, practicality, horse sense, and disinclination "to chase [men] around the block at every opportunity." (Any "flapper type of girl," as one recruiter said, was out of the question.)

A prerequisite for the job was having previously worked as a qualified nurse. At Boeing Air Transport, they wore matching hunter-green wool capes and berets, low-heeled black shoes, and knee-length skirts and, as nurses who had seen a lot in their time, were expected to be capable of taking care of both passengers and themselves.[27]

Behavior on an airplane was strictly regulated to avoid any hint of impropriety. Pilot-stewardess relationships were absolutely forbidden, partly to assuage the fears of pilots' wives that some saucy minx would steal their husbands on a layover in Kansas City. Stewardesses were also instructed to salute, not greet, pilots when they came aboard and were banned from entering the testosterone-drenched cockpit. (Somehow forgetting about Amelia Earhart and her "Ninety-Nines" club of accomplished aviatrixes, pilots believed female fliers were "impulsive and scatter-brained," rarely remembered to fill the gas tank, and habitually set off for out-of-range destinations because "that's where they want to go.")[28]

As for the passengers, stewardesses were instructed by the airline to "retain the respectful reserve of the well-trained servant. A ready smile is essential, but never permit yourself the intimate attitude of a traveling companion." Most important, only touch a male traveler when needing to wake him, in which case "tweak his elbow sharply."[29]

In the cabin, stewardesses had the power to remove abusive passengers—not that there were many. The most unruly person Olette Hasle ever had to deal with was a man who kept removing his shoes. There were occasional instances of groping, but stewardesses rarely put up with it. Harriet Fry Iden slapped a professor who lunged at her, as did Mary O'Connor when a passenger "got fresh" with her: Neither was reprimanded, Iden's supervisor adding that the brute had it coming.[30]

Smoking on board was encouraged, with United raising the bar by gifting each passenger fifty Chesterfields. Until the later arrival of sealed windows, a recurrent problem was passengers' habit of opening their windows and tossing the lit stubs out, much to the annoyance of farmers and forest rangers. The installation of ashtrays at every seat helped mitigate the issue.[31]

Playing board games was also popular.[32] More exciting, though, was the advent of the in-flight movie, the first of which—the title is, unfortunately, lost to us—was shown in the drawn-curtained cabin of a Universal Air Line plane flying from Minneapolis to Chicago on February 16, 1929. A projectionist suspended a screen at the front and, held spellbound for ten reels, "the passengers lost all sense of air consciousness," as *The New York Times* reported, equally spellbound.[33]

To erase the impression that passengers shared space with mailbags, airlines commissioned designers to come up with décors reflecting their brand. United Air Lines went with a traditional you-can-trust-*us*-to-get-you-there look. On their serene and sedate Boeings, one relaxed, as at one's club, amid leather seats, bronzed metal fixtures, muted lighting, and walnut paneling.[34]

A trendy upstart, Transcontinental Air Transport (TAT, soon to become TWA) prided itself on being more "progressive" than its competitors and hired an "artist in Greenwich Village" to "create an interior which would be modernistic, individual, and like absolutely nothing but an airplane," explained Velva Darling, who loved the lavender wall paneling, window shades with fluffy curtains, and cream seat covers.[35]

What passengers, whatever airline they chose to reward with their loyalty, did not know was that all the fancy seats and furnishings were designed to be easily removable so that cleaners could use a high-pressure hose to flush vomit off the floor. Non-flushing chemical lavatories, dubbed "slop boxes," were as hideous as their nickname implied, and stewardesses frequently had to sponge them down with a potent disinfectant during the flight.

Nausea was the bane of flying, and no matter how often executives claimed that only one out of a hundred passengers threw up, the true figure was much, much higher. An industry rule of thumb had it that 80 percent of passengers would heave when traveling over the Southwest owing to its unpredictable weather.

When an airplane hit turbulence, its tail would swing back and forth, and hundred-foot plummets were common. At those times, said Marcia Davenport, "your mouth flies open, your eyes goggle, you clutch the arm of your chair [and] your mental attitude is not helped much by the neat little cardboard ice-cream container fastened at each chair with the legend, 'Use in case of Air Sickness,' on its cover."

Most unpleasantly, when nauseated passengers sitting at the front opened the windows to unleash hell into the slipstream, high-velocity vomit would spatter those aft who'd unwisely opened theirs to enjoy the fresh air.[36]

None of this mattered, though, because during the great boom the aviation business, which in pre-Lindbergh 1927 had ranked 144th among manufacturing industries with gross revenues of $90 million (the corset trade was bigger), grew faster than any other. Until March 1928, only two aviation companies, Curtiss Aeroplane and Wright Aeronautical, had ever issued stock, but between that month and late 1929, twenty-one air companies went public, eventually reaching a paper value of $1 billion. Some speculators saw a 2,000 percent return, but Frederick Rentschler, a founder of Pratt & Whitney and a partner of William Boeing's, outdid everyone by turning his investment of $253 in the company that would become United into a fortune of $35 million.[37]

Twiddling his thumbs and cooling his heels after his ouster from Colonial, Trippe was desperate to get back into the game.

33. Terra Incognita

WITHIN WEEKS OF watching Lindbergh take off, Trippe formed a new venture with his wealthy ex-Colonial friends. They invested in him supreme power, giving Trippe carte blanche to make deals, hire staff, and buy whichever airplanes he thought best. Though grandly named the Aviation Corporation of America (ACA), it was, as yet, merely a holding company with nothing but $300,000 in a bank account to its name.

Trippe had noticed that all the major existing airlines were fighting for a share of the increasingly crowded U.S. domestic market; ACA's mission instead would be to conquer the potentially lucrative and virtually empty international routes to the Caribbean and South America.

It was time to use that letter from General Machado granting him exclusive landing rights in Cuba. In early June 1927, Trippe went to see a former navy pilot named John Montgomery. Back in March, he and a partner named Richard Bevier had founded a tiny airline optimistically called Pan American Airways, intended to fly between Key West and Havana. Like ACA, it had no actual airplanes, but it had something better: a fresh Post Office contract to carry mail internationally to Cuba. What they didn't have was the right to land there—a salient fact they might have forgotten to mention to the Post Office.

Having been burned before by overpromising entrepreneurs, the Post Office had stipulated that the first mail flight must take off on October 19 of that year. No exceptions would be granted. If a plane didn't depart that morning, the contract would be mercilessly canceled the same afternoon. Pan American had about four months, then, to be up and running.

Not only did he have the landing rights, Trippe told Pan American's owners, he was about to take delivery of the two long-delayed Fokker F-7s O'Ryan had allowed him to take from Colonial. He also had a team: Andre Priester, Hugo Leuteritz, and the grand panjandrum himself, Charles Lindbergh.

Priester was, like Eckener's man Ludwig Dürr, one of early aviation's more eccentric characters and difficult to deal with, but Trippe had quickly realized his potential importance. Short, bald, fiery, strange, remote, a roiling sun contained within a moon-sized body, Priester was a thirty-six-year-old Dutch immigrant working for his fellow countryman Anthony Fokker when Trippe met him at the latter's airplane factory. Trippe decided he would make an ideal chief engineer for his new airline.

Again, like Dürr with Eckener, Priester was willing to report only to the one man he respected. Anyone else was treated with disparagement, glared at "with eyes that never appeared to smile" (as one poor interviewee put it), or treated with icy formality. Only in letters to his wife, with whom he maintained amicable relations, would Priester relax enough to sign off with a "Best Regards."

Priester's major selling point was that he was a fanatic about process, doctrine, and planning. It would be he who ran schedules to the minute, produced a thick manual replete with checklists and stuffed with protocols explaining precisely what to do in any situation, and instructed pilots, a louche lot, that they would be fired if seen drinking or even smoking in uniform. Mechanics were dismissed on the spot if they were photographed wearing dirty overalls or if they failed to polish the brass fittings hidden deep inside engine nacelles—because a clean airplane was a tight airplane.[1]

Trippe's other acquisition was Hugo Leuteritz, a Brooklyn-raised technician who'd parted ways with Radio Corporation of America (RCA) when the company rejected his proposal to develop lightweight transmitters. He was necessary because, as Priester repeated like a mantra, "Der flying uff ninety miles uff vater iss no dchoke und iss not to be treated as such."[2] Ninety miles was the distance between Key West and Havana, a short journey but one fraught with danger.

On the mainland, pilots usually followed the "iron compass"—train tracks—between cities, and there were always landmarks and beacons to help guide them. But out there in the blue, Trippe understood, there was nothing, and if he ever wanted to expand deeper into the Caribbean, then pilots would have to find their way to isolated airstrips on tiny islands. As it was, even when flying the well-known area off the Florida coast, a pilot could easily drift as much as fifty miles off his planned course. In rain or fog, the odds rose considerably of missing as large a landmass as Florida or Cuba and heading either into the Gulf of Mexico or the south Atlantic, never to be seen again. As one journalist put it, an airplane lost at sea without a radio "had as much chance as a mouse among cats."[3]

Trippe and Priester agreed that any pilots they hired would have to be expert in blind instrument flying, which no other airline was demanding at the time, but it was Leuteritz's job to provide a fail-safe method of aerial navigation—one based on radio communication between an aircraft and new control towers in the departure and arrival airports. Weather reporting, too, would have to be timely, accurate, and constantly transmitted to pilots. Any man who balked at accepting instructions from ground stations, even if contrary to what his own senses and instincts were telling him, would be dismissed.[4]

As for Lindbergh, Trippe considered him one of the few people who grasped the immensity of what he wanted to do. After the aviator had returned from his U.S. tour, he'd agreed to meet Trippe but, being inundated with offers and honors, didn't know exactly when. The assistant's call came as Trippe was preparing to pick up Betty for a date.

Trippe had fifteen minutes with Lindbergh at the Commodore Hotel. Now. Don't be late.

Trippe stood Betty up, an act that would earn him an angry note from her insisting that "obviously, there are people you care about more than you do me." To Trippe, it was worth the flowers he'd have to send. At the Commodore, he outlined his dream of creating an airline that would eventually span the globe, starting with Central and South America. Impressed, Lindbergh accepted Trippe's proposal to serve as technical adviser.[5]

It was just this kind of farsighted thinking, expansive ambition, and studied perfectionism that distinguished Trippe from anyone else in the air business, as Bevier and Montgomery at Pan American were discovering. Like it or not, they needed Trippe more than Trippe needed them, and Trippe knew it. If they tried to go ahead without him, there was no way of making the October 19 deadline, Pan American would go bust, and Trippe would scavenge its carcass for the Post Office contract.

Trippe's terms were onerous. He wanted full control of Pan American, with Bevier and Montgomery as minor partners. Bevier and Montgomery in turn played their hidden ace: Pan American had another suitor— Atlantic, Gulf, and Caribbean Airways (AGC), yet another airline that didn't really exist in anything but name. AGC was financed by Richard Hoyt, a very sharp Wall Street investor who was not, to put it mildly, a man to be trifled with.[6]

Trippe did an end run around Bevier and Montgomery by having a secret meeting with Hoyt to propose an alliance. AGC and ACA would together take over Pan American and buy out Bevier and Montgomery for the air-mail contract.

Bevier and Montgomery were doomed. As the clock ticked down to October 19 and Trippe deliberately drew out negotiations to increase the pressure to make a deal, Bevier and Montgomery caved a week before the deadline. They handed over Pan American in exchange for $10,000 in cash and $45,000 worth of stock in Hoyt's AGC.

Within hours, as per their secret deal, Hoyt sold 52 percent of AGC to Trippe's ACA for $199,500—most of the cash Trippe had on hand—in exchange for becoming chairman of AGC and allowing Trippe to take over the Pan American subsidiary as president and general manager. From his chairman's seat, Hoyt would be able to keep an eye on Trippe, whom he regarded as being a little too clever for his own good, while making him do most of the work.

On October 13, Trippe summoned the new board of Pan American— packed, of course, with his friends—to outline his vision of its future. Given that he had six days to begin service to Havana and still had no airplanes—the Fokkers were *almost* ready—it was a remarkably ambitious one.

Pan American's headquarters would be in Miami, where he would build a world-class airport to serve as the base for two trunk lines to South America. The western line would extend to the U.S.-owned Panama Canal Zone and from there all the way down along the coast to Chile. The eastern line would run first to Cuba and then through the Caribbean island chain to the Dominican Republic, Puerto Rico, and Trinidad before reaching the mainland and tracing its way to Rio de Janeiro and, ultimately, Buenos Aires. A new airplane route across the Andes would then connect Buenos Aires to Valparaíso (Chile), completing the circuit.

A MINOR PROBLEM, unfortunately, was that none of these routes existed. When it came to airplanes, South America was terra incognita. To make it all work, Trippe would need to gain concessions and permissions from a dizzying list of countries and territories including Mexico, Colombia, Nicaragua, Uruguay, Ecuador, Costa Rica, Honduras, Guatemala, El Salvador, Peru, Venezuela, Chile, and Haiti. He would do it, he confidently promised as the board nodded along, but first, he needed to get to Havana inside of a week.

Aside from pestering Fokker to finish at least one of his damn planes, Trippe's priority was to get Key West up and running in time for October 19. Right now, there was no "airport" there, just a rock-strewn field filled with swamp pits. His contractors were urgently filling in the pits with the rubble left over from leveling two short runways, but on October 15, the manager there, a humorless, red-haired fellow named Captain Whitbeck, told him that they were ready.

And then the rain came. For two days, torrents fell and turned the dirt runways into thick mud. On October 17, Key West was unusable, Whitbeck sorrowfully informed Trippe. The only good news was that a Fokker had arrived in Miami safely late at night on October 18. Only, of course, there was nowhere at Key West to land.

Trippe pored over the air-mail contract, searching, as Eckener had done in the Treaty of Versailles, for some kind of loophole. Like Eckener, too, he found one. The Post Office had directed that a mail plane had to depart from Key West but had omitted to specify that it had to take off from an *airport*. The land-based Fokker was out of the question now, but a *seaplane* wasn't. The mailbags could be loaded onto one floating just off the coast of Key West to fulfill the terms of the contract.

All Trippe had to do was find a seaplane by morning. Late at night, he called Jacksonville, Chicago, New York, and Boston to charter one, but none were to be had. Meanwhile, the train from New York carrying the mailbags had already departed and was due to arrive by dawn. Then Captain Whitbeck had an idea: He called a friend in Miami, left off Trippe's list because no seaplanes were permanently based there. The friend looked out his window and, praise be, a Fairchild FC-2 seaplane had landed just a few minutes before for refueling and maintenance before heading to Haiti.

Whitbeck told his friend to get the pilot on the phone and found himself talking to one Cy Caldwell a few minutes before he was due to leave. The sum of $250 changed hands and Caldwell arrived in Key West a couple of hours later.

At 7 A.M. on October 19, he and Whitbeck loaded seven mailbags from the New York train onto the Fairchild, taking off at precisely 8:04 A.M. Sixty-two minutes later, Caldwell landed in the sea off Havana, and the local postmaster rowed out in a small boat to certify the mail's safe arrival according to the terms of the contract.

Trippe had pulled off a hell of a trick, with mere hours to spare, and Pan American was in business. A week later, on October 28, 1927, the airstrip having finally dried out, Trippe's Fokker departed Key West on the airline's first scheduled flight. Piloted by Eddie Musick, a meticulous aviator soon to become a legend in the air business, the Fokker—obsequiously named the *General Machado*—carried 772 pounds of mail. A jubilant Trippe cabled Betty: FIRST FLIGHT SUCCESSFUL. She burst into tears.

But now the real work would begin.[7]

34. El Dorado

Pan American proved a winner from the very beginning. In November 1927, it carried 19,946 pounds of U.S. mail and 877 pounds of Cuban mail; the following month, the numbers rose to 26,513 and 1,492 pounds, respectively. Thanks to Priester's beady eye, neither the *General Machado* nor the second Fokker, *General New* (after U.S. Postmaster General Harry New—Trippe laid on flattery with a trowel), suffered an engine failure, an equipment malfunction, or a late takeoff.

On January 16, 1928, Trippe announced the beginning of daily passenger service—a long-desired dream now, finally, fulfilled. Customers would board at Key West at 8 A.M. after the mail was collected, and the return from Havana would depart at 3:45 P.M. The eight-seat Fokkers were outfitted with wicker chairs, though they were rarely filled.[1]

The great aviation boom was taking off, but Americans were not yet persuaded that flying over water was truly risk-free. Trippe instructed his agents to rustle up passengers. Salesmen stood outside Pan American's office in Key West temptingly crying, "Fly to Havana and you can bathe in Bacardi Rum two hours from now." In Havana, they would roust inebriated tourists in bars and "persuade" them to fly home. Some of these passengers, shocked to find themselves waking up in midair, would panic and struggle so much they had to be physically restrained from opening the door.

Still, flying had some fans, particularly those who had some discreet business to conduct in Havana. One time, a short, squat fellow accompanied by four flashily dressed young men showed up in Key West. He gave his name as Al Capone, telling the ticket agent, "Better see it's a *safe* plane. If anything happens to us, remember, it won't be so healthy for you fellers." He paid $1,000 for a private flight to Cuba.[2]

With Key West–Havana up and running, Trippe set his sights on his next objective. From March 1928, the Post Office began to offer a series of contracts for the right to carry the mail to every foreign country south of Texas. This was the biggest game of all, and Trippe intended to win every hand.

Get an exclusive contract, make money on the mail, and open passenger service as an add-on. The plan was simple, if tricky to execute. A key point was that one success would lead to another, owing to a new provision in the Post Office rules that the lowest bidder would not automatically be awarded a contract. (The Colonial trick of underbidding to entice a competitor to buy it out had been played once too often.) Instead, "the lowest *responsible* bidder that can satisfactorily perform the services required to the best advantage of the government" would be the first choice. By "responsible" the Post Office meant the best-managed, best-financed, best-positioned airline, which generally ruled out a host of tiny, penniless upstarts.

The victorious bidder would also be expected to serve the U.S. government's interests in the region, which meant it not only had to be American-owned but also willing to work with the State Department to improve America's often prickly diplomatic, economic, and military ties with Central and South America.

Between 1898 and 1927, the *New York World* pointed out, the United States had intervened militarily there on thirty-one occasions and to a large extent regulated many countries' affairs through treaties and agreements, not all of them entirely voluntary. Suspicion of American motives was rampant, and any airline, unless it was careful not to be too closely associated with Standard Oil or the hated United Fruit Company, would raise hackles.[3]

Trippe was more than happy to make Pan American Washington's chosen instrument. If he served his master well, the State Department and the Post Office would smooth the path to his next desired destination by allotting him the relevant mail contract. One by one, inevitably and inexorably, the individual pieces would fall into place to form a structure.

That didn't mean that Trippe's local competitors would meekly walk away from the table. They had to be destroyed, undermined, outplayed—

all of them. A failure to subjugate one would mean the loss of a territory, and the loss of a territory meant a gap in the structure. Too many gaps and the whole thing would collapse.

For instance, Foreign Air Mail Route Number 6 (FAM-6), which began in Miami, stopped in Havana, proceeded to Haiti and the Dominican Republic, and ended in Puerto Rico, was a critical one for Trippe to nab. He needed FAM-6 to break out of Cuba and bring his planes closer to the South American mainland, but a tiny airline named West Indian Aerial Express (WIAE) was considered the front-runner.

Lo and behold, thanks to Trippe's friends on the Post Office committee, FAM-6 was awarded to Pan American, the more "responsible" airline of the two. With the loss of the contract, WIAE went bust, but Trippe had gained 1,930 miles' worth of routes and, more important, his entrée to South America. He acquired another 2,058 miles in Central America when FAM-5 (Cuba–Yucatan Peninsula–Panama) conveniently fell into his lap soon afterward.

The new routes brought in rivers of cash, courtesy of the Post Office. Whereas the Key West–Havana line earned $160,000 annually, Trippe calculated that FAM-5 and FAM-6 would together net $2.5 million per year. Since each contract ran for ten years, that meant guaranteed income of $25 million *on mail alone*. If he offered passenger service on top of that, well, the sky was the limit. His rinky-dink little airline had hit the jackpot.

Wall Street took notice. The stock price of ACA, which had gone public, rocketed from $15 a share to $50 and then to a high of $89; within a few years its cash holdings would rise from a few tens of thousands of dollars to $6 million, making it one of the richest predators in the world. Trippe went on a spending spree, buying up more airplanes than anyone in the business had ever seen and beginning construction on a long-desired airport on 36th Street in Miami.

Miami was intended to mark Pan American's arrival to the world. Designed by the famous New York firm of Delano & Aldrich, architects to the wealthy, the classy, and the clubby, 36th Street — or "Pan American International Airport," as Trippe preferred to call it — was one of the first truly modern air terminals in the United States.

A grand, light-filled two-story affair with a cantilevered roof, it was graced with a huge panoramic window overlooking the runways. There was a large waiting room with wicker furniture, palm fronds, and a huge wall map of North and South America. Spanish wrought-iron grillwork separated the terminal into "Arrivals" and "Departures," from which out-going passengers walked beneath canopied promenades to their airplanes waiting on the tarmac.

Everyone who passed through was, as Trippe intended, appropriately astounded at the Napoleonic scale of his ambitions.

As AMERICA'S MOST glamorous airline, Pan American made sure to look the part. Trippe instituted a rigorous selection process to find only the best cabin crew. They were expected to be "alert and good-looking youngsters" exemplifying the virtues he demanded of a Pan American STEWARD—"Service, Tact, Efficiency, Wisdom, Ability, Responsibility, Dependability"—and were required to speak three languages and have experience working in upmarket hotels and ocean liners.

Unlike U.S. domestic airlines, which had already begun hiring women trained as nurses, Pan American, like Zeppelin, remained stoutly all-male. After training in what would later be called the "Gable Routine"—in this exercise, stewards had to serve an imaginary high-maintenance Holly-wood celebrity while looking after the other passengers and supervising the mailbags—they were taught basic seamanship skills. Supremely con-fident and outfitted in smart uniforms of black trousers and white waist-length jackets over boiled white shirts and black ties, Pan American's cadre of stewards would become famed throughout the globe.

Trippe, who took only a small salary, owned fully a quarter of the com-pany. So, yes, one might say that he, a newly minted millionaire, was now gainfully employed—fully satisfying the demands of Betty's family. She was summoned home from her gilded Parisian exile and the wedding date was set. The ceremony on June 16, 1928, annoyingly, took place in the midst of yet another Pan American expansion, and the honeymoon was necessarily a brief one to the Catskills so that Trippe could get back to work.[4]

He did have a lot to do. Over the next two years, Pan American won every single one of the foreign air-mail contracts the Post Office put up for bid. Asked by an inquiring colleague how he had achieved this remark-able record, Trippe "just smile[d] his most guileless smile and shrug[ged] his shoulders."[5] As well he might. Nothing happened by happy accident during Pan American's conquest of the continent.

From each South and Central American government Trippe demanded a lengthy list of concessions in return for international air service. Trippe's men carried Pan American's fearsome "Form B" in their brief-cases, laying out his terms: twenty-five-year exclusive landing rights, customs and immigration privileges for his passengers and cargo, immunity from local taxation on his new airports and aircraft, and the right to carry mail to the United States.

Any complaints by existing local airlines about Pan American's muscling in would be assuaged in due course by cooperating with, co-opting, crushing, or buying them.

When potential resistance was too strong, Trippe always made a deal, his assumption being that he would be able to force out his partner later on. The giant conglomerate W. R. Grace & Co., for instance, had been operating in Peru for seventy-five years, but its roots also ran deep in Chile, Bolivia, and Ecuador. The company "owned practically everything on the west coast," a Trippe field agent reported—including, no doubt, the politicians.

For obvious reasons, Grace could make life difficult for Trippe, who urgently wanted landing rights in Peru so that he could drive toward Chile. So he set up Pan American–Grace Airways (PANAGRA), split equally between the two concerns but really run by Trippe. Within weeks, PANAGRA was awarded the lucrative new route FAM-9, Panama-Santiago. Another major piece—some 4,500 miles—of Trippe's planned western trunk line had fallen into place and ACA stock reached a new high in New York.[6]

If cooperation failed, co-optation was the fallback. Colombia was a particularly hard nut to crack, thanks to a German-owned airline already operating there: Sociedad Colombo–Alemana de Transportes Aéreos (SCADTA). Its boss, Peter Paul von Bauer, a courtly Austrian noble, was popular among Colombians and put up a fierce fight, but he was running out of money. Trippe made his offer: Work with us and get rich or die poor. For $1.142 million, half in bearer certificates (making it impossible to ascertain the identity of the owners) and the other in cash, the co-opted Bauer secretly handed over 84.4 percent of the company to Trippe and agreed to vote his remaining shares in Pan American's favor at board meetings.[7]

Sometimes, crushing the foe was the only option, as happened when a new airline, the New York, Rio, Buenos Aires Line (NYRBA), appeared. Its president was the juttingly jawed Ralph O'Neill, a man whose bravery was second to none (he'd won the Distinguished Service Cross and the Croix de Guerre as a fighter pilot in the war) but whose volcanic temper was ill suited to combating the likes of Trippe. The latter regarded O'Neill as someone "who thought that [he] would like to run an international airline. But [he] didn't really know what it was all about."

Thanks to Trippe's fix-it men, important shipments of NYRBA equipment vanished en route, construction work moved at a snail's pace, and usually efficient local officials repeatedly forgot to approve the necessary permits. By the time Trippe got through with O'Neill, NYRBA had blown through millions of dollars and had just $16,000 in the bank.[8]

Cooperating with Grace had brought a windfall, co-opting SCADTA had removed a nettlesome problem, and crushing NYRBA had been delightfully satisfying, but sometimes it was just easier to purchase an airline impertinently standing in the way, as happened with Compañia Mexicana de Aviación.

It was owned by George Rihl, a former Philadelphia banker with something of a checkered past. His right-hand man was Erwin Balluder, of German background and born in the Virgin Islands in 1893. They were the kind of colorful rogues with flexible morals ideally suited to the rough-and-tumble world of South American aviation.

Rihl and Balluder had made sure to bribe the Mexican postmaster into giving Compañia Mexicana the exclusive right to fly mail through the country, which meant that Trippe couldn't push them aside. For once, Trippe would have to pay over the odds. The pair made a killing: Their flyspeck of an airline, really a couple of ramshackle planes, was sold for $150,000 in Pan American stock. Trippe judged the transaction to be well worth the trouble, mostly because he was so impressed at Rihl and Balluder's cojones that he hired them as Pan American vice presidents, though their *real* job was to work as his dirty-tricks specialists.[9]

Trippe's methods were not usually outright illegal, but they certainly carried with them a whiff of the dark arts. He frowned on blatant bribery because, as John MacGregor, one of Trippe's leading fixers, wisely advised, "when you pay out graft, there's always a kick-back to it afterwards." Instead, the approved technique of greasing a local politician was to adapt Form B to include certain "donations" to his favored party newspaper or granting him free travel forever.[10]

But when push came to shove, cash was king, and Rihl and Balluder were the kings of cash. Balluder's specialty was to target the "people who knew people in the government": brothers-in-law of generals, the son of a major industrialist, former senior officials, and the like, sensing in them a refreshing willingness to exploit their influence in exchange for a payout.

As for George Rihl, he was even more cynical. During a stop to Guatemala City, he told a shocked Arthur Geissler, the U.S. envoy to the country, that "money will finally decide whether [we] will be permitted to fly in Guatemala." Indeed, the very purpose of his visit was to bribe the minister of the interior. To which poor, honest Geissler replied, "You seem to think that all public officials are either crooks or fools." "No, not all of them. Not quite all of them," chuckled Rihl, amused that Geissler was quite this naive.

A few days later, the minister of the interior announced that he was welcoming Pan American to Guatemala.[11]

TRIPPE ALWAYS KEPT himself aloof from such goings-on, preferring to conquer South America from his desk in New York. He never even bothered to learn Spanish, seeing in it a shameful reminder of his family background. Only after much pressure from his ace publicity man, William Van Dusen, who could scarcely believe that Trippe would *not* use such an obvious way to curry favor with Latinos, was he persuaded to sign letters (written by a Hispanophone secretary) with the detested "Juan."[12]

Trippe's ulterior motive for steering clear of his own dominions was that he could not afford to be seen by the scrupulously clean Lindbergh as being anything but aboveboard. Lindbergh, as Pan American's technical adviser at $10,000 a year (plus stock options worth $150,000) and the greatest advertisement for the airline one could imagine, knew nothing about the skulduggery involving his friend's conquistadors.

In the early fall of 1929, Trippe at last deigned to visit South America when Lindbergh mounted a tour there. Trippe brought along Betty; Lindbergh, his new wife, Anne. Trippe had told Lindbergh the tour was an opportunity to promote the cause of aviation and the beneficence of American business, but it more accurately marked Trippe's coronation as emperor—even if this emperor preferred to stay discreetly in the background to let the Sun King bask in all the adulation.

Lindbergh piloted the whole way as Trippe, invariably dressed in a scorchingly bright white suit, conducted business by radio in the cabin. The publicist Van Dusen, who later recalled that at Pan American "sometimes we had to invent" myths to obscure "truth and the absolutes," made sure reporters were given upbeat releases several times a day on their progress. Most of the resulting stories focused on the human-interest angle—Anne writing letters while airborne, Betty keeping a diary, Lindbergh flying alongside brilliantly colored tropical birds and flamingos—to generate huge interest in the voyage while Trippe got on with the real work.

At every stop, they were mobbed by well-wishers, garlanded with flowers, paraded through flag-draped streets accompanied by brass bands. Every night they attended dinners and dances in their honor. And every night, Trippe made time to meet State Department officials, Pan American agents, and presidents in palaces filled with copies of Louis XIV furniture who were waited on by servants costumed in tailcoats, brass buttons, knee breeches, and white gloves. And every night, a new deal was struck.

Over the course of three weeks, the quartet covered nine thousand miles, zigzagging their way through Pan American's vast, sprawling dominions. It wasn't, however, all business. Even the fastidious Trippe, who liked flying but hated traveling—he always brought his own linen sheets—was persuaded to let his hair down. Trippe, who treated people as interests, had few friends, but Lindbergh was an exception. One night, much to the amusement of Betty and Anne, the two of them got into a pillow fight. It was the first time their wives had seen Trippe and Lindbergh, two men of rigid emotional control, acting so boyishly. "The Trippes have been such fun, and wear so well," wrote Anne. "I think they're remarkable. The more I see of them, the more I think it."[13]

Trippe could afford to relax. He had accomplished something extraordinary, something, in terms of explosive growth, not only unprecedented in the history of aviation, and perhaps even the history of business, but unrepeatable.

At Christmas of 1927, Pan American had been a company flying a single ninety-mile route; two Yuletides later, its route network had risen to 15,575 miles; and by mid-1930, to 23,075 miles. By way of comparison, the *entire* U.S. domestic air network, serviced by *dozens* of airlines, amounted to 30,045 miles.

By the time he was thirty years old, Trippe the Great ruled an empire vaster than Pizarro, Cortés, or Columbus could ever have conceived. He oversaw an enterprise that encompassed twenty-nine countries, connected 83 percent of the South American population, employed 1,200 people, operated seventy-one airports on two continents, and owned forty-four multiengine aircraft that flew on so precise a schedule that you could, it was said, set your watch by them.[14]

Pan American planners forged an "air trail" between Buenos Aires and Santiago over the Andean wall, a place where winds could buffet an airplane so suddenly it shot from 13,000 feet to 22,000 feet in minutes, ice formed on the wings, pilots were blinded by snow, and each passenger nursed an oxygen bottle just in case.[15]

Trippe's men built airports in places deep in the interior once reachable only by riverboat, oxcart, or pack animal. His managers—many, incongruously, Yalies—braved poison arrows and crocodile-infested swamps and the green hell of deadly jungle to find promising locations, they called upon scout planes to drop bags of flour to mark in splotched white puffs where work crews coming by canoe or on burro would clear foliage, and they hired local tribesmen to haul five-gallon cans of fuel over peaks and through valleys.[16]

And the result of the blood and the sweat, the tears and the malaria? A letter that had taken six weeks to travel the eight hundred miles between, say, Lima and Iquitos, a jungle-bound city at the headwaters of the Amazon, now took thirty-six hours, and a passenger in Barranquilla wishing to visit Bogotá, who used to spend up to a month aboard a boat negotiating the treacherous river, could fly there in nineteen hours. Small wonder that an admiring journalist, astounded at the wonders wrought by Trippe, could only remark, "The Midas touch of the airplane is transforming everything it meets to gold."[17]

But Trippe had no intention of stopping at El Dorado.

While Trippe's idea of a first-rate Christmas party for his staff was to order in ham sandwiches and bottles of Prohibition-approved Coca-Cola, for the 1928 bash he had a surprise in store—two-week bonuses for everyone—and an announcement to make. Never a natural toastmaster, Trippe perched himself on the edge of a desk and gazed out the window, haltingly speaking of his plans for Pan American in the coming years: "We will be going across the Atlantic and after that, across the Pacific. We are going around the world."[18]

Hugo Eckener had had exactly the same idea. Two very different men with identical ambitions, one as implacable as the other in the pursuit of victory, meant a coming struggle for the mastery of the air.

Eckener mastered the trick of wearing a public mask—that of the avuncular, backslapping Zeppelin captain—to disguise a native cunning. In private or among a crew, though, he was much sharper and grouchier than Trippe: Heaven help you if you made a careless or unthinking mistake. Yet his men adored him, for his loyalty to the loyal was unshakable.

Eckener always made the Zeppelin project about himself. He and the airship were one and the same, its successes were his successes, and without him it could not survive. There was an air of gruff arrogance about him, but deservedly so: Nobody knew more than he about mastering the technical challenges of the weather, managing operations, planning voyages, and driving an airship.

The secretive, withdrawn Trippe, however, kept to the shadows, was hardly seen outside his private office, never piloted his own airplanes, and carefully came across as much less important than he was in order to disarm potential foes. When someone needed to be fired, the task was invariably delegated to an underling—Trippe hated to be seen as the bad guy. Lashings of charm and dollops of amiability covered his guile, and he tried to turn enemies into friends by offering them gifts and tokens.

Eckener took things personally that to Trippe were just business; Eckener bore grudges, Trippe had interests. Whereas Eckener told you bluntly if he disliked you, with Trippe you would never know where you stood until the moment you were ruthlessly dispensed with. Veterans of dealing with the Pan American chieftain learned to look out for the tells that he was up to something. His occasional stutter was a dead giveaway: "When Juan Trippe stumbles for words," advised one, "you have to watch out; he becomes more deadly than a seven-headed hydra."

Their strategies reflected their characters. Eckener favored frontal assaults by risking everything on mounting a spectacular feat to win the battle. Trippe preferred to stay on the flanks, covering his rear and advancing cautiously, step by step, to play for position.

How they worked was of a piece. Eckener spent little time in his office, which was in any case more of a study—book-lined, cigar-fugged, leather-armchaired—where, said a friend, "he'd talk with you for a few seconds, look you up and down with those all-seeing eyes of his, and know right away what you were thinking and what you really wanted of him." That of Trippe, on the other hand, contained only a giant globe festooned with red threads on which he would obsessively plot potential routes, as well as, oddly, two desks. The first was a regular flat table where he discussed various matters with visitors; the other, to the side, was a specially made roll-top desk that he locked whenever he left the room or someone entered. Secretive to a fault, he worked for hours at the rolltop, lost in his own world. When colleagues came to him to ask about some routine matter, he answered them brusquely, then turned back to his crystal-gazing. If they persisted, they received no reply, almost as if he had never heard them. Perhaps he hadn't.

Outside the office, Trippe, impeccably dressed in conservative suits, white shirts, and dark ties, his hair invariably parted and neat, camouflaged himself by vanishing into the background, but it was impossible to miss Eckener, bordering on six feet tall with a massive, deeply furrowed face, goateed chin, and severely crew-cutted hair, whose corpulence went undisguised by his cheap, wide-pinstriped suits. The jackets, often an odd shade of light blue, always seemed short and bulging, as if he were squeezing himself into them, and the trousers were covered in a dusting of old cigar ash that wafted in his wake.

Unlike Eckener, whose booming north German voice with its *verdammts* and *scheisses* could be heard from one end of a hangar to the other, Trippe let others do the talking for him. Eckener threw out quotes from Schiller, Goethe, the Bible, Shakespeare, and Mark Twain interspersed with analogies to Greek myths and Roman history to make his points, while in Trippe's home there were neither books on the shelves nor pictures on the wall, lest any visitors, of whom there were few, succeed in deducing his true interests from his literary or artistic tastes.

Really, there was not so much to learn of this Man of Secrets: He knew little of history, music, or philosophy, and his "religion," noticed one observer, consisted of worshipping the airplane: "He literally thinks and reads nothing but aviation." In public, he talked only of business, football, and golf—all safe, uncontroversial subjects—and restricted his swearing to "Oh, gosh" and "Darn" to avoid offending anyone.[1]

THEIR BUSINESS SITUATIONS, too, were very different. In one sense, Trippe had it easy: During the great aviation boom, everyone was making money, he had a complaisant State Department at his disposal, and South America was right on his doorstep. His problem was that there were many competitors in the airline industry, and breaking out of the Americas to get across an ocean was impossible, given the restricted capabilities of his airplanes.

Eckener, meanwhile, had to contend with the treacherous political and economic environment of postwar Germany, but he enjoyed a monopoly in the airship business. What he lacked was the money and the opportunity to exploit it.

Yet there were encouraging signs of a change in fortune. The Dawes Plan had amended Germany's reparations-payment schedule, much in Berlin's favor, and foreign loans had consequently flooded in, spurring economic growth. The funds built new factories and modernized old ones that had survived destruction in the war.

Better still, the Locarno Pact of October 16, 1925, brought Germany in from the diplomatic cold. Berlin and Paris mutually recognized their contentious border and pledged "that they will in no case attack or invade each other or resort to war against each other." Instead, they would bring disputes to the Council of the League of Nations for arbitration.

On May 22, 1926, infused with the "Spirit of Locarno," the Conference of Ambassadors, the body that had originally imposed the one-million-cubic-foot size restriction, lifted its edict. The "special exception" clause, too, was deleted, legally allowing Eckener to build as many Zeppelins as he wanted, of any size.[2]

The opportunity was now there, just not the money, thanks to the

Spende fund-raising debacle. To get it, Eckener appealed directly to Paul von Hindenburg, former commander of the kaiser's armies in the war and current German president, to save the Zeppelin. He sent a lengthy memorandum laying out the advantages of the airship over the airplane. There was no hope, claimed Eckener, that the airplane would achieve long-distance flight unless it was so colossal it could barely take off; further, it would never be as comfortable as an airship, and it was uneconomical to boot.

These were by now all old and much-rehearsed arguments, but Eckener more tantalizingly hinted that he had something big up his sleeve. "For the construction of the next airship in Friedrichshafen," he wrote, "a change in the design is intended, which has far-reaching significance, indeed, which can almost be called revolutionary with regard to the airship's navigation, safety, and cost-effectiveness; this change will improve the airship's situation compared to that of the airplane quite a bit more. Unfortunately, no further details about this design change can be disclosed here yet."[3]

Soon after, the government disbursed 2 million marks to Zeppelin to help pay for this mysterious airship. Eckener was still short, so, ignoring Colsman's heated objections, he bled the subsidiaries dry to fund the Zeppelin Company.

Everything, then, had to be sacrificed to the great god of Zeppelin, which manifested itself in Eckener's announcement in July 1926 that he was beginning work on LZ-127, an airship "in which one would not merely fly, but would also be able to *voyage*."[4]

THE PLANNING STAGE of LZ-127 proceeded remarkably quickly. By early 1927, shortly before Trippe was purged from Colonial and Lindbergh embarked on his Atlantic flight, Dürr and his team had completed the technical blueprints, and the duralumin needed for the rings and longitudinals was beginning to arrive. Friedrichshafen was more active than at any time since the war; Zeppelin City sprang to life again. For the first time, Eckener dispensed with the secrecy surrounding his hangars and offered guided tours and souvenirs. No longer had he anything to hide: LZ-127 was Germany's airship by hard-won right. No more intrusive inspectors from the Inter-Allied Commission or impertinent questions by American overseers.[5]

That LZ-127 was essentially a stretched copy of ZR-3 *Los Angeles* with new, improved Maybach VL-2 engines sped up construction. Eckener had wanted to stretch LZ-127 even more, but Dürr was inherently restricted by the dimensions of Friedrichshafen's main hangar, built in 1916. The hangar was 787 feet long and 138 feet wide, so LZ-127 was 776 and 100, leaving spare a mere 5.5 feet bow and stern and 19 on each side. Height-wise, with just 24 inches of space available, LZ-127's upper backbone, 110 feet in the air, barely scraped under the hangar's arches when walked in or out. A single minor mistake by the ground crew and they would have been tormented by the sound of shrieking metal and ripping fabric.

For the first time since the day-tripping LZ-120 *Bodensee*, Eckener was able to devote more attention to passenger comfort than to military demands. The new ship had a galley set behind the control cabin equipped with a stove, electric burners, a refrigerator, and racks carrying silver cutlery and white porcelain ringed in blue and gold and bearing the monogram "LZ" (for Luftschiffbau-Zeppelin).

The lounge, 16.5 feet by 16.5 feet, was brilliantly lit by four large, outward-slanting windows, two of which opened. Four dining tables could accommodate the twenty passengers on board, with a typical menu comprising creamed chicken soup, fried pork chops, red cabbage, mashed potatoes, cucumber salad, and ice cream sundaes for lunch, and a dinner of grilled fish with salted potatoes, accompanied by cold cuts and salad, and finished off with fruit compote. The lounge was covered with a wine-red carpet, the wooden chairs, cushioned and upholstered in chintz tapestry, complementing the dark curtains and mahogany walls. Eckener's interior design tastes remained staunchly traditional, and he preferred the clubby look of Pullman first-class trains.

Behind the lounge were two sets of five two-berthed cabins on either side of a hall, as there had been on the *Los Angeles*. They remained mostly the same, but the folding sofa was updated with a striped pattern and the walls were decorated with a pretty flower print. A closet, a folding canvas stool, and a small table had also been added. Down the hall were two modern toilets and washrooms with hot and cold water.

The most radical change—the one Eckener hinted at when he told Hindenburg that it would improve the airship's "navigation, safety, and cost-effectiveness"—was not immediately apparent.

A perennial problem of airship flying was that as the engines burned relatively heavy fuel during a trip, Zeppelins became lighter and so naturally tended to lift. In practice, the captain would valve out hydrogen to bring his ship back down into static equilibrium. In this manner, several hundreds of thousands of cubic feet were typically exhausted into the atmosphere on each flight. The *Los Angeles*, for instance, was no less than twenty-two tons lighter when it reached America in 1924 than when it departed Germany, and fully a quarter of its hydrogen had had to be vented to compensate.

The more hydrogen that could be preserved, then, the fewer times an airship would be obliged to land to replenish its stock of gas—translating into longer range in less time.

The solution was Blau gas, named after Dr. Hermann Blau. Used to light trains and heat stoves, it was, chemically speaking, quite close to the coal gas employed by nineteenth-century balloonists but, containing no carbon monoxide, was far less poisonous. Blau gas resembled propane in that it could be transported as a liquid but released in a gaseous state only slightly heavier than air. In the latter form, it could replace liquid gasoline as a fuel. So as it was consumed the airship would experience almost no lift because the overall weight remained virtually the same.

Some gasoline was still carried for emergencies, as ballast and for take-offs and landings, but a comparison between the two on a typical trip shows a dramatic difference in power and endurance: Had it not been for LZ-127's 918,000 cubic feet of Blau gas, which weighed thirty-three tons and was enough for one hundred hours of flight, a similar amount of gasoline would have kept the airship flying for just sixty-seven hours. In one stroke, Eckener increased airship range by a third *and* saved considerable hydrogen-replacement costs.[6]

On July 8, 1928—on what would have been Count von Zeppelin's ninetieth birthday and around the same time as Trippe was beginning to flex Pan American's muscles—his daughter, Countess Hella, christened LZ-127 *Graf Zeppelin*. Eckener chose the name to honor the count's risk-everything approach. When *Graf Zeppelin* soared into the heavens, Eckener wrote, "the result would be either victory and fame, or final downfall and the end of a defective concept."[7]

GRAF ZEPPELIN WAS put through its paces in a series of five trial runs over the summer and early fall.[8] On one of them, the airship flew over President von Hindenburg's hunting lodge to drop flowers and a congratulations card signed by Eckener and the crew to commemorate his eighty-first birthday.[9]

The airship experienced no major technical problems but caused two minor incidents. Both occurred on October 2–3, when Eckener decided to take the *Graf* for an evening flight over the English coastal towns of Lowestoft and Great Yarmouth. On the return leg, Eckener sailed toward the Rhineland, where it received an enthusiastic welcome.[10]

The French considered the overflight a challenge to the post-Locarno status of the Rhineland. They annoyedly directed, in the interests of keeping the peace, that henceforth Zeppelins fly over French territory only at night to prevent peeking at military bases, keep to scheduled routes, secure all photographic equipment, and stay at high altitude.[11] As for the British, the sight of a Zeppelin hovering over towns that had been bombed by them was needlessly provocative. Eckener claimed an honest mistake on both counts, which might have been true, but likely wasn't.

Eckener was in fact cannily playing to domestic politics. The Socialists, conservatives, and center-left Social Democrats were resolutely pro-Zeppelin, leaving only the Communists still denouncing the airship as a reactionary tool of capitalist imperialism. But Eckener's intention had been to throw a bone to the growing nationalist movement, which had previously attacked him for "giving" ZR-3 to America. The nationalists had, however, warmed to Eckener ever since he had announced that the *Graf Zeppelin* would be entirely German-made, German-staffed, and German forever.

Just a few weeks earlier, the extreme right had demanded a popular referendum to abolish the parliamentary system and bring back a form of monarchy, either by restoring the Hohenzollern dynasty or by electing a dictator. In the beer halls, too, there was talk of French perfidy and warnings that force of arms might be necessary to take back control of the Rhineland.[12] Complementing these dark rumblings was a burgeoning interest in the deeds of the airshipmen of the Great War.[13]

Erich Maria Remarque's novel *All Quiet on the Western Front* was appearing in serial form at the time, and nationalists loathed what they saw as its cynicism, defeatism, and tone of futility. They preferred a new genre of memoirs written by airshipmen and pilots, who had soared above the mud and blood of the trenches and fought a heroic, romantic war in the sky. Accounts written by Eckener's old comrades invariably deified the mad airship supremo Peter Strasser and gushed over his Wagnerian death in L-70 while fulfilling his duty—unlike the backstabbers who had betrayed sacred Germany in 1918 by surrendering. Even the destruction of the Zeppelins at Nordholz, a stupid act of self-defeating sabotage, was re-interpreted as a brave last stand against the Allied occupiers. Ernst Leh-mann, back from America, was a predictably keen proponent of this revisionism and helped popularize the erroneous belief that Strasser's air-ship fleet had forced the British to keep a "million" soldiers at home to defend their island.[14]

Eckener, who had known Strasser very well and had long been aware of both the Zeppelins' impotence and Lehmann's politics, could have quickly put this nonsense to rest. Instead, he placed his dream of airship domination above the truth. By compromising himself, Eckener was get-ting into bed with some very dangerous people, as he would eventually discover. Convenient allies have a way of turning into enemies at inconve-nient times.

To the outside world, Eckener presented a completely different face. America was his primary market, and it cared little about Germany's inter-necine politics. For its inaugural flight, Graf Zeppelin was turned into an international business machine. Advertisers in the American media pro-moted German products like Zeiss binoculars, Leica cameras, Krupp crankshafts—anything that was on board was packaged and sold as Zeppelin-approved. Eckener gave talks to leading German industrialists explaining how the airship, as an emissary of business, would help their bottom line, and he was gratified when the German-American Chamber of Commerce in New York reported widespread admiration for the Graf and a willingness to buy German products.[15]

Eckener always had a purpose. He never embarked on major undertak-ings like a transatlantic flight without a greater aim in mind. Yes, it was nice to advertise German goods and to bask in the warm embrace of na-tionalist feeling, but the real point of the coming voyage was to convince American financiers to put up $15 million to amass a four-dirigible fleet (with perhaps a fifth used as a spare), two of which would be built in Friedrichshafen and two in Akron by Goodyear-Zeppelin, in order to run weekly service back and forth across the Atlantic.[16]

In fact, Eckener was already sure that the Graf Zeppelin, being too small to be put to the test repeatedly over the Atlantic, was inadequate to the demands of regular New York service. To his mind, the Graf Zeppelin, as impressive as it was, was only a proof-of-concept device to demonstrate long-term commercial feasibility.

Timing was key. The *Graf Zeppelin*'s trip alone would probably not be enough to loosen American purse strings and make airships a viable business. In the midst of their aviation boom, the Americans were far too obsessed with airplanes to consider such a huge investment based on a one-time trip. Eckener had to create an unstoppable groundswell in airships' favor to make success inevitable.

Thankfully, in the fall of 1928, the winds were blowing strongly his way.

FOR ONE THING, the *Los Angeles*, commanded by Lieutenant Commander Charles Rosendahl, had been performing splendidly since the *Shenandoah* catastrophe. Rosendahl had been second officer on that doomed ship and was now counted — by Eckener, among others — as the country's leading airship expert. A broad-shouldered, blue-eyed man, born in 1892 as the son of Swedish immigrants in Chicago, Rosendahl attended the Naval Academy and later saw action in the war on destroyer duty in the Atlantic. In 1922, a call went around for volunteers for the navy's new airship program, and Rosendahl had put in his name.[17]

Over the previous few years, he had impressed his superiors by flying several reconnaissance missions over the Pacific in the *Los Angeles*. As a result, the navy decided to commission two new 6.5-million-cubic-foot helium airships, ZRS-4 (USS *Akron*) and ZRS-5 (USS *Macon*), in the week leading up to Eckener's departure from Germany.[18]

In a radical departure, *Akron* and *Macon* were intended to be airborne aircraft carriers. They would have to include a way of storing and launching at least four airplanes either to serve as fleet scouts or to defend the airship against attacking fighters. With an $8 million contract being offered, Goodyear-Zeppelin promised they would figure out how to include an internal hangar and an external "trapeze" to which to attach the airplanes hanging below.

Akron was to be delivered in thirty months, *Macon* fifteen months after that, and to build them Arnstein was charged with erecting the world's largest hangar in Akron, called the Air Dock. The height of a twenty-two-story building and enclosing an area of seven, maybe eight, football fields, the Air Dock could fit the Washington Monument, the Woolworth Building, *and* the *Lexington* and *Saratoga* aircraft carriers if tipped on their sides. Some visitors, peering into the depths of its cavernous vault, experienced a sense of vertigo or weakness in the legs, and it was said, somewhat plausibly, that a skilled pilot could take off, fly, and land a small plane inside. Its doors alone weighed 2,400 tons, each one so vast that, exclaimed an excited reporter, the task of opening one was "equal to swinging the side of an eighteen-story skyscraper on hinges."[19]

In his spare time, Arnstein had been advising Howard Hughes on the construction and interiors of wartime airships for a new movie being made, titled *Hell's Angels*. Hughes, wanting a big Zeppelin-on-fire scene for the pic, had no fewer than four colossal models built, putting him in the airship business almost as much as the navy or Eckener.[20]

Between keen anticipation for the epic movie, the announcement of the navy contract, and the Goliathan proportions of the Air Dock, better publicity for the *Graf Zeppelin* could scarcely be imagined. It was now up to Eckener to deliver a home run to win the game.

36. Survival of the Fittest

EARLY IN THE morning of October 11, 1928, *Graf Zeppelin* slipped its moorings and rose into the slowly pinkening sky. Aboard were twenty passengers and forty crewmen, including Eckener's three most experienced captains: Lehmann ("small, nimble, keen as a gimlet, always brimming over with good humor and dry wit," said one reporter), Schiller ("ever active, courteous, charming, with sea-deep eyes"), and Flemming ("tall and handsome"). There was also Knut Eckener, the great man's son, and not least Jacob Meyer, a junior Zeppelin machinist who'd begged Eckener for a spot in order to bring an engagement ring to Miss Dora Stoeckle, currently living in New Jersey, whom he'd earlier met in Friedrichshafen, "where he passed her home on his way to the hangars daily."[1]

Most of the passengers had not paid for their tickets. The free riders were a mixture of reporters, cameramen, government officials, and business partners, but it was reported that four people had paid $3,000 apiece, a vast sum, for the glory of being the first paying passengers to cross the Atlantic. One of them was Frederick Gilfillan, a mysterious American financier living in Switzerland who had recently survived a plane crash in the Azores (he also had two shipwrecks to his credit). He would prove such a handful that Eckener would have been better off paying *him* $3,000 to stay at home.[2]

The German press was well represented, but Eckener had separately arranged with William Randolph Hearst to cover the journey as an exclusive for his giant newspaper chain. As part of the deal, passengers and crew had to sign confidentiality agreements before they left to ensure they didn't leak to Hearst's rivals. The magnate sent two journalists: Karl von Wiegand, a friend of Eckener's who'd been covering airships since the war, and Wiegand's rumored lover, the only woman on board, Lady Grace Hay Drummond-Hay.

Lady Drummond-Hay, in her early thirties, was the recent widow of Sir Robert, a British diplomat in Lebanon older than she was by some half a century. Plump-cheeked, adventurous, witty, and vivacious, Lady Drummond-Hay had turned her hand to journalism and been talent-spotted by Hearst a year or two earlier. She had made quite a name for herself and was an ideal correspondent for the trip.

Drummond-Hay's dispatches overflowed with the kind of color, excitement, glamour, and brio readers hankered after. She brought more clothes than anyone else (on departure she wore a "black hat, white georgette waist, a black skirt, gun-metal silk stockings, and a gray squirrel fur coat with a bunch of white violets on her lapel"), made frequent changes (all the better to show her fans that flying was fashionable), and was the only one farsighted enough to bring woolen underwear, woolen stockings, wool-lined boots, and a jumper suit for the cold nights, made colder by the fact that the *Graf* lacked heating. It was she who seems to have given Eckener the *Graf Zeppelin*'s mascot—a canary named Hansy who sang for most of the trip and gave the reporters some excellent copy.[3]

The voyage began the same as so many others. As the *Graf* headed toward the Alps, passengers "could not tear themselves away from the windows, running from one side to the other, exclaiming at every new phase of the scenery," wrote Lady Drummond-Hay, or became "fascinated by the cabin arrangements, the charming little sleeping compartments . . . which contain almost every comfort and convenience one can wish."[4]

Two Americans had reportedly insured themselves for $1 million each (Eckener showed his confidence in his creations by always sticking to a basic $30,000 policy), and some of the Germans, who had evidently read too many westerns, had brought a chest of dollars with them in case the *Graf Zeppelin* became lost "on the plains of Arizona" and they were "stranded there among the cowboys and Indians."[5]

But any apprehensions they felt soon washed away as the airship passed over Gibraltar, gateway to the Atlantic, and life settled into a routine of chatter over bottles of wine, diary-keeping, and games of chess. A dance night was arranged, with Lady Drummond-Hay, as sole female, in high demand. She chose Captain Lehmann as her first partner, but the festivities were soon curtailed because the music, supplied by a lone accordion, "was not so good," according to one newspaper.

One morning, off the Azores, as the *Graf Zeppelin* was driving through the air at about 70 miles an hour at 1,250 feet, Wiegand and Lady Drummond-Hay were breakfasting when the sky darkened and "suddenly the bow of the ship dipt at a great angle as if the air-liner were about to dive into the cold waters, which at that moment looked perilously close." Drummond-Hay, ace reporter to the last, shouted, "Save my typewriters on the table in my cabin!" and Wiegand obediently ran to go get them but was thrown off his feet in the corridor when the nose sharply lifted. "There was a crash, jingle, and rattle as all the dishes on the three tables were piled up on the floor."

As strong and virile men "turned pale, with the thought of death in their eyes," Lady Drummond-Hay, in full Wodehousian Aunt Agatha mode, stoutly cried, "I've lost my coffee!" Her eggs landed in her lap, the plate in someone else's, and Gilfillan—who, said Wiegand, "had been panning everything on board the airship. The coffee was rotten, the air bad, the ship slow, and the service terrible"—suffered the indignity of a movie camera hitting his head, which put him in a mood fouler than before.[6]

Wiegand went to the control car to find Eckener in deep discussions with the other officers. What appeared to have happened was that a nearby squall had caused a *Shenandoah*-style vertical current and an inexperienced helmsman had overcompensated in bringing the nose down to regain trim. Then he exacerbated the problem by trying to ascend bow-first, making it feel as if "a giant hand had suddenly shoved up out of the sea below."

Eckener was again fully in command of the vessel, which had been righted, when the ashen-faced chief machinist arrived from the stern and told him that hundreds of square yards of fabric had been ripped from a stabilizer fin and entangled in an elevator hinge. Eckener ordered the engines stopped, turned to his guest Lieutenant Commander Rosendahl of the *Los Angeles*, and said quietly, "I want a ship to stand by." Rosendahl nodded and telegraphed the navy to send vessels to their position.

Eckener had to perform emergency surgery mid-flight. Knut, his son, and a handful of others volunteered for the task. As ominous clouds enveloped them, seven crewmen, including Knut and Chief Helmsman Ludwig Marx (with Zeppelin since the count's earliest days) undertook the perilous mission to crawl on all fours along the catwalk and then climb the girders until they were outside on the giant fin. The torn outer skin flapped in their faces as the rain got into their eyes; one poorly chosen move or an unlucky slip would result in a man falling a thousand feet into the churning sea. With shears and knives they cut away the cloth and painstakingly sewed it back together as the tail whipped fifty or a hundred feet this way and that, swaying them like sailors clinging to a yardarm. Eckener, in the control room, could not see or communicate with his son, but after five hours the team returned safely and Eckener told Rosendahl to call off the rescue. By midday, Eckener had restarted the engines and gingerly accelerated out of the storm at 71 mph.[7]

Among the passengers, excitement soon turned to boredom. The delay and a subsequent detour to avoid another storm resulted in food and drink running short. Guests took their lukewarm coffee in glasses as all the china had broken and, after the coffee ran out, had to make do with condensed milk because not enough wine and beer had been brought (and whatever drops were left were locked away as the *Graf Zeppelin* neared America, to conform with Prohibition laws). The electricity had not been turned back on while the crew checked for problems, so there were neither hot lunches nor hot water. Instead of cuisine prepared by Friedrichshafen's Kurgarten Hotel and sealed into vacuum cans for reheating, there was only smoked salmon on buttered bread to be had, and washing was done with wet paper towels that "melt[ed] to pulp in one's hand," Lady Drummond-Hay informed her readers.[8] Deepening the ennui was the ban on smoking, a source of considerable irritation to Frederick Gilfillan, who never shut up about it while others soldiered on "with unlit menthol cigarettes in our lips," as one passenger said, or "pulled at unlighted pipes."[9]

But finally, on October 15, land hove into view and everyone's spirits rose. The *Graf Zeppelin* took Washington, D.C., by surprise by floating over the capital for twenty minutes. As it dipped gracefully over the White House, President and Mrs. Coolidge, surrounded by nervous Secret Service men, emerged to gaze and applaud the apparition casting a shadow over them. Thousands of government personnel left their offices to watch the show, but, as the Supreme Court was in session, Justice Oliver Wendell Holmes and Chief Justice William Howard Taft were forced to miss the event. "I wish I could have seen that Zeppelin," complained Holmes.[10]

That afternoon, the *Graf Zeppelin* arrived over New York, hypnotizing the hundreds of thousands of spectators who gathered in parks or on rooftops. There were so many, judged one newspaper, that their multitude exceeded that which had congregated to watch the passing of Halley's Comet in 1910. During the *Graf's* half hour above the city, the courts of justice ceased dispensing, as did retailers and wholesalers; taxi drivers refused to pick up, mischievous office boys threw confetti everywhere, and elevator attendants quit their posts to go outside. Prison wardens even allowed their charges a little extra time outside to enjoy the sight. As the airship proceeded north along Broadway to turn west at 110th Street to cross the Hudson into New Jersey, bells jangled on streetcars, trains shrilled their horns, and factory whistles joined in with their baritones, adding to the deep-chorded notes of steamships greeting the mistress of the skies. A dozen biplanes cheekily circled the *Graf Zeppelin*, dodged over, and slipped alongside it.[11]

Shortly after, the airship loomed above Lakehurst and officially landed at 5:38 P.M., ending its 111-hour-and-38-minute voyage. The first visitors were Dr. Leon Van Horn of the Public Health Service, acting as quarantine officer, and Customs Inspector Theodore Morgan, whose task it was to transfer passengers, cargo, and luggage to the waiting room. This took an inordinately long time, partly because Eckener would not allow baggage to be unloaded until the ship was secured, and sparked a minor diplomatic incident.

Morgan's Customs officials examined the mail and goods before moving on to the luggage and only then allowed their Immigration counterparts to clear the tired passengers. Since the *Graf Zeppelin* carried nearly seventy thousand pieces of mail, as well as packages containing such things as five dozen pairs of women's leather gloves, Parisian hats, and precious jewelry, the process was a drawn-out one. Each and every suitcase was then inspected while the passengers stood around for hours.[12]

Eckener—who had slept for just eight hours in the previous four days—slipped out to smoke a cigar (the crewmen cadged lights from the base's American sailors), and was unaware that some passengers had begun to complain mightily of high-handed treatment by humorless Customs and Immigration officials. One German passenger shouted, "Is this the American freedom we hear so much about?" which somehow failed to move his interlocutors.[13]

The brouhaha, which eventually calmed down, paled beside the general atmosphere of excitement and anticipation outside. Some twenty thousand people had come to see the landing, and their numbers increased dramatically after carpenters built a wooden cradle to help rest the gondola. If they were willing to wait on line for a few hours, visitors were allowed to climb the stairs, peer in, and gasp at the control room, lounge, radio room, and passenger cabins.

Over the coming days, an estimated 150,000 people would come to Lakehurst, requiring the presence of seventy-six Marines, fifty sailors, and forty state troopers whose main jobs were to tell visitors to extinguish their cigarettes when anywhere near the airship hangar and to confiscate cameras (Hearst was determined to retain his exclusivity rights). With forty thousand cars on the roads leading to the naval base, ten special trains laid on from New York, Atlantic City, Philadelphia, and Newark, and hundreds of buses, the traffic for twenty miles around was sluggish indeed.[14]

Demand for Zeppelin news around the world was insatiable. "250 Reporters Held Lakehurst Sector," announced *The New York Times*, amusingly making it sound as if the Journalism Corps were stoutly defending the line on the Western Front.[15] In Germany, where people had listened in beer halls and at home for hourly radio reports during the flight, news of Eckener's triumph was greeted by cries of "God be thanked!" Hindenburg cabled his congratulations, and even the dour Dürr was more effusive than usual. In taking passengers by air, "the *Graf Zeppelin* has accomplished a deed heretofore unknown in the history of crossing the ocean," he announced before flashing a brief smile, a sight rarely glimpsed. Elsewhere in Friedrichshafen, where the airships' officers and crew lived, their fellow townspeople spent the night celebrating in the streets.[16]

The next day, Eckener got down to work. In New York, there were the usual ticker-tape parades, honorary luncheons, and celebratory dinners, as well as some amusing gossip items. One newspaper noticed that Eckener, who pronounced "exhausted" as *ex-howsted*, was unaware that he sounded like Lew Fields, a comic famed for his exaggerated German accent, while Captain Schiller was delighted that not only stores but even cab drivers refused to take his money.[17]

Less noticeable, but still detectable, was the lingering, low-intensity conflict between Eckener and Lehmann. When they were forced into close proximity for long periods, the normally ebullient Eckener would become "sardonic" while the chatty Lehmann turned "uncommunicative."[18] Disagreements over their relative status and fame lay at the heart of the matter. When Lehmann wrote a long opinion piece for *The New York Times*, Eckener was irritated to find that Lehmann had somehow omitted to include *his* name while mentioning his own numerous times—and had also supplied a very large photo of himself. In this case, Eckener gained a satisfying revenge for the slight when a subsequent article ran a picture of him and Lehmann, neither looking too pleased to be posing together, with a caption that ran: "Commander of the Zeppelin and His Aide."[19]

Lehmann, at least, could be dealt with later. Eckener's most urgent task was to sell the *Graf Zeppelin* as a business proposition. He was feted and lauded everywhere he went around the country, but the trip did not turn out to be the grand slam he'd anticipated.

No one committed to fund his $15 million airship venture. All his meetings went very well, but there were no signatures where it counted. What he heard repeatedly instead was, We're Interested But Let's Wait and See.

THE TRUTH IS, the 1928 flight had not decisively proven the airship's superiority. News that the *Graf Zeppelin* had needed urgent repairs mid-flight was damaging, as were the revelations of the troublesome American passenger, Frederick Gilfillan, who had waited for Hearst's confidentiality agreement to expire to give hostile interviews to every paper he could. Among his litany of complaints: running out of mineral water, lack of sleep caused by anxiety over the fin repair and the descent, the small cabins, the lack of room to exercise, the ban on smoking, a shortage of wine, the canned food, and the closed windows. "Which reminds me," he added, "I haven't had a bath for five days."

Though most of these could be written off as minor inconveniences, easily remedied if Eckener improved the in-flight entertainment and onboard dining selection, more harmful was Gilfillan's assertion that "the airship, as I see it, is too uncomfortable in its present stage of development to commend it to the traveler." He returned home by the Cunard steamship *Mauretania*.[20]

Gilfillan was a symptom, albeit a singularly annoying one, of a broader evaluation of the merits of Eckener's *Graf Zeppelin*. Now that the initial furor had died down, experts were taking a harder look at his plan to establish a regular transatlantic line. The general consensus was that, well, nobody quite knew since the results were so ambiguous.

On the plus side, the *Graf Zeppelin* had demonstrated to a nervous public that airship travel was safe (an airplane suffering a similar mishap mid-flight would have been doomed) and enjoyable (Gilfillan excepted), could traverse huge expanses, and could carry more passengers and freight than any conceivable airplane. To the man in the street, it definitely seemed to be the future. "In this vast monster of the atmospheric deeps," rhapsodized the *New York Herald Tribune*, "the public, despite the tragedies that have punctuated the history of the dirigible, feels a strange confidence." To that end, the *Graf Zeppelin* would be just "the first representative of that mighty argosy of air-liners with which the human imagination has for generations been painting the skyscape of the future."

The picture was less positive when it came down to brass tacks. In terms of cost, for instance, it was unclear whether airships could pay their way. Eckener claimed the flight had made a profit of $100,000 but remained opaque on details.

Before departing Germany, he'd said that operating costs for a one-way trip would be $54,000. Mail and cargo revenue was estimated to amount to about $70,000, but one had to consider the huge expense of building the *Graf Zeppelin* (thought to range between $840,000 and $1.4 million, though almost certainly considerably more), the gigantic $2 million hangars it required, and its ongoing maintenance (hydrogen, fuel, storage, parts, etc.). One also had to account for the large number of personnel required to fly a Zeppelin: If a ten-seater airplane employed two pilots and a stewardess to stay in the air, then the twenty-passenger *Graf Zeppelin* needed forty highly trained officers and crew to do the same.

All of which meant that if Eckener *had* turned a profit, it was probably owed to his selling of the media rights to German press agencies for $15,600 and to Hearst for another $67,400. But he could hardly count on doing the same for every trip.

Since only four people had paid full fare for the voyage, if Eckener could subsequently increase the number of paying passengers he might have a viable enterprise, but he would have to keep prices high and every cabin occupied on every trip to pull it off—and he'd need to make scores of trips to offset the sunk costs of construction.

Whether that was possible was another question. Eckener had always bragged that airships would beat ships handily across the ocean, but *Graf Zeppelin*'s travel time to New York had been 111 hours, unimpressive compared to a fast *Mauretania* time of 107 hours. Granted, Eckener had had to make a detour south and to contend with a lengthy repair, but North

Atlantic storms were inevitable occurrences, so how often would they delay flights? And assuming most flights did arrive a day or so earlier, was it really worth paying $3,000 for a ticket to take an airship across the Atlantic when a steamship would get you to your destination a little later but much more comfortably and cheaply?[21]

An aviation exposition in Berlin—Eckener's home turf, shockingly—took a dim view as to his chances. Even accounting for the challenges Eckener had had to overcome, reported *The New York Times*, among the experts present "there is beginning to be a doubt about the value in the advancement of commerce by these lighter-than-air craft." Rather worryingly, "aeronautical engineers of standing say their skepticism has been increased, rather than diminished, and they are more firmly than ever convinced the airplane will soon set other forms of air transportation in the background. Some even predict the Zeppelin will be as big a failure in the commercial world as it was in the World War."

The airplane-airship debate had taken on new vigor, as well as an alarming new aspect. In the early 1920s, airplanes and airships had been regarded as parallel technologies, neither impinging on the other, and existing in tandem. Everyone knew that airships handled long-range travel and airplanes short- and perhaps medium-range trips. But the calculus was changing as airplanes advanced in leaps and bounds. There were thousands of them now, and as Pan American was ably demonstrating, you could cover an entire continent in routes using these small, cheap, convenient, safe conveyances. Airships still had the advantage of range, comfort, and payload, but how long would that situation last? What would be their fate if a passenger airplane eventually flew across the Atlantic?

Observers now began to see the airplane-airship problem as a ruthlessly existential one when it came to long-haul travel. Inevitably, wrote Arthur Blessing in the *North American Review*, one would supplant the other to "square with the principle of utilitarianism, and this means the survival of the fittest."[22]

For the time being, whether the "fittest" would be the airplane or the airship was an open question, and at the exposition "what all agree[d] upon is that final judgment should not be passed until the *Graf Zeppelin* makes a round-the-world flight, which should be a crucial test of the worthiness of this type of aircraft."

One attendee in particular, as usual avoiding the spotlight, listened very carefully to the proceedings: Juan Trippe.[23]

· · · · ·

THANKFULLY, THE RETURN home helped put paid to the criticisms about the *Graf Zeppelin's* lackadaisical flight time. A strong tailwind bore the "silvery mammoth of the air" along at 88 mph, and Eckener and his crew arrived in Friedrichshafen in an amazing 71 hours and 51 minutes—far faster than any steamship.[24]

Eckener received the traditional greetings and torchlit processions, followed by the bestowal of three honorary degrees and a meeting with President Hindenburg, but once the festivities were over he was still faced with a major challenge: The *Graf Zeppelin* had to go around the world.

Eckener would have to make do with a ship he knew to be too frail for the task.[25] Neither was the irony of the situation lost on him: He had to circumnavigate a globe of nearly 25,000 miles in a small airship to prove that he could traverse an ocean 3,500 miles wide in a big one.

37. Around the World

IF AT FIRST the logistics and distances involved with girdling the globe appeared daunting, the endeavor, once Eckener sat down and began plotting the *Graf Zeppelin*'s course, was not so very different from taking the airship to America and back several times—and might even be easier than anyone expected. Traveling east across Europe was not hard, and much of Russia (including Siberia) was one gigantic plain; from the east coast of Russia to Japan was a piece of cake, and crossing the Pacific, though it had not been done before in a single leg by either airplane or airship, he did not consider a difficult proposition.

Of the two oceans, the Atlantic was the far trickier route, its storms and winds fiercer than in the Pacific. Crossing the United States from California to New York, Eckener believed, would actually be the hardest part of the voyage, but he had no intention of running the risk of combating the fearsome midwestern tempests that had doomed the *Shenandoah*. Eckener chose his route carefully to avoid the danger areas as best he could.

Of greater concern was isolation. Siberia was an unmapped region of trackless forest, oozing swamps, and algae-slimed lakes; it was a primeval place, virtually unedited by the hand of man, where very few of the very few inhabitants had ever seen a train or a car, let alone heard of an airship. If the *Graf Zeppelin* had to make an emergency landing, its passengers and crew would be beyond radio contact and stranded for months— assuming that they could ever be found. Overexcited reporters persisted in asking Eckener whether they would be eaten by "wild Kimucks and other bandits" in that event, to which Eckener always responded that he thought the inhabitants were "not so bad as they are painted."[1]

At least, though, if the *Graf Zeppelin* came down in Siberia it would come down on land; not so with the Pacific, where islands were in short supply and were invariably tiny, meaning that the *Graf Zeppelin* could easily vanish amid the blue waves.

But the greatest risk of all, even greater than crazed cannibals and sinking without a trace, was political. The *Graf Zeppelin* was entering a diplomatic viper's nest. Arranging a voyage to the United States had been a simple enough affair consisting mostly of submitting a request for a permit to land and agreeing to abide by customs and immigration regulations. But circling the world required gaining entry to two closed countries—the USSR and Japan—both thought to be as aggressive as they were secretive. In Europe, the insidious spread of Bolshevism was seen as a menace to existing liberal-democratic societies, whereas the Americans were particularly apprehensive of Japanese intentions in the Pacific. A *Graf Zeppelin* visit, then, came fraught with political tension, and Eckener had to be especially careful to avoid the slightest hint of favoritism or partisanship.

Financially, the costs associated with the Round-the-World voyage were prodigious: a minimum of $250,000 (shipping 900,000 cubic feet of Blau gas to Japan did not come cheap). Generating intense newspaper and newsreel interest, then, was again critical to the enterprise, and Eckener offered the reliable William Randolph Hearst a first-look deal to buy the exclusive media rights. Hearst responded with a colossal offer of $100,000 for the American and British reporting.

The Hearst deal came with conditions. The magnate insisted, for instance, that the voyage begin and end in New York, putting Eckener in the crosshairs of the German nationalists who saw him, yet again, selling out to the Americans.

To forestall criticism, Eckener took the expedient of simultaneously running dual around-the-world voyages. For the Americans, he spoke of the trip as New York to New York, but in Germany he referred to it as Friedrichshafen to Friedrichshafen. The extra effort required mattered little to him. If anything, it would help popularize airships in both his primary markets, the United States and Germany, but just as important, flying Friedrichshafen-to-Friedrichshafen would allow him to cover more costs by selling the newspaper rights for the European portions to German press agencies for $12,500.

The rest of the expenses were covered by a combination of Japanese newspaper fees, passenger fares (at $2,500 per ticket), and sales to stamp collectors of special issues. Eckener later claimed to have turned a profit of $40,000 on the flight. Maybe he did.[2]

ON AUGUST 1, 1929, Eckener lifted off from Friedrichshafen with twenty passengers, mostly reporters and American observers like the navy's airship expert Charles Rosendahl, but also present (though not given cabins) were a baby gorilla named Sue and Louis the chimpanzee, who had been ordered by an exotic-animal importer. Ninety-five hours later, they were at Lakehurst. The only kerfuffle happened after landing, when it was discovered that while everyone's attention had been diverted by the antics of Sue and Louis, a gang of thieves had taken the opportunity to steal the naval base's canteen safe containing $700.[3]

Late on the evening of August 7, the *Graf Zeppelin* was on its way back to Germany, where it would "restart" its voyage. The fifty-five-hour return trip was incredibly fast (there was no more talk of "slow" airships) but as boring as Eckener had hoped—all the better to prove that flying on an airship could be as routine as taking a train or a steamer. The greatest irritant was not the weather but the incessant, night-and-day clacking of typewriters in the lounge as reporters cranked out their stories.

Eckener, who rarely partook of alcohol in the air, was so relaxed he joined the other guests in a few glasses of wine and Champagne to celebrate his sixty-second birthday. In Friedrichshafen, Johanna, worried that her energetic husband was overexerting himself, was quoted as saying that the best birthday present for him would be an opportunity to sleep, a comment that led to rumors that the Round-the-World trip would be his last before retiring. Not by a long shot; he was only just beginning, Eckener reassured reporters.[4]

On August 15, when it was time to depart for the mysterious Land of the Soviets, Eckener crossed his fingers that the trip would prove equally dull. It hadn't been as hard dealing with the Russians as many had expected, though that didn't mean it was easy. They had insisted on having their own team of representatives on board; when Eckener had complained that they would weigh the ship down too much, the Soviets grudgingly reduced the number to one, a meteorologist named Professor Karklin, who spoke no other language than Russian. But he had gamely learned a sentence in German by rote, which he repeated at every opportunity: *Moskau Industrie, viele Leute, Sibirien nicht gut* ("Moscow industry, many people; Siberia no good").[5]

The *Graf Zeppelin* sailed across East Prussia and the Baltic coast, through Lithuania and Latvia, heading for Russia. Eckener was certainly fortunate in his timing. The Russians had ended their bloody civil war a few years earlier and Stalin was firmly in control, yet the horrendous impact of collectivization, the starving of Ukraine, the mass imprisonments in the Gulag, and the paranoid murderousness of the purges still lay in the future. The secret police were everywhere, of course, including on board the *Graf Zeppelin*, where Professor Karklin was expected to pull double duty as informant. Eckener noticed he was spending more and more time in the control car and growing increasingly nervous as they pushed on.

Eckener was not just using the USSR as a way to get to the Pacific, and his hosts had their own reasons for allowing him through their territory. For the Americans, the *Graf Zeppelin* may have been a business proposition, but for the Soviets, it was a symbol of a burgeoning politico-economic relationship.

Twelve months earlier, Stalin had inaugurated his first Five-Year Plan with the intention of modernizing a backward Russia to Western standards, and German firms had benefited tremendously from the crash program. In 1928–29, more than a quarter of Soviet imports came from Germany alone, and there were more German companies operating in the USSR than the rest of the world's combined. The Russians needed German industrial equipment for their modernization drive, and in exchange they played discreet host for armaments research and production still illegal under the Versailles Treaty. Grenades and artillery shells were secretly manufactured, and German "advisers" ran a tank school, a poison-gas production facility, and a pilot-training college, all well beyond the reach of Western inspectors.

Of equal or even greater value was German engineering, technical, and scientific expertise. Hundreds of specialized personnel had already been lent to the Soviet Union, and Moscow was eager to have Zeppelin staff join the ranks, for aviation was also part of the Five-Year Plan. Stalin had demanded that Soviet aircraft must fly "further, faster, and higher" than any inferior capitalist product, and airships were promoted as status symbols of modernity. If the Germans had them, so too must the Soviet Union.

What the airships would be used for was never clearly enunciated, but the main principle seemed to be that brandishing the hammer of revolutionary ideology in the service of state-run industry would only highlight the impotence of such mundane bourgeois concepts as profit and loss, supply and demand. To that end, Soviet proponents stressed quantity over quality to demonstrate superiority.

The head of the Dirigible Construction Trust, for instance, bragged of a fleet of 425 airships of varying sizes, the grandest of them to be inspiringly named *Lenin, Stalin, Old Bolshevik, Collective Farmer,* and *Pacific Ocean Proletarian.* None would ever be built, but at the time of the Round-the-World flight and the feel-good era of German-Soviet cooperation, Eckener was hoping to expand Zeppelin into Russia. Unfortunately, he had accounted for neither the singular nature of Soviet politics nor the vagaries of the weather.

Stalin, rather strangely, never spoke publicly about airships, and unbeknownst to the Germans, the whole subject was a contentious one within senior Party circles. True, the *people* were thoroughly excited about the prospect of a Zeppelin visit and their coming airship fleet, but in the topsy-turvy world of the USSR popular enthusiasm for something was counted as a black mark. In Stalin's eyes, homegrown, organic excitement was suspect; great projects had to be planted and cultivated by the Party and imposed on the people, like broccoli to children, for their edification and improvement.[6]

Innocently unaware of these internal Party struggles, Eckener had headed for Russia assuming it would be a normal trip. Hence he could not understand Professor Karklin's mounting anxiety when he announced in passing that he would have to skip the scheduled Moscow visit. High winds were heading toward the city and Eckener, wishing to reduce risk and fuel consumption, wanted to fly north around it. Karklin, growing ever more agitated, kept pointing at the map and signaling that Eckener *must* go to Moscow. "Siberia no good," he repeatedly insisted.[7]

Eckener ignored the warnings and pushed on. To him, this was a small matter—in America and Germany, he occasionally had to postpone an arrival and leave spectators disappointed—but he was taken aback by the hysterical Soviet reaction to his apolitical, weather-related change of course. *Pravda* devoted large amounts of space to attacking Eckener, German perfidy, and sinister capitalist influence. A long article by one Michael Kolzoff declared that the West was terrified by the "twelve years of proletarian dictatorship" that had resulted in Soviet superiority in automobiles and airplanes and that Eckener, owned by "insatiable Uncle Sam" and swayed by his "bourgeois foolishness" and "degenerate arrogance," had deliberately bypassed Moscow in order to avoid witnessing the achievements of collective farms and parades by the Young Communist League. He menacingly concluded that Eckener had better remember that the way to the Far East ran through Soviet territory, and "if it suits us—we'll allow the path to be used; if it doesn't suit us—we'll close the path."[8]

To help patch up relations, Eckener asked the German ambassador to convey his "deep regret" at having had to disappoint the hundreds of thousands of Muscovites who had gathered to see the *Graf Zeppelin* and to explain his reasons. As a result of his obeisance, the Soviet press was directed to tamp down the outrage and extend an olive branch.[9]

All's well that ends well, but the uproar over nothing in 1929 had convinced Eckener that there could never be a truly globe-spanning airship line, in the sense of a virtually nonstop voyage, at least not while Stalin remained in power. The thought of having to deal with Moscow's whims and tantrums on a repeated basis simply because the USSR happened to own some useful real estate was off-putting, to say the least. Instead, Eckener would stick with the Atlantic, where at least the Americans could be counted on to be sensible.

IN CONTRAST, THE flight to Japan was easy and the *Graf Zeppelin* was everywhere greeted with fulsome acclamation. Wrote a reporter rapturously, the airship "glinted in the sky over the land of cherry blossoms where geisha girls halted their clip-clopping to watch her pass, and where Fujiyama was mirrored like a tinted shell in the dark waters of Lake Hakone."[10]

To the Japanese, Eckener had brought their country closer to Europe, within four days of Berlin, as opposed to a month aboard a fast liner. Some four million people craned their necks at the sky to watch the heavenly chariot pass over Tokyo and rocked the earth with their cries of "Banzai!" while the newspapers proclaimed the flight the "greatest accomplishment mankind has achieved." Eckener, for having avoided a typhoon on the way over, was instantly dubbed the Storm King.[11]

Then it was on to Kasumigaura, where a quarter of a million people awaited the *Graf Zeppelin*'s arrival. From then on, Eckener scarcely had a moment to himself as the Japanese laid on an unending series of formal ceremonies.

First there was a visit to a teahouse, where Eckener, bemused at a geisha's gesturing at his shoes, experienced mild panic at the thought of taking them off and exposing a large hole in one of his socks. After several minutes' earnest discussion of the diplomatic ramifications of such an insult, the Japanese produced small woolen bags to protect his modesty. Then came the state banquet at Japan's grandest hotel with the foreign minister, the minister of communications, the minister of war, and the admiral of the navy in attendance, splendidly outfitted to the nines in formal tails or dress uniform. Finally, that rarest of honors: tea at the Imperial Palace with the recently enthroned 124th emperor, Hirohito, who presented Eckener with a pair of silver cups—which joined a ceremonial sword, silk embroideries, ornamental daggers, and porcelain vases in crates shipped back to Germany by freighter.

Back in her modest little house in Friedrichshafen, Johanna, busy making a sweater for Eckener, was told by a reporter of her husband's safe arrival. Accustomed by now to hearing about the accolades heaped upon his head and the medals pinned to his chest, she said modestly, "Isn't that wonderful? We go to sleep when the ship sails and awaken when it returns. Now I must get on with my knitting."[12]

Less discussed amid the excitement—the Japanese newspapers had printed more column inches about the visit than any other event in history—was that Eckener's stay coincided with the U.S. Navy's campaign to build up its own airship fleet; in Washington there was alarm at the enthusiasm shown in Japan for the *Graf Zeppelin*.

To glean some insight into Japanese designs, the State Department instructed its attachés in Tokyo to compile detailed summaries of the national press coverage of the visit and its possible implications. Much had been made of "German scientific progress," "German culture and civilization," and "German genius and technical skill," they reported back, adding that within Imperial Navy circles there was talk of buying several Zeppelins for oceanic scouting.[13]

The U.S. Navy's recent contracting for the *Akron* and the *Macon* airships had been predicated on the idea that they would spot the Japanese battle fleet, but if the Japanese also had Zeppelins, the U.S. Navy would be equally at risk of being spotted.

With this revelation, if Eckener had entertained any hopes of selling airships to Japan he would have to forget about them, at least if he wanted to avoid difficult questions in Washington about "trading with the enemy." In the event, the decision was made for him by the Japanese, who ultimately lost interest in airships and plumped for aircraft carriers instead.

Though the Round-the-World voyage was producing a huge amount of favorable publicity, it was also demonstrating, by dint of the loss of Moscow and Tokyo as potential customers, that the Zeppelin business would have to be confined to Germany and America.

ECKENER MADE HIS triumphant return to the United States at 6:06 P.M. on August 25. With his eye for a good publicity shot, he precisely timed his arrival and angle of approach over the Golden Gate of San Francisco Bay as the sun gloriously set behind him. "The oncoming monster of silver sheen," one newspaper reported, was "burnished at times to russet by the rays of the declining sun" as it entered the bay at an altitude of about 1,000 feet.[14]

Despite the tooting of thousands of car horns and the cheers of the multitudes below, Eckener tarried only an hour over the city, the *Graf Zeppelin* dipping and curtseying to the awe and delight of San Franciscans.

His ultimate destination was Los Angeles, but first he had to pay fealty to his benefactor. The *Graf Zeppelin* flew south along the coast and at about 11 P.M. passed over Hearst's famously gargantuan castle (fifty-six bedrooms) at San Simeon, the one later mocked for its vulgar megalomania in *Citizen Kane*. To Eckener's surprise, "everything was enveloped in deep darkness. Nobody seemed to be awake. But suddenly hundreds of lights went on, flooding with brilliant light the large mansion." Eckener telegraphed his thanks to the great man.[15]

They landed in Los Angeles several hours later for refueling and replenishing the hydrogen. With the fate of the *Shenandoah* in mind, Eckener was morbidly aware of the unpredictable weather lying ahead, but his flight across the United States was not half as bad as he thought it might be.

The desert of Arizona and New Mexico surprisingly proved to be the most difficult. As the crew of the L-59 had discovered in 1917 during their China Show cruise across Africa, the sun's heating of the air during the day caused the ship to lift 600 to 1,000 feet at a time and downdrafts pulled it earthward by the same amount. Quite a few of the passengers became seasick, and Eckener, himself a little green round the gills, resolved to never again cross a desert if he could help it.

His mood was not brightened by some aggrieved homesteader taking a potshot at the *Graf Zeppelin* with a rifle when they flew over Texas, but aside from that unwelcoming don't-tread-on-me gesture, the flight rapidly improved from there on out as the *Graf Zeppelin* soared diagonally from Oklahoma to Kansas to Chicago, and from there to Detroit and Cleveland and Akron (they waved hello to Paul Litchfield of Goodyear) and finally to Lakehurst.[16]

Thanks to his timing, Eckener avoided the midwestern storms in the fall, but between those and the southwestern desert in the summer, he was realizing that much of the continental United States was close to being a no-go area for Zeppelins. For the moment, that was a problem obscured by the tumultuous applause, the bedlam of whistles, and the cacophony of countless car horns he received everywhere he overflew. New York in 1929, for instance, surpassed even New York in 1928 when it came to the celebrations. The ticker-tape parade, Eckener noted, was satisfyingly bigger than previous ones and more telephone books, said the papers, had been ripped apart to make confetti than ever before recorded—more even than for Lindbergh.[17]

Eckener was invited to the National Geographic Society in Washington, D.C., to be awarded its Special Gold Medal, only the eleventh so given. Before an audience of six thousand, including cabinet secretaries, congressmen, diplomats from thirty-five nations, and gold-bedecked ranks of generals and admirals, Eckener was inducted into the pantheon of great explorers. According to Gilbert Grosvenor, the society's president, Eckener's voyage made him a fit companion to such immortals as previous medal winners Lindbergh, Peary, Amundsen, Shackleton, and Byrd.[18]

But these accolades were baubles compared to his greatest feat. In going around the world, one magazine said, Eckener had single-handedly revived the "airplane-dirigible controversy" and the Zeppelin had retaken the lead in the long-distance stakes. It was now up to the airplane "to prove its possibilities," not the other way around.[19]

38. The Monster

ECKENER HAD DEPARTED Friedrichshafen with a reputation as a flying Columbus, an adventurer who had crossed an ocean to the New World. He returned to New York hailed as an aerial Magellan, an explorer who had circumnavigated the orb of the world.

He had broken record after record—6,900 miles nonstop from Germany to Japan in 102 hours; the first Pacific crossing of 5,500 miles nonstop in 79 hours; the first American transcontinental nonstop voyage of 3,500 miles in 52 hours—with barely a mishap or more than a few days' delay.[1]

So swift had been his progress that passengers experienced a new and unexpected phenomenon: a discomfiting form of time dilation that had obliged them to repeatedly put their watches ahead an hour to keep up with the International Date Line. Some of them worried that they had mysteriously lost hours and days of their lives.

More immediately, however, the *Graf Zeppelin*'s success proved exactly what Eckener had set out to prove: Airships were a practicable proposition. To that end, he dispatched Lehmann home in the *Graf Zeppelin*—all the better to sideline him—while he toured New York, Washington, and Akron meeting officials and fielding calls from financiers eager to invest in Zeppelins now that Eckener had come through on what he had promised a year earlier.[2]

Now it was less "wait and see" than "hurry up and go." On October 22, 1929, the Pacific Zeppelin Transport Company (PZT) was formed to develop a line between Los Angeles or San Diego and Honolulu. Currently served by steamships that took up to five days, the route was ripe for exploitation. Some twenty-four thousand tourists had sailed from California in 1929, along with six thousand pounds of first-class mail each week. A Zeppelin could cut two to four days off going by sea, making the U.S.-held territory of Hawaii an enviable destination for affluent vacationers.

Jerome Hunsaker, a vice president of Goodyear-Zeppelin and a longtime airship advocate (he'd met Eckener before the war), became PZT's chief executive while Paul Litchfield of Goodyear served as chairman. Their business plan was a little vague, but the gist was that once Goodyear-Zeppelin had built the *Akron* and *Macon* it would begin work—scheduled for completion in 1932—on two or three giant airships for weekly runs to Honolulu.

Litchfield boasted that PZT's "Transpacific Zeppelins will have sleeping, dining, lounging, promenading, and other accommodations for eighty passengers" plus space for twenty thousand pounds of mail and packages. Fitting all of this would be a cinch with two decks: a lower, with a glass-enclosed promenade surrounding a large club room and the dining room; and an upper, an open-air promenade where travelers lolling in deck chairs could sip martinis—Litchfield suggestively hinted that "all the flying won't be done over Prohibition country"—in the balmy Pacific breezes or hang over the railings watching the waves pass by at 100 mph.[3]

Complementing the PZT was the International Zeppelin Transport Company (IZT), based in New York and ably managed by Willy von Meister, twenty-six years old, six and a half feet tall, and fluent in German and English. The son of a wealthy American woman and a German noble (the kaiser *and* the crown prince had stood as his godfathers), Meister had graduated from Darmstadt Technical University with a degree in mechanical engineering in 1924. A couple of years later, he'd sailed to America to set up a Maybach car dealership and later acquired a contract to service the *Los Angeles*'s engines. Meister had gone to Lakehurst on the day of the *Graf Zeppelin*'s arrival in 1928 to greet Eckener, only to find himself pressed into service as a translator during the fracas in the Customs Office. Eckener had been so thankful he invited the young man to accompany him on his business tour and to work for Zeppelin.[4]

Whereas PZT, essentially a Goodyear-run airship line, advertised itself as an "all-American company" and was associated with Zeppelin in Germany only by dint of its universally recognizable name—Eckener, for instance, was not on the board—IZT had, as its moniker indicated, more of an international flavor.

IZT announced that by 1934 it would assemble a four-airship line, each of 5.2 million cubic feet, with at least two built in Friedrichshafen. These "*Graf*-beaters" would have crews of 46 and accommodate 120 passengers in "luxurious staterooms . . . equipped with tubs and showers." Eckener had taken to heart criticisms about restricted room to move and boredom setting in, so when they weren't exploring the observation decks and promenades, passengers would enjoy a swing orchestra, a music room with a piano, a quiet writing room, a bar, and a solarium, which would serve as "a café-dansant by day, and night-club after dark." There would even be a radio room to listen to the wireless and, mirabile dictu, a room with television, then in an experimental stage.

Eckener had assembled an impressive roster of investors, first and foremost being Frederick Rentschler's United Aircraft and Transport Corporation, a sixteen-company conglomerate far larger than any other air operator, including Pan American.

When Trippe read the news, he couldn't understand why Rentschler, an airplane man through and through, was suddenly sinking millions into airships. What was his game?

Originally from Ohio, Rentschler was born poor but grew up rich as his father, an immigrant German ironworker, worked hard and became a major industrialist. While Frederick and Trippe were of course acquainted—the aviation world was a small one—and didn't actively dislike each other, they didn't exactly like each other, either. Rentschler, who had the hard-edged look about him of a hungry shark, contrasted markedly with Trippe's dolphinlike softness, but personal issues were in any case irrelevant, for their relationship was run on strictly business lines admixed with realpolitik.

A little later, Trippe himself explained how he regarded these sorts of things when he told a former partner he had just betrayed (and nearly bankrupted): "We're businessmen. We can't have friends. We can't be in the position of according favors. We have to look at each deal on a cold-blooded business basis. We've made some enemies."

Trippe assumed, as his own words indicated ("We've made some enemies"—maybe *he* had), that everyone else believed the same, and his career was and would be littered with examples of forming partnerships with enemies, making enemies of partners, and reversing himself when circumstances changed. Nothing, to men like Trippe and Rentschler, was permanent; everything was in flux. To Trippe, the mere fact that Rentschler was in bed with Eckener meant that Pan American was in their sights. That was just good business, after all.

Trippe quickly put two and two together. At that moment, United was poised to buy National Air Transport (NAT), which operated the New York–Chicago route. Since United already owned San Francisco–Chicago it didn't take a genius to work out that taking over NAT would create a single California–New York passage.

Still, United would remain only a domestic airline and no threat to Pan American, unless . . . Rentschler joined Zeppelin to go international. And not just across the Atlantic, where United planes would pick up and drop off Zeppelin passengers at Eckener's U.S. terminus, but potentially down in South America, too, where Trippe's exclusive landing-rights system would break down as governments realized they could play the rivals off against each other or insist on nonexclusive agreements.

Eckener had more friends in South America than Trippe had ever suspected. After the war, fleeing from economic and social strife, more than 130,000 Germans had emigrated there, and these *Auslandsdeutsche* formed a potent and talented population of industrialists, businessmen, and technicians who maintained bonds with the fatherland—not least of which was an attachment to the Zeppelin.

To maintain at least a semblance of the fiction that there was competition, Trippe had "allowed" a few tiny competitors to exist alongside his South American empire. First among these was SCADTA, the German-run (and now covertly Trippe-owned) Colombian airline, but there was also the Syndicato Condor, an operation based in Brazil that was quietly subsidized by Berlin through Deutsche Luft Hansa, the new German national airline. At the moment, Condor was nothing but a speck compared to Pan American, but it was gearing up to expand little by little.

The airline already had short hops from Rio de Janeiro to various Brazilian cities, as well as to Montevideo in Uruguay, and it was planning to open a line from Rio to Buenos Aires, then west to Santiago and on to Bolivia. There it would connect to Lloyd Aéreo Boliviano (another local German-run airline), which already flew to Chile.

Neither of these two tiny airlines would have mattered very much had it not been for the unseen hand, sinister designs, and ulterior motives of

Rentschler—as Trippe conceived them. He could see the terrible fate that awaited him if United invested in or bought Condor and Lloyd. After expanding United's presence in Brazil, Argentina, and Chile, Rentschler would inevitably push north into Trippe's fiefdoms in Central America and the Caribbean, then connect to his vast American network. That was bad, but it would get worse if Eckener decided to open a Zeppelin route from Europe to South America to link with United/Lloyd/Condor.

The hard truth was that Trippe's position in South America was rather more precarious than it looked on paper, especially once the stock market crash of October 1929 hit two months after Eckener's Round-the-World flight.

Pan American's revenue mostly came from business travelers paying top dollar for rapid access to South America's riches, but there were few businesses doing well in the aftermath of the smash. By January 1930, the froth had come off aviation stocks: In October 1929, there had been 278 airplane manufacturers, but three months later, all but a few of the very largest had gone belly-up.

Airlines exclusively dependent on fares and mail, like Pan American, were particularly vulnerable. Pan American was so big it would probably survive, but the stock price had plummeted 55 percent and there was no bottom yet in sight.

At the same time, soup-to-nuts conglomerates like United Aircraft and Transport Corporation, which made airplanes (Boeing), propellers (Hamilton), and engines (Pratt & Whitney) as well as ran airports, training schools, and of course its own airline (United), were, if not blossoming, at least in relatively better shape owing to the various subsidiaries balancing one another: United Aircraft was "only" down 35 percent, but Frederick Rentschler could count on a supply of cheap, in-house aircraft to overwhelm Pan American's weakened defenses.[5]

Trippe had arrived in South America like a Cortés, but now he was a Montezuma, trapped and trembling in his palace as the conquistadors battered the doors to steal his gold.

JUDGING BY IZT's rapid progress, Trippe was right to feel he was under assault. When you took a hard look at the state of transatlantic traffic, it was clear that IZT would make a fortune. Plying the Atlantic were currently eleven "monster steamers," of which five took six days to get across, the rest a full week. In order to justify its fares, which would be double those of a steam liner's first-class stateroom, IZT's business plan called for 2.5-day runs each way.

Operating costs would be heavily subsidized by cutting the number of passengers from the touted 120—a completely unrealistic number in the first place—to between 25 and 50, reserving the rest of the space for high-priced mail. Since the entire U.S.-originating mail (letters and postcards) to Europe amounted to thirty tons per week, if IZT could steal this time-sensitive mail from the steamship lines—which would still carry heavy packages and freight—on twice-weekly flights, a windfall would be theirs.

IZT had another ace up its sleeve. Senator Charles McNary of Oregon, a New Deal–friendly Republican, introduced the Merchant Airship Bill in April 1930 proposing that cheap federal funds be lent to Goodyear-Zeppelin to defray up to 75 percent of the costs incurred building airships. The loans would relieve Goodyear-Zeppelin from having to raise the colossal capital up front to finance its airships—a major boost as the Depression hit.

More excitingly for most people, IZT proposed to erect a mooring mast atop the Empire State Building, construction of which was due to be completed in 1931. Like so many other New York real estate moguls, Al Smith, the building's developer, had a penchant for exaggeration, in this case boasting, "[It's] on the level, all right. No kidding. We're working on the thing now."

Despite much heated speculation, there was never any intention of the Empire State Building serving as a Zeppelin main terminal. Owing to the need for wind protection, repairs, and refueling, the mooring mast would serve only to drop off passengers before the airship headed to a proper hangar, but still, what a thrill it would be to hitch to the tallest building in the world, cross a gangplank, and take an elevator down to the heart of Midtown, where cabs would whisk you to your hotel.

Eckener himself, however, regarded using a skyscraper mast as *any* kind of terminal as an inherently ridiculous idea. How would his airships get close enough to dock without endangering themselves or hitting other buildings? How were passengers supposed to disembark, more than one hundred floors up, amid howling winds? What would happen if they got vertigo looking down from their narrow walkway? Which fools would volunteer for the job of securing the airship as it swung round and round?

When quizzed as to his thoughts on the matter, Eckener confined himself to grunting that the subject needed further study, which meant "No"—but it was good publicity nonetheless. He preferred to focus on finding the optimal location for what would be the company's main hangars and mail facility. For the time being, Lakehurst would serve, but it was a military air base and once the navy got the *Akron* and *Macon*, there would hardly be room for Zeppelin's airships. His preference was Baltimore, nicely situated near United's airports in Washington, Philadelphia, and New York, and a major railroad nexus offering easy access to the West and the South.

Given all these factors, it is easy to understand why to Trippe, analyzing the problem from his office, the Rentschler-backed IZT looked frighteningly like a monster that would not only kill his ambition to start a transatlantic route but burn down his whole house.

ECKENER'S PUBLIC STATEMENTS did nothing to allay Trippe's fears that IZT was coming for him. The alliance between United and Zeppelin, said *The New York Times*, was proof that Rentschler "believes transoceanic traffic must be surrendered to the lighter-than-air branch of aeronautics," with Eckener himself adding, "This is of particular importance because it means that airplane and airship, long regarded as competitors, through two of their greatest companies are joining in a cooperative enterprise."[6]

Eckener also revealed his intentions, confirming Trippe's predictions that he would encroach on his South American empire: "We [Zeppelin] intend to undertake flights to Brazil by way of Spain," after which IZT would establish "a huge traffic triangle whose sides will comprise the United States and Germany, Germany and Brazil, and Brazil and the United States."[7]

"Triangle" was the key word. IZT could not be satisfied with flying back and forth along a single route. It needed an international network to funnel passengers and mail through hubs, which would connect with United, Luft Hansa, and Syndicato Condor for domestic destinations.

The American hub, Eckener had already said, would be Baltimore. More surprising were Eckener's choices for the European and South American ones. For the former, he had already decided that Friedrichshafen was too remote. Instead, it would remain Zeppelin's manufacturing center but the main terminal would be moved to Seville in Spain. Blessed with perfect weather and a colorful history, Seville was easy to get to from Berlin, Paris, London, and Rome, and still better, took a day off the travel time to South America.[8]

The South American terminal was, as Eckener indicated, to be in Brazil. He'd accepted an invitation to come to Rio de Janeiro from the Brazilian minister of commerce, Victor Kondor, who, being of German ancestry, was excited about Zeppelins. Kondor, unfortunately, had blanched at the cost when Eckener informed him that the government would need to build him a hangar. They compromised on nearby Recife, whose placid weather meant a cheap mooring mast would do. From Recife—already a Syndicato Condor connection—passengers could choose from a brief air, train, or boat trip to Rio.[9]

With Brazil confirmed, Eckener was ready for a demonstration flight to prove that the Triangle scheme was viable. With the wind at his back and the newspapers avidly following his every move, Eckener embarked on the Europe–South America leg from Friedrichshafen on May 18, 1930. After refueling at Seville, the *Graf Zeppelin* headed out to the Atlantic, the ship flying on "as softly and quietly as she always did over water." He set a course for Tenerife to await, as he said, "the big sensation, the north-east trade wind, which, like a helpful spirit of the elements, was to take us in his arms and bear us at great speed to the south across 20 degrees of latitude—a good 1,600 miles—in constantly fair weather."

Early the next morning, they approached the critical point where the cool trade wind mixed with the warm, moist air of the "calm zone," producing thunderstorms and rain squalls. The airship plunged into a giant cloud and water began driving through the chinks and crevices until it was inches deep on the control-cabin floor. The ship became noticeably heavier as water weighed it down, but Eckener shrugged off the danger. A few minutes later they were out of the squall and the eight tons of water that had seeped in began flushing out. There was nothing, Eckener realized, to be afraid of in a tropical cloudburst, and he was pleased to find, as he had predicted, that the winds had been relatively moderate and there were none of the hazardous vertical air currents that had doomed the *Shenandoah*.

By 8 P.M. Eckener had arrived safely in Recife. The next day, they made a ceremonial visit to Rio—the view of the most beautiful bay in the world was spectacular—and at 11 A.M. on May 28, they set out for Lakehurst, the second side of the Triangle, by way of Barbados and Puerto Rico.

Despite experiencing a severe squall, the *Graf Zeppelin* arrived safely in New Jersey. Eckener tarried in New York for two days to meet with the gentlemen of the International Zeppelin Company and was gratified to find that "confidence and interest" in the transatlantic project "had been increased by our flight."

The return trip via Seville was a jolly one, with Eckener pleased as punch by his New York reception. Or at least it was until they were in the home stretch just past Lyons in France and Eckener came within a hair's breadth of destroying the airship and killing everyone on board.

The trouble started when Eckener saw a thunderstorm forming but, perhaps a little cocky and certainly "foolish" (his word), he decided to enter it at the risk of flying blind. At first, it seemed just another lashing of rain—nothing to worry about—but the rain soon turned into hailstones the size of walnuts. Inside the *Graf*, the passengers could hear the hail drumming on the outer cover, but Eckener did not grow alarmed until Captain Flemming called out in horror that ice was forming and dragging the ship down. From 1,000 feet, it fell to 650, then 500, and then down to a bare 300 feet. Eckener charged up the engines, crying that at this point it wouldn't matter whether they smashed into the ground at 50 or 65 mph. But still the airship fell, and they were at a hair-raising 160 feet when the elevator man shouted, "I can hold the ship up!"

The *Graf Zeppelin* was finally holding steady thanks to Eckener's coaxing a few hundred more horsepower out of its trusty Maybachs to drive the airship out of the hailstorm. The terror had lasted all of ten minutes, and Eckener assured the worried journalists (untruthfully) that everything was perfectly fine and under control. All they had experienced was a sudden slanting, followed by the crashing of a few plates, but no pitching or rolling. The only passenger who suspected otherwise was the famous polar explorer Sir Hubert Wilkins, who'd been picked up in New York and had been in numerous life-or-death situations himself. He shot Eckener an "earnestly questioning glance," at which Eckener "nodded slightly to him, and he understood." Upon arrival in Friedrichshafen, the airship was carefully examined and more than fifty holes were discovered in the outer skin.

Only among his closest colleagues did Eckener ever admit the truth of the matter. Of all the dangerous situations he had ever been in, or would be in, "none of them affected my nerves" so greatly as that hailstorm. Had the *Graf Zeppelin* crashed, at one of the highest peaks of Eckener's career, the entire "Zeppelin enterprise would have come to an end." As it was, the close call was expertly covered up so that neither his investors nor the public ever found out about it.

As far as anyone knew, then, IZT was a go and Trippe a goner.[10]

39. Engage the Enemy More Closely

Eckener had managed to shift conventional opinion on airships. His Round-the-World and Triangle flights definitively proved, reported the Department of Commerce's Aeronautics Branch in a widely published statement, that the airship was superior to the airplane for transoceanic flying. Sweeter words Eckener never heard.[1]

Among the airshipmen and their backers, this judgment became an unwavering article of faith, but Trippe suspected they were praying to the wrong god. They saw only the present continuing forever, while he peered into the future. Trippe was not alone in perceiving this flawed assumption. Lindbergh similarly thought that the airship's superiority over long distances was true only in the "present-day," cautioning that "larger flying boats are being built" and that the airship had yet "to demonstrate its commercial practicability and economy."[2]

Never one to quit while he was behind, Trippe, intending to combat Eckener head to head by building a large, long-range airplane capable of challenging his commanding lead, called Lindbergh, Priester, and Igor Sikorsky in for meetings. Sikorsky, born in 1889 in Kiev and descended from a line of Russian Orthodox priests—he would remain observant until the end of his long life—had attended the Naval Academy in Saint Petersburg but became fascinated by the new field of aeronautics and resigned in 1906 to study engineering. Ultimately, he was a philosophizing dreamer with a talent for mathematics and a penchant for discussing the fantastical aerial machines of Leonardo da Vinci and Jules Verne as if they had really existed.

When he was eleven, he said, he once had a dream. He was walking along a long, narrow passageway with walnut doors on either side. As he padded along the carpeted floor, he noticed the corridor was suffused with a calm bluish light and he felt a vibrating in his soles. When he reached the end, he opened a door and entered the comfortable lounge of a "large flying ship of the air."

That dream formed the inspiration of his life. In the event, reality intruded. His first effort at building such an airplane in 1910 was successful in that it flew (at an altitude of four feet), but unsuccessful in that it couldn't turn and crashed soon after. He got better. During the war he constructed bombers for Russia but fled to France during the Bolshevik Revolution after being threatened with execution as an Enemy of the Proletariat. From there he emigrated to America in 1919, a former honorary tsarist army general now living on beans and toast and making 80 cents a day giving lectures to fellow expatriates on aviation and astronomy.

He set his sights on selling civilian planes and founded the Sikorsky Aero Engineering Corporation after raising a little cash from the likes of the composer Sergei Rachmaninoff and other Russian émigrés. Staffed by Russian mechanics, pilots, and engineers, the company managed to build a passenger plane he called the S-29-A by scavenging discarded hospital beds from junkyards and cannibalizing their iron to form its internal structure. In May 1924, on its maiden flight, it crashed, but after major repairs it went up again in September, only for Sikorsky to discover that he was early to the party and that no one, in those pre–Kelly Act days, was interested in a huge fourteen-person airplane when delivering the mail was all-important.

The Sikorsky Aero Engineering Corporation was based in Bridgeport, Connecticut, and Sikorsky approached Trippe, then running Colonial from its headquarters nearby, about whether he'd be willing to buy a few planes for his budding airline. Trippe had to turn him down as he'd already contracted for the Fokkers, but was left impressed, if nauseated, when Sikorsky unexpectedly "looped" him—Trippe's first time upside-down—to demonstrate his airplane's maneuverability.

Soon afterward, Trippe was out at Colonial and the nearly bankrupt Aero Engineering was reorganized as the Sikorsky Manufacturing Corporation. Sikorsky's next few years put him back at square one, with the S-30, the S-31, the S-32, the S-33, the S-34, the S-35, *and* the S-37 all flopping. Trippe, newly installed at Pan American, saved him from yet another financial drubbing by purchasing an eight-seat S-36 in December 1927 for Caribbean use.[3]

Sikorsky's next model was the S-38, an odd-looking seaplane with a long, protruding duck-bill nose, a boat-shaped wooden hull sheathed in aluminum, a small lower wing from which extended two pontoons, and twin booms or outriggers poking out from the upper main wing suspended above the fuselage, which weirdly lacked a tail. Everything was held together with struts and wires reminiscent of the Wrights' *Flyer*—but it worked and, crucially for overwater use, was reliable. Over the next few years, Trippe showed his faith in the design by purchasing no fewer than thirty-eight.[4]

The S-38 was exactly what it was: an ungainly 1920s aircraft. But now Trippe wanted one fit for the 1930s. Exactly what form it would take occupied Trippe, Lindbergh, and Sikorsky for much of 1929 as the shadow of the *Graf Zeppelin* loomed darkly over them.

Lindbergh was adamant that Pan American's next mainstay be a landplane. A pilot could fly at night to lighted airports, their smoothed runways guaranteeing safe takeoffs and landings; further, jungles and mountains presented no obstacles, and there was an established chain of radio communication and guidance stations to ensure accurate navigation. Seaplanes, he pointed out, could fly only during the day (to avoid hitting flotsam, jetsam, and other debris), could not be used in icy or rough waters, and were prone to corrosion, leaks, and salt caking the windows. In addition, radio signals at sea were subject to strange atmospheric phenomena.

Sikorsky, on the other hand, argued in favor of seaplanes. He conceded their liabilities but emphasized that for long-haul flights you needed an aircraft capable of operating far out to sea and free to choose from a myriad of ports, harbors, rivers, and bays as its destination. Landplanes were fine for short and medium distances over known terrain blessed with plenty of airports and frequent flights, but to transport a large number of passengers (many already worried about engines conking out) it was necessary to have a flying boat that could reassure them that in the event of a water landing all would be well. More passengers on board, too, meant a heavier aircraft, and it was better to glide in over soft water than to risk the undercarriage collapsing under the strain of hitting a hard runway.

That left Trippe in the middle. He could see Sikorsky's points, but

Lindbergh knew his onions. All Trippe was concerned about was getting across an ocean and beating Eckener. He asked Sikorsky to show them blueprints for his proposed seaplane to help him reach a decision.

What Sikorsky brought in to Trippe's office did little to change Lindbergh's mind. The S-40, as Sikorsky had named it, looked like the S-38's beefier big brother. At first glance, it seemed as if all Sikorsky had done was to more or less double everything: four 575-hp engines instead of two 410-hp ones, for instance, 77 feet long instead of 40, and a cost of $125,000 rather than $50,000.

The same oddly shaped wings and booms were still there, and a conventional tail had not yet made an appearance. But the S-40's larger dimensions and greater power had combined to give it geometrically enhanced carrying capacity and performance: thirty-eight rather than eight passengers, and a 50 percent increase in range, to nine hundred miles. If built, the S-40 would easily have been the largest commercial aircraft in the world.

Trippe was duly impressed, but the S-40 was not quite the futuristic aircraft he'd envisaged, and Lindbergh still held out. The latter's main concern lay with its "struts, wires, and huge pontoons," which added significantly to air resistance. "I objected to the awkwardness of design and said bluntly that it would be like flying a forest through the air," recalled Lindbergh. He was taken aback when Sikorsky, in his "delightful Russian accent," replied, "I agree with you, Co-ro-nel. The resistance is high. But to remove it is still another step."

What Sikorsky was saying was that the S-40 was, technologically speaking, an intermediate plane. He knew it was not perfect, but to attempt to skip a generation at this stage would be exceedingly hazardous. Airplane technology had not yet caught up with Trippe's (or Lindbergh's) ambitions, and while it was painful and risky to allow Eckener and the airship to keep their lead, Sikorsky needed *time*. With time would come, he promised, the opportunity to develop the aircraft he needed to take on Eckener.

Lindbergh understood, as did Trippe, that the desired "radical improvements" would have to wait. As Lindbergh recalled, "I was disappointed, but I had to agree [that] we were in desperate need of a better plane for Caribbean routes. The Atlantic could come later."

In December 1929, with Eckener and Rentschler gearing up for their assault on Pan American, Trippe made the final decision: Just as Admiral

Nelson had famously signaled to the fleet at Trafalgar, Pan American too would *Engage the Enemy More Closely*. He doubled down and ordered two S-40s with an option on a third.[5]

THE KEY FACTOR that Trippe grasped, as so few others did this early, was that the airplane was on the cusp of a major technical transformation, one that would overturn every precept of the decades-old airplane-versus-airship competition.

Beginning in 1925, the aviation boom had encouraged tentative steps forward in airplane design, but only by the decade's close were the pieces starting to come together to create entirely new types of aircraft.

As Lindbergh's complaint that the S-40 would be like "flying a forest through the air" implied, of central concern to aerodynamicists in this period was lowering air resistance through better streamlining. With reduced drag, an airplane could travel farther, fly faster, and carry heavier loads. Arnstein and Jaray over at Zeppelin had already concluded the same during the war and had altered airships' shapes from cigars to teardrops, but there was little more progress to be made in that respect. There was still much to do, however, when it came to airplanes, whose sharp angularities, ornamental touches, and cumbersome protuberances were begging to be smoothed out and cut off.

Scientists focused their attention on several key parts of the plane. Cowlings to cover open engines, for instance, had become a top area of interest after it was discovered that the exposed, spokelike cylinders of air-cooled radial engines alone were responsible for nearly a fifth of all drag. Indeed, on some models of aircraft, the addition of a cowling reduced drag by almost two-thirds, resulting in dramatic increases in speed at negligible cost. In one experiment, an open-engine Curtiss AT-5A, whose maximum speed was 118 mph, was modified with a cowling and hit 137 mph on its next outing. The cowling had cost a mere twenty-five dollars to build and install.

Similar advances were made in developing retractable landing gear (after takeoff, the wheels were mechanically tucked underneath the wing) to further reduce drag, supercharging engines, thickening and adapting the airfoils of the wings to improve lift, and inventing variable-pitch propellers (where the pilot controlled the angle of the blades to adjust speed and power during different phases of flight). Taken individually, each of these parts would have tweaked airplane performance; combined, they added up to far more than their sum, making for a revolution, or rather, an interlinked series of small revolts that overthrew the existing order. The airplane was being reinvented.

With an improvement in one part synergistically affecting the others, airplane technology accelerated in a matter of years, rather than decades. By using low-cost models in laboratories, aerodynamicists could run dozens of tests on minutely varying versions of, say, an airfoil or a flap, until they found a promising solution. Real-world trials took a matter of days or weeks, and if successful, were rapidly integrated into existing and fast-spawning new generations of airplanes.

Zeppelins, on the other hand, were the size of ocean liners, cost millions to build, and were too rare to risk introducing overly innovative improvements without huge amounts of study and a series of lengthy trial flights. Eckener's switch to Blau gas, for instance, had required years' worth of calculations to ensure it would not only work with the Maybach engines but also not throw off the delicate balance between the *Graf Zeppelin's* hydrogen and ballast. For each new Zeppelin, then, there would generally be only one or two enhancements, whereas each new type of airplane combined a fresh palette of technologies.

If lumbering Zeppelins bred like elephants and nimble airplanes like rabbits, then the airshipmen were ostriches with their heads in the sand. Even as the revolution was happening around them, they were curiously oblivious to it. Eckener, for instance, was *still* relying on the cube law to maintain what he regarded as the airship's superiority over the airplane.

Going back to the nineteenth century, the rule stated that an airship's lift—which determined how much it could carry—was based on its cubic feet of gas volume, whereas an airplane was regulated by the old square law, its lift being directly derived from the square footage of its wings.

Eckener was not wrong in believing that the cube law continued to hold, but he was blithely incognizant that the square law was in the process of being "annihilated," as one knowledgeable observer put it. In aviation circles the new buzzword was the wing-loading (W/S) ratio, defined as the total weight of the airplane divided by the square footage of its wings. Put another way, a higher wing-loading ratio meant that a smaller, better designed modern wing could lift relatively more pounds of weight (including fuel, passengers, mail, and so on) than an older, larger one could. The traditional assumption that to have any hope of competing against the Zeppelin airplanes would need wings so colossal they couldn't even take off would soon be rendered obsolete as ratios rose. Because you now needed a much smaller wing to lift disproportionately greater weight, ever-higher W/S figures meant that airplanes could get a *lot* bigger than anyone had ever thought.

So it was certainly true that by making airships bigger, Eckener would be able to accommodate more passengers, install bigger engines, bring additional fuel, and take more mail and cargo, all of which allowed his airships greater range, speed, and profit. But what he had not realized, even as Trippe, Lindbergh, and Sikorsky did, was that the airship's advantages were quickly eroding relative to the capabilities of the airplane. The ancient paradigm that airplanes were necessarily limited in size, range, and power was vanishing despite Eckener's reactionary insistence that he "had been hearing this tune for the last thirty years" and considered it nothing but a "vague hope" based on "figuring with miracles."

Taking their cue from Eckener, other airshipmen may have continued to claim that airplanes were "inherently restricted to moderate size and capacity" and that flying "over seas . . . is the [exclusive] field for the airship," but in just the eighteen months between launching the S-38 and presenting the plans for the S-40, Sikorsky had tripled takeoff weight, improved the W/S ratio by 25 percent, and increased range by half.[6]

In short, the cloud, in the fortressed minds of the airshipmen, remained superior to the bird—even as airplane designers were approaching or even exceeding the capabilities of birds, fulfilling the age-old dream of these machines replicating their ornithological equivalents in grace, form, and function.

FILLED WITH ILL-FOUNDED confidence that his lead was unassailable, in mid-1930 Eckener began work on LZ-128, a successor to the *Graf Zeppelin*, intended to be five million cubic feet, equipped with no fewer than ten Maybach VL-2s, and accommodating up to twenty-four passengers. LZ-128, Eckener bragged, would be the first of a new generation of giant airships for the Atlantic crossing, but he was quickly disabused when he showed the blueprints to his IZT partners and Goodyear.

The airship, they said, was already out of date. For one thing, it was designed to use hydrogen, and the Americans wanted helium airships for safety reasons. For another, whatever else LZ-128 was, or was supposed to be, it was hardly a wonder of the world, the kind of flagship that would wow in the looks department. The coming airship age was supposed to be one of awe, glamour, and excitement, but LZ-128 resembled a fat version of the aging *Graf Zeppelin*, being almost exactly the same length but containing half as much gas again. It showed a lack of imagination and demonstrated evidence of being a rush job—which it was.

To his credit, Eckener recognized his error and ceased further development of LZ-128. He needed something better. In his office, he started a new file for what he called *"Projekt LZ-129."*[7]

V OLUNTARILY MOTHBALLING LZ-128 had happened not a moment too soon, for Eckener would have been forced to do so anyway shortly after 2:10 A.M. on Sunday, October 5, 1930.

That was the moment a British airship, R-101, crashed and exploded near Beauvais in France on its maiden voyage. Of the fifty-four men aboard, all but eight were killed, including the secretary of state for air, Lord Thomson (who'd declared the ship "as safe as a house—except for the millionth chance"). In short order, the British grounded R-101's sister, R-100, steamrolled the infrastructure, and sold the remains for £600 (it was rumored that Eckener had purchased the scrap duralumin).[1] The British had tried time and again to launch an airship program, only to fail disastrously, and the latest catastrophic loss was insuperable.

Eckener had long been skeptical of R-101, which to his mind was a piece of junk, a sad relic of a saga that had been dragging on since 1924, when the British had approved the construction of the two five-million-cubic-foot airships. It had been a mess from the start, with R-100 being built by a private firm and R-101 by the government. Whereas Zeppelin (and Pan American) had a single Presiding Genius at the helm, the British employed different design and engineering teams for each airship, who competed for resources, never visited each other's workshops, and sniped at each other in the press. Work, accordingly, was not so much duplicated as multiplied.[2]

Even if its interior was quite nice, British aesthetic tastes leaning toward a look that called to mind a Ye Olde Tea Shoppe at a small seaside resort,

R-101 turned out to be a combination of a white elephant, a giant albatross, and a fat turkey. Aside from its ceaseless construction and mechanical problems, R-101 had been disastrously overweight by at least a quarter more than planned. R-101's underpowered engines alone had weighed seventeen tons, and it was questionable whether some homey touches, like a carpet that weighed a thousand pounds, were truly necessary.[3]

Eckener traveled to London to attend the state funeral at St. Paul's Cathedral and was asked to testify before the British Court of Inquiry, which judged that a forward gas cell had suddenly deflated, leading to a nose-down crash, with electrical sparks igniting a devastating hydrogen fire.

For Eckener, the destruction of R-101 was a Pyrrhic victory. Yes, it removed the British from the airship business once and for all, but it also made the business climate for airships uncomfortably chilly. In public he remained confident, saying that "the disaster will not affect the course of Zeppelin building any more than the sinking of a steamer causes an end of shipbuilding," but while his American partners did not walk away from IZT, they requested Eckener to slow down with his expansion plans until the dust settled and memories grew hazy.

A worrying sign of ebbing airship enthusiasm came in early December, when Senator McNary's revised Merchant Airship Bill dealt a blow to IZT's and PZT's fortunes by stipulating that instead of federal loans being extended to Goodyear-Zeppelin to defray the enormous cost of building airships, they would have to be self-financed. That meant that Goodyear could not enlarge its facilities without first raising money from its reluctant partners, but United had originally promised to supply any necessary capital only on the surety of federal backing.[4]

The Depression had hit United harder than anyone had anticipated. Company profits had dropped by nearly 70 percent since the 1929 crash, and its stock had fallen to all-time lows. Strapped for cash, Frederick Rentschler wasn't eager to sink millions of dollars into expensive projects until the storm passed.[5]

And it wasn't as if Eckener could help Goodyear. The Zeppelin Company was itself in terrible financial straits as Germany was suffering through its own terrible economic crisis. Bank failures were mounting, and by mid-1931 corporate bankruptcies were averaging more than 1,700 per month. Unemployment rose to 4 million, up from 2.6 million a year earlier, as the metal industry fired a third of its workers and construction firms nearly two-thirds. Fully a fifth of German shipping was laid up in dry dock, and exports and output in every sector were drastically down. Any hopes, then, that Eckener had of building more Zeppelins in Friedrichshafen for the IZT project were in deep freeze.

Eckener was disheartened by the developments, but the future still seemed bright. The Depression would eventually pass, and when it did surely the Merchant Airship Bill would be amended to restore the federal loan guarantee and money would flow from a reinvigorated United as Zeppelin geared up for production. In the meantime, the USS *Akron*—the first of the two navy airships—was due for completion, with the *Macon* soon following, thereby allowing Goodyear-Zeppelin to start building the PZT and IZT fleets when the funds arrived.

WHAT ECKENER *DIDN'T* know about was the extent and dexterity of Trippe's wire-pulling. It had been *his* Washington lobbyists who had persuaded Senator McNary to remove the federal loan guarantee to buy him time to organize his own Atlantic service.

Then he used an old trick: Beat them by joining them.

PZT had always been the smaller, poorer sister to the Rentschler-dominated IZT, which was forever off-limits to the likes of Trippe—Rentschler had the measure of the Pan American boss—but PZT was a soft target. Of course, Trippe had no great interest in airships, but he *was* greatly interested in knowing about Eckener's next move, so he pretended that Pan American *might* be thinking about launching a Hawaii route in collaboration with PZT.

The well-meaning, naive Jerome Hunsaker, in charge of PZT at the time, was only too delighted when a prestigious all-American company like Pan American came calling, and he invited the ever-emollient Trippe to join the PZT board without talking to Eckener first. Trippe quickly signed on as a minor investor, putting in just enough to indicate good intent but not so much as to trigger his name being mentioned in newspaper articles.

This was a grievous error on Hunsaker's part. Just having a place at the table, even if it was the children's one, was sufficient to allow Trippe access to confidential intelligence about Eckener's plans and progress, while Eckener had none about his. Playing both spymaster and spy, Trippe was soon in receipt of such valuable documents as PZT's annual reports, technical data, and financial records, allowing him to try to piece together exactly what Eckener was up to over at IZT/Zeppelin.

In return for allotting him the board seat, Trippe promised Hunsaker he would back an amended version of the Merchant Airship Bill reintroducing the federal-funds guarantee. It was then being examined by the House and Senate Committees on Commerce, and Trippe had only one small request: Would Hunsaker mind terribly if the word *aircraft* were substituted for *airship* in it to cover airplanes, as well?

Hunsaker did not mind terribly—another error. Paul Litchfield, chairman of PZT and Goodyear, compounded it when he was asked by a puzzled senator during the hearings whether he was amenable to the change. Not wanting to sound unappreciative of Trippe's efforts on his behalf, he replied, "We have no objections to having heavier-than-air included."

His gentlemanly sop to Trippe had just put airships and airplanes on an equal footing if the bill passed, thereby negating the original intent of giving Goodyear a financial leg up in its effort to build airships. Now it meant that Pan American would *also* receive federal loans to add to its own fleet if the bill passed.

At the time, Pan American was, against all odds, doing very well. In 1931, even after figuring in the costs of breakneck airport expansions, extending service throughout South America, and aircraft purchases, Pan American posted its first-ever profit. It wasn't a lot, a mere $105,452, but it was better than the $305,271 the company had lost in 1930 and, in relative terms, much, much better than any other major U.S. airline was doing. Indeed, at a time when the revenue of the domestic airlines was falling fast, Pan American's had increased by more than $2 million from 1930 to 1931.

Which raised a very good question: Why on earth would any politician want to give Juan Trippe *more* money when he was already flourishing? A natural reluctance to open the federal spigot when so many other industries were suffering from the Depression meant that the chances of passing the Merchant Airship Bill, even if it also helped Goodyear, fell drastically.

This was exactly what Trippe had intended. The full horror of what Trippe had done slowly dawned on poor Hunsaker after he was tipped off that, as he reported ruefully, "several Senators had been approached by Pan American agents to prevent action" on the Merchant Airship Bill. Trippe wanted it to die, and the change to *aircraft* from *airship* had been designed to ensure that. There was still a slight chance that the bill could be put to a vote the following year, 1932, but the best that Hunsaker could hope for was to cross his fingers. Trippe, flashing that guileless grin of his, pleaded innocence when Hunsaker confronted him.[6]

PLEASED WITH A job well done, Trippe turned his attention to the Atlantic while Eckener and his allies were on the ropes. He quickly discovered that breaking into the European market was much more difficult than cracking South America.

In South America, there were many small countries run by either strong men or weak governments open to various "inducements" to cooperate. Europe was a continent, conversely, of strong governments, a few large states, and many worldwide interests, all of which made Trippe the weak man. Like it or not, he had to make reciprocal deals, for each country had its own national airline and those governments considered their airlines instruments of statecraft. The kinds of tricks Trippe had played in South America were out of the question when negotiating with the mandarins at Whitehall, the Quai d'Orsay, or the Foreign Ministry in Berlin.

Great Powers had no truck with the exclusive landing rights and tax concessions Trippe demanded in Form B. They could at any moment withhold permission from Pan American to bring its planes in to their airports at home or at their possessions overseas.

Complicating matters still further was how the European flag-carriers were organized. Britain's Imperial Airways was particularly idiosyncratic and needed special handling. The clue was in the name.

In 1924, the British government had forcibly merged the country's major carriers into a single entity—Imperial—with an inimitably British muddled mandate: The new airline was a private monopoly with a public subsidy. Put another way, while it wasn't state-owned, it was state-funded, and its shareholders received dividends from taxpayers. In exchange for this beneficence, Imperial had to buy exclusively British airplanes, hire British pilots, and develop routes to India, Canada, the Middle East, Australia, and South Africa to service the Empire's needs.

Sir Eric Geddes, a former railway financier, was installed as part-time chairman, and George Woods Humphery, who'd helped run two of the merged private airlines, became his general manager.[7]

Both were highly capable executives, but Geddes suffered from poor health. Most of the heavy lifting, then, was delegated to Woods Humphery, a grammar-school boy from Glasgow who'd worked his way up from engineer's apprentice to the summit of Imperial. He did not, as the euphemism has it, suffer fools gladly and quickly made enemies of the airline's employees. Within weeks of being appointed to his position, he'd provoked a pilots' strike by refusing to call them "captains." He tolerated no indiscipline or work-shyness among the staff, having once sacked on the spot a man he found loafing about at the airport, only to discover that he was actually a passenger waiting for one of Imperial's perennially delayed planes.[8]

Trippe realized early on that Woods Humphery was the key man to talk to about a Pan American–Imperial arrangement. The two got along very well—they commiserated with each other about the *outrageous* salaries they had to pay pilots—and in May 1930 formed a working partnership.[9]

The Sikorsky S-40s Trippe had ordered had not yet arrived, but even when they did Trippe knew they would lack the range to make it across the Atlantic in a single leap. So how would he do it? Trying to head south through the British colony of Bermuda and the Portuguese-held Azores was his first choice, but the distances were too great and the Ditadura Nacional (National Dictatorship) government in Portugal proved unwilling to sell a concession in any case.[10]

He needed, then, Imperial's support to develop a potential "northern" route to Europe—one with pit stops. From New York, Pan American would head north up the coast to Newfoundland, then skirt the Arctic Circle over Greenland, get to Iceland, then make stops in the mountainous Faroe Islands (a Danish-owned archipelago) or the rugged coasts of the Shetland Islands, finally arriving in mainland Britain. From Imperial's point of view, cooperation with Trippe would speed along the development of what it called its "All-Red Route" to Canada to match those it was extending eastward to India, South Africa, and Australia to fulfill its government mandate.[11]

Eckener, though, happened to be thinking the same thing: Go North.

41. The Great Circle

ECKENER'S IDEA OF exploring what became known as the "Great Circle" northern route had first been broached in the aftermath of the *Graf Zeppelin*'s Round-the-World voyage two years earlier, in 1929. Then, a few pedants had pointed out that Eckener had not, in fact, gone "around the world," which to them properly meant following the 24,901-mile equator; he had "merely" flown a circle around the North Pole.[1]

Their salient point was that the shortest route between America and Europe was not directly in a straight line across the Atlantic, as most people, thinking of the traditional Mercator projection drummed into their heads at school, naturally assumed. Traveling globally meant using the earth's curvature to take advantage of shortcuts. With the advent of oceanic crossings, the possibility of cutting hundreds or thousands of miles off a flight by arcing northward toward the higher latitudes began to matter a great deal.

In this instance, Newfoundland, Greenland, and Iceland—once regarded as hinterlands—would be of potentially enormous importance. Eckener had already nailed down the best Europe–South America and South America–United States routes for IZT, but the optimal path between Europe and New York remained elusive. Could the Great Circle be it? In the summer of 1931 he embarked on three exploratory flights in the *Graf Zeppelin* to find out.

The first was a brief charter flight in the service of, of all things, the Swiss Automobile Club, which left for the Arctic outpost of Spitsbergen (today Svalbard) on July 9. The good bourgeois of the cantons had no inkling that Eckener was using them as guinea pigs to determine whether paying passengers could handle the cold weather. Along the way, he also found that Norway was often covered in cloud in the summer and that Finland experienced frequent heavy rain and storms. Neither was an encouraging indicator for regular service. Off Spitsbergen, for instance, they ran into deep fog banks and had to rise from 900 to 3,200 feet in order to make their way through, another bad sign. Machinist Wilhelm Fischer wrote that "it is not advisable to fly further, especially as this chain of mountains of perpetual ice and snow is supposed to be very high. . . . We would be in danger of colliding with the hidden glacier."

Eckener ordered a turnaround. After they turned south and headed for the Scottish coast, more thick fog enveloped them. The Great Circle route was appearing to be more hazardous than anyone had anticipated.

A week later, on July 16, with Ernst Lehmann now serving as captain, the *Graf Zeppelin* again set off north. Lehmann made for the Faroe Islands to check conditions. When a rough storm assailed them, Fischer recalled that "the ship was shaken throughout its whole structure, going up and down and rolling very severely." They continued until they reached the east coast of Iceland, where they finally encountered better weather, but the high, steep mountains were covered in heavy clouds and there were no lighthouses to use as guideposts. "It would have been very difficult to venture into these totally fog-covered mountains at night," noted Fischer.

When the *Graf Zeppelin* reached the southern plains of Iceland, he continued, "the few inhabitants stood around their wooden houses, astonished by this unknown wonder from the skies." To say that it would be troublesome to make an emergency landing in such a remote area with scores of passengers and crew aboard would be an understatement, and Lehmann was well aware of the dangers of mooring an airship out in the windswept open. As it was, another storm was threatening them from Greenland, and Lehmann decided to cut his losses and fly home.[2]

So far, the chances of successfully developing a Great Circle route looked pretty slim, but Eckener wanted to check whether going farther north into the Arctic itself might be easier than expected. An airship was primarily a seagoing vehicle, so if he could avoid Iceland and Greenland and stay mostly over water, perhaps the hard terrain and adverse weather could be overcome.

On July 24, 1931, the *Graf Zeppelin*, hastily converted into an Arctic exploration airship, departed for the Soviet Union. Out went the lavish multicourse dinners, and in came rations of pemmican (a beef and fat mixture), pea and potato meal, oatcakes, and chocolate; out went the fine china and the linen napkins, and in came aluminum plates and repurposed toilet paper.[3]

Eckener's first destination was Leningrad, the jumping-off point for the Arctic. In the late morning of July 25 they landed in the city with a crowd of a hundred thousand people cheering their greetings, the brief unpleasantness during the Round-the-World voyage evidently forgiven, if not forgotten, by Stalin. As the Red Army stood guard over the airship, there was a huge illuminated sign saying "Welcome" in German and Russian.

A decade or so later, Leningrad would not be so gracious to the Germans, but for now there was a gala held aboard the *Graf Zeppelin*. Local functionaries and regional dignitaries crowded the dining room and cabins, where they were serenaded by the Red Army Chorus and the tables "groan[ed] with caviar, while the clinking of glasses almost drowned out all other noises," said Fischer the machinist. The guards, he added, were roughly turning away hungry civilians at the door—the dire effects of Stalin's collectivization of agriculture were beginning to show—while guzzling vodka and food on the sly.

The next morning, they lifted off with a squadron of airplanes as escort (they were there to prevent any lingering over military installations); after a few hours, the *Graf Zeppelin* reached Archangel. Work had not yet begun on the dictator's White Sea Canal, built on the backs of a hundred thousand slave laborers, making for a more scenic journey than would have been the case just a few months later.

Once past Archangel, a squall arose and mist enveloped them for a time, but they reached the Arctic Circle without much trouble. Fischer wrote that whoever has not seen the frigid archipelago of Franz Josef Land "with its gleaming and transparent glaciers, in the fairy-like delicate tones, and the endless symphony of color of its ice-masses—its colorful beaches and the blue inlets between the fantastically shaped islands and foothills—has not known anything of the most beautiful thing which this earth has to offer to our eyes."

Romantic, yes, but Eckener, with his veteran airshipman's eye, looked down six hundred feet and saw nothing but mountains looming out of the fog—Danger. He noted that these were "critical areas to be carefully avoided" on a regular basis.

By July 30 they were on their way home, but a deep depression with strong winds and dangerous storms was approaching Leningrad, and Eckener had to cancel a planned landing—which of course was taken as another slight by the Soviets. They finally landed in Berlin at 6:30 that evening.

Eckener addressed the crowd that had gathered to welcome them home. He was, as ever, bullish, saying that "flight in the Arctic is as comfortable, beautiful, and the least dangerous that one could experience. It took place under a beautiful blue sky and calculating one's position was no problem."[4]

But the perpetual distinction between Eckener's public and private faces was in full effect. A true Arctic Circle flight would always entail dealing with Stalin on his own turf. The headaches, which would invariably include inexplicable changes of mind, outbreaks of paranoia, and minor extortions, were far more trouble than they were worth.

Then there were the weather problems. A constant menace would have been the formation of ice on the envelope and propellers, dragging the ship down to its doom. Atmospheric interference, too, had caused the *Graf Zeppelin* to lose radio contact with the outside world for an entire day at the 80th parallel, rendering it invisible to searchers if the ship had gone down.[5] Eckener privately conceded that an Arctic route was too unsafe and unpredictable to be viable.[6]

Yet the more southerly route via Iceland and Greenland was also out of the question, as he had learned from the two earlier flights. The treacherous winds past Greenland were too difficult for a ship the size of the aging *Graf Zeppelin*. Until he built bigger airships that could negotiate those winds and handle the longer sea voyage necessitated by avoiding Iceland and Greenland, he would have to restrict himself to South American flights, with a possible—if time-consuming—route north to Florida and along the East Coast to reach New York.[7]

That was the bad news. The good news was that August 8, a week after his return to Berlin, was the day the *Akron*, first of the two navy Zeppelins, was christened by the First Lady, Lou Henry Hoover, in Ohio, which greeted her with a children's choir, brass bands, and a trained-pigeon demonstration. Famed aviatrix Amelia Earhart attended the celebrations, as did 250,000 people from around the state.[8]

Meanwhile, work on the *Macon* was coming along nicely. Once it was completed, Goodyear-Zeppelin would finally be ready to build the IZT airships.

Until then, Eckener increased the pressure on Pan American by strengthening his connections with the Condor airline in South America and by keeping the *Graf Zeppelin* busy on flights to Recife and Rio de Janeiro. Shortly after returning from the Arctic, he embarked on three round-trips to Brazil in September and October, each one serving as a demonstration of the airship's long-haul viability.

There weren't any accidents, but in the depths of the Depression, passenger numbers were not high, and neither were the mail loads. On average, just eleven passengers paid their way along with 363 pounds of letters and postcards, making for half-filled flights, cavernously empty cargo bays, and bottomless losses.[9] Eckener sustained the short-term hit because he intended to begin regular, scheduled service to Recife and Rio the coming year, by which time, he hoped, the economy would have improved.

BETWEEN 1931 AND 1933, as the long struggle between the airship and the airplane heated up, Trippe's and Eckener's moves were closely synchronized: When one advanced, the other countered; when one countered, the other feinted—and both would suffer reverses.

Just a day after Eckener had departed for the Soviet Union to explore the Great Circle route on July 24, Trippe dispatched a spare Pan American Fokker F-10A to start doing the same from the opposite direction. The Fokker left Boston and headed to Bangor, Maine, on a survey flight, and was followed a few days later by a longer-range Sikorsky continuing the journey to Nova Scotia to open an unglamorous Boston-Halifax mail route.[10]

So far, so good, but Trippe's real objective was Newfoundland, the stepping-stone to Europe. From Newfoundland to Greenland to Iceland to the Faroe Islands to the Shetlands to England: an elegant geographical path with pit stops laid out by the God of Flight himself across the Atlantic.

Acquiring landing rights in the Faroe Islands, Greenland, and Iceland would be a simple matter, as they were all controlled to varying degrees by Denmark, which was happy to sell them for what amounted to petty cash.[11] That left the Shetland Islands, which belonged to Britain. With his Imperial Airways alliance in hand, Trippe would have no trouble getting permission to land there to refuel and refit.

It was Newfoundland that was unexpectedly causing problems. That island had since 1855 enjoyed self-government as an imperial dominion with the same status as Canada, but it was incompetently and corruptly run, so in Trippe's eyes it was as vulnerable as any small South American country to blandishments. And so it proved: His request for a fifteen-year concession in partnership with Imperial encountered no opposition in the House of Assembly in St. John's, the capital. When Trippe talked to Sir Eric Geddes of Imperial, he was told that "the chances are a hundred to one that we'll get final approval."

But the "one" came up. Newfoundland, it turned out, was also just as unstable as any small South American country. The Depression was hitting hard, public debt was spiraling upward, tax evasion was a habit, and fully a quarter of Newfoundlanders were on the dole. At one point, rioters broke into the Colonial Building and made heroic but unsuccessful efforts to beat up the prime minister.

Canada and Britain were forced to intervene, with a royal commission proposing that Newfoundland revert to the lower status of a Crown Colony and be temporarily governed by a committee of "six wise men" accountable to London to restore order.[12] All of which meant that when it came to asking for concessions, Trippe was no longer negotiating with St. John's but with London and Ottawa—tougher customers. The patriotic Canadians, in particular, started making noises about an Imperial Airways route through Newfoundland to Britain that omitted Pan American completely.

That rumor prompted a call to Geddes from Trippe, who pointedly asked, "Do we not have an understanding?" If Imperial went along with the Canadian plan, in other words, they would be breaking the gentleman's code. Woods Humphery agreed with Trippe, and in July 1932 he informed a conference of British and Canadian officials that Imperial must share its rights to the Newfoundland route with Pan American for the next fifteen years. With Trippe reassured that Imperial wouldn't stab him in the back, that summer he dispatched two expeditions to survey Greenland and Iceland on the ground.[13]

An "understanding," of course, goes both ways, and just as Woods Humphery had supported Trippe, he expected Trippe to support him. Part of the agreement with Imperial had always been that neither airline would begin operations before its partner was ready. And that meant Trippe had to wait for Imperial before opening a route through Newfoundland.

The problem with Imperial, as Trippe only now discovered, was that the British lagged far behind him. Their only "long-range" aircraft was the Handley Page HP-42, a slow, underpowered *biplane* introduced in 1931 that was obsolete even before it was delivered. Neither did Imperial operate a single route on which the hops were more than five hundred miles — less than an elderly Ford Trimotor could manage — and every stop between America and Britain was more distant than that.

Worse, a report from Lindbergh brought discouraging news. At Trippe's request, the aviator and his wife, Anne, had set off on a trip of several months in a specially adapted Lockheed Sirius seaplane to scout Greenland, Iceland, the Faroes, and the Shetlands. At the time, little was known of the flying conditions in these remote regions. Previous attempts to scout them by airplane had generally ended in disaster for the intrepid few, and the findings of the land-based expeditions Trippe had sent had been inconclusive.

Originally, Lindbergh had intended to do it alone, but his flight had been delayed by the kidnapping and death, probably murder, of his son, Charles Jr., in March 1932 — the "Crime of the Century." Since then, a second son, Jon, had been born, and Lindbergh sought an escape for Anne from the press attention. She left Jon, nearly a year old, in the care of her mother and traveled with her husband as the radio operator.

Lindbergh reached virtually the same conclusions Eckener had after his Arctic flight in 1931. The route was untenable for flying-boat operations. While Iceland seemed safe, the Faroes and the Shetlands would have to be avoided owing to their eternally terrible weather. It was possible to fly directly between Iceland and mainland Britain, but only by stretching a high-performance aircraft's capabilities to their maximum; a regular passenger plane could not yet hope to do the same.

Greenland presented the greatest obstacle: In the winter there would be too much ice for safe landings, its fogs and storms were outright hazardous, winds blew snow as high as a thousand feet up, radio communications could be cut off for more than a week at a time owing to atmospheric interference, and temperatures of minus 50 degrees Fahrenheit would wreak havoc on equipment and crew alike. "When you looked at a globe of the world, the Arctic routes were [so] tantalizing, [so] ideal for air routes," Lindbergh wrote, that it was "easy to forget [their challenges] in a New York office when you stretched a piece of string across the surface of the globe."[14]

In that New York office, recently moved to the fifty-eighth floor of the Chrysler Building, and fiddling with his reels of red string, Trippe could only gnash his teeth. How on earth was he ever supposed to get across the Atlantic with Greenland, the Faroes, and the Shetlands out of bounds?

He needed a new plane.

42. Master of Ocean Aircraft

ON OCTOBER 12, 1931—almost two months to the day since she had done the same for the *Akron*—Mrs. Hoover christened the first of Sikorsky's long-promised thirty-eight-seater S-40s with a bottle of Caribbean water as the navy and Marine Corps bands struck up rousing tunes. Trippe gave a nationally broadcast speech—*advertisement* would be a more appropriate word—extolling "America's Mightiest Airplane."

He went all out to celebrate this new member's introduction to the Pan American fleet. He hired Ethel Murchie, a society designer specializing in airplane interiors, to install dark-stained walnut paneling over sound-proofed walls as well as finished silver fixtures, blue carpets, reading lights, cigarette lighters, card tables, call buttons at each seat to summon the steward, a "ladies' lounge," drapes of gray silk, and Queen Anne–style chairs to achieve an "effect of cool spaciousness," in the words of one journalist. When Sikorsky first saw Mrs. Murchie's efforts, he realized he was living inside the "blue-lit airplane" dream he'd had when he was eleven.

Trippe was ecstatic about the plane. As a boy, he'd often traveled on Cunard liners, and his ancestors (from the more respectable side, that is) were Marylanders: Since Baltimore had built the original clipper ships in the nineteenth century, Trippe decided to dub his new airliners "Clippers" in their honor, thus instituting a long Pan American tradition (and trademark). This first S-40 was named the *American Clipper*, and it would be followed by the *Caribbean Clipper* and the *Southern Clipper*.

In keeping with the nautical theme, Trippe borrowed a number of maritime customs to reassure his clientele that what Cunard was to the waves, Pan American was to the clouds. Speed was calculated not in miles per hour, but in knots; time was denoted by "bells" and the crews' shifts were renamed "watches." Pilots' uniforms now resembled those worn by ships' officers—navy-blue serge with gold wings pinned to the breast, and rank indicated in stars (the gold rings on the cuffs and embroidery on the hat would come later). Captains were given the powers of a naval captain and enjoyed full control over their vessel as master and commander. Even a new elite rank would soon be introduced: Master of Ocean Aircraft.[1]

Using the term *Ocean* was a deliberate choice on Trippe's part. Pan American's S-40s made flying a seagoing enterprise—the Caribbean Sea—but he aspired to an *ocean*going one. To take on Eckener in the Atlantic, he needed bigger, more powerful airplanes. The S-40s may have been the largest commercial planes of their time, but *months* before the *American Clipper* even arrived in Miami to begin operations, Trippe was already looking past them to the future.

He wanted true transatlantic planes, and on June 26, about two weeks before Eckener embarked on his first Arctic flight, Trippe sent a letter to six major aircraft manufacturers asking them to draw up plans for Pan American's next airplane. The specifications were for "a high-speed multi-motored flying boat having a cruising range of 2,500 miles against 30-mile headwinds, and providing accommodations for a crew of four, together with at least 300 pounds of airmail." Of the six, four replied to say that such a plane was an impossibility. The two that didn't were Sikorsky and the Glenn L. Martin Company of Baltimore.[2] Sikorsky, of course, had long had the inside track with Trippe, making him the favorite.

On November 19, the very day the *American Clipper* was scheduled for its first commercial flight, Lindbergh was in Miami. There, he had two jobs: first, to captain the *American Clipper* on what would be one of the most important flights of the era; and second, to discuss with Sikorsky the shape and specifications of Trippe's ideal ocean aircraft.

LINDBERGH WOULD HAVE plenty of time to do both at once because the *American Clipper* would be embarking on the longest passenger airplane flight across open water in the world. Until now, the Pan American route from Miami to Panama (the ultimate destination) had been an arduous one requiring multiple planes, many short hops, and numerous overnight stays: From Miami, one flew to Havana, then to the Yucatán Peninsula, then Belize, Guatemala, Honduras, Nicaragua, Costa Rica, and finally, Panama. The S-40 had the range to skip most of that. Travelers would henceforth depart Miami, head for Havana and then Kingston, Jamaica; from there, it was a straight shot of 660 miles over water to Barranquilla on the Colombian coast and thence to Panama. One journalist trilled that "you will breakfast in the West Indies, lunch in South America, and dine in Central America." Less romantically, Trippe saw it as an "Atlantic Prep School," a laboratory for the conquest of the Atlantic.

When Lindbergh arrived in Miami, he was put through his paces by Priester, whose iron law stated that no pilot could take command of a new airplane without having first practiced ten takeoffs and landings. Lindbergh, the greatest aviator in the world but also knowing what Priester was like when rules were flouted, good-naturedly performed the maneuvers in quick succession and was given the go-ahead to take the *American Clipper* on its maiden flight.

With four passengers on board, including Sikorsky, who'd paid full fare (Trippe was not one for freebies), the S-40 set off for Havana. Soon afterward, Lindbergh left his copilot in charge and came back for the first of many conversations with Sikorsky. Usually meeting over lunch on the plane or dinner at their hotel, Lindbergh and Sikorsky sketched out on the backs of menus the plane they were already calling the S-42.

The S-42 was to be "the next step" Sikorsky, Lindbergh, and Trippe had agreed on after the intermediate-stage S-40. The aeronautics revolution of the late 1920s was by now paying rich dividends, and the time was right to move to a true 1930s airplane. Sikorsky and Lindbergh went back and forth on every point, giving a little here and sticking a little there. Lindbergh still preferred a landplane but conceded that only seaplanes had the capabilities required, so Sikorsky won that battle. But in return, Lindbergh insisted, he wanted the ugly outriggers to vanish, the plane should have as few struts as feasibly possible, and external wiring must be kept to a minimum to improve streamlining. The wing-loading ratio, both agreed, must be higher, and range should be about twelve hundred miles to allow for direct, nonstop Miami-Panama service.

The *American Clipper* arrived in Panama on time. Lindbergh and Sikorsky talked more on the way home—the former even allowed Sikorsky to pilot occasionally, but not too often, for as a colleague said, "Igor flies like a professor with a textbook in his hand"—and waited to see what Trippe would make of their handiwork.[3]

THE BOSS WAS both pleased and disappointed. The S-42 certainly would make a wonderful addition to the Caribbean and South American fleet, but it was not, frankly, the transoceanic carrier of his dreams, the truly breakthrough airplane he needed to beat Eckener. It was a next step, but he wanted a longer stride.

Sikorsky was delighted when Trippe nevertheless told him he would order at least three of them at $242,000 each once preliminary design work was completed the following year but shocked him (and Lindbergh) with the news that he would *also* be ordering three Martin M-130s at $417,000 per plane. (To put those figures into context, at that time a domestic airliner usually cost about $78,000.)[4]

Trippe explained to a confused Sikorsky why he was accepting offers from *both* applicants to his June letter. Put simply, he needed the Martins for the Atlantic route. That part was true, but as always with Trippe, the layers of truth were multiple and nebulous.

In this case, Frederick Rentschler at United had purchased the Sikorsky Aviation Corporation in July 1929, shortly before the stock market crash. United was still financially weak, but it would eventually strengthen and rouse its alliance with Zeppelin from dormancy. Trippe feared that Rentschler, once he got back on his feet, could cripple Pan American by slow-walking or even canceling his S-42 order at a critical moment. Hence Trippe had to cover his bets by contracting out to an independent manufacturer.[5]

Glenn Martin was *very* independent. Like Trippe, Lindbergh, Priester, and Sikorsky, he was highly idiosyncratic, a man who never quite fit in, an island unto himself. In Martin's case, his closeted homosexuality—in media reports, cagey euphemisms like "never married," "shy bachelor," "monastic," "women alarmed him," "feminine intuitiveness," and "lives with his mother" allowed sophisticates to read between the lines—made him shrink from public exposure even as his brilliance and spectacular appetite for risk attracted the spotlight.

Born in 1886 in Iowa, and now in his late forties, the long-legged, thin, and bespectacled Martin was fascinated by kites as a boy—he turned a

profit selling them for a quarter to the local kids—and built his first working airplane in 1909 out of bamboo and silk strung together with wire.
Propelled by a four-cylinder engine, the aircraft covered a hundred feet at
an altitude of two feet. Three years later, he became an early barnstormer,
notching up various records and competing in aerial races. By the Great
War, he was in the bomber business and employed 150 men in his factory,
but the post-1918 slowdown hit him hard. In 1929, with the aviation craze
in full swing, he moved his plant to Baltimore and looked around for orders.

Martin had a reputation for innovative designs. During the war, he'd
drawn up plans for an armored bomber, a torpedo-armed plane for sinking
ships, and an airplane that carried a motorcycle, to be used for scouting
behind enemy lines after landing. He was also famous for thinking very
big. As early as 1918, one of his bombers had a wingspan of seventy-one
feet and carried nearly a ton of bombs, four machine guns, and a cannon.
From there, they only got bigger.[6]

So in 1931, when Trippe was calling around asking for a large flying
boat, Glenn Martin seized his opportunity. He offered something that
Sikorsky didn't: a genuine leap forward. His planned M-130 dwarfed the
S-42. With a length of 91 feet, a wingspan of 130, and a height of 25, the
plane would deliver, Martin promised, 3,320 horsepower (520 more than
the Sikorsky) and a takeoff weight of 52,250 pounds (nearly 30 percent
more than the S-42).

Most astoundingly, owing to the sheer volume of fuel on board, the
M-130 would enjoy a maximum range of 3,200 miles, or more than two
and a half times that of the S-42, already considered a long-range aircraft,
and substantially greater than the 2,500 miles Trippe had so impossibly
proposed. Of course, the range was an as yet theoretical figure, subject to
significant dropoff once you added in forty-one passengers, but even so,
the M-130 would be like nothing else on earth. At 3,200 miles, the Martin
airplane was distantly approaching Zeppelin-airship numbers.

The only downside? Martin was currently working on a new bomber,
the B-10, one far in advance of anything else on the market. The proposed
M-130 would benefit from B-10 technology, but Martin could not develop
both simultaneously. He told Trippe that the very earliest he could foresee
for delivery was sometime in late 1934, three years hence, but later was
likelier.

Since the first S-42 was also due for delivery in early 1934, Trippe was at an impasse. For the next few years, he would have to rely on the S-40s, which meant he was locked into the Caribbean and South America. In the meantime, rather than twiddle his thumbs, his task was to set up the future conditions of success by chivvying Imperial along so that they could launch Atlantic service simultaneously.

He fed Woods Humphery details on the upcoming S-42 and M-130, but Major Robert Mayo, Imperial's technical expert, laughed at Trippe's claims of their payload, range, and capability. Those kinds of figures, he said, "could not possibly be achieved." Instead, Imperial would build a flying boat of his own design, but *when* was left infuriatingly opaque.[7]

All of which left Juan Trippe with $2 million worth of Sikorskys and Martins on order—and nowhere special to fly them when they came.

The only bright spot was that Eckener wasn't going anywhere, either.

AFTER THE *AKRON* had been christened in August 1931, technical problems and seemingly unending malfunctions in the navy's mandated scout-plane trapeze kept it in its Ohio hangar until October, and even then the planes themselves weren't due to arrive for nearly a year after that. Making do nevertheless, the airshipmen strove to put on a good show for the upcoming fleet exercises in March 1932.

There were high hopes for the *Akron*. The exercises were to take place off Hawaii—fine hunting grounds for a long-range airship—and simulate an attack on Pearl Harbor by bombers launched from the "Japanese" car-riers *Lexington* and *Saratoga*. *Akron*'s task would be to scout far ahead of the fleet and alert the friendly cruisers and battleships to the enemy's loca-tion. Admiral Moffett, the head of the navy's airship program, could only rub his hands with glee at the thought of teaching his carrier adversaries a lesson in humility.

An excellent performance at Fleet Problem XIII, as the war games were called, would form a perfect backdrop to Hunsaker's great coup: finally getting the Merchant Airship Bill before the House of Representatives for a vote, scheduled for July 16. If the Senate followed suit, then IZT, PZT, Goodyear, and Zeppelin would be able to establish Pacific and Atlantic routes in short order, notwithstanding Trippe's perfidious acts of sabotage a year earlier.

As always seemed to happen with airships at inconvenient moments, there was an accident. On February 22, *Akron* was being brought out of the hangar to begin its journey to the Pacific when a gust of wind tore its tail loose. No one was hurt, but the runaway airship hit the ground several times, breaking the tail fin. This happened right in front of four members of the House Committee on Naval Affairs, invited by Moffett to witness the beginning of a new era of American airships. For the next two months, *Akron* underwent repairs, missing Fleet Problem XIII.

To help Hunsaker whip up support and divert attention from *Akron*'s woes, Eckener swung into action and with scant notice scheduled flights on the *Graf Zeppelin* between Friedrichshafen and Recife. He put a brave face on the effort, as he always did, but his plan to begin regular service to South America had been predicated on the assumption that by 1932 the Depression would have lifted—only it hadn't, and Germany was in straits still more dire than even the year before.

The truth lay in the numbers. The year 1931 had been a terrible one, with an average of just eleven passengers and 363 pounds of mail to and from Recife, but now those numbers fell to nine people and 262 pounds. On some trips, just seven or eight pounds of freight was shipped, meaning that Eckener was devoting a colossal airship and scores of crewmen to transporting packages that together weighed less than a carry-on bag across an ocean. On one of Trippe's old Caribbean S-38s, it's sobering to realize, nine passengers, 262 pounds of mail, and a couple of packages would have been counted as a full load—but at a minuscule fraction of the cost.[8]

Trippe could only wonder how on earth Eckener was staying in business. The short answer is that he was receiving large amounts of cash from philatelists. He would arrange with the German Post Office to issue special sets of Zeppelin stamps, which were purchased by collectors at a significant markup. Eckener once mentioned that from stamps alone he'd made no less than $100,000 on the Germany–Brazil–United States triangle flight. The back-and-forth runs to South America earned far smaller but not inconsiderable amounts. Still, there was no getting around the fact that by 1932 Eckener was managing an international airship operation heavily reliant on the continuing interest of stamp collectors.[9]

This was not, to put it mildly, a sustainable long-term business model, and Eckener covered up the South American flights' dreadful figures. He stressed instead the *Graf Zeppelin*'s undoubted safety and regularity. The trick allowed Hunsaker the time to redouble his lobbying efforts, which paid off dividends in the July 16, 1932, vote. The troubled Merchant Airship Bill passed in the House with a margin of seventeen. This was a fairly narrow squeak and by no means the overwhelming vote of confidence the airshipmen had been hoping for, but it would serve.[10]

The next step was the Senate, though as Hunsaker already knew, Trippe had lots of friends in that august body. But 1932 was a presidential election year, and the Democratic candidate Franklin Roosevelt was likely to prevail against Herbert Hoover. Hunsaker counted on the Democrats being more open than the Republicans to the kind of grand spending project and high-employment plan a home-grown airship industry would exemplify.

When Roosevelt won a landslide victory in November, the Senate vote was scheduled for debate on March 2, 1933, two days before Roosevelt's inauguration. It was sure to go favorably, as many of Trippe's pals had lost their seats, but early that morning Senator Thomas Walsh of Montana died of a heart attack and the vote was adjourned out of respect. Upon reconvening the following day, the Senate, faced with a crowded New Deal agenda, decided not to set back the calendar to take care of previous business and the Merchant Airship Bill vanished into the legislative ether.

Through sheer luck—bad, in the case of poor Senator Walsh—Trippe had stopped Eckener and his allies in their tracks.[11]

More bad news for the airshipmen was on the way. A month later, on the night of April 3–4, the *Akron* was making a routine flight along the New England coastline when a thunderstorm unexpectedly veered in its direction. Commander Frank McCord ordered *Akron* up and farther out to sea to avoid it, but the storm caught up.

As lightning flashes filled the sky, McCord could see the altimeter reading eight hundred feet—safe enough—only to feel the ship rapidly descend, followed by a shuddering shock and the loss of all control over the lower rudder as it bruised the sea. A few moments later, an officer cried, "Stand by for crash!" and then the control car was submerged in the chilly water. A lack of life rafts brought death quickly to seventy-three of the seventy-six aboard, including Admiral Moffett.

The *Akron* had set a new and unenviable record: Its loss had caused more fatalities than any other air vehicle in history.

The only silver lining Arnstein and the design team at Goodyear-Zeppelin could see was that as a helium airship, the *Akron* had not caught on fire—an achievement of scant consequence when exposure and drowning had killed 96 percent of those aboard. Few in a grieving nation paid much attention to the distinction.[12]

The navy had lost one of its two Goodyear-built airships at the worst possible time, and the death of their biggest cheerleader, Moffett, was an almost irrecoverable blow to the airshipmen's hopes for further expansion.

The long, grueling match between Eckener and Trippe had ended in a temporary stalemate, but a new game would soon be afoot.

43. The Hooked Cross

FOR MUCH OF 1932, Eckener's eye had been not on America, but on Germany, where the rise of Hitler had dragged him, unwillingly, into politics.

Like Trippe, he was concerned above all with the preservation of his great air project, and like his American counterpart, he presented himself as a moderate-in-all-things patriot, a staunch defender of the middle-ground consensus, in order to ensure widespread support for Zeppelin. For that reason, in January 1932 he made a radio speech in support of re-installing the universally respected Field Marshal Paul von Hindenburg, now eighty-four years old, as president.

Hitler, whose Nazi Party enjoyed more than a third of the vote in some places, was mulling whether to run against the old soldier, but even he was wary of taking on the Victor of Tannenberg in an open fight. Hitler proposed instead that he would not oppose Hindenburg on condition that he dismiss the beleaguered chancellor, Heinrich Brüning, a bespectacled academic, devout Catholic, and leader of the Center Party group in the Reichstag (Parliament).

If Brüning did not go, he threatened, blood would run in the streets and flow more freely than in even the previous year, when more than eight thousand paramilitaries from all parties were injured or killed in brawls. "It is the duty of all citizens to back up the government without qualification," Eckener thundered in response, and not to indulge such "demagoguery and incitement."

Eckener's was a powerful and influential voice. In the days after his radio speech, which was reprinted in newspapers around the country, veterans' societies representing three million members and the Federation of German Industries announced their support for Hindenburg.

Still, his speech was just the kind of thing that tended to get one's name added to the Nazis' ever-lengthening enemies' list. In the aftermath of the address the Nazi press published a number of, as Eckener said, "malicious pamphlets" about him, and he received a heap of abuse from correspondents calling him a "despicable, narrow-minded person" and pledging never to give another pfennig to the Zeppelin Company. Hitler, it was rumored, was incensed and swore to punish Eckener when his time came.

The outbursts were an unpleasant wake-up call for a man who had always tried to play both ends against the middle, his only goal being to further the interests of the Zeppelin. Try as he might, Eckener could no longer be friends with everybody.

But still he tried, by establishing a pressure group that would eschew the extremes of both left and right. This organization, he pledged, would "put aside all partisan and personal special interests and take a united approach . . . in order to protect the people and the Reich from extreme economic and cultural upheaval." The German National Association (GNA), as he called it, would be committed to freedom and humanism, the kind of values Eckener had long claimed were represented by the airship, that engine of commerce and international understanding.

It went nowhere. For his pains, the *Deutsche Zeitung* commented that "Dr. Eckener builds castles in the air. . . . The middle is politically dead. And not even a name like Dr. Eckener's will perform the miracle of reawakening it." Caricatures were printed of Eckener as a hobbled old man staring grimly at an airship and saying, "Great—my entire new party will just barely fit into the Zeppelin." And there was another torrent of abusive letters telling him he should stay out of politics and stick to flying airships or accusing him of Jewish sympathies or of selling out Germany.

Eckener was out of step with the times—which were growing ever more extreme. Hitler was publicly pledging to have "heads rolling in the sand" when he came to power, while his henchmen talked excitedly of "the thousands of Marxists to be fusileered and of Jews' heads lining the way from Munich to Berlin." Over the summer, more than a hundred Nazis and Communists were murdered in an escalating series of clashes and riots.

Eckener soon gave up his efforts in view of what he called the "total hopelessness" of the political situation and devoted himself to flying the *Graf Zeppelin* to South America to wait out the storm—but it only worsened in the closing months of 1932.

Though he had forced Brüning to resign, Hindenburg continued to hold out against Hitler, asserting that he would never allow a "Bohemian corporal" so enamored with violence to become chancellor. Instead, Hindenburg appointed Franz von Papen, a vastly wealthy aristocrat married to the daughter of an industrialist. Reactionary, lightweight, and charming—some might say oleaginous—Papen was liked by the country's big businessmen, and he packed his cabinet with his noble friends, moneyed pals, and Hindenburg loyalists.

Then on November 17, following a Nazi-organized vote of no confidence in the Reichstag, the cabinet was forced to resign, leaving the control of Germany in the hands of two men: Hindenburg and an isolated Papen, now ruling by emergency decree over a country in fast-motion economic and social collapse.

By January 1933, unemployment stood at around six million, but the real figure was closer to nine million if one includes part-time workers, meaning nearly half the German workforce was either unemployed or underemployed. The stock market had fallen by two-thirds since 1929, and industrial production had been halved. The pensions of disabled veterans and war widows had been drastically cut, and every day brought news of farmers and small-business owners declaring bankruptcy. Suicide and crime rates were rocketing, and apathy and hopelessness sought an avenue in extremism, not a refuge in Eckener's form of centrism.

Papen was soon forced out by a court intriguer, General Kurt von Schleicher, whom nobody liked, least of all Hitler. So greatly did the Nazi leader detest him that he actually consented to a series of secret meetings with the slightly less detested Papen.

At their first meeting, on January 4, 1933, Papen promised Hitler that he would back him for the post of vice chancellor if Hitler agreed to his having the chancellorship. Papen still had the ear of Hindenburg and thought that he could get the Grand Old Man to agree to the idea. That would mean knifing Schleicher, which was a pleasurable thought for the both of them.

Then came a front-page story in the *Neues Deutschland* newspaper predicting Eckener as the next chancellor. The report was obviously nonsense—the editor was a keen astrologer and published Eckener's horoscope prophesying his destiny to lead Germany—but Eckener later said that this one article provoked more Nazi hostility to him than anything he had so far done, including forbidding the use of the Friedrichshafen hangar for a Hitler rally.

He quickly disavowed the story, and in order to allay any dangerous suspicions that he was in fact secretly trying for the chancellorship, Eckener publicly announced that he was burying himself in preparations to work on *Projekt* LZ-129—stalled ever since his blueprints for the LZ-128 had been roundly panned by IZT and Goodyear three years earlier.

In the intervening time, Dürr's design team had completely rethought what Zeppelin's next big ship would encompass. They knew from the start that it would have to be able to handle helium, which they weren't enthusiastic about, but they were marching to an American drum now. The performance hit on lifting ability that using helium entailed meant that LZ-129 would have to be much, much larger than previous German airships. This also made it more expensive to build, of course, and Eckener had no money, leaving *Projekt* LZ-129 a kind of imaginary airship—but at least it signaled he was harmless and harbored no political ambitions.

With an Eckener candidacy no longer an issue, Hitler persuaded Papen that it would be in his best interests to walk away while he still could. Ultimately, January 28, 1933, brought news of the mass resignation of Schleicher and his cabinet, and two days later, a tired Hindenburg finally agreed to appoint Hitler and summoned him to the Reich Chancellery to swear his oath to uphold the constitution he had long pledged to tear up. "Hitler is Reich Chancellor. Just like a fairy-tale," gushed Joseph Goebbels, his propaganda master. Privately, Hitler assured his comrades that heads would soon roll.[1]

ONE OF THEM might well have been Eckener's. His name may not have been at the top of the enemies list, but it was on there. At first, he was left alone, but on February 27, a fire burned down much of the Reichstag building. Hitler spuriously claimed that it was ignited by Communists, and the next day the Reichstag Fire Decree nullified most remaining civil liberties, allegedly to forestall a Red uprising. Arrests began immediately, accelerating after the passage of the Enabling Act of March 24, which granted Hitler the power to enact laws without the consent of the Reichstag.

Tens of thousands of Communists, Jews, democrats, centrists, old-line conservatives, and Catholics began to be held under what was euphemistically called "protective custody." What this really meant was that they were kidnapped by Hitler's stormtroopers, beaten up, whipped, and tortured in old military barracks and *Heldenkeller* (secret dungeons). A suspiciously large number were "shot while trying to escape."

Eckener later found out that he was to be rounded up, too, but claimed Hindenburg had personally intervened with Hitler to stop the order. This possibly might have been true, though a little far-fetched; more likely is that the new Nazi leadership's will-they-or-won't-they policy on arresting Eckener reflected their own confused and ambiguous views of the airship.

To some senior figures like Hermann Göring, a former fighter ace and currently the aviation minister (along with other posts, such as Prussian minister of the interior), the airship was a relic of the imperial past and an obsolescent has-been compared to the modern, technologically advanced airplanes he wanted to build. On the other hand, the airship was famed abroad for its ability to forge new air highways and develop trade links with the Americans, northern and southern.

Owing to his own uncertainty, then, while Göring initially rejected a funding request from the Zeppelin Company for 2 million marks to keep it afloat, he reversed his refusal a few weeks later. Cutting *Graf Zeppelin* service to expatriates in South America when the regime was claiming to stand up for German rights abroad would have been hypocritical, and throwing hundreds of employees onto the welfare line when Hitler had repeatedly promised to put Germany back to work, embarrassing—especially when Zeppelin was saying it would forge ahead with LZ-129 after the bailout. Not that Göring really wanted a new airship—it was just that *not* having it was worse than having it.

Over at the Propaganda Ministry, Goebbels, a pronounced rival of Göring's, was just as conflicted, but for different reasons. For him, the Zeppelin symbolized Germany's national rebirth out of the struggle of the 1920s, and its sheer gigantism awed and fascinated those below. Eckener had, despite the odds, heroically succeeded in keeping the airship cause alive in the face of British and French opposition, much as the Nazis had, so Goebbels figured, fought valiantly against their enemies to bring the German people out of the darkness and into the light. Balanced against that, of course, were Eckener's dubious politics, so for Goebbels it was critical to make clear the distinction between the machine and the man, the technology and the cultural icon who personified it. If the world had learned to identify Eckener with the Zeppelin, the world now had to un-learn it.

So it was that in April 1933 Goebbels demanded, in the manner of a creditor calling in his debts, that in return for the two-million-mark fi-nancing Eckener release the *Graf Zeppelin* for a flight over Berlin on May 1 to mark the Nazi version of May Day—a celebration of German brotherhood via *Gleichschaltung* (which can be translated in various ways, such as "coordination" or, less positively, "bringing into line").

The *Graf Zeppelin*'s reputation as an elitist, rich man's means of travel was turned on its head so that the airship displayed the proper *Volksge-meinschaft*, or the soul and spirit of united national community, by invit-ing guests of *all* classes aboard. "Next to the general stands the unskilled worker from Lake Constance, next to the son of the postal official, the deputy switchman, and next to the airship captain, the salesman," cooed one tame journalist. "A single feeling fills us: gratitude to the unknown German worker whom we celebrate today, and who helped create the wonder of German technology."[2] Eckener was not mentioned.

A couple of weeks later, Goebbels, after offering an additional 2 million marks, took the *Graf Zeppelin* to Rome to visit one of his heroes, Musso-lini. Their romance was characterized by mutual flattery—Goebbels gushingly declared him a "political genius," and Il Duce played to his admirer's intellectual vanity by calling him Il Dottore (the Doctor).

Unfortunately, who was in command of the *Graf Zeppelin* during the May Day or Rome flights is difficult to say. Eckener was certainly not, not when the regime wanted less publicly visible faces. Of the five captains available at that time, at least two—Max Pruss and Anton Wittemann—were or would shortly become Nazi Party members, and Ernst Lehmann, by far the most experienced, was closely identified with the Party. It was probably Lehmann, then, assisted by one or more of the others.[3]

In the meantime, Eckener was not playing along as well as he was ex-pected to. On April 26, Göring, alarmed at the scale of the ongoing vio-lence inherent in the "protective custody" scheme, had created the Geheime Staatspolizei (Gestapo), or Secret State Police, to investigate and neutralize political opponents. In other words, the terror was to be-come more *orderly*, even if the intent remained the same. Its first chief was a (relatively) urbane and (objectively) philandering member of the Prus-sian Political Police named Rudolf Diels.[4]

In late May Diels, no doubt on Göring's say-so, invited—or, perhaps, ordered—Eckener to dinner at the Hotel Esplanade in Berlin. After some small talk, Diels said that he'd heard some "complaints" about Eckener and produced a file containing enough material to brand him an enemy of the state.

In Friedrichshafen, apparently, he'd been overheard talking to an Austrian friend, to whom he had exclaimed, "My dear Count, you're forgetting that Germany is now a nation governed by criminals!" Another time, while walking through the lobby of the Hotel Bristol in Berlin to meet a colleague, he'd seen some SS officers drinking coffee and muttered loudly, "Look at these gangsters. They would have locked me up long ago had they any guts." A phrase Eckener liked to bandy around—"Nazism brings to the surface all the evil qualities of human character"—was particularly troubling. He'd also been rather foolish by expressing these views in front of the workers and crews at Zeppelin, most of whom ignored them, but it seems that a few had not. What was certainly clear was that Eckener was being watched and informed upon.

"We don't expect you to become a member of the Party," Diels smoothly went on, "but it would be to your best interest to stop your remarks and to retract those that you've already made." At this, Eckener looked at his watch, said gruffly he'd think about it, and left, telling Diels that "I've got to get back to Friedrichshafen for a flight to Brazil."[5]

In spite of his later claim that he marched out, head held high, Eckener had been given a scare. If he didn't clean up his act, there was a good chance, no matter how internationally prominent he was, that he'd be spending some time in one of the new concentration camps. Eckener subsequently made himself scarce in the first two weeks of June by undertaking the voyage to South America he'd mentioned to Diels, and then doing the same in July, only to find himself having a chance encounter with Hitler himself several days after returning home.[6]

On July 19, according to Eckener's daughter Lotte, the family was on vacation near Berchtesgaden, Hitler's favored resort area. They were staying in a rural hotel, quietly admiring the views of the mountains, when a wave of great excitement rippled among the guests: Hitler was coming! Eckener headed for his room, grumbling something along the lines of, "Don't jump off the roof unless you have to," but he was duly summoned to the verandah, where a long table had been prepared for Hitler's entourage, including Rudolf Hess, Julius Streicher, and Eva Braun. Remembered Lotte, "My mother and I could now look through the window and curtains and observe how the two stood opposite each other as if no connection at all existed between them."[7]

What exactly was said is uncertain—Eckener mentioned later that Hitler issued a torrent of words—but the meeting, and the mere fact that Hitler, who usually remained at a discreet distance when he wished someone dead, wanted to see him in the flesh, indicated that Eckener could count himself safe, at least for the time being. He was lucky, too, that the wave of arrests was subsiding and that the Nazi leadership had decided that the Zeppelin was worth saving.

For Hitler personally, airships as a whole were dangerous and awkward, and he refused to ever travel in one. Asked about them, he once declaimed, "The whole thing always seems to me like an inventor who claims to have discovered a cheap new kind of floor covering which looks marvelous, shines forever, and never wears out. But he adds that there is one disadvantage. It must not be walked on with nailed shoes and nothing hard must ever be dropped on it because, unfortunately, it's made of high explosive."[8]

Yet even Hitler recognized airships' power. In a signal that Eckener might rehabilitate himself if he kept his views to himself, in September 1933 Hitler approved an appearance of the *Graf Zeppelin* over that year's Nuremberg rally. Filmed by Leni Riefenstahl, Hitler's court director, the airship passed slowly over the parade and, in the words of a Nazi propagandist, "traces her majestic route over Germany's youth. . . . A frenetic jubilation shakes the whole stadium. Over one hundred thousand voices ring as a single immense scream of joy and pride at the giant of the air."[9]

Though, again, it was likely that Lehmann was captaining that key flight, Eckener was permitted, perhaps as a loyalty test, to take the *Graf Zeppelin* to Chicago later that month for the World's Fair. The Nazis, who had not yet established themselves incontrovertibly as the legitimate government of Germany, obsessively monitored the foreign press and were eager to make a good impression after enduring any number of hostile articles reporting on the bloodletting against their domestic enemies. A visit by the *Graf Zeppelin*, Goebbels thought, presented an ideal opportunity to show off the New Germany, and having Eckener in command would prove that the claims of tyranny were meritless calumnies.

Eckener, however, was aware of the ploy and told his wife, Johanna, "I'm beginning to think that they feel I'm more valuable to them alive.

They know the propaganda value of having German airships flying the world's air routes. They also know that I enjoy many overseas friendships, particularly in America, and that I can succeed in obtaining foreign cooperation and support whereas they cannot."[10]

What should have been an easygoing and uncomplicated visit turned out to be a political minefield. The trouble began even before the flight, when Eckener, following precedent, requested that the U.S. Post Office issue a special Zeppelin stamp to bring in much-needed revenue to help pay for the trip.

When the green 50-cent stamp was issued, it caused an uproar: The *Graf Zeppelin* had been depicted without a flag on its vertical tail fins.

The flag was a fraught issue both in Germany and among German expatriate populations at the time. The traditional imperial flag, with its three horizontal stripes of (from top to bottom) black, white, and red, had been in use until 1918, when it was replaced by that of the Weimar Republic (a tricolor of black, red, and gold). Throughout the 1920s, the Weimar flag had been abhorred by the ultra-right as the hated and humiliating symbol of democracy and reparations, so on March 12, 1933, it had been outlawed in Germany and replaced by two official flags: the old black-white-red imperial one as a temporary sop to Hindenburg and the conservative traditionalists and the flag of the Nazi Party, a black swastika inside a white disk on a red background.

A year after Hindenburg's death in August 1934, Hitler would scrap the imperial flag, but during the World's Fair the law was that the Nazi flag must be painted on one side of all aircraft and the imperial tricolor on the other. In the *Graf Zeppelin*'s case, twenty-foot-high swastikas were added to the port side of the upper and lower rear fins, with the imperial stripes on the starboard.

Yet on the stamp, which showed the *Graf Zeppelin* with its port side plainly visible, there was no swastika. Clearly, someone at the Post Office had decided to depoliticize the problem by omitting the flag—but had succeeded only in politicizing it.

In Chicago, where there was a 600,000-strong German community, the absence of the swastika on the stamp proved particularly divisive. The German Group of the World's Fair, the official organization overseeing the German exhibitions and comprising mostly older-generation Germans, worried that the Nazi flag symbolized anti-Semitism and flew only the imperial colors, while young pro-Nazi activists belonging to the

Friends of the New Germany, usually called the Bund, made great sport of tearing down the imperial flag and replacing it with the Nazi one.

So far there had been just a few scuffles over the issue, but the police feared that the *Graf Zeppelin*'s visit would lead to a riot, bomb threats, or even a sabotage attempt. Hans Luther, the German ambassador, was so worried he warned Cordell Hull, the U.S. secretary of state, of rumors that Communist militants were planning on blowing up the ship.

Eckener and Willy von Meister, running the American arrangements, had early on been notified that trouble was brewing, but were promised that three hundred deputies, a hundred police officers, and two companies of soldiers would be on hand. "Bring that Zepp to Chicago," growled William Meyering, the Cook County sheriff, to a relieved Meister.

Thus reassured, Eckener went ahead with the visit on October 26 but made some discreet changes to the schedule. Usually, Eckener tried to arrive at the most propitious moment for news photographers, who loved taking shots of huge crowds surrounding the airship. Accordingly, he announced that the *Graf Zeppelin* would be landing near the World's Fair in downtown Chicago at 9 A.M. — perfect for the late-morning editions. But in fact he touched ground about twenty miles away at Curtiss Field, in the suburbs of Chicago, three and a half hours earlier.

That decision alone was very strange, but then there was Eckener's odd choice of approach direction. He had been coming from Indiana, southeast of Chicago, and the most expeditious route would have had him heading into the city with Lake Michigan to his east. Yet he went far out of his way to bear west and approach Chicago from the north, whereupon he'd circled the metropolis in a clockwise direction until landing at Curtiss. When Meister noticed the detour, he asked Eckener about it, who replied quietly, "And let my friends in Chicago see the swastikas?"

In an act of small rebellion, Eckener had deliberately allowed only the *starboard* imperial tail markings to be seen easily from below and then had made sure to land in such a way that the swastika side was mostly obscured from view.

Unfortunately, stills snapped by an airplane-borne photographer from the sunlit east over Lake Michigan with the fairgrounds as a backdrop were more picturesque than those shot from below or from the west. So when the newspapers printed them, millions more Americans saw the swastikas than Eckener had expected. Unaware of the ruse, the Propaganda Ministry in Berlin was thoroughly delighted with the result, and graded Eckener accordingly, if undeservedly.

The rest of Eckener's time in Chicago was occupied with similar acts of muted resistance. These consisted of restricting the content of his speeches and interviews to boilerplate rhetoric about the *Graf Zeppelin*, its technical achievements, and airships' contribution to world commerce. Eckener may have been obliged to collaborate with the regime for the sake of his own safety and the good of the Zeppelin Company, but he would serve neither as its propagandist nor as its apologist.

At one reception, ostensibly to celebrate "Zeppelin Day," a supposedly nonpolitical event, Eckener discovered that the venue, the Medinah Temple, was festooned with Nazi flags—a gigantic one hung behind the podium—and the ushers were dressed in stormtrooper uniforms. It turned out the Bund had arranged the affair, and its members occupied themselves making stiff-armed salutes. Eckener, clearly unhappy, gave a brief speech yet again extolling the technical abilities of the *Graf Zeppelin* and sat down to near-silence, after which a pro-Nazi speaker roused the crowd into singing militaristic choruses of "Deutschland Über Alles."

Eckener, who had heard the anthem in New York in 1924 after his first triumphal flight, was brought almost to tears. It was a pitiful reminder of how greatly its meaning had changed—from one of Weimar republicanism and liberty to that of racial and ideological domination—in the intervening decade.

On his return home, Eckener was told that the Chicago Bund leaders had informed Berlin that they had been disappointed by his lack of political enthusiasm, though Eckener received a note of appreciation for the voyage from Göring, granting him a modicum of protection from further investigation.[11]

After all, he still had his uses.

44. The Ledgers

ECKENER'S RELATIONSHIP WITH the regime would remain ambiguous: He never knew for certain on which side of the Nazi ledger he was. Debit or credit? Maybe he was on both—or neither.

On the plus side, in 1933 and 1934 the *Graf Zeppelin* was turning a corner with its South American service, partly owing to Germany's slow emergence from the Depression but mostly because Berlin had decided to funnel a million marks a year into the Zeppelin coffers as a subsidy.

By 1934, the *Graf Zeppelin* was transporting an average of nineteen passengers—almost a full house—415 pounds of mail, and 352 pounds of freight. It would be going too far to say it was a blazing, profitable success, especially since the entire edifice would have collapsed without the government cash and there was still a lot of empty cargo space, but after a dire 1931 and 1932 at least the numbers were now heading in the right direction.[1]

On the minus side, alas, the government was taking an increased interest in the Zeppelin Company. It was noted with dismay that Eckener had not yet purged his Jewish employees and moreover had hosted the Goodyear (and former Zeppelin) designer Karl Arnstein, Jew, in late 1933 for a lengthy visit to discuss LZ-129.[2] Just as the universities and scientific institutes were being pressured to fire their Jewish professors, the so-called Jewish Influence on Germany's newest and proudest airship would not be tolerated, and it mattered not that Arnstein happened to be, even more so now that the elderly Ludwig Dürr was nearing retirement, the world's most experienced designer.

With this in mind, Eckener's position as the head of Zeppelin was increasingly precarious. A secret memorandum of March 30, 1934, marked for highest eyes only (among the recipients were Hitler, Göring, Goebbels, and Heinrich Himmler), judged that airships were economically, culturally, and politically important as a means of international transportation and should be exploited as such. More Zeppelins should be built to take advantage of Germany's near monopoly on the airship industry and more agreements negotiated with foreign countries to allow more frequent flights.

The memorandum conceded that Eckener was "an excellent airship operator and a man of unusually outstanding merit," but ultimately he was too much in love with the Americans, too involved with the Jews, too globalist in his outlook to be permitted to represent Germany. Henceforth the Reich should take over all dealings with Washington and fund the Zeppelin Company to whatever amount ("forty, thirty, or even twenty million" marks) was necessary to ensure success. In short, the memorandum recommended that Eckener be ousted at some point and the Zeppelin Company fully Nazified and nationalized.[3]

That this had not already happened—political parties, companies, trade unions, sporting associations, social clubs, and the civil service had all been shut down, crushed into submission, or ruthlessly cleansed—was testament to Eckener's world fame but not exclusively so. Other noncompliant world-famous figures, like Albert Einstein, the playwright Bertolt Brecht, the conductor Otto Klemperer, and the novelist Thomas Mann had already exiled themselves or been forced to emigrate, but unlike these subversive parasites, Eckener was considered necessary to the airship effort, at least for the time being.

Until he exhausted his utility, Berlin, keeping Eckener on his toes and his back foot, gave with one hand and took with the other.

So, sometimes he received the kid glove treatment, such as when he was requested to visit the Propaganda Ministry, where Goebbels, seemingly casually, asked him how he was coming along with LZ-129. When Eckener replied that he needed the equivalent of at least half a million dollars to reach the next stage of construction, Goebbels "said with great composure, 'Is that all? Why, that's a mere bagatelle! I will approach some big industrialists and suggest they contribute what you need.'" When his rival Göring heard about Goebbels's generosity with other people's money, he too disgorged a nonrepayable "loan" from the Air Ministry.[4]

And then at other times, he was shown the mailed fist. An arrest order was issued for Eckener during the Night of the Long Knives between June 30 and July 2, 1934. This was the operation in which Hitler decapitated the brownshirted leadership of the SA militia, rid himself of troublemakers, and avenged himself on old enemies. In Eckener's case, two Gestapo men knocked on his door, and upon being told he wasn't there, departed without saying a further word. By good fortune—or maybe the Gestapo just wanted to give him a scare in absentia—Eckener was at that moment on board the *Graf Zeppelin* in South America. He only arrived back in Germany on July 7, after the murders and beatings had ceased.[5]

This unpredictable combination of punishment and reward was cleverly thought out, and Eckener was by no means the only victim. It was a common enough strategy to bring the recalcitrant into line once they realized the world could be theirs if only they behaved. The Nazi leadership knew perfectly well that Eckener was addicted to airships and was likely to do anything in order to be allowed to finish LZ-129. An opportunity to test Eckener's loyalty soon came along; if he passed, perhaps he could stay as the head of Zeppelin for a time longer.

Following the death of Hindenburg in early August 1934, a national plebiscite was ordered to approve the bestowing upon the former Herr Hitler the title of *Führer und Reichskanzler* ("Leader and Reich Chancellor"), thereby making him supreme commander of the armed forces and a divine entity to whom unconditional obedience was required.[6]

In the days leading up it, Goebbels asked Eckener "in a somewhat threatening tone" to make a radio speech supporting the measures. It was not a matter of whether or not to accept the request, for as Eckener recalled, "it was immediately clear what a refusal of this imposition . . . would mean to me." His concern was, rather, what should he say and how should he say it?

Eckener's speech was an extremely careful and most cunningly worded model of balance, in and of itself a subtle sign of resistance at a time when lickspittles and toadies were falling over themselves to praise the Führer. Much of it stated the obvious and recited banal truisms, but other parts could be understood in different ways depending upon the listener's perspective.

The Propaganda Ministry, which had a sensitive nose for these sorts of things, allowed the speech to be broadcast—so Eckener had passed the test—but only after one of the censors had made changes to the "significant final sentences" that had, Goebbels thought, called a touch too much attention to the late Hindenburg's "integrity, sense of duty, and love for his country." If one didn't know better, it might be thought that Eckener was discreetly contrasting Hitler to his predecessor.[7]

ECKENER'S COOPERATION HAD bought continued government support for the airship program, his primary concern. Construction work on LZ-129, which had proceeded in fits and starts for years, now advanced at a fast clip.

Harold Dick, a young Goodyear engineer, received the rare prize of a yearlong secondment to Friedrichshafen at the personal request of Paul Litchfield. Not since the early 1920s and the arrival of the U.S. Navy inspectors had any foreigners been permitted to visit Eckener's Chamber of Secrets, but he graciously welcomed this one.

Eckener wasn't being entirely altruistic: He knew Dick had an inside line to Litchfield and would be reporting on LZ-129's progress. It was critical for the future success of IZT that Litchfield be kept apprised of the marvel that he was planning so that Goodyear could get ready to build its own in Akron.[8]

Explaining Dick's visit was that the IZT scheme had very recently, and entirely unexpectedly, been given new life after years in the doldrums—but it would be without United's Frederick Rentschler, its leading backer. He'd become a major person of interest during a recent Senate investigation into the Post Office's cozy awarding of airmail contracts in the 1920s and early 1930s. When his staggering salary was revealed, he had been roundly condemned as the worst of the worst, a man who grew rich on stock pyramids, market manipulations, and the lamb-shearing of innocents. As a result, the giant United Aircraft and Transport Corporation was broken up into three companies and Rentschler was banned from running an airline.[9]

The ouster of Rentschler, who'd long since lost interest in airships, had been greeted with sighs of relief in Akron. At last free of United, once seen as their savior, Goodyear and IZT could pursue their own ambitions without having to worry about pleasing the airline.

The bullish Litchfield now wanted to go full speed ahead on transatlantic passenger service. Commercially successful flights by Eckener's next-generation LZ-129 would lead to increased American interest in IZT, which would enable Goodyear-Zeppelin to quickly build LZ-129 copies in the currently empty Air Dock.

DURING THE AIR Mail scandal that had taken down Rentschler, Trippe, too, had been asked a series of uncomfortable questions once Post Office auditors found that Pan American, like United, had long been benefiting from favoritism and backroom deals to receive maximal subsidies. Between 1927 and 1934, Pan American had pocketed no less than $35.7 million from the government, but it proved impossible to distinguish between what was proper and what was padding since Trippe, more adept than Rentschler in covering his tracks, had made it diabolically difficult for the investigators to understand Pan American's bookkeeping.

Any number of key accounts were dispersed among Alaska, Texas, and Florida, while others were scattered among the many operating divisions in South America. Internal reports carelessly (or carefully) obfuscated distinctions between types of expenses. Breakdowns of the sums for overhauling aircraft, records of passenger miles, and other data critical for cost analysis went mysteriously missing or were sent to the wrong office.

The upshot was that very little, if anything, could be pinned on Trippe personally. Certainly, there were raised eyebrows—why, pray tell, were Pan American's administrative expenses quadruple those of other airlines? And why did the New York headquarters charge its overhead to the Caribbean division?—but, legally speaking, Trippe apparently knew nothing of his own company's financial operations even if he obviously knew everything. Rather like Eckener, he was on both sides of a ledger and on neither.

In the event, he escaped from the scandal barely scathed: Mail payments to Pan American would be cut by 25 percent. Trippe wasn't bothered one way or another because he'd already prepared for the worst by reorienting Pan American toward the less politicized, less regulated tourist trade—and now he had the perfect plane for the job.[10]

Sikorsky had delivered the first S-42, the long-range successor to the S-40, right on schedule. In early August 1934, Trippe's chief pilot, Edwin Musick, took it for a faultless four-circuit run totaling some 1,200 miles along the Manhattan riverfront, Long Island Sound, and coastal Connecticut, serving in effect as an eight-hour-long advertisement for Pan

American. Immediately afterward, the *Brazilian Clipper*—as Trippe had dubbed the plane—was assigned to the Miami-Rio route. Travel time between the two cities was cut at a stroke from eight days to five, and the *Brazilian Clipper* was soon joined by the *West Indies Clipper*, *Jamaica Clipper*, and *Antilles Clipper*.[11]

The Sikorsky S-42s scored a huge success for Trippe, who brilliantly exploited a growing interest in Brazil among Americans. One of his board members was the extraordinary adventurer Merian Cooper. In his time he had chased Pancho Villa in Mexico, served as a bomber pilot in France in the Great War, flown for the volunteer Kościuszko's Squadron in the Polish-Soviet War of 1919–21, been shot down and interned in a Soviet POW camp, escaped from that POW camp, and been pursued by pirates off the Abyssinian coast. In the late 1920s, he had joined Trippe's merry band at Pan American, but board duties took up little of his time: He otherwise devoted himself to—why not?—the movie business, where he pitched an idea about a giant ape ravaging New York that had come to him in a dream.

King Kong came out in 1933 to great acclaim, and on the back of its success Cooper became production chief of RKO, where he oversaw the making of an ambitious "aerial musical" called *Flying Down to Rio*.

If ever a movie served as outright propaganda for a company, *Flying Down to Rio* was it. As the film went into production, Cooper—still a Pan American board member, but not a word was heard about the conflict of interest—and other RKO executives became frequent visitors to Trippe's offices in New York and enjoyed themselves "scouting" locations in scenic Brazil courtesy of the airline.

Released in January 1934, the film was a major blockbuster, though audiences paid less attention to the romantic triangle among the leads (pulchritudinous Dolores Del Rio, beefcake Gene Raymond, sizzling Raul Roulien) than to the dance numbers performed by a scene-stealing duo named Fred Astaire and Ginger Rogers, together for the first time.

The film's spectacular chorus-girls-dancing-on-a-wing flying sequences and its invention of a "new Brazilian dance craze that is driving the world Melody Mad" called the Carioca drove American tourism to lush Rio in 1934—with Pan American, featured heavily in the movie, reaping the benefits. That year, the airline reported a 44 percent increase in South American passenger miles over 1933, as tourists flocked to experience, as one magazine put it, "palm-fringed beaches fanned by the southern trade-winds, exotic scenery, Latin folk, brilliantly plumed parakeets, and sloe-eyed señoritas."

To accommodate them, Trippe built the Dinner Key Terminal in Miami, intended to complement the city's 36th Street Airport in handling Pan American's growing fleet. Designed in the cool Streamline Moderne style, Dinner Key was the most modern seaplane terminal in the world and could control up to four aircraft landing or taking off simultaneously.

The main entrance doors were of bronze, like some Homeric palace, topped by a frieze of winged globes, the symbol of Pan American. Inside, visitors were greeted by the hulking prominence of a three-and-a-quarter-ton rotating globe ten feet in diameter that showed Trippe's boundless empire.[12]

Or so he wanted people to think. In fact, the boundary stopped at the Atlantic—but he didn't care anymore.

THE ATLANTIC OCEAN was still a mire, thanks to Imperial Airways, Trippe's ball and chain of a partner. Granted, Glenn Martin was late in delivering his M-130s—he'd asked for an extension until the fall of 1935 to deliver the first one—but Imperial hadn't even *started* to build its own transatlantic plane, and, thanks to his "understanding" with Woods Humphery, Trippe couldn't go anywhere in the meantime.

Imperial had been bragging to Trippe that its contracted seaplane, the Short Brothers "Empire" S-23 flying boat, would be a world-beating aircraft. But then Oswald Short, co-founder of the firm, and Arthur Gouge, its chief designer, attended a lecture by Sikorsky at the Royal Aeronautical Society on Pan American's S-42 and were flabbergasted by what they heard.

Before a stunned audience Sikorsky spoke of such proven numbers as a top speed of 182 mph, a payload of 8,363 pounds, and an astounding wing-loading ratio of 28.58. To give some idea of the scale of Sikorsky's achievement, consider that the S-40's respective figures were 137 mph (25 percent less than the S-42), 3,200 pounds (62 percent less), and 19.5 (32 percent less).

In the question period afterward, a shocked Gouge confessed that the superiority of the S-42 was such that Short Brothers would have to "re-cast their ideas" and redesign the Empire from scratch.[13]

The very earliest it could be delivered, Woods Humphery told a horrified Trippe, was the summer of 1936—fully two years away—by which time Sikorsky would no doubt be generations ahead technologically even as Trippe's expensive M-130s languished in a hangar until Imperial could get its act together.

TRIPPE SOUGHT A way of turning the Imperial liability into a Pan American asset. He had the glimmerings of an idea, but it was a hell of a long shot.

He stopped by his old haunt, the New York Public Library at 42nd Street. In the mid-1920s, when he'd been running Long Island Airways for Coney Island thrill seekers, Trippe had spent countless hours in the library studying the history of trains and shipping as he tried to understand the arcane science of airline economics. Now he lodged himself in the Maps Room, peppering, or pestering, the librarians with requests to bring old, yellowed maritime charts up from the archives.

He was looking for one tiny speck in an endless ocean, an island in the *Pacific*. But none was to be found, no matter how closely he held his magnifying glass to the page. He had another idea: the logs of the clipper and U.S. Navy ships that had once plied the blue. One brittle manuscript after another was brought to his desk and Trippe diligently deciphered the crabbed, handwritten numbers and notes within. And all of a sudden, there it was: a single mention of a place named Wake Island, located roughly halfway between Midway and Guam.[14]

Little was known of Wake. First discovered by a Spanish navigator in 1568, who dismissed it as a waterless, barren island inhabited by nothing but seabirds—an accurate appraisal—it was rediscovered by Captain Samuel Wake of the *Prince William Henry* in the 1790s, but nobody cared. The place was so forgotten and forlorn that it had seldom been added to maps since.

Trippe investigated further and ordered up the five volumes of Lieutenant Charles Wilkes's *Narrative of the United States Exploring Expedition During the Years 1838, 1839, 1840, 1841, 1842*. While surveying the Pacific Ocean, Wilkes had made a brief stop there on December 20, 1841, and had written of the place: "Wake's Island is a low coral one, of triangular form, and eight feet above the surface. It has a large lagoon in the center, which was well filled with fish of a variety of species; among these were some fine mullet. There is no fresh water on the island [and] from appearances, the island must be at times submerged, or the sea makes a complete breach over it. . . . The reef around this island is very small in extent."[15]

That was the sum total of knowledge of Wake Island. Judging by Wilkes's description, it did not sound very promising, but the lieutenant had spent less than a day there and had only guessed that it was sometimes underwater. Maybe it wasn't. How large and placid was the lagoon? How small and dangerous the coral reef?

Trippe needed to find out more because Wake was so perfectly situated. If a flying boat could land in the lagoon and the island was supplied with fuel and water by ship, then Wake was an ideal spot for an island-hopping Pacific route from California to Hawaii to Midway to Wake to Guam to Manila, and from there China or Japan was an easy ride.

In June 1934, Trippe called a trusted manager named C. H. "Dutch" Schildhauer, a former naval aviator, into his office with the rolltop desk and asked him to go to Washington and look into Wake Island. "Do it quietly," said Trippe as he ushered Schildhauer out. "You don't need to bring our name into it."[16]

Schildhauer made straight for the Navy Department, which was sure to have more up-to-date information available. It had some, but not much. In 1898, Hawaii had been annexed; soon after came the Spanish cession of Guam and the Philippines to the United States as it expanded across the Pacific.

At that point, there had been a thought, Schildhauer learned, that Wake Island would make a good coaling station for the fleet. To that end, in 1899 Commander Edward Taussig of the *Bennington* had dropped anchor and claimed "the Atoll known as Wake Island" as American territory. But it proved unsuitable for a naval base (Midway was better), and for the next couple of decades the island had mostly played host to Japanese pearl hunters, feather poachers, and sharkfin cutters.

A follow-up report in 1922 noted that USS *Beaver*, under Lieutenant Commander Sherwood Pickering, had visited the main island for a few hours and discovered that the lagoon was just fifteen feet deep, making it unsuitable for anything other than small boats. He had also measured the total land area, much of it lying only ten to fifteen feet above sea level, at about 2,600 acres (the three-island group surrounding the lagoon was roughly 4½ miles long and 1½ wide).

Pickering's report confirmed the existing impression that Wake was useless for naval operations, but it nevertheless contained the information Trippe was looking for. He didn't need a deep lagoon to land a seaplane, just one sufficiently calm to allow a smooth set-down and takeoff, and Pickering had determined that the island remained above water.

There was one sentence that particularly excited him: Pickering, rather farsightedly considering it was 1922, had remarked, "If the long-heralded trans-Pacific flight ever takes place, Wake Island should certainly be occupied and used as an intermediate resting and fueling port."[17]

Trippe ordered Schildhauer to find out who was in charge of administering Wake: the State Department or the navy? Schildhauer asked State's historical adviser, who hadn't the foggiest idea but called the navy's Hydrographic Office. The navy didn't know, either. Wake Island existed in bureaucratic limbo. Eventually, the navy got it.[18]

Trippe was also in limbo. He had to decide what to do next. A major strategic dilemma faced him: Pan American's goal had always been to cross the Atlantic, but Imperial's backwardness was preventing that, and there were no guarantees that Imperial would fix itself. On the other hand, a Pacific route lay open and beckoning, though whether it was possible to fly it was as yet unknown.

The distances alone made it seem out of the question. California to Hawaii, at roughly 2,400 miles, and Hawaii to Guam, at about 3,500 miles, were the two longest overwater jumps in the world and made the Atlantic look like a pond in comparison. But add in Midway and Wake Island as pit stops and the calculus changed: Hawaii to Midway was 1,149 miles; Midway to Wake, 1,034 miles; and Wake to Guam, 1,334 miles. From Guam, it was about 1,600 miles to Manila.

As for the California-Hawaii leg, a stripped Sikorsky S-42 loaded up with nothing but fuel could theoretically get to Honolulu, albeit with zero safety margin and fumes powering the engines. The distances thereafter were within reach, but then the other problem was that Wake, Guam, and Midway were impossibly small tufts of land; the tiniest navigational error would doom a hopelessly lost pilot to a watery death.

Commercially, too, flying the Pacific was unlikely to make much money. China and Japan were by no means major markets, and relatively few people wanted to vacation in the Philippines when London, Paris, Rome, and Berlin were far more attractive destinations.

Then again, Hawaii had proved highly lucrative for the steamship companies, and since PZT was a bust, Trippe would be assured of tourist dollars for that leg, at least. If nothing else, since Pan American's longest water route at the time was between Kingston and Barranquilla, a mere 660 miles, any lessons learned from mastering the Pacific could be applied to the Atlantic when the time came. And anything was better than having his airplanes languishing in hangars for years.

Trippe rarely blazed trails without official backing, and in the Pacific he could count on strong support in Washington. In 1934, the State Department and the navy were confronted by an expanding Japanese empire in the Pacific. As State sought to buttress American allies there, the navy was trying to fortify its chain of bases against potential attack, though this was loudly denounced in Tokyo as a treaty violation.

But what if a *private* company developed the islands with radio stations, fresh water, and seaplane docks? A company like Pan American, for instance, that agreed in the event of war to hand over its facilities at Midway, Wake, and Guam to the navy. With the State Department's blessing, Trippe and the navy made what they called an "informal working arrangement" to that effect.

Trippe made the final decision in the early fall of 1934. In the coming year, Pan American would pivot its efforts to the Pacific, he told his disbelieving staff in New York. Despite everything he had said previously, the Atlantic would become the airline's secondary concern and Glenn Martin's delayed M-130s would be based in California instead of on the East Coast.[19]

When the news was announced on October 14 that Pan American was heading to Asia, Eckener could scarcely believe his luck.[20] It meant Trippe was abandoning the Atlantic enterprise and leaving the field wide open for Zeppelin domination. But he had to work fast.

GOODYEAR'S EMISSARY TO Friedrichshafen, Harold Dick, accordingly reported to a rapturous Litchfield that Eckener was driving his workers hard to finish LZ-129 so as to begin Atlantic service as soon as possible.

Dick was astounded by what he saw in the hangar. It was not so much the size of LZ-129, for Dick had been up close to the *Akron* and *Macon*, and LZ-129 did not dwarf them, being only slightly larger.

No, for Dick, it was the scale of the ambition, for here at last was the grand transoceanic airship he and so many others had long dreamed of building. While the *Akron* and the *Macon* had been utilitarian military vehicles that devoted significant space to their internal (and eternally troublesome) scout-airplane hangars, LZ-129 was intended for between fifty and seventy-two passengers. Eckener had finally overcome one of the *Graf Zeppelin*'s drawbacks—lack of room for travelers—and the plans he showed Dick depicted luxury in the air far beyond anything else available.

Dick, however, was a jaded rarity. To any regular observer, LZ-129, even in its then-skeletal state, made the *Graf Zeppelin* look puny, though it was only thirty feet longer. It was its width and height that exponentially multiplied its dimensions and boggled the senses—as they still do. Whereas the maximum diameter of the *Graf Zeppelin* was 100 feet, LZ-129's was a third more, making it nearly *five times* the height of what is today the world's largest passenger airliner, the double-decked Airbus 380 (subtracting the upward-protruding tail). Its length was 804 feet, more than twice that of a football field. One of LZ-129's four tail fins alone measured 105 feet long and 49 feet in breadth, and was 12 feet thick at the base.

There was still a lot of work to do, Dick told Litchfield. The passenger cabins were mostly complete, but there were no walls, fittings, or water pipes in the crew and officers' quarters. None of the outer cover was in place and no gas cells had been installed, nor had any of the complex control mechanisms or mooring equipment.[21]

He also revealed the "next stage of construction" by adding that a new type of engine was undergoing testing in Stuttgart. This was a reference to a coming changeover from Maybach, which had supplied Zeppelin engines since the count's day. It was not a sad goodbye. Karl Maybach had lost interest in the low-margin business of manufacturing airship engines and had turned instead to making an exorbitantly expensive automobile—one could purchase five large houses for the same price—named the "Maybach Zeppelin," fittingly so considering the car's gigantic dimensions.[22]

Replacing Maybach was Daimler-Benz. The Nazis had lately instituted a "motorization" policy to increase car production (and to build military vehicles for the semi-covert rearmament program).[23] A prime beneficiary was Daimler-Benz, whose board of directors was accordingly packed with Nazi Party members and fellow travelers.[24] They were no doubt some of the "big industrialists" Goebbels had casually referred to when he once spoke with Eckener about raising money, and they had been prevailed upon to supply Zeppelin with the engines he required.

The new DB 602 engines were specifically designed for airship use and to Eckener's specifications. Airship engines were different from airplane engines in that they were expected to run at cruising power for days, rather than for hours. That made them heavier than airplane engines, by up to five times, but fuel consumption was the more important measurement, as a thriftier engine allowed less fuel to be taken—useful, when one had to budget for about sixty-five tons of the stuff for transoceanic voyaging.[25] It was time, Eckener had decided, to switch from gasoline and Blau gas to diesel. Not only was it cheaper, but it was a great deal more energy efficient.

Harold Dick also reported on something else, of greater moment, at Friedrichshafen: the imminent switch to helium.[26]

In the aftermath of the R-101 tragedy, Eckener's American partners had made it very clear that for safety reasons he must use helium instead of time-honored hydrogen. Like Trippe with his turn to the Pacific, then, Eckener had had to calculate the debits and credits involved with such a radical pivot.

There were a lot of minuses. As Trippe had for years assumed that the alpha and omega of Pan American was to get across the Atlantic, hydrogen had from day one served as the basis for Zeppelin's entire operating system. For the Zeppeliners, switching to helium was not as simple as choosing a diesel engine over a gasoline one, and neither was it a matter of stupidly adhering to a gas known to be dangerously flammable.

It was the one thing everyone at Zeppelin, past and present, agreed on. The ousted Colsman had recently published his memoirs, in which he had warned that the company should never rely on a vital resource, helium, available exclusively from the Americans.[27] When Willy von Meister had brought up the touchy subject with Ernst Lehmann, he was told bluntly that "we have been operating our commercial service with hydrogen for years" and that he saw no reason to change to an unfamiliar gas. Hydrogen, Lehmann added, was perfectly safe when handled expertly, as only the Germans knew how to do.[28]

Then there were the technical factors involved, including adverse changes in performance, weight, and capability. And cost, of course. Hydrogen was cheap, plentiful, and easy to acquire, but Eckener estimated that purchasing helium in America and then shipping it to Germany would be eight times as expensive as buying locally produced hydrogen.[29] The price for the *seven million* cubic feet of helium needed just to inflate

LZ-129, and heaven knows how much more to replenish it regularly, would be staggering.

But Eckener, like Trippe with Imperial, was forced to face the reality of his situation and make a drastic decision. If he wanted to conquer the Atlantic, he had to have helium, like it or not, and that meant asking the Americans to sell it.

On October 11, 1934, three days before Trippe announced that Pan American was heading across the Pacific, Eckener boarded the liner *Albert Ballin*, bound for New York.[30]

45. The Medusa

ECKENER WOULD ONLY be in New York for a short time, and he no doubt stopped by for talks, as he always did, with Adolph Ochs, the publisher and proprietor of *The New York Times*. His meetings with Ochs were highly confidential, but Eckener served as an unofficial source for German political news. In return, recalled Captain von Schiller, whenever he was "in trouble with the Nazi regime politically, the *Times* [went] to bat for him editorially or [by publishing] significant news stories" to relieve the pressure.[1]

Then Eckener headed to Washington, where he had two related aims in mind. The first was to arrange a helium purchase, and the second, to lay the groundwork for starting transatlantic service in 1935 under the IZT umbrella. The newly formed Federal Aviation Commission was currently examining the latter issue, and Eckener and Goodyear's Litchfield had been invited to testify.

He optimistically expected to be able to kill both birds with one stone, but Eckener, who may have been accustomed to dealing with Nazi bureaucracy—a Hydra to be sure—was innocently walking into the lair of Medusa, she of the many-snaked hair. Washington was a far more complicated, if less personally dangerous, place to navigate than Berlin.

Having been advised by officials that "this government would not aid any foreign country to set up a permanent air service," Eckener agreed with Litchfield that in their testimonies on October 29 they would instead talk up "the inauguration of American-built and American-operated Zeppelins in intercontinental service."[2] Litchfield assured Eckener that since the Merchant Airship Bill debacle he had been working his contacts at the Department of Commerce. They were now recommending, he said, $17 million in funding for two giant Akron-built airships, a smaller one for South American trips, and the long-desired Atlantic terminal in Baltimore.[3] Once this was approved, it would be a simple matter for Zeppelin to acquire its own helium as a courtesy to American-owned Goodyear.

Unfortunately, Litchfield had gravely miscalculated the balance of power in Washington by proposing that the $17 million come as a grant from the new Public Works Administration (PWA), a New Deal initiative dedicated to building large-scale public works that was headed by Harold Ickes, the secretary of the interior. The Commerce Department, then whose purpose was to promote business interests, could certainly make recommendations, but the decision was in Ickes's hands.

In approaching the PWA Litchfield assumed that an airship line would be right up its alley, and in thinking this he may not have been entirely wrong: The PWA *did* fund airport construction, runways, mapping, and other programs it believed were part of the national infrastructure. These were, however, for public use and the public good; funding a privately owned airship business was quite a different matter, especially when traveling aboard those airships was advertised as a privilege reserved for the affluent few.

Still, the PWA was supposed to involve private enterprise in its projects, and Goodyear-Zeppelin could justifiably claim that federal financing would boost employment and spur consumer demand. But Litchfield did himself no favors by asking for so much money. Compared to other PWA air projects it was an absurdly extravagant amount. Between 1933 and 1939, the PWA distributed grants for 384 airport-related projects totaling nearly $20 million—which means Litchfield was requesting almost as much for a single project as the PWA would spend nationally over six years. At a time when the PWA served millions of regular Americans by allotting $18 million to help build Chicago's subway system, Litchfield's $17 million proposal to help speed the elite to Europe for their vacations had not a hope of gaining Ickes's approval.[4]

And despite Litchfield's sunny assurances, the chances of Eckener's acquiring helium were equally dismal. Since the Helium Control Act of 1927, the Bureau of Mines—responsible to the Commerce Department— held jurisdiction over the production, storage, and sale of America's helium, but the land the gas lay under and its exploitation rights were controlled by Interior. Further complicating matters, the export of helium, a strategic resource, was subject to the unanimous recommendations of the Departments of War, Navy, *and* Commerce as well as presidential approval.

Arguing that this system was somewhat unwieldy, Interior Secretary Ickes, a very argumentative fellow, had won a bruising fight with his colleague Daniel Roper at Commerce in April 1934 to wrest control of the Bureau of Mines from him, though Roper retained a say in whether to approve export. As Ickes was also the PWA head *and* a noisy critic of Hitler, it meant that Eckener and Litchfield were contending with a man not overly inclined to favor either an American airship plan or, to say the least, the sale of helium to a German firm he thought served only to further Nazi propaganda.

As for the other players, the army was neutral, but Navy Secretary Claude Swanson was dead set against selling helium to the Germans, saying that "to give our commercial competitors [Zeppelin] the advantage of using helium in the future development of commercial lighter-than-air transportation seems inadvisable."

Even so, Eckener made an attempt to convince the government to make an exception. Eckener—as evidenced by his careful reading in 1919–20 of the Versailles Treaty that permitted him, or so he had thought, to build small civilian airships despite an Allied ban—was adept at exploiting pinprick holes in legislation. He believed that he had spotted one in the Helium Control Act.

In Section 3, there was a line stating that "any surplus helium produced may, until needed for Government use, be leased to American citizens or American corporations." When helium had been in short supply, the provision hadn't mattered, as there was no surplus—but now there was, thanks to increased production and less demand from the navy for its diminished airship fleet. Since Goodyear was an American corporation and Zeppelin was in a partnership with that American corporation, all it would require was an exceedingly minor tweaking of the act to allow him to purchase helium. To that end, he testified that it was "imperative that in a future combined service of domestic and foreign airships helium be made available to *all companies* who will participate in this service."[5]

It was a good try, but a hopeless one: Ickes had the ear of President Roosevelt, and Swanson remained unmovable on the subject, pointing out that the navy had in fact recently discovered a myriad of nonairship uses for some of its surplus helium. The gas, added to pressurized air, mitigated cramps and nausea in divers, as well as reducing decompression times. Used in recently developed diving bells, helium could help rescue trapped submariners. It could also be employed to treat asthma, croup, and diphtheria, and larger quantities aided the laying of bridge foundations and tunnels—making it also a PWA concern.[6]

At this, Eckener realized there was no point pursuing the idea, and during a subsequent friendly meeting with Roosevelt at the White House he never raised the subject. Instead, fully aware of Eckener's disappointment, the president graciously offered to let him use the navy's facilities at Lakehurst if LZ-129 visited in the near future.[7]

The voyage home in mid-November was a depressing one. With helium out of the question, months of testing had been wasted and LZ-129 would have to be reconfigured as a hydrogen airship. The only silver lining in this particular cloud was that, quite unexpectedly, he would have a giant seven-million-cubic-foot vehicle capable of lifting fifty tons of weight rather than the thirty-two tons achievable with helium. But on the downside, there was no possible way LZ-129 could be ready to fly by the summer of 1935, not with all the changes it was necessary to make.

WORSE, ANY LINGERING hope that the navy would order another Goodyear airship vanished on February 12, 1935, when the USS *Macon* encountered a storm off Point Sur, California, causing structural failure in the ring securing the upper tailfin. Harold Miller, one of the scout-plane pilots off duty in the smoking room, remembered a feeling "as though a giant hand had swatted us." The bow suddenly dropped, then the airship swerved wildly to starboard, and the *Macon* started sinking.

If one had to experience an airship crash, being aboard the *Macon* in the Pacific Ocean would be most people's choice. The crew had twenty minutes to prepare themselves for collision as the ship gently deflated; the navy had this time provided sufficient life rafts; the water was relatively warm and calm; and there was plenty of time to send rescue vessels. After a soft, stern-first landing, *Macon* sank within a few minutes. Only two men died among the eighty-three-strong crew.

The loss nevertheless left the navy's airship program with neither airships nor friends. President Roosevelt declared he would spend not a penny more on airships, and Carl Vinson, who chaired the House Naval Affairs Committee, and after whose home city *Macon* (Georgia) had been named, said he recognized a "death-knell" when he heard one.

Press reaction was predictably hostile. As the usual investigations got under way, the *New York Post*, with all the nuance for which it is famous, shouted that "no more funds of American taxpayers [should] be squandered on these useless gas-bags."

The *Post* was right. The American record with airships had been disastrous. ZR-2 (R-38): destroyed. *Shenandoah*: destroyed. *Akron*: destroyed. *Macon*: destroyed. The U.S. government had spent, or wasted, more than $16 million on its airships and now-forlorn hangars. The only one left was the venerable *Los Angeles*, which had recently been decommissioned after eight years of faultless service—and it, of course, had been built by the Germans.

As always, the Germans were the exceptions. Eckener's products experienced minor problems here and there, but so too did any aircraft, and his had never, in more than three decades, suffered a fatal peacetime accident—a safety record unequaled by any other form of transport. Between September 1928 and December 1934, the *Graf Zeppelin* had made 423 individual trips covering a distance of 630,000 miles; it had carried 10,500 passengers and spent a total of 10,005 hours in the air; and it had sailed the world and had crossed the South Atlantic no fewer than sixty-eight times.

American opinion was not against airships per se—there was due excitement about the prospect of a visit by LZ-129 and the start of transatlantic service—but was directed against navy airships in particular. The leading American airshipman Charles Rosendahl fought a rearguard action by claiming that since 1918 trains had killed "300 times," and cars "1,000 times," more Americans than airships ever had, but the navy airship program was dead. In its place, the navy commissioned aircraft carriers like the USS *Ranger*, *Yorktown*, and *Enterprise*.[8]

Eckener at this point considered approaching the Americans once more about purchasing helium. After all, the navy now *definitely* had excess capacity, but he wisely refrained. With Ickes in the saddle, he would have been turned down, and in any case, within a couple of weeks he was in no position to.

· · · · ·

THERE HAD BEEN a coup at Zeppelin, and Eckener, a past master of internal dogfighting—as his struggles against Lehmann and Colsman attested—now found himself the loser.

The government was following through on its secret memorandum of a year earlier, which had laid the groundwork for Eckener's ouster and the Nazification of the company. To Eckener, the move came as a surprise, but he could not have been shocked. Other large companies were gradually being brought under state control or had Nazi loyalists appointed to their boards. Now it was Zeppelin's turn. An early warning sign had come at the beginning of 1935, when he'd been advised to prepare to move Zeppelin's operating base (though not its manufacturing facilities) to a terminal being built outside Frankfurt. Deutsche Lufthansa, the government-backed airline, would share the new airfield.

Perhaps Eckener had missed the flashing danger signal because Frankfurt, as an airport, made a lot of sense: Unlike Friedrichshafen, it was near two new national highways and a major city, and coordinating flight schedules with Deutsche Lufthansa would make connections for airship travelers to or from Germany that much easier.[9] But it also meant that Deutsche Lufthansa would have increasing say in Zeppelin affairs, which really meant Berlin's voice would be amplified.

In early March, the blow fell. Eckener was informed that a new, virtually nationalized company was being formed, to be called Deutsche Zeppelin Reederei (German Zeppelin Airline Company, or DZR), with one half owned by Deutsche Lufthansa and the other by Göring's Air Ministry, and headquartered in Berlin. Until then, Eckener had ruled over both the Zeppelin Company (which built the airships) and the DELAG (which flew them). The DELAG, which Eckener had birthed in 1909, was wound up and its duties transferred to DZR, while the Zeppelin Company would continue construction work on LZ-129 and its successors.

Under the new arrangement, Eckener was sidelined as head of Zeppelin building in Friedrichshafen and was "promoted," or so he was told, to become the toothless figurehead chairman of DZR and a member of its Supervisory Committee, where he was joined by several regime stalwarts to keep an eye on him.

No longer serving as king and pope, Eckener would remain the outward face of Zeppelin in its dealings with the Americans, but real power—management of flight operations and oversight of commercial affairs—would no longer be in his hands. That job fell to his rival, Captain Ernst Lehmann, who had lusted after it since being outmaneuvered by Eckener in the early 1920s. Lehmann, a friend of the regime (Eckener subtly remarked that he was "more acceptable than I"), would be a far more pliant and pleasing chief executive officer of DZR than the irksome Eckener.

Eckener suffered no illusions as to what was actually happening. The Nazis, he said, had pushed "the old nuisance out of the control car and behind a desk where he won't be so much in the way."

There was no doubt that Göring had organized the "putsch," and it was of a piece with a related plan to turn Germany into a great air power. A week or so later, on March 11, he revealed the secret existence of the Luftwaffe to a stunned world. Then, on March 22, Göring commemorated the formation of DZR with a speech. In it, he was rather gracious to Eckener, considering that he'd just knifed him, thanking him for his "great endeavors" in representing "the German engineer on the one hand and the German worker and spirit of enterprise on the other."

Then came the big change. The airship "does not have the exclusive purpose of flying across the Atlantic, but also has a responsibility to act as the nation's representative." The pursuit of service to New York remained *a* goal, in other words, but it was more important to fly the flag, exemplify German power, and "make known to the world the healthy and workmanlike spirit of our Fatherland." Zeppelins would henceforth serve the Reich, not just Eckener's dream of creating a transatlantic airline.

A tangible sign of Eckener's diminished power came after the *Graf Zeppelin*'s first flight under the new DZR flag several days later: Not only were *both* sides of the tail fins carrying the swastika but a photograph of the company leadership and captains showed Lehmann front and center, with Eckener humiliatingly off to the side. Snappy Nazi salutes were now expected, though Eckener, recalled a colleague, used to make minimal efforts with "a somewhat lackadaisical upward swinging movement of his right arm." ("No one advances here without these acrobatics!" murmured Eckener to a friend.)

By altering Zeppelin's mandate, Göring was throwing a bone to Goebbels at the Propaganda Ministry. Göring himself had little interest in airships, as Eckener well knew. If anything, he viewed them, as Eckener said, "with contempt, if not actual hostility." When Eckener once asked him whether he'd like to take a ride, the air minister "vehemently declined and declared that he had no real confidence in the 'gas bag.'" Goebbels, however, keenly understanding the spell they wove on Germans, wanted to employ them as propaganda tools for the upcoming Olympics and for other nationalist purposes.[10]

With government control came a fraught discussion over what to name LZ-129. Some of the more servile flunkies at the Führer's court "suggested" it be christened *Adolf Hitler*, but on second thought, it was hardly likely that the Great Leader would want his name associated with the project. How his critics would snicker if the *Adolf Hitler* went down in flames over one of his own ridiculous rallies.

What the Hitlerites didn't know is that Eckener already had his preferred name: *Hindenburg*. Because Hitler brooked no competition to his own personality cult, *Hindenburg* might be a tough sell, but on the other hand, the president, safely dead, was regarded as a symbol of Germany, and Hitler had a weakness for naming big things like battleships after statesmen, admirals, and generals (*Bismarck*, *Tirpitz*, and *Scharnhorst*, for instance). *Hindenburg* could be pitched not as a political statement—even if it was, in Eckener's eyes—but as honoring Germany's most esteemed field marshal.

Hindenburg it was, then. Bella Fromm, a prominent society reporter with an excellent ear for gossip, bumped into Eckener at a party and noted in her diary that "good old gruff, upright, dry-witty bear Hugo Eckener" had been incredibly relieved: "I couldn't think of going to the United States with *Adolf Hitler* painted on the ship," he told her.[11]

FOR THE REST of 1935, Eckener busied himself constructing the *Hindenburg*. At sixty-seven, he could very well at this point have retired honorably, basking in the grand title of DZR chairman and the acclaim of a grateful nation, but he decided to stay on, even if only, as he said, as a "fifth wheel" to annoy the Nazis.[12] Not only was he concerned that the enterprise would come to grief without him, but he still needed to fulfill his life's ambition of inaugurating a transatlantic passenger service. He would not, could not, let Lehmann steal the laurels due him after so many decades of work and sacrifice.

Progress was quick. By the summer, the engine gondolas were complete and the skeleton was virtually done. The outer skin, with new, specially developed doping and tougher fabric to withstand the harsher weather of the North Atlantic, had been stretched over the majority of the airship by September, though its belly remained open. Installation of the fuel tanks, steering mechanisms, and signal and electric equipment were all on schedule, and the gas cells had been made but not filled by December. By then the engines had arrived and were close to being fitted.[13]

By February 1936 most of the exterior work had been finished, aside from a few patches here and there, but the interior decorators were holding things up. They would be worth the delay, though. Eckener had hired Fritz Breuhaus and Otto Arpke to do, respectively, the furnishings and wall decorations. Both artists had worked for the North German Lloyd shipping company, on whose vessel *Bremen* Eckener had sailed home from America in late 1934. Perhaps it was during that depressing voyage that he first admired their handiwork and thought it would suitably complement his hopes for the *Hindenburg*.

In choosing these two artists, Eckener was firmly distancing himself from the now old-fashioned Pullman look of the *Graf Zeppelin* and embracing a modern, streamlined aesthetic more in keeping with the 1930s. No more heavy wood and garish brass, no more old-fashioned bric-a-brac and the busy chintz so characteristic of the passé Edwardian era. All was sleek, vibrant, and gracefully curved.

For the dining room and writing room, Arpke contributed twenty-one lovely Japanese-inspired watercolor murals on silk-and-cotton wallpaper illustrating the development of mail, from hand delivery to airborne, and depicting the *Graf Zeppelin* in flight to Rio, at Friedrichshafen, above Lake Constance, and over Spain, Africa, the Cape Verde Islands, and the Brazilian coast. A giant world map in the lounge depicted famous voyages of the past. Alongside those of Magellan, Captain Cook, Vasco da Gama, and Columbus were the Round-the-World journey and the crossing of the *Los Angeles*. In the smoking room, the yellow pigskin-leather walls featured episodes from the history of lighter-than-air flight, from the Montgolfier balloon to the *Graf Zeppelin*.

Breuhaus, an architect famed for his functional yet sophisticated eye who also produced lamps, furniture, rugs, and silverware, designed the *Hindenburg*'s light, tubular aluminum chairs (upholstered in brown in the writing room, red in the dining) and tables. The chairs, marveled one guest, could be lifted with a finger. Breuhaus also almost certainly created the *Hindenburg*'s china. His plates were sternly stamped "Property of the Deutsche Zeppelin Reederei" to discourage passengers from stealing souvenirs and bore a gold-and-blue band around the rim with the DZR crest—a white Zeppelin, outlined in gold and superimposed on a blue globe with golden meridians of longitude and parallels of latitude.

Like the name *Hindenburg*, the selection of Breuhaus and Arpke bore a hidden dual meaning known only to Eckener. On the one hand, they had done fine commercial jobs for the North German Lloyd line and could then be sold to the likes of the untutored Lehmann. But on the other, Arpke had declared himself a pacifist after the Great War and his favored school of style, New Objectivity, would soon be declared "degenerate" by the Nazis; as for Breuhaus, the government later banned him from working as an architect for political reasons. So these two were by no means approved Nazi artists, just as Eckener was not quite an approved Nazi businessman.[14]

MARCH 4, 1936, was set as the *Hindenburg*'s first test flight. Lehmann wasn't there, being engaged in Berlin, but Captain Max Pruss, his ally, was in command. Eckener took advantage of Lehmann's absence to make sure he was present on this historic day.

After taking off (*"Luftschiff Hoch!"*), there was a ripple of fear among the crowd below as what appeared to be smoke trailed from the *Hindenburg*'s stern, but it turned out to be merely dust, accumulated during the ship's months inside the hangar, blowing off. The engines were tested at increasing speeds over the Bodensee and encountered no problems; likewise, turning maneuvers proved no obstacle. In calm air the ship was so stable it almost seemed to steer itself, and there was a notable lack of vibration. The outer cover flapped in places, generating some noise, but that was easily rectifiable. Otherwise, it was almost dead silent inside, the engines being placed far enough away from the passengers so as not to disturb them with their roars and rumbles.

Over the next few days, *Hindenburg* was taken out for more trials. There were a few kinks to iron out (a smell of fuel oil in the keel, it was chilly in the smoking room, a floor needed replacing in the control room, and so on), but nothing serious. On the afternoon of March 18, *Hindenburg* officially passed muster.[15]

A week later, Georg Wagner, a leading typographer, arrived from Berlin, having been commissioned to design the lettering for the *Hindenburg*. The obvious color for long-distance recognition was vermillion red, and Eckener had no say in the choice of font: It had to be Fraktur, regarded since the nineteenth century as *the* national German typeface and fully Nazified. Wagner painstakingly sketched the ornate individual letters—all in lowercase, as Fraktur capitals were even harder to read—that were then painted onto the hull.[16]

And with that, *Hindenburg* was ready for service.

46. The Labyrinth

A LOUDSPEAKER SYSTEM HAD also been installed, though it required days of trial and error before anyone on the ground could possibly hear what was being broadcast over the noise of the engines.

Eckener would have been happy never to have been able to fix the problem, but fixed it had to be. Goebbels had ordered the loudspeakers to be used on the *Hindenburg*'s important upcoming flight. Lehmann buzzed around repeatedly demanding progress reports from Eckener, who now cordially detested him. The two men had long distrusted each other, of course, but most of their disagreements had simmered rather than boiled. Eckener, as the boss, had admittedly always been arrogant with his subordinate, but now the shoe was on the other foot. Lehmann made sure that Eckener knew who was in charge, who had the ear of the government, and who made the decisions.

After a great deal of sniping, they arranged an armistice. If both were on the *Hindenburg* at the same time, then one would act as commander while the other had only the status of a passenger with no operational control. In this case, Eckener was relieved to step aside at Lehmann's insistence. What Goebbels had dubbed the *Wahlfahrten* ("plebiscite flights") was to him a ludicrous and thoroughly dishonest endeavor, having been prompted by Hitler's movement of troops into the Rhineland, a clear breach of the Treaty of Versailles, on March 7, 1936.

Since the 1920s, the Rhineland had been a bone of contention, and Hitler's remilitarization was intended to show France that the province was German territory, not a French buffer zone. Goebbels had accordingly arranged a plebiscite, or national referendum, for March 29 to further demonstrate that the German *Volk* firmly approved of the Führer's action. Just to make sure, the voting slips bore no NO box, only a big YES one.

Goebbels requested that Eckener make a speech, as the latter said, hailing "the occupation . . . as a brilliant deed of Hitler's," but this was too much for Eckener, and he avoided having to do it by pleading that he, a mere worm, was unworthy of the honor. Once again, his name was written down in a little book for being unhelpful. Privately, Eckener objected to the exploitation of his *Hindenburg* for nefarious political purposes when the Atlantic and his dream of establishing a true passenger airship line still beckoned.

Eckener was not invited on the cruise. Lehmann took instead sundry Party hacks, Air Ministry representatives, some Luftwaffe officers, a number of reporters, and members of the Plebiscite Commission. Captain von Schiller, meanwhile, was in charge of the *Graf Zeppelin*—this was to be a double flight.

The two airships took off at dawn, the *Graf Zeppelin* first. Lehmann prepared for liftoff but was warned that the wind had picked up. If Eckener had been captain, he would have waited a short time for it to die down, but Lehmann—impatient, anxious to show the pride of the fleet to the dignitaries, and eager to distinguish himself from his overly cautious predecessor—decided to leave anyway. In his defense, the *Hindenburg* could handle such gusts but Lehmann inexplicably chose the most difficult method of lifting off: idle the engines, lighten the load aft, and allow the wind to pick up the tail so that it pushed the airship upward under its belly. It was an impressive trick to show off but a foolish thing to do under the circumstances.

When the wind unexpectedly slammed the elevators down, the horrible sound of crushed metal and tearing fabric was heard as the lower fin hit the ground and the *Hindenburg* went sailing over the Bodensee for a time until Lehmann regained control.

Upon landing, Lehmann assessed the damage. It was not severe, at least not as severe as the chewing out he received from the furious Eckener. "How could you, Herr Lehmann, order the ship out in such wind conditions? You had the best excuse in the world to postpone this shit flight

[*Scheissfahrt*]. Instead, you risk the ship merely to avoid annoying Herr Goebbels."

Eckener's long-held suspicions that Lehmann was an excellent pilot but habitually made careless mistakes was confirmed that afternoon when Lehmann again took off and made the identical error, though at least this time he avoided a mishap. Eckener could only shake his head. Being an airship captain, he wrote, was a matter not merely of being capable and experienced, as Lehmann was, but of character. He must be able to say no frequently, which is harder than saying yes, and Eckener knew Lehmann worried that if he said no it would be interpreted as a lack of confidence and courage. Lehmann lacked the strength to stand up to authority, and his moral cowardice, Eckener believed, would reap a harsh reward.

A day after the accident Eckener circulated a memorandum to the airship captains, now seven strong, which, though it avoided mentioning his name, deeply humiliated Lehmann. It read as if he were being chastised like a naughty schoolboy. Eckener said that "one of the gentlemen" involved had not even "recognized" that he'd made a "disastrous mistake" and had blamed the crew for the error. Waspishly, Eckener added that *he'd* never experienced any such problem with *his* crews.

Lehmann was furious at the dressing-down, especially because the plebiscite flights had been a great success. The airships visited nearly a score of cities around Germany, all the time blaring martial music from the loudspeakers, interspersed with uplifting electioneering messages ("The Führer's purpose is peace and honor. He will build a united nation. Give your vote to the Führer!"), and dropping small swastika flags and millions of pamphlets. The Zeppelin officer Albert Sammt recalled that the Propaganda men on board insisted on playing Hitler speeches even when they were flying over cemeteries, as if summoning the dead to vote.

Over the Tannenberg Memorial, resting place of Hindenburg, Lehmann broadcast the national anthem and telegrammed: "From the proud flight of our two ships over the free German Reich, the crews and passengers of the airships *Hindenburg* and *Graf Zeppelin* think of their Führer with gratitude and in loyalty." Hitler cabled back that he wished them a continuing good flight.

Goebbels had thoughtfully provided a voting booth on the *Hindenburg*. The results were not overly unpredictable: 104 votes in favor, 100 percent of those aboard. When he heard the tally, Eckener puffed on his cigar and amusingly remarked he was surprised there weren't more, considering that it was a Nazi election.[1]

ASIDE FROM THE politicization of his airships, Eckener was angry at Lehmann's carelessness because repairing the *Hindenburg* threatened to delay an upcoming flight to Rio that he intended as a trial run for beginning New York service, planned for early May. The trip would give him a chance to iron out any issues at an early stage.

In the event, Eckener's flight of March 31, *Hindenburg*'s first overseas voyage, which took thirty-seven passengers astounded at the accommodations and the freshly prepared food, was a routine one. Or at least it was until April 2, when Eckener received a radiogram from Reuters in London asking him to confirm a curious report they'd heard in Germany.

Aboard the *Hindenburg*, Eckener couldn't make head or tail of the message. What was Reuters talking about? The mystery was soon solved. It turned out that a Propaganda Ministry official named Berndt had overheard Eckener's chewing out of Lehmann ("*Scheissfahrt*"), and someone else had found his little joke about the "Nazi election" offensive; they had both reported him to Goebbels, who went to see Hitler, who (according to William Dodd, the American ambassador) "ranted and shouted that the man must be dismissed."

Goebbels, preferring a quieter method, summoned a press conference, where he declared that "Dr. Eckener has alienated himself from the nation. In the future, his name may no longer be mentioned in the newspapers, nor may his picture be further used." In these situations, the reporters present were directed to keep their mouths shut. The victim would not even know he had been "unpersoned"; he would simply vanish from the papers, newsreels, and radio waves. Such a sentence was usually the prelude to forced exile or confinement in a concentration camp.

It so happened that in this case one of the journalists at the conference had mentioned it in passing to a British pal, who published the nugget in his own paper at home. Hence the Reuters transmission. But Goebbels, said Eckener, "had made the mistake . . . of overlooking the interest in foreign countries in my flights and my person." Eckener, put simply, was too famous to unperson, quietly or not.

This did not mean that Eckener was not in danger. The Nazis were no strangers to arranging convenient car accidents. As it was, there were rumors circulating that Eckener had already been arrested and executed for high treason (the charge: selling secrets to the Americans) and others claiming that he had committed suicide. Eckener also received word from a sympathetic *Berliner Lokal-Anzeiger* correspondent on board the *Hindenburg* that his newspaper was already reporting that "Lotte's father is seriously ill"—never a sign of continuing good health in Nazi Germany. Lotte was Eckener's daughter, and the omission of his own name meant the press was obediently toeing the Party line.

A friend, Lieutenant Commander Scott Peck of the U.S. Navy, happened to be on board as an observer, and Eckener went so far as to ask him whether he should go to the American embassy in Brazil and seek emergency refuge in the United States.

The dilemma facing Eckener at that moment had already been or soon would be experienced by any number of nonconforming, non-Jewish luminaries psychologically manipulated by the demonic Mephistopheles (as Eckener called Goebbels): You could leave Germany and start again from nothing in a foreign land or you could stay, go along to get along, and be honored, have baubles showered upon you, and enjoy favors aplenty. In creating this addictive acquiescence to the regime, Goebbels had constructed a labyrinth of moral compromise and dependence masquerading as free choice and autonomy.

Eckener, like so many others, chose to stay. If he had gone into exile, the chances were that Johanna and their children would eventually have been allowed to join him, but emigration would forever have prevented him from taking the *Hindenburg* to America on the very eve of his greatest success, the very culmination of his life's ambition.

But if he stayed, it would be on *his* terms, Eckener calculated. Upon returning to Germany on April 10, Eckener used his fame as a lever. Goebbels no doubt wanted to replace him as captain of the *Hindenburg* with Lehmann, but he was determined to give the propaganda minister no other choice than to let him fly the *Hindenburg* to America.

It was a dangerous game to be playing with the likes of Goebbels, but Eckener had two trump cards: Roosevelt and Göring.

The president and Eckener had met thrice before, the first time after the Chicago World's Fair visit in 1933, when the White House had called him and extended an invitation to visit, and then again in October 1934, after Eckener's testimony to the Federal Aviation Commission, when Roosevelt had said he could use Lakehurst if the *Hindenburg* came to visit.

And finally just a few months earlier, during a brief visit to Washington in February 1936. At this third meeting, Roosevelt spoke to him warmly about the achievements of the *Graf Zeppelin* before getting to the subject at hand: "Well, now you want to make regularly scheduled flights over the North Atlantic?" Eckener confirmed that, yes, this was his intention, only for Roosevelt to laugh: "I must tell you frankly, I don't believe you can do it." "I believe we certainly can," replied Eckener. Roosevelt smiled and asked how he could help. Eckener seized the opportunity to ask permission to lease Lakehurst—the property of the U.S. Navy, after all—for the ten flights he intended to make that year. Roosevelt mulled it over briefly before replying, "Good, you shall have it! The question interests me."

Now he needed Roosevelt to come to the rescue. On April 16, a week after returning home from Rio, Eckener had lunch in Berlin with his old friend the journalist Karl von Wiegand and William Dodd, the U.S. ambassador. There, Eckener outlined his predicament to Dodd, who promised to get a message to the White House.

Two days later, Dodd wrote to Robert Walton Moore, the assistant secretary of state. He explained the pickle Eckener found himself in and suggested that he thought it appropriate "for the President to give him some attention when he goes to Washington. The whole outside world respects Eckener, and it would seem to me to be nothing but fair if our Washington authorities could even give him a luncheon and allow the press people to know it."

Lo and behold, the White House soon lodged a formal invitation with the Air Ministry in Berlin for Eckener to have luncheon with the president "upon his arrival in the United States aboard the new airship *Hindenburg*."

In other words, the president of the United States was *expecting* to see Eckener, and not Lehmann, when the *Hindenburg* came over. The gambit stirred Göring into action, not least because he saw an opportunity to twit Goebbels over this burgeoning diplomatic imbroglio his rival had caused with his idiotic unpersoning scheme.

Eckener was invited to Berlin to see Göring in an effort "to clear up the difficulties . . . in which he finds himself." In their chat, Göring did not bother bringing up the tired subject of whether Eckener had made the disrespectful comments during the plebiscite flights. Of course he had. He was more concerned with why Eckener had refused to, as Eckener put it, "sing a hymn of praise" for the reoccupation of the Rhineland, and why he'd heard reports that Eckener had made fun of the Hitler salute at various times. Eckener assured him, not entirely believably, that all that was idle gossip, and Göring seemed content to let it go.

Then the air minister brought up one other little matter. "They say you would have liked to have succeeded Hindenburg as president!" To this exhumation of the bizarre, astrology-imbued *Neues Deutschland* story from January 1933, Eckener could honestly reply that this rumor had been started by a gutter rag, but it was evident "from Göring's dissatisfied manner that this point carried a certain weight in high Nazi circles."

Moving on, Göring suggested that Eckener write to Goebbels to say that his seeming lack of enthusiasm for the plebiscite flights had been misinterpreted; his misgivings had actually been due to his "understandable anxiety about the ship being endangered." Eckener would pretend to apologize; Goebbels would pretend to accept the apology. Once done, Göring would talk to the Führer about having the unpersoning edict lifted.

The intervention worked, and Eckener was quickly rehabilitated. To save face, Goebbels demanded that Eckener Avenue in Berlin be renamed Adolf Hitler Street, but Eckener's was only a tactical victory. It was actually Goebbels who had won the battle: The master of propaganda had kept Eckener in the country working (albeit unenthusiastically) for the regime, and Eckener's apology made it look as if he had thrown in his lot with the Nazis.

So intent had Eckener been on gaining the eternal glory of taking the *Hindenburg* to America and spanning the Atlantic with a passenger airline that he failed to notice that he had trapped *himself* in the Mephistophelian labyrinth.[2]

47. Here Be Dragons

TRIPPE HAD NOT been idle since he'd announced the pivot to the Pacific in October 1934. Six months later, in March 1935, the same month Eckener was ousted as head of Zeppelin, Pan American began its mission to Mars, or at least so it might have seemed, judging by the variety and quantity of the equipment Trippe's men were taking to Midway and Wake to build his bases. Everything was thought of, nothing forgotten.

Alongside four landing barges and four motorboats, there were water tanks, cisterns, five tractors, two ten-ton generators, windmills, electrical fixtures, mobile floating docks, and sufficient prefabricated, anti-vermin-treated plywood buildings to erect two villages plus huts for radio equipment, storage needs, and medical facilities. There were also 250,000 gallons of aviation fuel in barrels and forty large antenna masts. Creature comforts included a movie projector, a "library" of forty-five books (there would not be, it was hoped, much time for reading), ten decks of playing cards, board games, and tennis equipment.

Meat and vegetables were packed into dry ice and stored in giant refrigerators. Since no one knew what, if anything, would grow on Wake and Midway, there were seeds to cultivate royal palm, coconut, papaya, beets, onions, radishes, spinach, tomatoes, cucumbers, cauliflower, and peas. There were even some for assorted flowers, in case anyone wanted to make the place look a little less Robinson Crusoe–like. Ink, chewing tobacco, pillowcases, coat hangers, eggbeaters, sugar bowls, pie tins, and, rather optimistically, a dozen iced-tea stirrers—all were counted and packed. One could say that they brought "everything but the kitchen sink," but there were a few of those as well. All in all, Trippe spent $500,000 outfitting his expedition, which, according to its manifest, contained exactly 1,018,897 items and weighed some six thousand tons.

Pan American had chartered a freighter, the *North Haven*, to depart San Francisco and return in six months. Its 118 men had no idea of what they would encounter. Anything could be out there: Maybe the islands would be ravaged by merciless tsunamis; maybe the ferocity of the storms would put those of the North Atlantic to shame; maybe fierce and strange tribes would shrink their heads or chew their flesh. Who knew? Truly, as the medieval maps had marked of regions unknown that existed beyond the realm of civilization: Here Be Dragons.

The men Trippe sent were not the laborers New Dealers sought to employ in the PWA and the Works Progress Administration. Aside from the specialist Pan American staff attached to the project, most were fresh-faced, clean-limbed volunteers scarcely out of their teens, students or graduates of Harvard and Yale (or Stanford, at a pinch) up for an adventure. They were also cheap to hire.[1]

Meanwhile, Trippe didn't even know whether a Pacific run was feasible. Someone needed to fly to Hawaii and back; if that proved impossible, he could still recall the *North Haven*. Lindbergh had originally been slotted to make the exploratory flight, but the trial of Bruno Hauptmann, who had kidnapped and killed his child, had begun in January and for obvious reasons the aviator was ill equipped at that moment to undertake such a dangerous task.

The job fell instead to Pan American's "Pilot Number One," Ed Musick. Like Lindbergh, Musick was a former barnstormer who had learned to be exceedingly careful. Before every flight, every detail, every part was fastidiously triple-checked, then triple-checked again. Few knew much about him, but there was little to know.

Aged forty, the son of a hardware salesman, Musick was tight-mouthed, five-o'clock-shadowed, round-shouldered, and maintained a permanent squint, perhaps acquired from staring so intently out of a cockpit window at the horizon. With ten thousand hours' flying time under his belt, Musick was one of the very few pilots in the world qualified to fly any type of airplane. He was habitually silent in the air, uttering little else but necessary commands, his face otherwise a wooden mask. *Time* magazine ran a cover story on him, and its researchers sought in vain for "color" vignettes to liven up its profile; the best they could do was, "He lives quietly with his blonde wife, Cleo, has no children, likes baseball, Buicks, apples, ham and cheese sandwiches, [and] vacations in Manhattan."[2]

In preparation, an S-42 had been taken off Caribbean duties and sent for modification. The luxurious interior was torn out and replaced by fuel tanks with pipes leading to the wings, resulting in an ineradicable smell of gas inside the plane. The tanks contained 4,500 gallons of the precious stuff—enough for an extra three hours in the air, or 500 miles, over an already bone-stripped S-42's absolute maximum of 2,500 miles.

Musick took the plane, now named the *Pan American Clipper*, up over Miami to fine-tune the ideal carburetor settings, engine speeds, fuel mixture, and altitudes for stretching its range to the very last quarter mile. He had to know with scientific precision what the point of no return was, for he would have absolutely no room for error. Weather reports from that deep in the Pacific were sparse and untrustworthy, and there were few radio stations to help fix his location. Hugo Leuteritz, Trippe's radio expert, had managed to extend coverage to 1,200 miles, but after that Musick would be on his own.

Fortunately, he had a fine navigator, Fred Noonan (later to vanish alongside Amelia Earhart), assigned to assist him. He specialized in finding coral atolls in remote places in the Caribbean, and it would be his eyes (and celestial measurements) that would spot the tiny, greenish islets hiding amidst the camouflaging waters.

On April 16, 1935, just before 4 P.M., Musick took off from San Francisco with no fanfare (which he would have hated anyway). Every half hour he radioed in a progress report, frustratingly dry. Hoping for a vivid description to feed hungry reporters, William Van Dusen, Trippe's public relations man, asked Musick to "send something about sunset over the Pacific," and received in reply: "Sunset, 6:39."

The overnight flight to Hawaii went so astonishingly smoothly even Musick cracked what passed for a joke when the crew spotted Molokai: "We can swim to port from here." The return home was more troublesome as the S-42 bucked head- and crosswinds the entire time. Noonan noted with alarm at one point that they were far west of where they should have been and hundreds of miles behind, but it was impossible to measure their angle of drift at night.

By 10 A.M. the next day, Musick was still five hundred miles from the California coast, the plane was flying at less than 100 mph, and their fuel was running worryingly low. Trippe sat in his New York office listening to updates. He could do the calculations as well as anyone: The estimated travel time from Hawaii was at most eighteen hours, no one had ever flown a S-42 for more than twenty-one and a half hours straight, and yet there was no sighting of Musick—twenty-three and a half hours after he'd set off.

On board the plane, Musick saw that he had only thirty minutes of fuel left, and California was not in sight. But then the clouds parted and there, in the shimmering distance, it was. He steered straight for San Francisco Bay. At 5:21 P.M., he smacked the water and taxied slowly in to the dock. One of the engineers on board later checked the tanks with a gauge stick: They were "just about damp on the bottom. I don't think we could have made it once around the bay."

It was the closest of close shaves, but it was enough for Trippe to announce that "the results justify early inauguration of through service to the Far East."

PAN AMERICAN'S WHEELS started turning. The trusty *Pan American Clipper* was dragooned into further exploratory flights over the coming months to survey Guam, Wake, and Midway as the *North Haven* and its crew completed its assigned tasks.

At Wake, the freighter unloaded three thousand tons of cargo using the barges and motor launches; it was then hauled over the sand piece by piece on the men's shoulders until a primitive railway was rigged using spare lumber and girders.

Sunburned and steaming hot, the teams began to erect their village. During the Age of Exploration, the first building had always been a church; in the Age of Pan American, it was the airline business office. Then came the antenna masts for Leuteritz's radio boys, and only after that, accommodation for the work crews.

The lagoon, where the flying boats would land, was quickly discovered to be too shallow for easy use. The coral heads lurked menacingly just under the surface and would tear the guts out of any plane that landed. They had to be blasted away using dynamite to form a channel a mile long, three hundred yards across, and at least twelve feet deep. Over the next three months, at least a hundred reefs were blown up, and they were barely halfway done.

Work at Midway was faster. By mid-August there was even a concrete tennis court, a nine-hole golf course, and a baseball diamond. Unwilling to be held up any longer, Trippe ordered the S-42 (under Musick's former copilot Rod Sullivan) to land at Wake to test conditions. It would carry morale-boosting salads—nothing grew on Wake, it turned out—and a cage of canaries to cheer up the lads.

It was touted as a "historic" visit. Van Dusen unimprovably wrote in the company magazine that "crimson fire flooded the glassy surface of the ocean as [Sullivan] took off from the shelter of the Midway base at dawn today." Later, he encountered clouds like "towering battlements that dissolved into rainbow-silver mist as the Clipper flashed through. Coming out of these misty baths, the sun, sparkling on tiny drops of water, turned the great wing into a brilliantly glittering strip of jewels."

The reality was that Sullivan's flight had been so featureless—eight hours of flying over nothing but water—that he'd felt as if he were in suspended animation, as if the S-42 weren't moving. Tedium, however, turned to horror when he eventually reached Wake and saw that the water runway was still too short for a safe landing. He was furious when he eventually taxied into the dock, telling the base manager that he wouldn't have come if he'd known the conditions were that bad. A celebratory cake that had been made, with "Welcome to Wake Island" inscribed in icing, went uneaten.[3]

THE WAKE FACILITIES would need to get better, faster. On October 9, 1935, Trippe accepted delivery of the first of three Martin M-130s in Baltimore after a brief ceremony, broadcast nationally. In his speech, Trippe announced that "this flying boat will be named the *China Clipper*, after her famous predecessor which carried the American flag across the Pacific a hundred years ago."

Glenn Martin, however, wasn't there. He claimed he was sick and couldn't attend, but really he just didn't want to see Trippe. Martin's M-130s may have been the best in the world, but Trippe, annoyed at the production delays, had told him that he was already searching for a replacement and that his services were no longer required. Martin was crushed by the betrayal of trust: He had invested a huge amount in the factory infrastructure to build M-130s and needed a large order to cover the $850,000 he'd put up. Martin was almost bankrupted—"Sure, Martin lost money, but he didn't have the next step," Trippe casually remarked—and never forgave the Pan American boss.

Another casualty of Trippe's decision to bridge the Pacific was Sikorsky. The S-42 was at heart a Caribbean and South American plane and could probably have served in the North Atlantic, but it was unsuited to Pacific use. All Pan American operations in the latter would henceforth be based on the Martins. Lindbergh pleaded with Trippe to give his friend Sikorsky a chance to produce a larger plane, but the boss was unmoved. This was business, and Sikorsky had nothing in the works.

Lindbergh, for that matter, was a diminishing force at Pan American. Suffering under the harsh spotlight of press attention during the Hauptmann trial during the first few weeks of 1935, Lindbergh had become quite reclusive and moved his family to Britain to escape its blinding glare. He would remain close to Trippe, but their once-Siamese relationship became one of separated twins. Trippe was coming into his own as lord and master combined, even as Eckener had lost his status as king and pope after the creation of DZR that same year.

The inauguration of the Pacific route was arranged for the afternoon of November 22, 1935, and Trippe turned it into a truly international affair. In America, CBS and NBC carried the ceremony live while seven other networks in Europe, South America, and Asia rebroadcast it. Something like a hundred thousand people gathered to watch Musick take off from San Francisco Bay.

Trippe was given the microphone to talk to Musick. "*China Clipper*, are you ready?" "Pan American Airways *China Clipper*, Captain Musick, standing by for orders, sir." "Stand by, Captain Musick, for station reports."

One by one, nicely choreographed by Van Dusen, the Pan American radio stations crackled their call signals: "KNBF, Honolulu. Pan American Airways ocean air base No. 1—Honolulu, Hawaii. Standing by for orders." Then the same from KNBH (Midway, No. 2), KNBI (Wake Island, No. 3), KNBG (Guam, No. 4), and KZBQ (the Philippines).

Trippe: "Captain Musick, you have your sailing orders. Cast off and depart for Manila in accordance therewith."

The *China Clipper*'s engines roared in response as a rousing rendition of "The Star-Spangled Banner" played. Car horns blared as twenty-two cannons exploded in salute and ships' whistles tooted. A fireboat loosed streams of water from its hoses to wish it farewell.

As the *China Clipper* vanished into the distance, the reporters moved into a lobby equipped by Van Dusen with a giant Pacific map on which he studiously marked Musick's position every hour. In the back room, he went to work, churning out copy for the press ostensibly written by the seven crewmen. "Within a few short weeks," ghostwrote Van Dusen in *Cosmopolitan,* "Mr. Manhattan Business Man will be flying into Manila Harbor with a briefcase full of papers five days after he leaves his home office."

Even he, though, could not compete with Musick's typically succinct quote upon reaching Manila on November 29 after 59 hours and 48 minutes of flying time. When asked what it was like, he replied it was "without incident, an uneventful trip." It was exactly what Trippe wanted to hear. Tourists and businessmen alike could fly across the Pacific in the safe and capable hands of Pan American.[4]

MUSICK HAD BROUGHT no passengers, just mail, which could be lodged in sacks and required no food. Before he could fly people, Trippe first needed somewhere to accommodate them; a shack or tent on an isolated island was romantic but an unaccustomed hardship for wealthy Americans.

The work teams on Wake, in particular, were exhausted, ailing, bored, and homesick. Trippe pushed them harder. The *North Haven* was again dispatched, this time bearing the contents of two complete forty-five-room hotels, bound for Wake and Midway. He'd commissioned Delano & Aldrich, who'd designed Miami's 36th Street Airport, for these as well. Two wings emanated from a central lobby, with each room, equipped with coat hangers and ashtrays, enjoying a wide verandah, screen netting, and a bathroom with running hot water. There was even an aquarium in the lobby. The construction crews must have been envious.

For a time, it looked as if the hotel's occupancy rate would be zero percent. Over the three months between December 1935 and February 1936, Pan American made one successful round-trip to Manila but suffered three false starts. On December 22, for instance, Musick, after unhappily posing in a Santa Claus costume for publicity reasons, took off, encountered a cyclone seven hundred miles out and cautiously returned to San Francisco. On January 5, the *China Clipper* again left the ramp at Alameda, taxied across the harbor, and hit a floating log, which punctured its hull. The *China Clipper* finally got away again in mid-February. For thirteen hours it butted headwinds, and at the thousand-mile mark Musick cocked an ear at a bad report from Hawaii and scooted home.

Airship advocates hooted at the delays. These "great air cruisers" of Trippe's were kept in their hangars by the weather, an enthusiasts' magazine named *The Airship* laughed. "What sort of weather?, you ask. Tempest, blizzards, squall lines, ice? Not one of them." The answer was wind. A "situation that would have been no deterrent to the airship"—the magazine seems to have forgotten the travails of Zeppelin's early days—was proving, yet again, that airplanes were no match for Eckener's airships on oceanic routes.

Much to Trippe's relief, however, these proved to be teething problems, and thereafter the schedule rapidly improved. Passengers stepping aboard one of the M-130s would have been amazed at their surroundings. Reporters certainly were. Trippe had commissioned Norman Bel Geddes, a former theater designer, to prepare the interiors. Soft gray fabric covered the walls, disguising the cork soundproofing, and depending on which clipper one was on, there was either the Green Scheme (green seats with gold piping, corresponding with the curtains), the Brown Scheme (brown seats with eggshell piping), or the Yellow Scheme (yellow with green), all in appropriately subdued shades. The seats could be converted into full-size beds. In the separate lounge, seating twelve, there were clubby easy chairs, sofas, and card and magazine tables with rubber or felt surfaces to prevent slippage.

China Clipper, a Warner Bros. movie, appeared in August 1936 and only further served to advertise Pan American's Pacific service (Van Dusen was the film's technical adviser). It broke no new ground in storytelling technique, though it had some fine footage of the *China Clipper* and contained a potted, if sanitized, history of Pan American. Humphrey Bogart starred as a kind of Ed Musick, and indeed quite looked like him, but was more talkative than the real one. More interesting for some of the audience, however, was Pat O'Brien, who played a Trippe-type figure, a visionary dreamer who sacrificed his personal relationships to forge a new airline. At one point, Bogart's character says of O'Brien's, "I was just thinking how swell it would have been if he'd said thanks," which must have drawn knowing laughs from Trippe's long-unthanked staff.

By the time the movie was released, however, Trippe had already moved on. The Pacific had been conquered, and in spite of the tremendous achievements it had encompassed, for Trippe that ocean was a side-show to the main event. He *had* to get to Europe.

Boeing, which had been split off from Rentschler's United conglomerate after the Air Mail scandal, was at the time developing a new flying boat, the Boeing 314, with nearly double the M-130's horsepower and, it was claimed, a range of 3,500 miles with up to seventy-four passengers. One of the most amazing features advertised was that the traditional chemical toilet would be replaced by a "flushing" one—actually, a rotating drum rolled 180 degrees and emptied its contents outside, then righted itself.

Lindbergh considered it a mediocre plane and advised his friend that Boeing's promise to deliver 314s in December 1937 was utterly fanciful (it would take a team of engineers a year just to design that toilet, as it turned out). But Trippe, swollen with vainglory after his Pacific success, ignored the cautions and spent more than half of Pan American's treasury ($3.3 million) on six of them in July 1936.

Status, if nothing else, demanded an Atlantic route. Trippe controlled the largest airline on earth. Pan American's routes now extended over forty thousand miles, nearly double the range of its nearest competitor, Air France. The airline's passenger traffic far exceeded anyone else's; it owned or leased 202 airports and 129 radio/weather stations around the world; and it had the highest percentage of on-time departures and arrivals, the fewest fatal accidents, and the most advanced directional radio system.

To Trippe, it was frankly offensive that the mightiest airline king of all time could not fly to Europe, a doddle compared to the obstacles that had been faced in the Pacific, but it rankled still more that there *was* a way one could cross the Atlantic by air—according to one journalist, "the greatest, most lucrative, most important . . . blue-ribbon route of trade, diplomacy, and society" in the world": Eckener's Zeppelin.[5]

48. Master and Commander

MAY 6, 1936, was a dark day in the history of aviation—if you were Juan Trippe—but if you were Hugo Eckener, it was an epochal one. At 9:27 P.M., the *Hindenburg* lifted off from Friedrichshafen bearing fifty-one passengers and fifty-four crewmen. Destination: New York.

Eckener had beaten the unbeatable Trippe and his Pan American colossus. His *Hindenburg*, said one journalist, was about to "[steal] the show" and win the race across the Atlantic.[1] In spite of the dozens of attempts by others, after decades of single-minded struggle, after overcoming obstacles that would have felled a lesser man, it would be he, Hugo Eckener of Flensburg, who would go down in history as the titan who conquered the Atlantic Ocean.

Yet it would not, at least officially, be him captaining the *Hindenburg* to America. Eckener had expected the honor thanks to his Roosevelt Gambit, but the prickly Lehmann had raised a ruckus about it with Goebbels. *He* was the head of DZR, not Eckener, and this was yet another instance of Eckener's grandstanding. Eckener had no other choice but to acquiesce.

So, keeping to the arrangement that when he and Lehmann were both aboard, only one would serve as captain, Eckener was merely cited in the officers' list under his own name, whereas Lehmann received the coveted rank of commander. But in an admirably diplomatic solution, possibly suggested by Göring, Eckener's name was positioned above Lehmann's.

When Louis Lochner, an American journalist based in Berlin filing stories for the Associated Press, asked Eckener about it, he just laughed good-naturedly and "said resignedly, 'Anyway, I head the list. That's something!'" But in his diary of the flight, Lochner wrote that "I had the impression nevertheless that he felt hurt."

To Eckener, fussing over who was in command was insignificant compared to the Big Thing he'd achieved. For just $400 for a one-way ticket ($720 round-trip), a mere 10 percent more than the price of a first-class steamship fare and less than a fifth of what it had cost passengers on the Graf Zeppelin in 1928, a person could cross the ocean in three to three and a half days from Germany to the United States, or two and a half back.

Not that he would make a profit. Keeping in mind Eckener's habitual fuzziness with numbers, each round-trip voyage allegedly cost the company $80,000, though it was probably more. With every berth sold out and taking a heavy load of mail, he would be hard pressed to do better than break even, at least in the first year of operation with ten return trips to Lakehurst planned.

These were no longer the days when Eckener could sell exclusive press rights to Hearst for many tens of thousands of dollars to subsidize his costs, but he got something better in return: international blanket coverage of the flight.

The New York Times, for example, hired ex-Hearst correspondent and indomitable Zeppelin fan Lady Drummond-Hay as its onboard reporter; Karl von Wiegand, her "companion," worked for the International News Service; Sir Hubert Wilkins, the explorer and another old friend, was there courtesy of Hearst; Ralph Barnes of the New York Herald Tribune was also present, as were Webb Miller for the United Press and writers for the Staats Zeitung, the Paris-Soir, the Völkischer Beobachter, the National-Zeitung, London's Daily Mail, and a Roman Catholic paper. Making their first appearance on a Zeppelin were two gentlemen of the radio, Dr. Max Jordan (NBC's European chief) and a correspondent for Germany's Reichsrundfunk, as well as a cameraman for Paramount News Reel and a shutterbug named Franz Gayk working for Heinrich Hoffman, Hitler's official photographer and unofficial matchmaker (his assistant was Eva Braun).

All in all, a fun, hard-drinking, chain-smoking group. The number of women aboard—ten—soon prompted jokes about hanky-panky in the cabins, and the journos entertained themselves discussing "a suitable name for the first child conceived in mid-air aboard a Zeppelin—a possibility nowadays," with Webb Miller getting some laughs by suggesting "Helium if it were a boy, and Shelium, if a girl."

Their fellow passengers were a mixed bag.

There were several businessmen, like William Beckers, a director of Goodyear-Zeppelin; Carl Bruer, fountain-pen king; and Hans Hinrichs, a grain dealer and brewer.

There were some professionals: Harold Dick from Goodyear and Lieutenant Commander Scott Peck of the navy. They spent most of their time in the control car.

There were the glamorous, such as the enigmatic, dark-eyed beauty "Madame Titayna" (actually Elisabeth Sauvy-Tisseyre), the *Soir* reporter who turned up, said Lady Drummond-Hay, "in a symphony of royal blue and red." She had recently published a fawning interview with Hitler ("I was astonished and surprised by the bright blue of his eyes. . . . His face radiates intelligence and energy and emits a special glow when he speaks") and would become an enthusiastic collaborator during the German occupation of France.

There were the famous, like Leslie Charteris and his current wife (the first of four), Pauline. In 1928, Charteris had created the character of Simon Templar ("The Saint") and was currently working on his seventeenth adventure involving the proto–James Bond battling sundry evildoers. In full thriller-writer mode, he quickly announced that "I think it would be actually possible for a small group of men to overpower the crew and take possession of the ship."

There were the fliers. Miss Clara Adams, a Pennsylvania bluestocking and inveterate world traveler, was on board—as was Erla Parker, the long-widowed spouse of a Cleveland physician and a friend of Adams's. "Traveling [by airship] is a wonderful beauty secret," said Mrs. Parker. "It is so absolutely calm and effortless. There's no nervous strain—now any woman knows what that does for her appearance." Frederick Murray Simon, too, was present. He had been the navigator on Walter Wellman's misbegotten *America* dirigible flight back in 1910. "I vowed at the time that I'd fly across the Atlantic yet and now that moment has come," he said.

There were the special guests. One, the Reverend Paul Schulte, was known as the "flying priest," as he had established a service to bring missionaries via air to the remoter parts of Africa. Pope Pius XI had given him dispensation to conduct Mass on board the *Hindenburg*—a first in aviation—on the grounds that it was "a place of dignity and history and that the great airship is so steady that sacramental wine will not be spilled." Schulte had agreed, however, not to light any candles.

Another was Dr. Franz Wagner from Dresden. As Lehmann's previous attempts with the accordion had not been crowd-pleasers, he'd been hired to play a specially commissioned aluminum piano. A small grand, the piano weighed just 356 pounds and was covered in thin pigskin, presumably to reduce its tinniness.

And then there were the Nazi Party hacks and hangers-on. Wilhelm Traupel, for instance, was an SS officer who, when he was not traveling in comfort across the Atlantic, kept himself busy liquidating the mentally ill. Joseph Berchtold was the deputy editor of the *Völkischer Beobachter*, the official newspaper of the Nazi Party. As Party Member number 964—indicating *serious* commitment—Berchtold had participated in the Beer Hall Putsch and had commanded Hitler's bodyguard, which later evolved into the SS. His comrade from the *National Zeitung*, Eberhard Graf von Schwerin, was a Nazi of the aristocratic variety. The trio don't seem to have chitchatted with the other guests much.

More entertaining was Dr. Karl Ritter, chief of the Economic Department at the Foreign Ministry. Lochner the journalist found him a "jolly soul." After Lochner said he'd mentioned his name in a news report, Ritter "exclaimed in mock distress, 'Good heavens, did you do that? Now all [my] American sweethearts will be at Lakehurst to meet me.'"

Lochner, who was sharing a cabin with Ritter, thought the response all the more humorous because he had assumed Ritter to be a "confirmed bachelor," despite Ritter's having an affair, or at least being thought to be having an affair, with another passenger on the *Murder on the Orient Express*–like *Hindenburg*: the redoubtable Countess Rosie Waldeck.

Born into a banking family, she had briefly married Ernst Gräfenberg, a high-society gynecologist (the G-spot was named after him) before moving on to Franz Ullstein of the Ullstein publishing fortune. According to an American intelligence report, while married to Ullstein she had embarked on "an intimate personal relationship with RITTER for many years, and she was, according to one source, deeply in love with him." That at least explains her presence on the *Hindenburg*, though Dr. Ritter was perhaps not as smitten with her as she was with him.

She and Lady Drummond-Hay—the two people who brought the most clothes, commented Lochner—were assigned a cabin together. They got along well enough, thankfully, and took turns at the tiny mirror. The countess said that the cabins had been designed by a misogynist (Ludwig Dürr, more of a misanthropist, was responsible) and a "very mean one" for not including enough wardrobe space (there was room for six suits) and no drawers. "I always dress for dinner and no airliner can stop me," Waldeck determinedly added, yet her gabardine suit "was breathlessly un-wrinkled," remarked Lady Drummond-Hay (who was taken aback to learn that unstylish Pauline Charteris had brought only the dress she was wear-ing and a small suitcase).

The flight itself passed uneventfully enough. Up there in the clouds, it seemed as though the Continent, a happy and contented place, was at peace. Most of the passengers did not realize that the French had banned the *Hindenburg* from their territory owing to Hitler's provocations; that the *Hindenburg* could not cross Spain, where a civil war was brewing; or that the British had ordered the airship not to fly over London because Oswald Mosley's fascist Blackshirts were causing riots in the streets. The last thing the authorities wanted were giant floating swastikas further inspiring them.

Instead, the air-pressurized smoking room was always full—the bar-man's responsibilities included stopping anyone walking in with a forbid-den box of matches or walking out with a lit cigarette. Inside, there was an automobile-style electric lighter, which was tough on cigar aficionados and even tougher on pipe smokers. With a modicum of charm, it was pos-sible to persuade the steward to provide, "with a rather reluctant mien," a single match, but, said a traveler, he "retains hold of said match with great tenacity, from ignition to charred cinder. For the passenger to do his own lighting is *streng verboten* [strictly forbidden]."

The food was not bad but was (wrote one correspondent) "neither in-spired, nor much varied," and sometimes a little *too* German for Ameri-can tastes: heavy on the meat and heavier on the cream sauce. Frankfurt sausages were served for breakfast, Bavarian fattened duckling for lunch, and smothered venison cutlets for dinner. Some 250 bottles of wine were carried aboard, the great majority Rhine and Moselle whites with a leav-ening of French reds. Heinrich Kubis, the head steward, who had been with Zeppelin since before the war, performed up to his usual standards.

Belowdecks, as they say, the fare was rather more monotonous. To pre-vent tiffs breaking out between the south German and north German members of the crew, they alternated between fish (for the northern-ers) and pasta (for the southerners), the pattern broken occasionally by chicken—the crew suspected the cooks of giving them vulture—as a kind of mutually acceptable dish.

Many passengers repaired to the writing room to compose postcards and letters for hours at a time; others napped or admired the wonderful sights—cruising over an iceberg was a highlight—from the windowed promenades; they also played cards and munched on sandwiches. At night, the stewards drew the curtains separating the public rooms from the promenades to prevent the light pollution from inside ruining the vista of moonlight glistening on the blacky waves amid the unparalleled constellation of stars.

If a passenger hadn't brought something to read, he or she was out of luck: The airship's "library"—all in German—consisted of a copy of Hindenburg's tedious war memoirs (good for insomnia, though), a dictionary, a few other random tomes, bound volumes of the newsweeklies, and a novel by John Knittel, a Nazi-approved Swiss writer. All were primly locked away in a cabinet to discourage theft. (The reading matter on board Trippe's Pacific M-130s was better: Each carried four copies of Margaret Mitchell's new novel, *Gone with the Wind*; there was talk of a movie adaptation.)

One day, searching for something to report on, the journalist Webb Miller asked for a tour, a request readily granted by the ever effusive Lehmann. On their way to see the engine gondolas, he brought up the touchy subject of the wartime Zeppelin attacks on London. "I was two miles below you dodging your bombs," Miller said. Lehmann laughed: "Well, that was a long time ago."

Only then did Miller realize the catwalk they were walking along was but a foot wide. "I clutched nervously at struts and girders, fearing that a misstep would plunge me through the thin fabric into the ocean half a mile below." Lehmann waved off his concerns. "That fabric is strong enough to bear the weight of a man. You wouldn't go through if you slipped off on it." Miller was, understandably, incredulous. "Here, I'll show you," said Lehmann, as he jumped off the catwalk and trampolined on the outer skin, just a fraction of an inch thick.

When they arrived at the engine gondolas, Lehmann passed him off to Fritz Sturm, the chief engineer. The gondolas were suspended fifteen feet away from the side of the airship; the only things holding them secure were, now that Miller was looking at them up close, some alarmingly spindly-looking struts. Sturm tied a helmet to Miller's head and pointed to a foot-wide ladder slanting down from an opening in the envelope into the egg-shaped gondola. He showed Miller, now very worried about what he had gotten himself into, "how to clutch the frail ladder on two sides, crooking my elbow around it to the windward and clutching the other edge with my fingers. This precaution was necessary to prevent the eighty-miles-an-hour wind from tearing me bodily off the ladder. I found it a ticklish, frightening business; each time I raised a foot the wind wrenched it away from the ladder rung and flung it back toward the stern of the ship. Nothing in the world could save you if the hurricane-like wind tore you off the ladder." After a few steps, Miller wished he were back in the bar knocking back martinis, but he couldn't appear a coward.

Inside the gondola was a colossal diesel engine driving the twenty-foot propeller that "deafened me with its thunder in spite of my padded helmet." He felt as if he were "being shot through the air inside a huge artillery shell with open windows." A mechanic was permanently stationed beside each engine for a few hours at a time. The scenery was splendid, of course, but Miller longed for the whisper-quiet, virtually vibrationless environs of the lounge and cabins tucked inside the ship itself.

HIDDEN FROM THE passengers was the muted tug-of-war going on between Eckener and Lehmann, always over small things. The tussle over titles, for instance, continued to rankle Lehmann. For much of the flight, the American newspapers inadvertently insulted him by naturally assuming that Eckener, because he was Eckener, was the captain and Lehmann his underling. Eckener did not trouble to correct the stories.

Lehmann's willingness to take reporters on tours of the ship was partly inspired by his desire to set the record straight out of Eckener's hearing. The ploy paid off—partly—when on May 9 an article finally, and correctly, described Lehmann as the "actual master of the *Hindenburg*." Unfortunately, a different article that same day irritatingly added that he served "under the general direction of Dr. Hugo Eckener," the "Commander of the *Hindenburg*."

Eckener was always aware that it was *his* name, not Lehmann's, that sold newspapers, and whenever occasion demanded it, Eckener came first. So when a special dinner for all the guests was held one evening, it was in Eckener's honor rather than in Lehmann's, despite the latter having spent almost as many years at Zeppelin. Eckener was accordingly placed at the top of the head table with the most prominent guests, and Lehmann had to settle for the second table.

Then there was the matter of the radio. NBC had conceived the idea to broadcast a half-hour show from aboard the *Hindenburg*. At 8:30 P.M., when the *Hindenburg* was a hundred miles south of Newfoundland, Dr. Max Jordan, head of the network's European office, radioed in to check reception: "This is the airship *Hindenburg* calling New York." And heard back: "Go ahead, signal excellent."

Adding to his annoyance, Lehmann had not been asked to make a speech, an odd omission considering that his English was excellent, but Eckener, of course, headed the program—so perhaps not so odd, after all.

Millions of Americans tuned in to hear Eckener. "Many passengers told me today all this seems unreal," he said. "How long will it be until all this will have become just [an] ordinary thing, that will cause fear to nobody, and will hardly be mentioned in the newspapers any more. I really wonder when that will be." Then he announced the imminent inauguration "of regular airship travel across the Atlantic."

Afterward, Sir Hubert Wilkins, Lady Drummond-Hay, Lochner, and Wiegand gave addresses enthusiastically backing Eckener's dream of oceanic travel. To round out the show, Lady Wilkins trilled a current hit, "I'm in the Mood for Love," accompanied by Dr. Wagner on the piano, who then played Schubert's "Serenade" and Strauss's "Blue Danube" for an inoffensive touch of the light-classical.

Father Schulte's Mass was equally uncontroversial, if attended only by about a quarter or perhaps a third of the guests (the Church was not in Hitler's good books) and crew. Neither Eckener nor Lehmann, who weren't the believing sort, appears to have been present, though Lochner, a Lutheran, went out of curiosity and a nose for a good color story. By fitting happenstance it was the Feast of the Apparition of St. Michael the Archangel (the patron of mariners), and the priest conducted the service without spilling a drop of consecrated wine, using a portable altar equipped with a chalice, host, and crucifix (the lit-candle problem was solved with electric ones). Max Jordan, who later became a monk, served as choirboy to Wagner's accompaniment.

The service was interrupted by the incessant whirring of cameras, prompting the good-natured Schulte to say that "never before have I conducted a divine service in the presence of so many camera clickers and so few members of my faith." But, he added, he had experienced "the most impressive feeling I ever had. It was all quiet up there and one seemed near heaven."

AROUND DAWN AT Lakehurst on May 9, Commander Charles Rosendahl, now in charge of the naval air base, waited anxiously for the *Hindenburg*'s arrival.

He had nearly three hundred sailors and soldiers serving as ground crew. Separately, Willy von Meister had arranged for four small American Airlines planes to be on hand to whisk passengers to Newark airport, thirty minutes away, whence it was a brief drive into the city. Passenger processing through Customs and Health had also been vastly improved: No one wanted another embarrassing holdup like the one in 1928.

Then the siren atop the great hangar sounded. *Hindenburg* was here.

The crowds awaiting its arrival were noticeably smaller than in previous years. Owing to the hour, many of Rosendahl's men were not quite ready, so several journalists and German diplomats volunteered to help guide the airship in with its ropes. One Jewish reporter dryly wondered, "Hell, who would ever have thought I'd help land a Nazi airship?"

That was the problem right there. One might partly explain away the diminished crowd of only 8,000 to 20,000 waiting at Lakehurst as a consequence of airship flight becoming more routine: "Zeppelins were no longer the fabulous aircraft they were a score of years ago," remarked one report.

But the primary reason was that the *Hindenburg* was a "Nazi airship," and could not help but be identified with the regime. That uniformed members of the pro-Nazi Bund would frequently try to barge their way close to the airship for photos did nothing to improve its image. And the impact of the recently passed Nuremberg Laws on the Protection of German Blood and German Honor and the Reich Citizenship Law was felt particularly keenly in New York, with its large Jewish population. The laws forbade marriage between Jews and ethnic Germans and stripped non-Germans of their citizenship rights. Boycotts and vandalism of Jewish-owned stores were common, and Jews were banned from practicing medicine and the law, purged from the universities, and forbidden from employment in the civil service.

In the little borough of Lakehurst (pop. 968), near the naval air base, which had been hoping for a bonanza—stores had stockpiled $5,000 worth of sandwiches, ice cream, peanuts, and pennants (inaccurately) reading "Hindenburg—Maiden Flight"—the mayor, Harold Fuccile, despondently said that local merchants had sold just a thousand dollars' worth. He blamed "Hitler" for the "financial fiasco," with others adding that a lot of potential attendees were put off by all the Bund's swastikas (and the giant ones on the tail fins).

Eckener put a brave face on the slight turnout. He was after bigger game, but first he had to deal with the Lehmann Issue. During the flight, Lehmann had confronted him to complain about his second-class status, and Eckener had promised to fix the problem.

During the press conference afterward in the hangar, Eckener made a point of standing Lehmann right next to him as he clarified the situation. "It was quickly established," wrote a reporter bemused at why he had opened the session on such an inside-baseball note, "that Dr. Eckener had relinquished the post he has held for so long as an airship captain." Eckener made sure that everyone knew "it was Lehmann who wore the uniform today and it was Lehmann who gave the commands from the bridge of the great airship."

Journalists are sometimes like cats in that they sense the slightest movement in the dark. A few quickly realized that there was something else going on here, one asking Lehmann point-blank, "Are you in command?" which he headed off with "a ready smile" but did not answer. Then came the inevitable follow-up: "Then is Dr. Eckener's position advisory to you?" "Oh yes," answered Lehmann, who then "turned and smiled affectionately" at Eckener, as if there were an understanding between them.

The press gang suspected the change in leadership was related to the reports of Eckener's rift with the Nazi government and asked him what the situation was. Eckener had his answer down pat. "Oh, that was all a misunderstanding and . . . is all quite settled," he said casually, waving away the problem. "I went directly to Berlin and saw General Göring and we straightened [it] out."

With the unpleasantries out of the way and Lehmann's hurt feelings assuaged, Eckener changed the subject to a happier one. "I think that, with the completion of a dozen trips on schedule, we can show the American people what the airship can do, and that is the purpose of our flights this Summer." When they were done, he said, he expected to be able to raise private capital for Goodyear and Zeppelin to form an airship line.

Eckener spent the remainder of his visit hammering home this point and trying, not altogether successfully, to avoid airship politics. Before a dinner—which included such delights as Chilled Melon Cocktail *à la Eckener*, Okra Soup Friedrichshafen, and (recognition at last), Filet of Bass Almondine *à la Lehmann*—at the Waldorf-Astoria that evening organized by the Board of Trade for German-American Commerce, the orchestra had played the "Horst Wessel Song," the Nazi anthem, and about half of the thousand attendees stood and made the required salutes. As there were too many people watching—including German ambassador Hans Luther and Lehmann, plus who knows how many informers—to not follow suit, Eckener was reported as reluctantly holding "the salute by lifting his arm slightly above the level of the table."

Eckener was relieved to be out of there to catch the 1:40 A.M. train to Washington to see Roosevelt in the White House—his "Get Out of Jail Free" card. Unfortunately, he had two undesired escorts, Lehmann and Luther, so the meeting the next day was amicable, but nothing important was discussed. Old fishing and sailing stories seem to have been the main topics of conversation. Nice and safe, in other words, and guaranteed to bore Lehmann and reassure Luther that nothing untoward was going on. Sensing Eckener's discomfort at not being able to talk freely, Roosevelt claimed he had some pressing engagements and had to cancel luncheon, but the large photograph of Eckener and FDR that appeared in the newspapers the next day stood guarantor of Eckener's safety.

Afterward, the trio rushed back to New York as Lehmann and Eckener were due to depart later that night. The *Hindenburg* took off at 11:27 P.M., carrying forty-eight passengers and 150,000 letters.

There had been complaints about the lack of fruit and vegetables, and Kubis, the steward, had made sure to include more "typical American edibles" for the return journey. A thousand pounds' worth of turkeys, lobsters, ice cream, potatoes, lettuce, spinach, grapefruit, oranges, bananas, and apples (along with American whiskey, ginger ale, and sparkling water) had been loaded into the storerooms.

There were far fewer journalists this time, their places being taken by an interesting variety of regular Americans, such as eighty-seven-year-old Harriet Hague, a longtime air enthusiast, and Billy Googan of Cleveland, aged thirteen. He was a Boy Scout, was wearing long trousers for the first time, and when asked for his views on airship travel, said "I dunno" a lot. Paul Mack-Hale, a restaurateur of Worcester, Massachusetts, officially became the fattest man to ever fly in a Zeppelin, tipping the scale at 235 pounds. The newspapers, pretending this was a serious issue, solemnly calculated that three thousand cubic feet of hydrogen would have to be reserved to lift him. A not-quite-yet-famous passenger was Dr. William M. Scholl, described as a "foot specialist and manufacturer of orthopedic appliances." And for unknown reasons, there were three people from Peoria, Illinois, where the *Hindenburg* evidently played well.

The airship passed over Manhattan and traffic came to a halt as it made its way up Broadway, where the post-theater crowds cheered. The *Hindenburg* acknowledged their applause by dropping below a thousand feet, with the blaze of lights along the Great White Way illuminating its path. Then past the Empire State Building, where late-night office workers watched agog from the windows as it cruised at eye level and headed toward the bay. There, the ninety-two floodlights shining on the Statue of Liberty went dark for a moment and lit up again, blinking their salute to the queen of the air.

Hindenburg would not be returning to Friedrichshafen; it was bound for a new home, the "Airdrome" at Frankfurt, to share the facilities with Deutsche Lufthansa. There, the new airship hangar was the largest in the world at 900 feet long, 170 feet high, and 170 feet in width—a benefit of being a "Nazi airship" was that the Nazis had virtually limitless funds.

There was another reason, too, for the *Hindenburg* to be granted such largesse: The Berlin Olympics was coming up in early August, and Goebbels had ordered the *Hindenburg* to put on a good show.

Eckener was not invited to participate, much to his relief, and he was pleased when Lochner informed him on the way home that, because Goebbels's "unperson" edict had been lifted, his name was again being mentioned in the press, albeit without the usual glowing plaudits and only *after* his meeting with Roosevelt. To Lochner, whom he knew he could trust, Eckener growled, "*Die Giftkröte* [the venomous toad] has seen after all that I am not without friends."

Lehmann remained a problem, despite the choreographed press conference at Lakehurst. The dynamic between him and Eckener was becoming increasingly testy and tense, though they tended to keep their arguments out of the public eye. Once back in Friedrichshafen, unable to contain his anger anymore, Eckener wrote him a politely nasty letter, informing Lehmann that he'd decided not to come on the next scheduled *Hindenburg* flight. In saying that "I'm withdrawing willingly, and your ambition shall be fully satisfied," Eckener touched a nerve: He'd long accused Lehmann of being too obsequious to the Nazis for his own good.

And then, in a prickly reminder that Lehmann should know his place, Eckener added, "I reserve the right to take part in future flights, when it suits me." The next day, Lehmann dashed off a curt note saying that in fact it was Eckener's vanity that had put him on the outs with the government and forced him to step aside.

The sniping continued. In June, Lehmann made sure that a twenty-four-page special supplement included with the leading Frankfurt newspaper celebrating the new airport and the history of Zeppelins kept Eckener out of the main section and lauded his own contributions. Eckener was relegated to providing a brief word buried in the back pages. A subsequent letter from Lehmann to Eckener demanding a higher salary to suit his new status didn't even greet him with a pro forma salutation but was addressed instead to "The Supervisory Board," as if Eckener no longer counted.

Aside from these unpleasant clashes, however, the American voyage had achieved Eckener's intention of stirring up interest for a full-fledged Atlantic line. When Meister released a schedule of that season's flights— one every ten days or two weeks—that would end on October 7, every cabin was fully booked well in advance.[2]

Most important, for early October Meister was arranging a special domestic flight from Lakehurst to Boston that he had taken to calling the "Millionaires' Flight." This was intended to be the main sales pitch for American financiers and businessmen to fund new Goodyear-Zeppelin airships.

One invitee in particular must have attracted Eckener's eye as he glanced down the impressive roster: a certain Mr. Juan T. Trippe of Pan American, apparently still on the board of PZT.

IN THE MEANTIME, DZR had work to do. On August 1, shortly before 7 A.M., the airship, now bearing on its sides the Olympic rings, departed Frankfurt carrying sixty-five passengers, mostly Nazi officials and their wives, some officers, and various journalists.

Eckener was not on board, for obvious reasons, but neither was Lehmann, who was overseeing operations from Berlin, and so the *Hindenburg* was instead commanded by Lehmann's ally Captain Max Pruss, a Nazi Party member.

Pruss had to get to Berlin on time. Arriving at 1:40 P.M., the *Hindenburg* crisscrossed the flag-bedecked city, enthralling the two to three million people congregated in the streets. The *Hindenburg* thrice traveled the length of Unter den Linden, Berlin's grand thoroughfare linking what had been the royal palace in the east to the Brandenburg Gate to the west. By then it was three o'clock and Pruss directed the *Hindenburg* to the new Olympic Stadium, the largest in the world. There, a hundred thousand people, an orchestra, a half-dozen military bands, and a thousand-strong choir awaited its arrival.

To mark the opening of the games, the *Hindenburg*, trailing the Olympic flag, slowly passed over the stadium, to the adoration of the handkerchief-waving, cheering crowd. Task completed, Pruss sent a telegram to the Führer thanking "with respect and gratitude the leading sponsor of the Berlin Olympic Games, which bring peoples together." Hitler replied by wishing the crew and DZR "good wishes for a continuing good flight."

With that sideshow out of the way, the *Hindenburg* could return to its American service in earnest.[3]

49. Mr. and Mrs. Brown

O N October 5, captained by Lehmann, the *Hindenburg* set off on its final flight to Lakehurst for the 1936 season. By most yardsticks, it had been a tremendous success. On the ten westward crossings from Germany, passenger numbers had averaged 49; on the eastward, 54. Given the full houses, Eckener had already approved expanding *Hindenburg*'s cabins to accommodate 75 passengers for 1937. In the following year, he wanted a four-berth "family" stateroom (with an in-flight nanny thrown in).

Over the course of the summer, the *Hindenburg* had transported some twelve hundred passengers (including flights to South America), and even the airship's adversaries had to admit that the ocean weather had proved no hindrance; the *Hindenburg* had stayed almost exactly on schedule, and it had taken off and landed in the type of conditions (fog and rain) that invariably grounded airplanes.

Average westbound travel time was 64 hours, handily beating by *days* even the newest generation of liner, such as the British *Queen Mary*, and to a still greater extent going east, at 56 hours. Strikingly, the *Hindenburg* and the *Queen Mary* could leave their home ports within hours of each other, and the *Hindenburg*—despite having to fly across Germany and even after an entire day spent at Lakehurst for maintenance—would be halfway back across the Atlantic by the time the *Queen Mary* arrived in New York.[1]

But in terms of luxury, the *Hindenburg* was finding it hard to beat its maritime competitors. The dining room of the French *Normandie* alone, for instance, was longer than the Hall of Mirrors at Versailles, and almost as opulent, while for virtually the same price as a ticket on the airship one could spread out on the *Queen Mary* in a first-class suite including a spacious bedroom, a separate living room, and a private bathroom. Ludwig Dürr's monastically ascetic cabins and shared bathrooms could not compete. Neither could guests enjoy the children's playroom, barbershop, squash court, sauna, and swimming pool that came with the *Queen Mary*. But if you wanted to get to and from Europe quickly, preferred to avoid seasickness, and didn't care about having a swim or a haircut, the *Hindenburg* would admirably serve your purposes.[2]

Financially, Eckener had performed quite well, claiming that he had broken even, which might possibly have been almost true. To bring in extra revenue, he wanted to raise ticket prices from $400 to $450 for the coming year.[3] The picture was not so rosy at Goodyear-Zeppelin, where Litchfield had been forced to cut jobs after the destruction of the *Macon* and the loss of navy interest. Where once there had been more than eight hundred men working at the Air Dock and the engineering and design shops, there were now just ten. Apart from Arnstein, who occupied himself building mockups of his dream airship (it would travel at 100 mph, have forty-three cabins, and feature a dining room adaptable into a movie theater), most of the technical staff had already been transferred to other departments, returned to Germany, or joined other aeronautical firms.[4]

Litchfield reported that even if he got a commission for a passenger airship, Goodyear would need at least two years to build it.[5] Eckener was disappointed but reassured his friend that help was on its way.

It was now that Eckener revealed to Litchfield that he had a secret airship on the books: LZ-130. This ship had been commissioned in late 1935, in the full bloom of DZR's romance with the government, and was essentially an updated version of the *Hindenburg* intended to better compete with the ocean liners' standards of luxury (some cabins would have space for a large wardrobe and armchair). But construction work had only just begun, and its first flight, he said, would not be until 1938.[6]

The plan was for the *Hindenburg* and LZ-130 to run simultaneously—by which time DZR planned to have at least two more airships under construction. Lehmann claimed that by 1940 Germany would have no fewer than four Zeppelins operating between North and South America, and he spoke enthusiastically about expanding to India and Asia.[7]

In which case, Goodyear could still be in the game, so long as Eckener and Litchfield could get the process moving *now* to ramp up development. If they could raise the necessary funds, Akron could be put back to work and at least one Made-in-America airship would be ready by 1938, just as Zeppelin came into its own.

And it was for that reason that Eckener headed to New York. He had a flight to catch: the "Millionaires' Flight."

EVERY MAN ABOARD the flight, more fully described in the prologue to this book, had been carefully selected. This was by no means a mere sightseeing cruise over New England, optimally timed for the changing of the leaves.

Each guest had a purpose: He was needed to pony up money, to provide political backing, to supply material for construction and operations, to form a business alliance, to popularize the cause of airships, or to grease the wheels of statecraft.

So why was Trippe there? As a PZT board member he was probably eligible for an invitation, but Eckener could easily have crossed his name out. And Trippe could just as easily have sent a polite note saying he was busy.

But there was a mutual curiosity between the two long-standing rivals. Each had read and heard so much about the other, yet they had never met.

Each wanted to check out the competition, to size him up, to take his measure. Neither believed the other had a chance, and each was confident in the superiority of his chosen instrument. Each knew, too, that the next couple of years would decide the question of airship versus airplane, the cloud or the bird.

Nothing is known of what they said to each other. Very little, most likely. Eckener was engaged with his other guests, and Trippe was always quiet. Probably there was a shaking of hands, some small talk, a few insincere congratulations on how much the other had achieved, and then the moment passed, just as it once had when Zeppelin met Orville Wright in 1909.

Yet Eckener's hooded blue eyes and Trippe's inscrutable browns must have met over the course of the day. Maybe a nod was exchanged, or glasses were raised, in acknowledgment of two old foes at last meeting on the field of battle.

TRIPPE HAD BEEN appropriately impressed by the *Hindenburg* but was also nonplussed by the Germans' assumption that airships would continue to have the Atlantic to themselves. Eckener's remark in a newspaper interview—"he could see only [airships] regularly transporting passengers across the North Atlantic"—struck him as obnoxious, considering that Pan American was already regularly crossing a much bigger ocean.[8]

The stomach-patting smugness of the Zeppelin people seemed bizarre to Trippe. True, he was still stuck with Imperial, but time, money, and effort would eventually pry open the Atlantic to Pan American. With that in mind, shortly after the Millionaires' Flight, Trippe and his wife, Betty, embarked on an around-the-world trip of their own. It would be a chance to fly, for the first time, on Imperial and compare it to Pan American.

Starting in San Francisco on an M-130 and stopping along his chain of Pacific island bases, the Trippes stayed each tranquil night in one of Pan American's hotels. He had never visited his own properties, now equipped with electricity, private baths, a small movie theater, cultivated gardens, paths of crushed coral, and bamboo furniture.

From Asia to Europe, they were on Imperial turf and Trippe was eager to see what was on offer. It proved disappointing: Imperial's planes were slow, old, and uncomfortable. While in Karachi, Trippe noticed that KLM, the Dutch airline, was making the trip from Baghdad in one day on its American-built Douglas planes, but Imperial needed two to cover the same route. It was not a pleasing revelation, and neither was discovering that Imperial's fifteen-passenger flying boat from Alexandria to Rome flew "like an old tug" (said Betty) at a stately 90 mph, compared to the *China Clipper's* zippy 140.

There was better news in England, where Woods Humphery took them to Rochester to see the factory where the much heralded, completely re-designed Short Empire S-23 flying boats were being manufactured.

The Empires were not as good as the Martins or the Sikorskys, but they would be good enough. They were, at least, notably spacious and comfort-able. One engineer had spent two years fine-tuning the reclining seats, which on American planes were a constant problem and had prompted innumerable customer complaints. But on the Empires, said one awe-struck journalist, all you had to do was press a knob "and the chair stretches you out to any length within the limits of decency."

Just a few weeks earlier, Major Brackley, Imperial's air superintendent, had flown the first Empire off the production line to Marseille in France with no problems. And Woods Humphery told Trippe that a new plane was coming off the production line each month. In the coming new year of 1937, two more Empires were set to fly nonstop from England to Egypt as a test.

Best of all, Woods Humphery expected an S-23 to cross the Atlantic—from Shannon in Ireland to Botwood in Newfoundland—nonstop on an experimental basis sometime in early July 1937. By the end of that summer, Trippe realized, the unsuspecting Eckener might have some unwelcome competition on his precious Atlantic route.

Soon afterward, Trippe and Betty visited their old friends the Lindberghs, who'd decamped to England, where the two men discussed aviation problems. Anne Lindbergh noticed that Trippe, now older and wiser, was not as "close" as he had been several years earlier, by which she meant he "did the right things, but he was not suave, he was very closed in, shy." Yet she also saw he was markedly ebullient now that he'd been to Rochester.

This was turning out to be a fantastic business trip. Trippe was in such high spirits that he decided to surprise Betty: They would not, as she had assumed, be returning to New York by ship. How would she like a detour to South America aboard the *Hindenburg*?

After flying into Berlin, they were met by Lufthansa officials and stayed at the famous Hotel Adlon. "Our German friends," noticed Betty, "were tight-lipped concerning any reference to Hitler and the political situation." At dinner parties and lunches, whenever she asked about "anything relating to German politics, the subject was immediately changed, by asking me if I liked to swim or did I care for music." A visit to the Air Ministry set off further alarm bells when she saw that the windows were equipped with automatic sliding steel blinds to provide protection against bombing raids.

The Trippes took the train to Frankfurt, where they had breakfast with Martin Wronsky, who was on DZR's supervisory board. Many years earlier, they had known Wronsky, then a "bright, ambitious boy," when he had briefly worked for Pan American in New York, but he had since changed, said Betty, into an "overbearing, conceited, opinionated man [who] was almost rude in his controversial attitude about the future disadvantage of airplanes versus airships."

Trippe kept his own views to himself, finding it more interesting to listen to what he regarded as typical Zeppelin hubris brought on by the success of the 1936 flying season and the Millionaires' Flight than to bother arguing the point. They would be humbled soon enough.

Trippe was concerned with keeping his upcoming flight secret. A newspaper story about the head of Pan American taking a Zeppelin across the Atlantic would be too juicy to spike, so he and Betty traveled under the aliases "Mr. and Mrs. Brown."

For Trippe, it was a chance to do opposition research, to see where Eckener was weak, to understand his enemy. For years, he had closely studied Eckener, and though he rarely admitted it, he feared the Zeppelin. William Van Dusen, his publicist, later recalled that Trippe was intensely concerned about Eckener's plans for the Atlantic. In fact, the year before he had cabled Van Dusen not to travel on the *Graf Zeppelin* "because it would be misunderstood over here as favoring lighter-than-air if I did that." Asked why Trippe was so worried, Van Dusen replied that "it was a pretty hot issue [and] so many people [and big wheels in this business] favored airships over airplanes at the time. . . . These damn things would fly—and they kept on flying. They were big and impressive."

Betty enjoyed her trip on the *Hindenburg*, spending much of it chatting to her fellow passengers. The flights to South America tended to be more diverse than the higher-status and more expensive ones to New York. On board their flight were forty-eight men, seven women, and a three-year-old. Germans, British, and Americans predominated, but there were also Argentinians, Brazilians, Uruguayans, Chileans, and a lone Turk. Zeppelin travel, to many of them, had become routine.

For first-timers like Mr. and Mrs. Brown, though, the sights were breathtaking. At night, black as pitch, the sea would sparkle owing to a multitude of phosphorescent creatures, and during the day shoals of flying fish skimmed over waves and dived like torpedoes. You could see rays, looking like butterflies, patrolling and playing beneath the water, as well as whales coming to the surface.

Trippe had no complaints, and there were many things to be said in the *Hindenburg*'s favor, to be honest, but the trip had confirmed his own view, as he told Betty, that the end of the age of the airships was nigh. Within the next five years, he predicted, airships would be no more.[9]

ECKENER WAS JUBILANT when he returned home. All the talk about the Millionaires' Flight had been positive. Many of the participants, Eckener reported to DZR, had "expressed lively interest" in launching an American-based Zeppelin business.[10]

In New York, Willy von Meister got to work. Since IZT had been dormant since Rentschler's departure, he rebooted it with new directors and renamed it the American Zeppelin Transport Company (AZT). AZT would play a major role in the reinvigorated Zeppelin enterprise. First, it would handle all the investment capital that would doubtless flow into Goodyear-Zeppelin to build new airships. And second, and more important, it would serve as an independent partner to DZR. Henceforth, all Zeppelin flights to and from the United States would necessarily be a joint DZR/AZT venture.[11]

That AZT *had* to be the fulcrum of operations was a key consideration too often lost, downplayed, or ignored by DZR in Germany. Lehmann could crow all he liked about running four German airships across the Atlantic, but success hinged on using American facilities, constructing an American terminal, getting permission to enter American airspace, and buying American helium—all of which required federal approval. And to get that you needed an all-American company—AZT.

The real heart of the matter was that without Washington backstopping such an ambitious project with subsidies, no bank would lend additional funds or extend credit, no matter how much the millionaires had enjoyed their flight. Without that government money, there could be no Goodyear airships, and without American airships to disguise Zeppelin's German origins, there would be no federal support.

Eckener understood that this particular circle could become either virtuous or vicious, and warned his colleagues about the risks of the latter, but amid the celebratory mood at DZR's annual meeting in November 1936, no one was listening. Nor did they heed his caution that they *must* obtain federally approved helium to reassure American investors of the airships' safety. Lehmann had instead sharply shut down any further discussion.[12]

• • • • •

COINCIDING WITH THE DZR festivities, U.S. commerce secretary Daniel Roper received a letter. It was from a Michael Weinstein of Asbury Park, New Jersey, probably a real estate broker, who was serving as the "National Junior Vice-Commander-in-Chief" of the Jewish War Veterans of the United States (JWV).

The JWV had been established in 1896 and now comprised around twenty-six thousand members. Since 1933 it had joined its Catholic partners to organize anti-Nazi marches and boycotts of German goods in order to highlight the persecution of religious minorities in Germany.

Little could he have known it, but Weinstein's unassuming letter would be the proximate cause of the downfall of Eckener's entire Zeppelin scheme in America.

In it, Weinstein mentioned that the AZR/DZR arrangement was no innocent "commercial treaty" between two companies. In fact, because DZR was a Nazified entity, each visit by the *Hindenburg* would serve Berlin's purposes by "promot[ing] Nazi principles," while Goodyear's proposed American ships would in effect be German propaganda fronts.

Until that moment, Roper had been encouraging the DZR/AZT scheme—he was the commerce secretary, after all, and his job was to foster such business deals—but now its dangerous implications had been rammed home. If the Department of Commerce backed the DZR/AZT venture, financially or otherwise, then it could look as if Roper had officially endorsed a *political* alliance with Berlin—and one, moreover, that stipulated no provision for ensuring religious freedom in Germany. The White House, Harold Ickes over at Interior, and the State Department would have a collective fit. (Roper also may have been worried about the reaction of his son-in-law, Frank Bohn, a senior figure in the fiercely anti-Nazi group the German-American League for Culture.)

This was not, to put it mildly, the most auspicious moment to be suspected of pro-Germanism. To much American disgust, news had just arrived of the new Nazi Criminal Code, according to which individual rights were to be relegated to the bottom of the pecking order, below the rights of the nation, the state, the regime, the Party, and the Führer. It meant that anyone, especially Jews, would be an open target for Nazi judges.

Within a day or two of receiving Weinstein's letter, a shaken Roper wrote directly to his correspondent and distributed the text of his reply to *The New York Times* to make sure that he was on the record. Roper hastened to assure Weinstein that he had never committed "ourselves to Germany in lighter-than-air development plans." Neither had there been any "treaty or tentative treaty discussed [with] the Department of Commerce."

He affirmed that the United States would not "form an alliance with Germany" in this matter and that in his casual conversations with AZT about building American airships he had been interested solely in "keeping the United States in the forefront of modern transportation, of which lighter-than-air service is a segment."

Roper was essentially saying that the matter would go no further. In Ohio, Paul Litchfield could recognize Washington code for killing a proposal as well as anyone, and he understood there and then that the dream was over.

Akron would never build another airship—though it would later, famously, manufacture blimps—and Goodyear-Zeppelin closed its doors. Arnstein and the remaining Germans who had come over in the 1920s were transferred to a new company: Goodyear Aircraft Corporation.[13] The name, carefully chosen to avoid mentioning either airships or Zeppelin, said it all.

ASIDE FROM ECKENER, who understood the gravity of what had just happened in Washington, life went on as before at DZR, where Lehmann and his allies seemed excited, rather than disappointed, by the prospect of a necessarily all-German airship line. Dealing with Goodyear had long been an albatross around their necks, and kowtowing to American bigwigs for cash was never their style. That had been Eckener's obsession, not theirs, and everyone knew that Eckener was damaged goods in the eyes of the regime.

Lehmann and DZR were entertaining the delusion that they could go it alone with the *Hindenburg* and its successors with only Reich support behind them. When Goodyear's Harold Dick returned to Germany after several months away, he detected a "confident, even arrogant, attitude" at the DZR about its prospects.[14]

Indeed, in early 1937 one might easily have been misled into thinking that nothing of any import had occurred, so steadfastly did DZR ignore the fact that the coming flight season would again have to rely on American goodwill to allow landings at Lakehurst with hydrogen airships.[15]

Now that the chance to build a private terminal at Baltimore had vanished, all it would potentially take for the navy to withdraw official permission to use Lakehurst was a diplomatic crisis, public revulsion at a piece of Hitler thuggery, another embarrassing revelation à la Roper, or some other mishap.

Instead of trying to find a solution to this Big Problem, Lehmann busied himself with fixing small ones, mostly stemming from customer feedback collected by Meister, who had little to do now that AZT had been relegated to the status of a remote back office.

On the grounds that tipsy revelers had kept other passengers awake tinkling "Chopsticks," the aluminum piano would no longer travel with the *Hindenburg* and would be replaced by a gramophone imperiously controlled by the steward. The paltry library would have more magazines and newspapers in English, and there would be two backgammon boards and some jigsaw puzzles to stave off boredom.

Menu cards were to be printed in both German and English. Too many times, it seems, American passengers had been unexpectedly greeted by such German culinary delights as boiled meatballs in creamy anchovy sauce with beetroot—the feared *Königsberger Klopse*.

Beer, too, had to be served colder, and coffee, hotter. The barman urgently needed training: He'd never heard of a Manhattan. Meister went so far as to send the puzzled DZR board of directors a cocktail stick he'd pilfered from the Waldorf-Astoria, along with a note explaining that Americans liked to impale olives and cherries with such implements.

Other vitally important details were capably handled. The inkwells in the writing room were considered out of place with the décor and were replaced, though a suggestion to change the jam pots appears to have been passed over. Requests for a better selection of onboard gifts and souvenirs were easily fulfilled. Henceforth, there would be chocolates, eau de cologne, cigarettes, Havana cigars, and Champagne available.

Americans like clean bathrooms, and the *Hindenburg*'s standards had not been up to par. Inadequate ventilation in the lavatories had resulted in "odors"—the *Königsberger Klopse* had struck again—the hand towels went unchanged, and there had been smeared soap on the basins. Meister was particularly exercised by the hand towel issue, informing his mystified

German counterparts that Americans preferred single-use paper tow-elettes and regarded "general-purpose ones" as fit only for barbarians.

Meister also insisted on encasing the feather mattresses in fitted sheets and providing two or three blankets as a "concession to [American] living standards" to replace the German-style eiderdown duvets, which some Americans, like Lester Gardner, the publisher of *Aviation Magazine*, had complained left "parts of his anatomy exposed to the cooling blast of the ventilator" at night.

It was a long list of improvements, certainly, but, added Meister, "such things do not cost very much and are material to the creation of an overall good impression."[16]

And with those changes made, the *Hindenburg* would have all the trappings of a most luxurious air yacht. In January 1937, DZR began advertising that the upcoming season's schedule had been expanded to no fewer than eighteen round-trip voyages. The first was slotted for May 3.[17]

MEANWHILE, TRIPPE RACED to launch his own transatlantic service to compete with Eckener's. He cared about nothing else, not even the mounting financial losses in the Pacific, where Japan and China were gearing up for war. Business and tourism, as one might expect, were disastrously low. In 1937, many of Trippe's clippers would travel between San Francisco and Honolulu with just two passengers, and from Hawaii to Manila empty. Pan American would lose $504,000 on its Pacific expansion alone that year, while its rather less ballyhooed South American network would make $1.035 million—90 percent of the company profits.

The Pacific was turning out to be Trippe's grand folly, but at least there was progress, finally, being made in the Atlantic. After years of wrangling, the rights and concessions he desperately needed to get across the ocean were beginning to fall his way.

On February 22, 1937, the British Air Ministry finally granted Pan American permission to land and take off in the United Kingdom and Newfoundland; on March 5, the Canadians followed suit, as did the Irish Free State on April 13.

With the signatures still wet, Trippe set the Pan American machine in motion. First came the creation of the Atlantic Division, 113 men under the direction of Colonel J. Carroll Cone, and then the beginning of construction on a city-owned flying-boat terminal at North Beach, Long Island, later to become LaGuardia Airport. Afterward, Trippe dispatched teams to Newfoundland to cut down a square mile of forest at Botwood to make room for hotels and offices while construction crews worked on a marine base. Mooring cables, buoys, a radio station, and a new access road appeared at Foynes on the river Shannon in Ireland, seemingly by magic. Before the work had been completed, cadres of Pan American meteorologists arrived to take their measurements.[18]

Pan American had Zeppelin in its sights.

EVEN IF AIRSHIP travel was becoming routine, it was still a thrill to watch the *Hindenburg* land.¹ On May 6, five newsreel companies sent camera crews to Lakehurst to capture the airship inaugurating the 1937 season. They were accompanied by about a dozen photographers from the major news agencies as well as several reporters from the dailies.

The *Hindenburg* remained newsworthy, though it no longer dominated the headlines. To a third-tier radio journalist like Herbert Morrison, who hosted a midday farming news and light entertainment show, *The Dinner Bell Hour*, on WLS in Chicago, the *Hindenburg*'s coming to the United States presented an opportunity to showcase one of his new sponsors, American Airlines.

WLS's fare generally consisted of husk-by-husk coverage of regional corn-husking competitions, up-to-the-minute livestock reports, and crowd-pleasing fare like *National Barn Dance*, all avidly listened to by midwestern farmers. The previous year, American Airlines had been impressed by Morrison's reporting from an airplane about local floods and had offered him a chance to take one of its airliners gratis to Newark. As a bonus, he and his recording engineer, Charlie Nehlsen, would be taken to see the *Hindenburg* land at Lakehurst. (American's planes connected *Hindenburg* travelers to nearby airports.) Journalists working in local radio like Morrison rarely traveled outside their areas, so his plan was to interview any Chicagoan airship passengers he could find for some hometown interest, then return to Illinois that night.

· · · · ·

MORRISON AND NEHLSEN were just happy to be there, but few of the more veteran New York newsmen were. For much of the day, there had been persistent rain, and a thunderstorm hadn't improved their moods. One cameraman, from Universal, had already given up and gone to see a movie.

The *Hindenburg* was late. It had been due to land at about 6 A.M., and many of the journalists had risen before dawn to get to Lakehurst on time. It was now about 4 P.M. Where was that damned airship?

On board the *Hindenburg*, Captain Max Pruss was even more annoyed than the inconvenienced press corps. On the way over, he had run into headwinds and fallen some eight hours behind schedule. When he radioed Commander Charles Rosendahl at Lakehurst to tell him of the delay, Rosendahl moved him provisionally up to 6 P.M. and promised to have his landing crews ready earlier if need be.

A half-day's delay, though irritating, might not otherwise have mattered too much, but Pruss was scheduled to depart Lakehurst at 10 P.M. with sixty-five passengers, now at the Biltmore Hotel "laughing and joking while photographers took their pictures," said one reporter. Ten of these planned to connect with Deutsche Lufthansa at Frankfurt to attend the coronation of King George VI in London on May 12.

It would have to be a fast turnaround to prepare for the return leg, but Pruss thought he could make up for the delay by exploiting the favorable tailwinds. One thing was certain: *Hindenburg* needed to land at Lakehurst as soon as possible.

Thankfully, he wouldn't have to do it alone. Joining Pruss in the control car was Captain Anton Wittemann, his fellow Nazi Party member, though he was there only as an observer. Lehmann, too, was right there alongside him, ostensibly serving in an advisory role.

As the head of DZR, Lehmann captained few flights these days, but he had come along on this one as part of his eternal quest to replace Eckener as the Face of Zeppelin. Once in New York, he planned to give newspaper interviews to publicize the *Hindenburg*'s first voyage of the 1937 season, but he also had other business to attend to. One of the passengers aboard was an aviation journalist named Leonhard Adelt, who had co-authored Lehmann's memoirs. Together they were working on the final chapters and hoped to whip up interest among Manhattan publishers. (Abiding by his arrangement with Lehmann, Eckener had decided to absent himself from the flight by embarking on a lecture tour in Austria.)

Pruss was known to be a good airshipman. He had years of wartime Zeppelin experience under his belt, and had served as the elevator operator—the most demanding post for any crewman—during Eckener's voyage to America in LZ-126 *Los Angeles*. In 1934, he was promoted to commander of the *Graf Zeppelin*, and then became watch officer on the *Hindenburg* for much of the 1936 season. Harold Dick of Goodyear remembered that when he "was in command of the ship, every landing was perfectly executed."

Still, as Lehmann's protégé, Pruss naturally tended to look to the senior man for guidance in difficult situations like this one. On the thoroughly Nazified *Hindenburg*, it was critical to Lehmann, Pruss, and Wittemann to represent the "new" Germany and the "new" DZR by distinguishing themselves from the "old" Zeppelin ways, as represented by the habitually ultra-cautious Eckener. This mentality, combined with Lehmann's propensity for what Eckener regarded as unnecessary risk-taking and cutting corners, would help determine their subsequent decisions.

It HAD BEEN a strange day, weather-wise, at Lakehurst. Shortly after 4:30 P.M., a cold front advanced through the air base, lowering the temperature to the mid-60s, followed by a local thunderstorm. The *Hindenburg*, after a tour of Manhattan to wait out the poor weather, was now nearing Lakehurst, but the storm had not yet passed. Exasperated and powerless to do anything else, Pruss turned the airship in the direction of eastern New Jersey to kill time.

At 5:42 P.M., Rosendahl radioed him, in the telegram-ese of the era: "Conditions still unsettled recommend delay landing until further word from station advise your decision." Ten minutes later, following a discussion with Lehmann and Wittemann, Pruss replied: "We will wait till you report that landing conditions are better."

At 6:12 P.M., another thundershower followed on the heels of the storm, but Rosendahl believed that by the time the *Hindenburg* arrived it would have passed on, so he radioed an update: "Conditions now considered suitable for landing ground crew is ready." Hearing nothing back, Rosendahl pressed *Hindenburg* for instructions ten minutes later: "Recommend landing now."

Oddly, silence from Pruss. Perhaps he was discussing the matter with Lehmann. Finally, at 7 P.M., Rosendahl advised: "Overcast moderate rain diminishing lightning in west ceiling 2,000 feet improving visibility." Then, eight minutes later, he radioed once more: "Conditions definitely improved recommend earliest possible landing." Finally, Pruss acknowledged the message at 7:10 P.M., then shut down the ship's transmitters and reeled in its trailing antennae. Rosendahl went outside to the edge of the landing circle, where he met Meister, there to take care of passengers.

The *Hindenburg* began the approach maneuvers. It was coming in from the southwest at an altitude of about six hundred feet, with the wind blowing from the east. Pruss checked the landing circle below: There was a light rain falling, and he could see that the ground crew was not quite in position, despite Rosendahl's assurances.

Airship captains usually liked to land heading into the wind to help slow them down, and that is exactly what Pruss now did. He overflew the landing circle and turned to port, intending to describe a large oval that would bring him back pointing to the east.

So far, everything was by the book, but in the control car there were several curious departures from the norm. Standard textbook practice, as laid down by Eckener, was for the captain of the ship to stand aloof from the watch officers and let them perform their duties, supervising them closely, of course, and intervening only if a disagreement or a sudden problem should arise. Thus, one officer was in charge of the rudder and engine power, while another looked after the elevators, ballast, and gas.

On *this* flight, however, there was a more specialized division of labor. Pruss himself took over the rudder and engine position, while First Officer Albert Sammt supervised trim and altitude and Eduard Boetius manned the elevators. Sammt also directed Watch Officer Walther Ziegler, charged with overseeing the gas and valving, and Second Officer Heinrich Bauer, handling the ballast.

Lehmann's role was vague. As he was the senior officer on board, perhaps Pruss had pliantly deferred to his wish to command the landing, or perhaps Lehmann had long hankered to get rid of what he saw as Eckener's outmoded policy and had ordered a change in the hierarchy.

The *Hindenburg* continued turning left at full cruising speed—about 73 mph—and Sammt ordered Ziegler to valve hydrogen for fifteen seconds to reduce altitude and bring the ship level. A minute later, the *Hindenburg* had come around the bend almost 180 degrees and was about to straighten and head directly east to the landing circle. Pruss idled the engines to reduce speed to 33 mph, and a minute after that, Ziegler valved gas again from the bow cells to bring the nose down.

The *Hindenburg* was closing in on the landing circle very nicely, but then the wind shifted direction, from the east to the southwest. Now, at 393 feet, the *Hindenburg* was forced to change course at short notice and tacked east-northeast. Pruss put the rear engines into full reverse to slow down, then stopped them to allow the forward engines to draw the ship around to the northeast. Within a minute, he idled the forwards, too.

About three-quarters of a mile from the landing circle, Sammt noticed that the tail seemed a little heavy. To bring the ship back level, at 7:18 P.M. Bauer dropped 661 pounds of water ballast, and then a thousand feet later another 661 pounds, followed finally by a third drop of 1,100 pounds. Six men were sent to the bow to equalize the weights. Sammt confirmed that the ship had been righted.

Trimming an airship like this on its approach was a common occurrence, and nothing of the procedure so far was worrisome. Most likely, the rain the *Hindenburg* had recently passed through hadn't yet had time to roll off the back of the ship and the elevator fins.

More serious was the track the *Hindenburg* was pursuing. Lehmann and Pruss had opted to make a partial S-curve, beginning from the earlier port turn to the northeast and then swerving sharply starboard in a clockwise semicircle to bring them in to land. This turn was performed at a speed of about 27 mph, a third as fast as *Hindenburg*'s cruising speed.

By Eckener's standards, Lehmann and Pruss were coming in way too quickly, as they themselves seemed to realize about a thousand feet from the landing circle, when the engines were ordered into full reverse for an entire minute.

This kind of stylishly fast, aggressive maneuver bore every hallmark of Lehmann, and it's possible that he was also showing off a little for the benefit of Adelt.

Veteran American airshipmen watching outside noticed that there was something abnormal about the tight S-curve, and they all professed surprise at the rapidity of the *Hindenburg*'s arrival.

Chief Boatswain's Mate Willie Bishop said that usually the Germans "would come in slowly and gradually. From my observation I saw the *Graf Zeppelin* last year and they always came in very slowly and made their descent very nicely, while on this night it seemed to me [the *Hindenburg*] was in a great hurry to get down for some reason or other." Lieutenant Richard Andrews, in charge of taking hold of the control car when it descended, confirmed that "as the ship came in, she was coming, in my opinion, much faster than usual and backed the engines much longer and harder than I have [ever] seen."

It was either Lehmann or Pruss, probably a bit of both, who decided on the "high landing" that came next. Eckener relied on a technique known as a heavy landing, in which he would bring the airship in as low as possible by valving gas so that it would more or less settle into the hands of the ground men, who would then walk the ship into the hangar. It was efficient but required a large, experienced ground team. The Americans, whose ground crew consisted mostly of part-time volunteers, had preferred the high-landing or flying-moor method, meaning that ropes would be dropped from several hundred feet up and winched in by the mooring mast.

Neither Lehmann nor Pruss had done many high landings, and the reason it was chosen now was that they wanted not only to set themselves apart from Eckener but also to habituate themselves to American practice.

At 7:21 P.M., the *Hindenburg* hovered at two hundred feet, and the crew dropped the bow starboard rope, followed moments later by the portside one. The latter was soon coupled to the line attached to the mast as the ground men began working on the starboard rope. A light rain fell upon them as the *Hindenburg* pulled hard on the port rope, moving up and astern and swinging slowly to the right.

Somewhere near Cells 4 and 5, below where the top vertical fin connected to the upper spine of the *Hindenburg*, a thin shear or bracing wire, there to help strengthen the fin, had become overstrained by the earlier, high-speed S-curve. Eckener later said that "the stress on these wires is especially high in such a sharp turn," though none had ever snapped before on the *Hindenburg* (at least under his command, he subtly implied).

But now, tautened beyond endurance by the pull of the airship, one did.

BY SUPREME ILL luck, the wire's tortured metal strands happened to rend a modest tear in one of the cells, releasing around 1,500 cubic feet of hydrogen a second—an amount as yet imperceptible to the pressure gauge connected to Ziegler's board in the control car—into the space between the cell wall and the airship's skin.

On the ground, R. H. Ward, in charge of the port bow ground team, looked up and saw a weird fluttering of the outer fabric in the area over Cells 4 and 5. The cover seemed to be rippling, like a flag waving in the wind, as if gas were "pushing up" against it from the inside.

Invisible to nearly everyone, something even more curious happened. Lieutenant Richard Antrim, stationed on top of the mooring mast— allowing him to see from a greater height than those on the ground— noticed a "small flame at the base of the leading edge of the upper fin. This small flame licked forward for a distance, I judge to be 20 to 30 feet. This flame died out."

Professor Mark Heald, a history professor at Rutgers University, who had taken his wife and eight-year-old son to Lakehurst, also saw it. His family were standing some distance outside the main gate of the naval air station, giving them, like Antrim, a different view from anyone else. According to Heald, about a minute, likely somewhat less, after *Hindenburg* dropped its bow lines, he watched a dim "blue flame" flickering above the backbone about a quarter of the length of the airship from the tail.

At that precise moment, Herbert Morrison was doing his best to describe the landing as sound engineer Charlie Nehlsen recorded his words. "It's starting to rain again; it's . . . the rain had, uh, slacked up a little bit. The back motors of the ship are just holding it, uh, just enough to keep it from. . . . It's burst into flames! Get this, Charlie; get this, Charlie!"

Morrison was describing what Heald, Ward, and Antrim had that instant seen for themselves. Antrim confirmed that after the "small flame" disappeared a few seconds passed and "then [there was] a puff of flame on the after end of the ship just forward of the spot where I first saw a flame." Ward said that the puff suddenly erupted into a mushrooming flower of fire about ten feet in diameter, which Heald estimated burst a third of the length along the airship, so slightly closer to the bow than before, as Antrim had mentioned.

Within moments of the mushroom appearing, the entire aft of the *Hindenburg* was consumed as the fire feasted greedily on millions of cubic feet of hydrogen. Rosendahl, himself seeing the first small burst of flame, felt "at once that that spelled the doom of the ship." He was right. The time was 7:25 P.M.

•••••

THE HINDENBURG, LIKE every Eckener airship, had been built with safety in mind. Each part of the system—for instance, the gas cells, the outer cover, the wiring, the infrastructure, the hydrogen—could not be the *sole* cause of a disaster. They *all* needed to fail to destroy a Zeppelin, and such an outlandish scenario had never happened on Eckener's watch. But at Lakehurst that evening, starting with the S-curve, the snapped wire, and the torn gas cell, a cascading series of minor, madly improbable events did happen—and led to catastrophe.

Even then, had it not been for the weather, all would have been well: The tear in the gas cell would have quickly been found and patched and the wire replaced with a spare. No one would have given it a second thought.

The atmosphere above Lakehurst had become highly electrically charged by the thunder front earlier, and though it had dissipated some-what, the air around the *Hindenburg* had been reenergized with voltage by the arrival of a second, less detectable, and less severe front over Lake-hurst just before the airship approached.

In itself, this was not at all dangerous—every single metal part on the *Hindenburg*, as on its predecessors, was bonded to another to distribute electrical charges evenly—and airships had been struck dozens of times by lightning with no harm coming to them. Neither was it dangerous to drop bow ropes to ground the airship's charge with the earth; the ropes, when they fell, were bone dry and as such, poor electrical conductors.

But because of the drizzle and the decision to use the high-landing method, the ropes had had time to become damp—and thus better elec-trical conductors. The *Hindenburg's* accumulated electrostatic charge, then, was *quickly* equalized with that of the earth, which in turn created a sudden, drastic, and dangerous differential between the *Hindenburg's* framework charge and that of the surrounding, still highly electrified air.

The airship's outer skin, too, contained its own charge and was natu-rally less conductive than the metal skeleton. To reduce chafing between the cover and the framework, thousands of thin (less than half an inch) wooden dowels had been inserted between the fabric and the outer edge of the duralumin girders. Again, this was perfectly safe, but it did mean that the cover retained its charge for longer than the frame.

Normally, *none of this would have mattered*. But in this unique instance, the conditions were ripe for a small electrostatic arc discharge—of the same sort that happens when you shuffle across the carpet and shock your friend—to leap the doweled gap between the wet cover and the framework. And such a spark would otherwise be harmless but for the leakage of hydrogen caused by the broken wire tearing the gas cell.

This gas combined with the oxygen in the space between the top of the cell and the inner surface of the cover to form an oxyhydrogen mix. With the *Hindenburg* stopped and little wind present, there were none of the usual drafts that blew through the body of the airship in flight to disperse any stray gas through the vertical vents that poked through the airship's skin to the outside. As a result, the oxyhydrogen had begun to accumulate and lurk there.

The arc discharge ignited this relatively small volume of oxyhydrogen. The small flame that Heald and the others had seen, then, was the first, pale-blue burst of lit oxyhydrogen as it exited a wood-and-fabric vent-shaft hood on top of the *Hindenburg*. It continued to burn inside, which no one could see, acting like a pilot light for the second, larger burst witnessed by ground observers.

In the *Hindenburg*, the heat was melting the cover of Cell 4. Helmut Lau, who was stationed near the bottom vertical fin, then saw a "bright reflection" *inside* Cell 4 above him. "It was at first red and yellow and there was smoke in it," he said, and "the cell suddenly disappeared."

The flames had met highly purified hydrogen, which burned exponentially faster than the diluted oxyhydrogen, and Lau heard a *whoosh* sound similar to that of lighting a kitchen gas range left on for a time.

This almost instantly resulted in a colossal, high-velocity chain reaction to the other cells, one after another progressively rupturing and releasing their hydrogen to feed the insatiable maw of the fire.

In roughly half a minute, this succession of wildly remote possibilities, none inherently dangerous in and of itself, turned into lethal certainty.

INSIDE THE *HINDENBURG*, many people, like the cook Albert Stöffler, radioman Herbert Dowe, Third Officer Max Zabel, Boetius the elevator man, and Watch Officer Heinrich Bauer, heard either a "metallic tearing" or a "dull thud." Max Henneberg, a cabin steward, who had "heard so many explosions in my life during the War," likened it to sounding "like a big gun, you know, boom."

When a strong vibration subsequently quaked the ship, those in the control car assumed that the port rope had snapped. Annoying, but fixable. Sammt looked out the window and was confused to see that the rope was fine.

On the ground, Morrison the reporter had seen the burst of flame and cried: "It's fire . . . and it's crashing! It's crashing terrible! Oh, my! Get out of the way, please! It's burning and bursting into flames and the . . . and it's falling on the mooring mast. And all the folks agree that this is terrible! This is the worst of the worst catastrophes in the world! Oh, it's [unintelligible] it's flames. . . . Oh, it's crashing, oh, four or five hundred feet into the sky, and it . . . it's a terrific crash, ladies and gentlemen."

The ferocity of the combustion of millions of cubic feet of hydrogen cracked the rear quarter of the *Hindenburg*, arching the bow sharply upward. Inside, Stöffler recalled that in the kitchen all the dishes fell down, and hot water streamed all over the floor, probably from a broken pipe. Xavier Maier, the chief cook, fell over, while Severin Klein, one of the stewards, was thrown into a corner, with the nearby passengers also tumbling. Fritz Deeg, a cabin steward standing in the dining room, managed to hold on to a windowsill but everything else, "including chairs, everything not fixed on the ground, fell to the back." In the control room, too, the officers slid about and the log book, drawers, and other objects crashed to the floor.

The airship fell in slow motion, its back agonizingly broken, as fire billowed out from its nose.

Back to Morrison: "It's smoke, and it's in flames now; and the frame is crashing to the ground, not quite to the mooring mast. *Oh, the humanity!* And all the passengers screaming around here. I told you; it—I can't even talk to people, their friends are on there! Ah! It's . . . It . . . It's a . . . ah! I . . . I can't talk, ladies and gentlemen. Honest, it's just laying there, a mass of smoking wreckage. Ah! And everybody can hardly breathe and talk and the screaming. I . . . I . . . I'm sorry. Honest, I . . . I can hardly breathe. I . . . I'm going to step inside, where I cannot see it. Charlie, that's terrible. Ah, ah . . . I can't. Listen, folks; I . . . I'm gonna have to stop for a minute because I've lost my voice. This is the worst thing I've ever witnessed."

From the first glimpse of the flame to the time the immolating *Hindenburg* settled on the ground, just thirty-two seconds elapsed. As the holocaust petered out, the molten duralumin structure crumpled inward as black clouds, caused by the burning of the diesel fuel, swirled and enveloped the stripped, glowingly red carcass.

In the control car, everyone knew enough about airships to know that the best thing to do was to get out. Those present found the closest window and jumped. Lehmann, Sammt, and Pruss leapt from the starboard side while Boetius chose the port. Ziegler followed the first three out, but girders fell and a curtain of flame rose before him, so he ran back to the control car and jumped through a window in the navigation room when he saw a gap in the flames to the left.

Heinrich Bauer, too, had discovered that in the forward part of the ship the port side was the safer option, and he managed to sprint twenty yards away. Likewise, Anton Wittemann was hoping to follow Lehmann and the others, but the window collapsed. He looked out and saw Lehmann, Pruss, and Sammt cut off by flaming framework crashing into the ground. Desperately, he looked for another avenue of escape and noticed that the wind was blowing smoke and flame starboard, and "then I saw a clear opening shortly after on the port side, and with a quick decision I ran there and without difficulty [charged] through a short streak of fire." He would suffer nothing but a slight ankle sprain.

Herbert Dowe was also extraordinarily lucky. He had been on a stairway next to the control car when one of the *Hindenburg*'s gigantic water ballast tanks burst above, drenching him. When the infrastructure collapsed on top of him, "I threw myself on the ground, and the heat became so intolerable, so great, that I started to burn on the head and the face and the hands, then I crawled into the wet sands . . . and I waited until the outer cover burned off." Then he stumbled around until two ground men brought him to the dispensary for treatment.

Elsewhere, crewmen sought the closest exit—a door, a window, a hatch—it didn't matter where. The nearer you were to some kind of exit the higher the likelihood of surviving. Of the eleven mechanics in the outboard engine cars, for instance, nine survived, and ten out of twelve men in the control car would live, but of the nine souls forwardmost in the inescapable bow, none did. There was also a higher fatality rate among those personnel deeper inside the ship or on the starboard side, onto which the *Hindenburg* now leaned.

There were exceptions, such as chief electrician Phillip Lenz. He was trapped in the switch room amidships and "could not get out, because the doors were distorted." The fire started burning above him so "I found a metal cover and held it as a shield against my face. Then the air got extremely hot, so that one could hardly breathe anymore. Then I thought, 'This is the end.' And then the room was hotter and hotter and I turned around, and all of a sudden, standing on the outside" was a fellow crewman who took hold of him. It seems the fire had burned a hole for him to crawl through.

Helmut Lau, in the lower fin, was also trapped but saw mechanic Richard Kollmer climbing out of a hatch, and Rudolf Sauter, the chief engineer, yelled at him to move. Hans Freund, one of the riggers, helped pull Lau out. Sauter had blood streaming down his face, and together they extracted him as well. Freund had burns to the back of his head and one of his cheeks, Kollmer was limping, but otherwise they were safe, though they could hear dreadful screams coming from within the wreck.

All seven stewards and five cooks plus messboy Werner Franz had been waiting in the passenger service areas (lounge, dining room, kitchen, cabin corridors) when the *Hindenburg* sharply inclined. Heinrich Kubis, the chief steward, had just finished setting up a table for the Immigration and Customs officials who were due to board to check passports (Meister had really upgraded the arrivals process), when he was thrown off his feet and knew instantly "there was something out of order." Keeping his head, he opened the center window and ordered the other stewards to keep the rest open. Eugen Nunnenmacher, a dining room steward who had been preparing a platter of sandwiches, called out to him, "Mr. Kubis, jump out," only to be told by Kubis not to let anyone jump—they were too high up to survive.

Then a passenger jumped. Kubis looked down and saw that at about fifteen feet it was dangerous but necessary. He shouted to anyone who could hear, "Get out!" and then followed him, as did Nunnenmacher and Max Henneberg, who momentarily clung to the window frame, scared, before letting go.

WHEN THE AIRSHIP had lurched upward, many passengers hadn't the presence of mind to grab onto something sturdy. Those who did and then headed straight for a window survived, for the most part. George Grant, for instance, saw the flames to starboard and urged his fellow passengers to jump alongside him through a port window, but some had not realized the severity of their predicament, or had frozen in fear, or were too slow.

Anyone who went deeper into the ship to fetch a loved one died. John Pannes, the New York manager for the Hamburg-America steamship line, was in the dining room and at a window about to jump when he discovered that his wife, Emma, had returned to their cabin to get her coat. He went to find her, and they weren't seen again. Hermann Doehner, who was traveling with his family, had returned to his cabin shortly before the fire, and Irene, his fourteen-year-old daughter, left her mother and two brothers to go find him. Neither survived.

Those passengers who did not jump tumbled down the steep incline. In the dining room, Margaret Mather, fifty-eight years old, was "hurled a distance of fifteen or twenty feet against an end wall" and was pinned there by "several Germans who were thrown after me. I couldn't breathe, and thought I should die, suffocated, but they all jumped up." Then "the flames blew in, long tongues of flame, bright red and very beautiful," but the lurching of the ship threw everyone "repeatedly against the furniture and the railing, where they cut their hands and faces against the metal trimmings." She saw several men jump from the windows "but I sat just where I had fallen, holding the lapels of my coat over my face, feeling the flames light on my back, my hat, my hair, trying to beat them out, watching the horrified faces of my companions as they leaped up and down."

Just then, she recognized William Leuchtenberg—she had met him, a "red-faced elderly man who had evidently been celebrating his departure with something stronger than Rhine wine," briefly at the beginning of the flight—and she recalled that he was a German-American importer. When they had taken off, he had admiringly exclaimed, *"Mein Gott! Mein Gott!"* but now he had thrown himself "against a railing (arms and legs spread wide) with a loud terrible cry of *Es ist das Ende* [It is the end]."

Leuchtenberg himself remembered that while he was badly burned "I immediately knew I had to do something to save myself." He clutched a "girder"—he probably meant the railing mentioned by Mather—but could not see as his eyelids were swollen owing to burning. After Mather heard him cry out, Leuchtenberg let go and slid down toward the fire, expecting to die. But then he heard a voice asking in English: "Can you jump?" Leuchtenberg said he couldn't, "I am almost all in." Then he slid down farther toward the bulkhead, where Mather had fallen—but no one was there anymore.

That was because, according to Margaret Mather, "I heard a loud cry! 'Come out, lady!' I looked, and we were on the ground. Two or three men were peering in, beckoning and calling to us. I got up incredulous and instinctively groped with my feet for my handbag, which had been jerked from me when I fell. 'Aren't you coming?' called the man, and I rushed out over little low parts of the framework which were burning on the ground."

Then it was the surprised Leuchtenberg's turn. He heard someone tell him, "You crawl through here quickly, it is afire, but it won't burn you." So he did. One thing he remembered doing was removing his artificial teeth and putting them in his pocket. A member of the ground crew was waiting to pull him out and said "I cannot hold you, I will have to drop you" and let him fall what he thought was ten to sixteen feet (it was actually just a couple) into the arms of a sailor, who bore him away from the wreck.

Their saviors were the ground parties and naval personnel at Lakehurst, who had rushed to the dying Zeppelin. "Those boys dived into the flames like dogs after rabbits," said Gill Robb Wilson, the New Jersey aviation director.

THE MAKESHIFT HOSPITAL in the base dispensary was bedlam. Rosendahl had already organized a morgue in the hangar to avoid inflicting further unnecessary trauma, but the dispensary was overwhelmed with casualties. Of the ninety-seven passengers and crew aboard the *Hindenburg* that evening, there would be thirty-five killed (plus one member of the ground crew). Virtually everyone else was wounded, many grievously.

Until doctors and ambulances could arrive from surrounding areas with proper equipment, burn victims like Margaret Mather were given bottles of picric acid to swab over their injuries. American Airlines stewardesses—trained nurses, fortunately—rushed from Newark to help. One, a Miss Tyler (or Pyler), was briefly interviewed by Herbert Morrison, who was diligently reporting on the aftermath. She didn't say much, so busy was she.

Mather had been among the first to arrive, but within a few minutes more carloads of people came in, including one bearing Lehmann's co-author Leonhard Adelt and his wife, Gertrude. The dispensary, he said, "swarmed with excited people like a disturbed ant heap. In the corridors on tables, stretchers, and chairs lay the seriously wounded. An ambulance orderly with a morphine syringe the size of a bicycle pump ran about to give everyone an injection. In one room a dying young mechanic called from his stretcher alternately for his bride in Germany and a priest. A neighbor on board was led by. He was badly burned and hung more in the arms of his companions than walked. Men with bloody burns strode searchingly through the rooms."

Both Adelt and Mather encountered Captain Lehmann, in shock and suffering from second- and third-degree burns to his face and body. During his appearances among the passengers when the ship was in flight, said Mather, "he always looked alert but genial, with keen blue eyes," but "now his face was grave and calm, and not a groan escaped him as he sat there, wetting his burns. His mental anguish must have been as intense as his physical pains, but he gave no sign of either, and when my burns became intolerable and I would reach for the bottle [of picric acid] he would hand it to me with grave courtesy, wait patiently while I wet my hands and receive it back with a murmured 'Danke schön.' It was a strange, quiet interlude, almost as though we were having tea together."

Shortly after, Adelt found Lehmann "crouched upon a table. He was half unclad and sat bent over." Adelt thought he'd broken his back escaping from the wreck. "I went to him. 'What caused it?' I asked. 'Blitzschlag [lightning],' he replied. These were the only words we exchanged. . . . We looked into each other's eyes; when I could no longer contain myself, I left."

Still later, Rosendahl visited him, later claiming (according to an FBI report) that Lehmann had told him that an "infernal device"—a bomb— had been planted in the airship and "that the accident would have happened, no matter whether they had utilized helium gas or hydrogen in the superstructure of the Hindenburg."

51. The Infernal Device

THIS REVELATION OF a bomb on board, rather than Lehmann's earlier, somewhat more accurate statement, prompted an FBI investigation into the possibility of sabotage, as well as myriad conspiracy theories. The *Hindenburg* did, after all, have many enemies. Communists and anti-Hitler activists may well have wanted to strike a spectacular blow against such a powerful symbol of Nazi power.

In a memorandum to J. Edgar Hoover, director of the FBI, dated the day after the disaster, Percy Foxworth, one of his assistants, notified him that agents would be checking out tracks and paths near Lakehurst to elicit whether a shooter—there was already a rumor that someone had fired phosphorus-filled incendiary bullets at the *Hindenburg*—had left behind empty cartridges. Over the coming weeks, the FBI and police would devote, or waste, a lot of time making plaster casts of footprints in the area.

Meanwhile, FBI Special Agent G. N. London reported on recent articles in *The Daily Worker* boasting that Communist sailors were doing "perilous underground work" aboard several German passenger liners. They spotted the Gestapo agents aboard, held secret meetings, and circulated Communist propaganda by means of tiny parchments rolled inside cigarettes. Perhaps a similar cell had been operating on the *Hindenburg*, he asked? But no evidence could be found of Communist infiltration in Friedrichshafen or Frankfurt.

The Red angle wasn't working out, but there were so many others to choose from. Someone suggested that "photographing racketeers" had set up the fire to make money selling pictures and had warned their newspaper chums to stay off the *Hindenburg* that day.

From there it got only more bizarre. Someone wrote a misspelled, strangely ungrammatical letter to Senator Royal Copeland of New York saying that with the destruction of the *Hindenburg*, "Zeppelin competitions in America against my airship invention are now out of my way." He promised to "construct an real American helium airship, better & satftier than a Zepplin, it will cost lesser money, useful for war as well, as commerce." A tired FBI agent wrote on the back of the letter, "I am trying to find out all info about such nuts. This guy might actually have done the dirty work, or some other disgruntled inventor like him."

That unpromising lead went nowhere, but then another nut emerged. This was Hans Omenitsch of Jackson Heights, New York, who claimed that he could read secret codes in newspapers giving orders to sinister operatives to destroy the airship: "The Codes appear to be operated by an invisible super-government and they are decidedly anti-American."

Even the Nazis themselves, who many thought (accurately) had burned down their own Reichstag in 1934, were not above suspicion. Destroying the *Hindenburg* could have been part of a diabolical ruse to generate sympathy in the United States for Germany. An anonymous note that Special Agent in Charge Whitley forwarded to Hoover asserted that *Lehmann*, of all people, had planted a time bomb with the approval and knowledge of Eckener, who "got Berlin O.K." Alternatively, suggested Dr. Nathaniel A. Davis, president of the "Planet-Aryans" in Los Angeles, a sexy female Nazi agent placed the bomb in the *Hindenburg* so that Hitler could benefit from the insurance money.

Given the tenor of the times, it didn't take long, of course, for the usual scapegoats to be targeted. Within twenty-four hours of the disaster, one correspondent informed Senator Copeland that "the *Hindenburg* was fired at in lower New York—I saw it from a skyscraper window. . . . Some hireling of anti-nazis [*sic*] did this job and was paid by the jews [*sic*], [including the "half jew, half wop Mayor La Guardia"] . . . And it won't be long now, Senator, before Americans start massacring kikes just as other nations have done."

An FBI man later reviewed all the correspondence received, finding that 30 percent of the letter writers believed the destruction was caused by mechanical defects, half by static electricity, and fully 20 percent by Jewish sabotage. Within the bureau, there was a lot of eye-rolling at the patent craziness on display, but Lakehurst's Charles Rosendahl's suggestions were taken more seriously.

On May 19, W. S. Devereaux, Special Agent in Charge, informed Hoover that the evening before he had received a phone call from Rosendahl asking him to come to Lakehurst. He implied that he had something important to impart that only could be done in person.

When Devereaux and Special Agent Lee Malone arrived, Rosendahl said that Eckener had stated to him that "the first opinion that he had formed of the disaster of the *Hindenburg* was that her wreckage was caused by sabotage . . . either on the part of Communists or on the part of sympathizers with the Anti-Nazi movement." As for Rosendahl, he was, said Devereaux, convinced it was sabotage and specifically mentioned Lehmann's opinion that there had been an "infernal device" on board.

Rosendahl, unfortunately, gave no indication as to exactly *when* Eckener had said these things, and another key point was that Rosendahl did not state that Eckener had said that this was his *current* opinion, just that his "first opinion" was that it sounded like sabotage. There was, it would turn out, a large difference between what was then and what was now.

ON THE NIGHT of May 6 Eckener had been asleep in Graz, Austria, where he had been scheduled to give a talk the following evening. That afternoon he had visited the well-known Austrian sculptor Gustinus Ambrosi, who had shown him a new work, of Icarus plunging aflame from the sky.

The Berlin correspondent of *The New York Times*, having enterprisingly discovered his whereabouts soon after hearing about the disaster from the New York office at 1 A.M., woke Eckener up at 2:30 to relay the news of the *Hindenburg*'s end. Few details were available and Eckener was groggy, but he had expressed "deepest grief over the loss of so many lives." As his voice "shook with emotion," he added, "I am shocked and very sorrowful for all on board. I can scarcely say more . . . until I have learned the circumstances of the disaster."

At 8 A.M., Eckener, still in the dark as to what exactly had happened and fretting, despite his atheism, that Ambrosi's Icarus had been some kind of divinely inspired premonition, was driven to Vienna to fly to Berlin on a 1:30 P.M. private flight for emergency meetings with Göring at the Air Ministry.

In view of his "impaired political health," as he put it, whenever Eckener traveled abroad he was assigned an official custodian to make sure that he behaved himself. In this case, it was his friend, a wartime Zeppelin captain, now Luftwaffe colonel, Joachim Breithaupt of the Air Ministry, and it was now that the system failed. At the airfield in Vienna, before Breithaupt could stop him, Eckener mentioned offhandedly to reporters that the company had for some time "received threatening letters, and it is just possible that one such threat materialized in the brain of some fanatic."

It was this ill-considered "first opinion," combined with Lehmann's alleged assertion about an "infernal device," that became the root of Rosendahl's sabotage theory.

IN THE MINUTES, hours, and days after the disaster, Eckener, Rosendahl, and Lehmann experienced varying forms and degrees of cognitive failure when confronted with the shock and magnitude of what had just happened—*which could not have happened*.

Rosendahl, who had served aboard the ill-fated *Shenandoah*, the *Los Angeles*, and the doomed *Akron*, was severely confused, bouncing off one theory or another as he tried to come to terms with the seeming end to his decades-long dream, so often disappointed, of making airships an American success story.

In the aftermath of the explosion, for instance, he had pestered FBI investigators about a harmless German-American passenger named Joseph Späh, an acrobat/contortionist who performed under the stage name of "Ben Dova." Rosendahl said one of the stewards believed Späh had acted oddly during the flight. He had "appeared to him [the steward] to be unsympathetic to . . . airship travel and impressed him as being a peculiar type of passenger; also that this individual remained aloof from the other passengers and was not at all responsive to the explanations of the crew regarding the various technicalities of the airship itself."

Ignoring the otherwise simple explanation that Späh might just not have been interested in an airship's "technicalities," Rosendahl helpfully added that "it would take a fairly active man to climb the rigging separating gas cells 4 and 5 in order to gain access to the gas shaft" and plant a bomb. Späh was an acrobat, was he not?

When that didn't pan out, he switched to a different tack and wanted the New York Police Department, the FBI, and the Bureau of Mines and Explosives to test a mysterious yellow substance allegedly found smeared on a valve. He suspected it was sulfur and therefore proof of sabotage, but it turned out to be nothing of any interest, as Devereaux noted to Hoover.

Rosendahl sometimes wasn't even quite sure whether he himself believed his own conspiracy theory. Two days *before* he had called Devereaux to his house, Rosendahl had told the special agent that "there was no evidence of sabotage." After the Devereux visit, when he had changed his mind yet again, Rosendahl was regarded as making a fool of himself: Nobody authoritative was by then citing sabotage as anything but the remotest of possibilities.

Lehmann, likewise, could not admit the reality, but for different reasons. The *Hindenburg* was the pinnacle of German engineering, German technique, German airpower, and German prowess: It was fundamentally inconceivable that human error, an act of God, pure accident, or a simple defect of any kind could have murdered it. Hence his conviction that some sinister external force—an infernal device—must have been the cause of the disaster. Lehmann would die in terrible pain at 5:55 P.M. on May 7, the day after the disaster, at Paul Kimball Hospital in Lakewood, New Jersey, so he never was given the chance to change his mind.

Eckener, however, after his initial suspicion of foul play, faced the truth squarely and recognized that the *Hindenburg* had been destroyed *partly* by human error: that is, Lehmann's.

He never stated this openly, because he couldn't. The ghastly death of his old comrade, rival, and foe made it impossible to point a finger directly, and he was surely affected by the distress of Frau Lehmann. The fact that their only child, two-year-old Luv, had died five weeks before of complications arising from an ear infection only doubled her sorrow. Blaming her husband would have been an uncharacteristically cruel act on Eckener's part.

When all was said and done, Eckener and Lehmann had known each other since 1909, had worked together, had stuck together, had intrigued together to make the Zeppelin the marvel of the age. They had fought and bickered and backstabbed each other, too, which could not be forgotten, but it could be forgiven, and maybe it didn't seem so important now. "For nearly thirty years we worked together, always in perfect harmony and as close friends," Eckener announced with "unrestrained grief." This was untrue, but it seemed the right thing to say, given the fate Lehmann had met on the very cusp of their success.

A relationship as complex as that between this old married couple could not be explained in a few cheap words or glib sentiments, as Eckener well knew. The Zeppelin Project had to continue, and Eckener understood that the truth was paramount, even if it had to be shaded.

To investigate the causes of the disaster, the Department of Commerce organized a Board of Inquiry, which held hearings between May 10 and May 28 in the Lakehurst hangar. There were two components to the inquiry—the American and the German—and two subsequent reports. The American "accident board" comprised members of the department's Bureau of Air Commerce, with Commander Rosendahl, among others, serving as a technical adviser. The "German Commission," which included Eckener (who told Rosendahl about his "first opinion") and a reluctant Ludwig Dürr (his first time off German soil), arrived four days after the hearings began. The Board of Inquiry interviewed scores of witnesses, crewmen, officers, and experts.

In his own lengthy testimony, Eckener spoke very carefully to obscure but not hide the truth. He placed special emphasis on the wire-breaking "sharp turns" that had immediately preceded the disaster and highlighted the decision to land quickly amid the storm-heavy atmosphere. At no point did he comment on who bore responsibility for these actions.

Many years later, Eckener did name names during a chat with his old friend Karl Arnstein. Eckener reproached himself for not having piloted the *Hindenburg* himself that day, as he would not have made the S-turn that had sparked the fatal chain of events. He remarked that Lehmann "almost always obeyed commanded instructions, though he also put safety on the line once in a while"—as he had done in 1936, when he carelessly damaged the airship during Goebbels's plebiscite flights. Eckener felt, perhaps a little unfairly, that Rosendahl had been trying to rush the *Hindenburg* into a dangerous landing, but Lehmann, as he had earlier warned, lacked the mental fortitude to say no.

At the Board of Inquiry, the Zeppelin officers also closed ranks to protect Lehmann, not only out of respect for a dead comrade but to defend the reputation of the company and of Germany.

To a man, the surviving officers of the *Hindenburg* swore to the Board of Inquiry that all was "absolutely, perfectly normal" on the approach and landing, that nothing was "unusual," and that everything was exactly as it should have been. Yet their statements were contradicted by numerous American observers not beholden to Lehmann's memory, and, more circumspectly, by Eckener himself.

The universal agreement in the officers' testimony that nothing was out of the ordinary seems likely to have been the result of gathering ahead of time to get their stories straight. Heinrich Bauer, one of the watch officers, came close to giving the game away in an odd exchange with his examiner:

Q: Do you believe that the landing maneuver that you have described was proceeding satisfactorily up to the time of the fire?
A: Yes.
Q: Do you believe that this view was shared by all of the other officers with you in the control car?
A: Yes, we are all of the same opinion, that the approach was good.

Perhaps there was a mistranslation on this point, but how would Bauer have known "all" were "of the same opinion" unless they had already discussed the matter and reached a consensus?

What the officers had deliberately left out in their testimonies inadvertently revealed more than what they included: None of them, not a single one, cited Lehmann's name with respect to his role in the control car. It was as if he, the most senior captain, the veteran commander of the *Hindenburg* and *Graf Zeppelin*, the head of DZR, were a ghost.

He had apparently said nothing, done nothing, made no decisions, and given no guidance in the *Hindenburg*'s final hours, minutes, and seconds. His first appearance is invariably *after* the fire broke out, when he was leaping from the window in the company of Captain Pruss. Eckener, however, judging by what he said in his later talk with Arnstein, was well aware that Lehmann must have been giving the orders.

Speaking of Pruss, where was he? He sustained grave injuries and was hospitalized, but he never testified before the Board of Inquiry—which conducted several interviews in hospitals—so the primary witness to the events in question and Lehmann's role in them was not put on record. The simplest explanation is that Pruss, who would be disfigured for life with facial scarring from his burns, was not able to speak owing to his injuries, yet the omission was nevertheless crucial to keeping Lehmann's reputation burnished.

As a consequence, Lehmann was instantly elevated to the status of hero cut down in his prime, a fallen warrior of Nazi Germany. On May 11, a few days after the crash, a funeral service for the German passengers and crew was held at Pier 86 at West 44th Street—the pier used by the Hamburg-America line. Attended by ten thousand mourners and organized by the German legation and the Bund, the ceremony was a solemn mini-rally replete with swastika-draped coffins and the singing of the "Horst Wessel Song" and "Deutschland Über Alles." Lehmann's coffin, bedecked with wreaths like the others, was placed in the center, *Last Supper*-like, of the single row of twenty-eight.

Brown- and white-shirted members of the Bund carried Nazi banners in double file, gave salutes, and formed a "guard of honor" over the coffins as an orchestra played a traditional hymn. Father Schulte, who had held the airborne Mass, bade a personal farewell to the captain, and both Hans Luther the German ambassador and Rosendahl made speeches on a stage draped in swastika flags.

Lehmann's coffin remained in New York while the others were loaded onto the *Hamburg* liner. Frau Lehmann and Martin Wronsky of the DZR would accompany it home a couple of days later.

When the dead arrived in Cuxhaven, Germany, at 11 A.M. on May 21, the ship slowly made its way up the Elbe estuary toward the quays as thousands of spectators doffed their hats in silence. All flags were lowered to half-mast and airplanes circled, dropping wreaths onto the deck. For the official ceremony, the great hall on the pier had been draped in black, and there were eight giant pylons with lit torches. After Chopin's funeral march, the Air Ministry's General Milch (representing Göring) gave a speech unctuously thanking the Führer for ordering a state funeral.

52. Resurrection

THERE WERE SOME naifs who believed the destruction of the *Hindenburg* brought the story of the airship to a close. Despite the initial shock, this was far from the case.[1]

When Herbert Morrison and Charlie Nehlsen returned to Chicago later that night carrying their precious recordings on four sixteen-inch disks, the only news released had been a brief NBC radio bulletin at 7:45 P.M., followed by two more live updates at 9:07 P.M. and 9:15 P.M. that included interviews with some of the survivors and personnel.

Morrison had tried calling the main NBC newsroom line in New York from Lakehurst, but the operator told him she was putting no one through. He explained who he was and she said, "WLS? That's in Chicago. What would someone from WLS be doing in New Jersey?" And then she hung up.

In fairness, news organizations had been slammed with calls about the disaster—*The New York Times* received more than five thousand in two hours—and Morrison's tale that he had gold was scarcely believable. Even if he had been taken seriously, his report would not have been broadcast in any case, as the radio networks at the time banned recordings. A radio play, for instance, would be broadcast live on the East Coast and then, for the West Coast stations, the actors would reperform it live.

But Morrison's astounded managers at WLS decided to break the rule by playing short excerpts early in the morning. Now that the dam had been broken, NBC followed suit with a portion of Morrison's recording about twenty minutes before noon on May 7 and then another portion, adding some live interviews from Lakehurst, in the late afternoon. These were the only occasions on which Morrison's as-it-happens reports were ever played on the radio, but they marked the first time that recordings of a breaking news event were ever broadcast.

Listeners were horrified by his famously urgent narrative of the disaster. In truth, Morrison had been more restrained in his description, but Nehlsen's portable Presto 6D device had recorded the event slightly slowly so that during normal playback it made Morrison's mellifluous, deeper "radio voice" sound higher-pitched, faster, and more excited than it actually was.

By the weekend, the newsreel footage, narrated by studio announcers using prescripted lines, reached movie theaters, causing further tremors among audiences. Everyone had heard and read of previous airship disasters, but these had always occurred at sea, at night, or in remote locations. Never before had a disaster as epic and as spectacular and as close as the death of the *Hindenburg* been caught on film. (Only decades later, it's worth mentioning, would Morrison's "Oh, the humanity" voiceover be dubbed onto newsreel footage; at the time, no one heard and saw them together, notwithstanding any number of instances of false memory.)[2]

Frank Nugent, the film reviewer for *The New York Times*, judged the newsreels to be "the most dramatic spectacle ever shown on screen." There they were, "on record, changeless as long as film endures—a picture of disaster, death, and human heroism and a mighty testament to the force of that strongest instinct of all—self-preservation. No Hollywood fiction will ever equal it, and automatically it becomes the most dramatic picture of the year."[3]

Audiences may have stiffened in their seats and clutched the armrests as the *Hindenburg* went to its fiery doom, but airshipmen were galvanized into action by the footage and newspaper coverage. In the end of the *Hindenburg* they heard not the tolling of a bell but its ringing. This crisis was the opportunity they had been waiting for.

Eckener, upon first visiting the site of the *Hindenburg*'s grave, had seen nothing but a "disorderly tangle of girders, wires, and crumpled sheet metal." It appeared to be nothing less than the "hopeless end of a great dream, a kind of end of the world, a mournful symbol of what I, proscribed myself, expected to be the final outcome for Germany." But on second thought, though admittedly suffering from "emotional depression," he confided to Johanna that "maybe the accident, which could have been worse, will quickly help us along."[4]

Eckener now openly and repeatedly admitted that airships were dangerous—but only, he added, because they were filled with hydrogen. Replace the hydrogen with helium and they would be 100 percent safe. Had the *Hindenburg* been a helium dirigible, as it had originally been intended to be, the "disaster" would have ended with a gentle landing at Lakehurst with the passengers none the wiser.

In short order, this argument became immensely influential in both Germany and America.

IN GERMANY, THE immediate reaction to news of the *Hindenburg's* destruction was one of disbelief. Only a few minutes earlier DZR had received a telegram from Lakehurst stating that the airship had arrived safely. The earliest editions of the morning's papers had accordingly gone to press carrying the happy news, only for later ones, now black-bordered, to be rushed out relaying the tragic update. The official German news agency, which had initially dismissed the news as a monstrous hoax, finally admitted that "the German people . . . have not been spared" the kind of airship disaster which had "[befallen] other nations."

There were few details available. Many employees showed up for work at Friedrichshafen unaware that anything had happened. They were greeted by a notice at the gate and a directive to gather in the hangar, where Ludwig Dürr informed them the *Hindenburg* had been lost.

Dürr, usually so reticent, drew deep on Zeppelin mythology to allay their fears. The German people, he said, had found themselves in a similar position in 1908 after the Echterdingen accident, but they had rallied behind the count and Zeppelin had emerged only stronger from the test. They would do so again.

Ministry of Propaganda policy followed a similar logic. The official line, Goebbels had quickly decided, must be "Forward, despite everything." He had noticed a disturbing sense of national sadness at the loss of the *Hindenburg* and wanted to curb such enervating weakness. Subversive sentiments like that found in the *Frankfurter Zeitung*—"the children went to school sad and gloomy-faced . . . and have aged from the experience"— were to be crushed.

Government officials were soon urging Germans to "stand up under the blow," for "the young and strong nations" can bear such challenges and emerge more resilient than before. Rest assured, a new ship, LZ-130, was already partly completed and would "take the place of the *Hindenburg* as ambassador from continent to continent, carrying the German flag over the ocean."

The official newspapers obediently followed Goebbels's cue to steer public reaction in the approved way. *Der Angriff*, the Nazi Party organ, reasoned that "we would feel only dark despair this morning if we were not Germans who have conquered forever the desire to capitulate. The king of the air is dead; we will provide another king." The SS paper, *Das Schwarze Korps*, optimistically cried, "Whoever thinks that the crash would mean the end of the Zeppelin idea doesn't know the Germans!"[5]

Meanwhile, Göring had called an urgent meeting with DZR and Eckener at the Air Ministry. In the late afternoon of May 7, it was decided, first, to appoint Eckener as managing director of DZR in Lehmann's place, and second, to send him and a small party to the United States to serve as a German Commission to investigate the causes of the disaster alongside the American Board of Inquiry being formed.

During the meeting, Göring had said (as Eckener told his wife) that "he had never thought much of airships, but now, it was necessary to persevere." In another instance of the regime's cynicism, Goebbels noted in his diary on May 10, "The Führer is right. The future belongs to the airplane."

A key point emerged in the course of these discussions at the Air Ministry: Though Göring and Goebbels themselves credited the idea of a bomb being placed on board, the German Commission *must* conclude that sabotage was nothing more, despite a sleep-deprived Eckener's careless speculation earlier that day, than a very distant possibility.

The Nazis had a congenital inclination toward conspiracies—both mounting and theorizing about them—but in the case of the *Hindenburg*, they conspired to tell the truth rather than the usual lies. If the *Hindenburg* had been blown up by evildoers, as Göring and Goebbels believed, it would indicate that Hitler was not universally beloved, undermine the image of Germany as a placid and law-abiding society under Nazi rule, and give the enemies of the Reich, like the Jews and the Communists, hope that the regime was vulnerable.

Thus, in order to explain the discrepancy between Eckener's initial, ill-formed "first opinion" in Vienna and the new official line, Goebbels directed him to make a brief radio speech for American listeners confessing that "of course a question of sabotage . . . occurred to me in the first moment" in Graz but that he had since come to the conclusion that this was "only a very slight probability."[6]

German representatives abroad were then given their marching orders to quash any rumors of sabotage. When asked about the subject the day after the disaster, the German Embassy in Washington curtly replied that there was "no possibility" of foul play. Fritz Kuhn, the national leader of the Bund, who usually saw sinister Jewish puppet masters behind even the most minor events, said he could scarcely believe "hatred could possibly run so deep" as to invite sabotage.[7]

During the Board of Inquiry's proceedings, the German commissioners accordingly showed a marked lack of interest in investigating sabotage leads relative to their American counterparts (who, as noted, would turn up nothing). The only reason the final German report mentioned, briefly at the end, the possibility of sabotage was to keep the door open in case Berlin ever decided a charge of foul play would come in useful for propaganda purposes.

Instead, the German Commission worked on the assumption that the disaster was due to hydrogen and asserted that had helium been used, the *Hindenburg* would never have come to grief. Eckener went along with this version of the truth because, put simply, it *was* the truth and happened to comport with his own views.[8]

In short order, helium had become the key to the future of the Zeppelin Project and transatlantic travel.

IN THE AFTERMATH of the disaster, the Germans could count on a sympathetic hearing in America, that they knew. The year 1937 was a relatively quiet one in U.S.-German relations, and the loss of the *Hindenburg* brought a groundswell of condolences and support for Zeppelin and Germany.

Roosevelt and Hitler had exchanged friendly messages of sympathy and thanks, while Göring expressed to Rosendahl his gratitude "for the heroic rescue of the survivors of the German airship *Hindenburg*" and composed a lovely letter to Mrs. Rosendahl for "car[ing] for my fellow countrymen with unending kindness." He enclosed a handsome photo of himself for some reason.[9]

So confident were the airshipmen that the *Hindenburg* disaster marked

A Fokker F-7 with Juan Trippe on board. The plane would later survey the
route from Miami to Havana.

Trippe (right), with his wife, Betty, Charles and Anne Morrow Lindbergh, and
two mechanics during their South American tour, with one of Pan American's
Sikorsky S-38s

Building the *Graf Zeppelin* in Friedrichshafen, with its intricate system of wires and girders visible

Hugo Eckener watching Karl von Wiegand playing chess during the *Graf Zeppelin*'s Round-the-World flight. Note Eckener's beloved Beethoven records.

Captain Ernst Lehmann, looking debonair

A vacationing Eckener accidentally meets Hitler in Bavaria. Their conversation was an awkward one. Rudolf Hess (standing, left) takes a photograph as the deranged Julius Streicher, publisher of the luridly anti-Semitic *Der Stürmer,* nosily listen in.

The new and the old: a Pan American flying boat passing over a sailing ship on the Spanish coast.

Inside Trippe's Martin M-130 *China Clipper*

Trippe, ever watchful, in his austere office with the rolltop desk—
where he locked away his secrets

Trippe and Paul Litchfield of Goodyear watch the passing cavalcade on
Eckener's "Millionaires' Flight."

Interior of the
Hindenburg

SAN DIEGO AIR & SPACE
MUSEUM/HENRY CORD
MEYER COLLECTION

A stateroom on the *Hindenburg*

EVERETT HISTORICAL/SHUTTERSTOCK

The dining room of the *Hindenburg* (Eckener in background, marked by arrow)

EVERETT HISTORICAL/SHUTTERSTOCK

The "unpersoned" Eckener with Roosevelt in the White House meeting that may have saved his life. Lehmann and Hans Luther, the German ambassador (left), were undesired presences.

LIBRARY OF CONGRESS

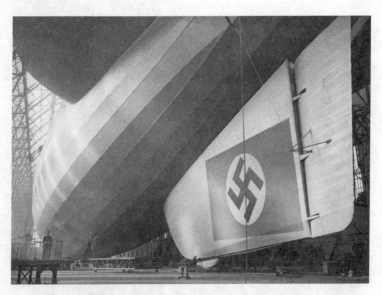

The colossal tail fin of the *Hindenburg*. Note the tiny figure standing next to it.

GETTY IMAGES

The *Hindenburg* serenely cruises over New York a few hours before its destruction.
GETTY IMAGES

May 6, 1937: the death of the *Hindenburg*
EVERETT HISTORICAL/SHUTTERSTOCK

The Nazified funeral service for the victims of the disaster at Pier 86 on West Forty-fourth Street in New York

EVERETT HISTORICAL/ SHUTTERSTOCK

Three years to the day after the *Hindenburg* disaster, the Zeppelin hangars at Frankfurt were unceremoniously dynamited.

SAN DIEGO AIR & SPACE MUSEUM/HENRY CORD MEYER COLLECTION

Eckener, indefatigable

GETTY IMAGES

not an end but a new beginning to their efforts that in August Meister industriously produced a long and detailed study of their bright future. In the weeks following the disaster, no fewer than 4,144 editorials, letters, opinion pieces, and articles in American newspapers appeared, of which 59 percent were favorable, 19 percent unfavorable, and 22 percent neutral toward airships. It appeared, he concluded, that "the airship has many friends in the United States."[10]

The effect of the *Hindenburg*'s destruction on the American public's view of airships was much less dire than people today tend to assume. Air disasters were, after all, a frequent phenomenon at the time. In the sixteen months between January 1, 1936, and May 1, 1937—a few days before the end of the *Hindenburg*—American domestic airlines killed an average of six people a month, and there had been no fewer than *ten* major crashes in the same period.

In one particularly dreadful two-week period (December 15–30, 1936), five airliners had gone down, killing forty-one people, soon followed by another spate between January 25 and March 25, adding twenty-nine to the pyre. (This was just the scheduled airlines; private planes were veritable death traps, with 541 dead in 1936–37.)

When an airplane crashed, it was usually the case that everyone died horribly. In that light, that "only" a third of the *Hindenburg*'s passengers and crew had perished in a cataclysmic blaze seemed a testament to the airshipmen's familiar argument that airships were actually safer than airplanes in an emergency.

The question of air safety happened to be of utmost concern among Americans that summer. The *Hindenburg* was destroyed amidst this greater crisis, caused primarily by an antiquated safety and navigation system struggling to keep up with flying's increasing popularity. In short order, updated air-traffic control measures, mandatory filing of flight plans, increased reliance on instrument flying, and more federal oversight of inspectors were introduced. As a result, passenger fatalities dropped by 85 percent between 1937 and 1939, with only nine people killed in the latter year.

As one magazine remarked a week after the *Hindenburg* disaster, now "there is a general expectation of tighter safety rules, more gadgets, improvements all along the line" for airplanes and airships alike. In other words, the common reaction to the loss of the *Hindenburg* may have been one of horror, but also there was an assumption that the introduction of helium, as part of the new push for "tighter safety rules," would render future disasters impossible.[11]

It was imperative that the Germans be allowed to have it.

AS IT HAPPENED, shortly before the *Hindenburg* fire, the Business Advisory Council (an arm of the Commerce Department) had reported that it saw no danger in selling surplus helium to Zeppelin through AZT. The House Military Affairs Committee had broadly agreed. Public opinion, too, was in favor, with fifty-seven newspaper editorials arguing that helium should be made available to the Germans, with only three opposing.[12]

In the days following the disaster, Roosevelt established a Special Cabinet Committee composed of the secretaries of the interior (Harold Ickes, serving as chairman), commerce, the navy, war, and state to advise whether helium should be sold to the Germans. It met for the first time on May 19 and was heavily in favor of export.[13] Even Ickes, that most unrelenting of anti-Nazis, wanted to extend the hand of sympathy to a bereft nation while improving safety. As he told Roosevelt, "It would appear to be the duty of this country as a good neighbor to share any unneeded surplus [helium] it may have with other countries for the promotion of commerce and science, alleviation of human suffering, and safeguarding the lives of passengers on airships."[14]

His backing came with one caveat, as Ickes made clear: Under no circumstances would helium be released if it was to be used for "military purposes." If a permit was issued for helium exports, it would be canceled at the first hint of the gas being used for anything other than Zeppelin transportation to the United States.[15]

With the Board of Inquiry winding down at Lakehurst in late May, Eckener traveled to Washington to make a statement before the Senate Military Affairs Committee.

In his opinion, Eckener definitively stated, Zeppelins "would never again be used by Germany as an instrument of warfare," adding that "it is quite impossible to use airships for military operations in Europe." When pressed as to what assurances he could provide that the German government would not confiscate the helium for military use, "Eckener said he felt sure such guarantees, if desired, would be forthcoming."[16]

In a meeting with Eckener on May 25, Ickes specified what kind of guarantee he wanted. It had to be a written promise not to use helium for anything else but civilian airship flights signed, preferably, by Hitler, though Göring would do in a pinch.[17]

Ickes's insistence on an iron clad guarantee was motivated by his knowing something that other people didn't. In 1933, the Bureau of Mines, under Ickes's jurisdiction, had discovered that when welding was done in a sealed helium atmosphere, it did not corrode light metals. This was not of particular interest in 1933, but after 1935 and the announcement of the Luftwaffe's existence, the fact that one could use helium to dramatically improve the weatherability, finishing, speed of construction, and durability of welded airplane wings became of great importance.[18]

To Ickes, this was a potentially dangerous military application for helium that would enable a rapid expansion of the German air force, already performing deadly service as the "Condor Legion" in the Spanish Civil War. The bombing of Guernica, which devastated the town, had recently occurred (April 26), and Ickes was horrified at the deliberate use of terror attacks against civilians.

In a mistake that would come back to haunt him, Eckener assumed that Ickes's "written guarantee" condition was just a bit of political puffery he could ignore. After all, Cordell Hull, the secretary of state, Daniel Roper over at Commerce, Harry Woodring at War, and even Claude Swanson at Navy, who'd been so adamantly against the idea a couple of years earlier, favored selling helium, believing that it had no military purpose, and President Roosevelt had raised no objections. Eckener returned home, pleased as punch.[19]

Helium would soon be on its way to Germany. LZ-130 construction, which had been paused since the disaster, recommenced with renewed effort to get it ready for the 1938 flight season.

Time was of the essence. It was alarmingly clear that Trippe was gearing up for war. He already had all his landing permissions for Britain, Newfoundland, Canada, and Ireland in hand; the Atlantic Division was close to finishing the airport infrastructure; and even Imperial was almost up to snuff. In a nicely choreographed arrangement on July 3, one of its new S-23s tested out the westward route by flying from Southampton in England to Ireland, then on to Newfoundland while a Pan American S-42 simultaneously flew the opposite direction. The Atlantic, finally, could be crossed by airplane, even if only on an experimental basis for the time being.[20]

It was the following year that would be critical in the fight against Trippe. The first of his Boeing 314s was scheduled to be delivered in December 1937, and sometime in 1938, it was reported, Pan American would have fully six of them operating across the Atlantic.

Eckener was nevertheless optimistic that Trippe was liable to bite off more than he could chew in the Atlantic. Had he not similarly tried to conquer the Pacific? Look how disastrously that was turning out. Neither was it certain, given the production delays and cost overruns, that the Boeing 314 would make a successful airplane. All it might take for Trippe's Atlantic campaign to come to a sudden halt was a single Pan American or Imperial plane crashing or disappearing amid the gray-blue ocean owing to an error of navigation or a technical malfunction. . . . No, the race was still on.

Eckener's bullishness was tempered, however, on September 1, 1937, when the Helium Act was amended to take account of a potential German sale. There was a surprising change in its wording. In the new version, the State Department had to provisionally grant an export license, *then* it had to be unanimously approved by a new State committee—the National Munitions Control Board (NMCB)—and *then* the secretary of the interior, Ickes in this case, was granted the final say for approval. In short, ultimate power over the survival of Zeppelin had fallen into the hands of one man, Harold Ickes.

There was another niggling, minor, tiny detail that would become of major importance—or at least, Ickes would make it so. The amended Helium Act spoke, in a provision admitted even by Ickes to be "very unusual," of the export of helium to foreign countries being prohibited if it was believed to have "military *importance*," not if it were known to have "military *purpose*," the phrase everyone, including Ickes, had been using.[21]

The distinction between "purpose" and "importance" was a subtle one and had probably been lost upon the drafters. Everyone knew that a Zeppelin filled with helium that set off on a bombing mission to London or Paris would be shot down by a fighter. Thus, it had no *military purpose* in that no rational general would ever employ Zeppelins as bombers. But Ickes decided that "it didn't matter whether [the airship] could get back or not if it could be used for the purpose [e.g., bombing a city] admitted."

To the interior secretary, if a German airship lifted off in any kind of military capacity, rational or not, it meant the helium inherently possessed some *military importance*. Hitler had already shown himself to be mad, in any case, so what was to stop him from dispatching suicide Zeppelins if he so desired, or from finding some as yet unsuspected use for a helium airship?[22]

Ickes and Eckener had a cordial relationship, and they had met on several occasions. They respected each other, shared hostile views of the Nazi regime, and were both canny political operators. So Ickes's position was motivated not by a dislike of Eckener but more an antipathy toward Cordell Hull, the secretary of state, who he believed was not sufficiently anti-Nazi, and thus the once-obscure topic of helium exports turned into a cabinet power struggle.

In October 1937, when Meister submitted a request to purchase nearly 18 million cubic feet of helium to cover the initial inflation of LZ-130 plus frequent replenishments over the coming seasons, Hull approved a first installment of about 2.5 million cubic feet and sent the request along to the NMCB, as the new Helium Act required. The NMCB saw no issue and approved the export license, due to begin January 31, 1938.[23]

As Eckener waited anxiously for news of the helium delivery, Ickes sat on Meister's application. Finally, in mid-March, he said that he would approve it only if three conditions were met. First, he still needed that written guarantee; second, he increased the price of the helium from an already expensive $8.50 per thousand cubic feet to around $10; and third, on top of the payment for helium he required a "penal bond" for poor behavior of about $500,000 up front. If a government inspection detected any sign of military use, Germany would forfeit the money. To put that colossal figure into perspective, since Ickes wanted to charge $10,000 per million cubic feet of helium, and the Germans wanted 18 million cubic feet of it, the penal bond alone amounted to nearly three times what the gas was worth.[24]

Germany at the time lacked the currency reserves to pay that kind of money, and the thought of being dependent on a foreign raw material—helium—would be too humiliating for the Nazi leadership to accept, as Ickes well knew. He was also sure that there was no way that Göring, let alone Hitler, would demean themselves further by signing "I promise to be a good boy" letters at the whim of an American cabinet secretary.

All in all, the newly stringent provisions were designed to stop any chance of helium exports while at the same time embarrassing Hull, who had long supported a liberal trade policy and a neutral position toward Germany. Hull retaliated by prescribing a limit of $32,000 for the penal bond and arguing that Washington had a "moral obligation" to honor the helium contract. As for Ickes's insistence that U.S. inspectors be allowed to visit the airships whenever they felt like it, the State Department commented that "it would not seem to be in good taste for one country to express its distrust of another in such a manner."[25]

Ickes suspected that if he just delayed enough, Hitler would reveal his true colors and give him a "good excuse" to quash the helium deal altogether.[26] The warning signs of a more bellicose Germany were already present: On February 4, 1938, the aristocratic, relatively cautious foreign minister, Konstantin von Neurath, was dismissed, ostensibly for being too dovish, and was replaced by the servile toady Joachim von Ribbentrop, who could not have been more hawkish (or more stupid). Ribbentrop wanted a war, the sooner the better.

After that, Ickes didn't have to wait long. On March 12, Germany annexed Austria, not only for racial-nationalistic purposes but to gain control over its raw materials and factories to accelerate arms production. Almost immediately, Jews were attacked in the streets and their homes and stores plundered. Already in force in Germany, Aryanization, the policy of driving Jews out of public life, the trades, and the professions, began in earnest within weeks. In his diary, Ickes, who had for years been warning of Nazi anti-Semitism, wrote that he was glad he had "refrained from signing the [helium] contract as amended. In view of Germany's ruthless and wanton invasion of Austria I doubt whether it is right for us to sell helium gas to Germany under any pretext."[27]

Ickes then stirred up public opinion against the sale through the friendly press. The words of the airshipmen were used, or twisted, against them to prove that selling helium to Germany was dangerous. In a press release, Ickes helpfully compiled a catalogue of their apparently compromising statements.

Charles Rosendahl's recently published book *What About the Airship?* proved a rich seam for quote-mining. In it, he'd written that at the end of the First World War "there was being constructed a type of airship capable of bombing New York and returning to Europe nonstop!" He added that "an airship like the *Hindenburg*, equipped as a bomber, could have a nonstop cruising ability of 9,000 miles with a bomb-load of 25 to 30 tons."

In the latter instance, poor Rosendahl had, of course, been referring to building new *American* airships to use as bombers, but Ickes then quoted Lehmann's posthumously published memoirs, where he had remarked that Germany's airships "are under the administration of the National Air Ministry." As such, they were not civilian vessels, suggesting (or at least as Ickes suggested) that Göring must therefore consider airships militarily important.

In his diary, Ickes, pleased with his handiwork, recorded that the labor unions had come out against selling helium, editorials now favored a refusal, and that "ninety-nine per cent" of the letters he was receiving backed his position. Much of Congress, too, had swung around to withholding helium. It was quite a turnaround from the situation of the previous year, though support of his position was not as uniform as Ickes flattered himself to think. *Scientific American*, for instance, thundered that Ickes's antagonism was a "crime against science [and] a shameful commentary on human nature that [his] obvious personal prejudices should be permitted to shut the door against development of a useful science."

The most important person, President Roosevelt himself, was "quite open-minded" on the subject, said Ickes, but still refused to be drawn out one way or another. The president knew perfectly well that Ickes was being sanctimonious about a negligible point—the distinction between military purpose and importance—but he could neither fire nor prod into resigning one of his most loyal supporters and effective fundraisers months before the midterm elections. Roosevelt seems to have treated the whole affair with some humor, telling Ickes he could drag out the issue a while longer, probably to avoid antagonizing the German-American vote with an outright denial.[28]

At the end of April, Ickes heard that Eckener was planning to come to America in a few weeks to get a definitive answer on the helium question. Eckener needed to *know* one way or another.[29] For Zeppelin, it was make or break: LZ-130 was virtually completed, and there was still just enough time to have the helium shipped before the flight season began.

But Ickes was not going to budge. "I believe that our decision should be against selling this helium to Germany," he wrote in his diary. Hitler was behaving in an increasingly threatening way, and everywhere Ickes could see Fascism marching victorious over the dead bodies of the democratic countries. With Austria gone, he believed that Czechoslovakia would fall next and that General Franco in Spain would eventually defeat the government forces and install a pro-Hitler regime.

Ickes, as interior secretary, had no international duties, but on this one exceedingly minor issue, of the sale of helium for an airship or two, he could insert himself into foreign affairs. Cordell Hull complained mightily about this, noting that in Berlin there was angry talk of the dishonorable United States reneging on contracts and that a very annoyed Göring had summoned the U.S. ambassador, Hugh Wilson, to have him explain the meaning of the diplomatic insult.

Göring, speaking "with deep emotion and bluntness," had told Wilson that German-American relations were reaching a "low point" over this ridiculous matter and warned of repercussions. As a warning, Germany announced on April 26 that it was withdrawing from the 1939 World's Fair in New York. Hull and Wilson worried that the next form of repercussions would involve shutting down talks on aid to Jewish refugees, the very people Ickes claimed he wanted to protect.[30]

The "steady flow of letters" he was receiving, however, as well as increasing congressional support, only encouraged Ickes's intransigence.[31] By this stage, the aggravated Hull was beginning to believe that, in the words of a State official, it was "easier to deal with an irate and distant Germany than with an irate Secretary of the Interior at hand."[32]

On May 11, by which time Ickes had bombarded Roosevelt with memoranda on the subject, he got the indirect backing he needed from the president after a discussion at the White House to solidify the policy before Eckener arrived "to lobby helium out of us," as Ickes put it. During the meeting, Roosevelt argued that helium had no military use as far as he could see, but Ickes countered that the whole decision process was going to have to be restarted from scratch in any case.

It turned out that, legally speaking, the NMCB meeting that had approved the sale months earlier had to be attended by no fewer than six cabinet secretaries, but in the event only their representatives had been present. So its decision was invalid.

The solicitor general, Robert Jackson, who had told a State Department official only the day before that Ickes was full of "legal rubbish," piped up: "Mr. President, under the law I do not think anything can be done so long as you have such a stubborn Secretary of the Interior with so much information in his briefcase."

"At this point," wrote Ickes, "the President gave up." Roosevelt did not say no outright and block the helium deal—that was never his way—but instead stated officially that according to the terms of the amended Helium Act, "the President is without legal power to override the judgement of Secretary Ickes and to direct the sale of helium for export."

Roosevelt didn't much care about helium, but he had realized, wrote Ickes, that saying yes to the Germans, given the growing opposition to dealing with Hitler and the upcoming midterm elections, would be "a bad political move" and that by piously quoting the law "he had a perfect out in letting me carry the responsibility."[33]

AT THIS STAGE, the one person who stood even the slightest chance of persuading Ickes to release helium was Eckener, of whom Ickes still had "a very high opinion." The interior secretary, an untrusting man, trusted Eckener's word and he wanted to speak to Eckener one-on-one. If Eckener crossed his heart and said that the helium was and would always be strictly for his Atlantic airships, then he, Ickes, could overlook the possibility that the gas would be exploited by Hitler if it was found to have any military importance or purpose.

On the morning of May 14, 1938, Eckener came to Ickes's office. He was older and in worse health than ever before. The incessant struggles of the previous three decades had taken their toll, and for more than a year he had been suffering from "chronic catarrh of [the] large intestine."

It had been hard for Eckener, and he was not in the best of spirits. Indeed, in a recent article for the Zeppelin Company's internal magazine, Eckener had, for the first time since he had joined the count's jolly band, expressed some hesitations about the future. After reciting the occasions when prospects had looked almost as bleak—the yawns that had greeted the advent of LZ-1, the destruction of LZ-2 and LZ-4, the loss of so many airships during the war, the shuttering of the DELAG after Versailles, the rise of the airplane—Eckener hailed the spirit of his airshipmen to fight against the odds and to never give up. Only this time, he conceded, the "Zeppelin Crisis" caused by the loss of the Hindenburg and the need for helium was by far "the gravest."[34]

So, when Eckener entered Ickes's office, he knew that his, and the Zeppelin's, fate lay in the balance. This was it, the final chance. A yes from Ickes would mean LZ-130 would be flying to America that very summer, with the prospect of more Zeppelins to come. The Atlantic would be bridged permanently.

In the meeting, Eckener repeated the point that airships had no discernible military value, which Ickes listened to with the utmost courtesy, without changing his position. Finally, Eckener asked him to explain how a helium airship would be used militarily. Ickes stumbled slightly and suggested that they could be employed to bomb London. Eckener laughed, telling the secretary that Zeppelins in their hangars would be bombed into nothingness within ten hours of war being declared. Ickes retorted that the Germans would have foreseen that scenario and already moved their airships out to sea.

That was impossible, Eckener snorted, but Ickes insisted that it "could" happen while Eckener argued that it "would" not. Ickes, now on his back foot, closed down this meandering *could-would* syntactical debate by saying that so long as anything was within the realm of possibility, he was legally bound to refuse the helium request.

Then the secretary, seeing Eckener so despondent, spoke frankly to him. He said, in substance, "We know you, Dr. Eckener, your accomplishments, and your reputation for integrity. We implicitly trust you. We are convinced that you are asking for our helium solely to make future Zeppelins for peaceful purposes as safe as is humanly possible. There has, however, as you are well aware, been a complete change of government and political ideology in Germany, with Adolf Hitler at the helm. What guarantee can you give us that the Nazi regime will not seize the helium gas for other than peaceful purposes?"

Eckener paused, weighing his words. "At that moment I knew that my life's work was at stake. If I assured [Ickes] that in my opinion there was no danger of any seizure by order of Adolf Hitler, I would get the helium; if not, my dream of a lifetime, which now at last seemed destined to come true completely, would be shattered."

In that one clear, shining moment, Eckener found himself unable to obfuscate. Over the decades, he had sometimes played fast and loose with the facts, he had fudged the numbers, he had cooked the books, he had hedged his words, he had suppressed adverse information, he had exaggerated for effect, he had morally compromised himself—all for the greater good of keeping airships alive. But he could do so no longer. The Nazis, and Germany itself, did not deserve the Zeppelin. Such mendacity and cruelty and brutality could not be rewarded, not by an honest man, even if he knew helium was militarily useless.

Eckener made the moral choice to *not* fight for what he had so long stood for and told the fatal untruth: "I can give you no such guarantee; in fact, my only fear is the same that you entertain."

Ickes could see how "painful" the meeting had become and how "terribly disappointed" Eckener was. To console him, "I told him that Congress might amend the law or I might have a successor who would have a different point of view." But they both knew that was a lie.

As Eckener recalled, "I returned to Germany empty-handed. We obtained no helium. The death knell of the Zeppelin had sounded."[35]

So HAD ECKENER'S. He had never stopped loathing the Nazi regime, and now that there was nothing to fight for anymore, it was time to go. The regime agreed with him.

Questionable actions, like removing the official swastika flag from the hood of his car and pointedly rejecting a dreadful book (*The Pioneers of the Third Reich*) by the head of the Hitler Youth sent to him for the Zeppelin Company library, had repeatedly called Eckener's loyalty into question.[36]

There were also suspicions that Eckener hadn't tried his utmost with Ickes to acquire helium, though Eckener was careful not to drop any hint that he had helped to sabotage the effort. Still, he was scapegoated for the failure, because someone had to be.

About six weeks after returning home, he announced to a meeting of the DZR on July 18 that he was stepping down from the post of managing director. He would retain the chairmanship in an honorary capacity and be available for consultation, but he would no longer be an active part of the enterprise.[37]

That month, and it was no coincidence, happened to be the hundredth anniversary of Count von Zeppelin's birth, and Eckener bade both him and the company farewell by publishing an admiring, if somewhat hedged, biography of the count. It was a small consolation, too, that a poll of 259,000 workers designed to test their political and historical knowledge revealed that 78 percent of respondents—almost four in five—were aware of who invented the rigid airship. To put that in context, the swastika, among the most famous, or notorious, symbols on earth, was recognized accurately by 78.4 percent, and the year the Nazis seized power was correctly answered by 78.7 percent. The count, it seems, was still loved, as were his airships.[38]

Also that July, *Fortune* magazine appeared on newsstands carrying an authoritative, lengthy article about the future of long-range air transportation. Crammed with interviews with the titans of American industry, like

Glenn Martin and Igor Sikorsky, and filled with detailed analyses of airline financing and engineering, the article featured a splendid, full-color foldout illustration of LZ-130, "Germany's Bid" for supremacy in the Atlantic.[39]

The artwork had no doubt been commissioned months earlier, before Ickes's decision was publicly known, and it was a sad reminder that the struggle for the mastery of the sky was over. The bird had conquered the cloud.

Eckener was happy to leave. He had beaten Trippe across the Atlantic, just as he had beaten everyone with the DELAG all those decades ago, and that would have to be enough. A little less than a month later, he celebrated his seventieth birthday (August 10), receiving, much to his surprise, a leather portfolio from Hitler himself containing a congratulatory letter.

As soon as the delegation from Berlin left, Eckener handed the letter to an aide and told him to burn it.[40]

53. King Across the Water

HELIUM WAS THE airship's Achilles' heel, but even if Harold Ickes had never existed or Eckener had said that he trusted Hitler or Goodyear-Zeppelin had received a slew of airship commissions, the Zeppelin business would nevertheless have been financially ruinous if it had continued.

In mid-1937, Meister conceded in a highly confidential memorandum that two unsubsidized, Akron-built airships would together cost $11.4 million, with a terminal adding nearly $4.9 million, leading to a predicted loss of about $1.2 million annually when regular passenger service began. Even if the American government were persuaded, miraculously, to fund construction, the airships would eke out a profit of a mere $8,150—a sum hardly worth the effort to earn it.

Operating cost per mile, a key yardstick in the air business, painted an equally horrifying picture. Each of the two American-built airships would cost $9.97 (including equipment, handling, fuel, terminal use, running expenses, and interest charges) per mile. In comparison, each of Trippe's Boeing 314s (priced at $550,000) were calculated to operate at $1.10 per mile. Yet fare prices were more or less the same, and even taking into account that the airships would carry twice or maybe thrice the passengers, the extent of the disparity was staggering.[1]

That, of course, was if one assumed that the airships were even allowed to fly. The destruction of the *Hindenburg* had been the costliest insurance loss in aviation history: the equivalent of roughly $40 million today, much of it borne in the London market among a number of chastened underwriters, very few of whom would in future be willing to cover the risk of another failure without helium available.[2]

The picture in Berlin was just as bad. If a helium-filled LZ-130 made fifteen voyages to the United States in 1938, DZR reported, the total income would be 1.3 million reichsmarks, but costs would amount to 5 million reichsmarks. Worse, if this tentative LZ-130 plan proved to be a success, it would bankrupt the company since DZR would then have to build LZ-131 (3.5 million reichsmarks) to alternate with LZ-130 plus another shed at Frankfurt (6.5 million reichsmarks) to keep it in.[3] Not even the Reich could afford to foot bills of that magnitude when it was spending everything it had on rearmament.

It was, ironically, helium, the very stuff that might have saved airships, that was also their killer. Subsidized construction or not, helium still cost at least ten times more than hydrogen. For Zeppelin, a concern already trembling under the weight of its finances, it was too much to ask.[4]

THAT RAISED THE question of what to do with LZ-130, the last of its kind. For reasons of national prestige, Berlin could not do the obvious thing and scrap it; if anything, LZ-130's very existence rendered it more necessary than ever to keep it for fear of losing face. Breaking up the airship because of a slap in the face by Ickes might be taken to imply that Germany was weak—an intolerable idea.

The policy became, then, one of redoubling efforts to keep LZ-130 flying until such time as it could quietly be put out of Berlin's misery. In a couple of years, once the point had been made that it was *Germany's* choice to kill the program, few would notice LZ-130's vanishing from the scene.

Goebbels proposed reverting to the tried and true—propaganda flights—to invent a reason for LZ-130 to exist. Göring, however, thought that the airship might have some marginal utility by deploying it as a mobile laboratory to carry out radio experiments. But otherwise he agreed with Goebbels in that this would merely help run out the clock.

In the meantime, the two Nazi princes were adamant that under no circumstances was LZ-130 to be risked. Eckener, the honorary chairman of a company that existed virtually only in name, was lured, either through duty, pride, or a desire for a last time at the controls, into captaining three brief test flights of LZ-130.

September 14, 1938, launch day, was a somber one. The only dignitaries present at Friedrichshafen were Eckener, Dr. Issel from the DZR board, and Eckener's friend Colonel Joachim Breithaupt from the Air Ministry. No one from the government attended, though Göring sent a formal telegram extending his "heartfelt greetings for an auspicious first flight."

Eckener made a speech extolling the workers for their labor over the years and their dedication to creating the "best of German worksmanship" to show the world. Then he revealed the name of the new ship, which he had been keeping tightly under wraps. Since LZ-130 had been completed in the centenary year of the birth of the founder, it should carry his name: Graf Zeppelin.[5]

It was an appropriate sendoff, Eckener thought. Initially, there were rumors that LZ-130 was to be called Grossdeutsches Reich, after the pan-German empire the Nazis envisaged, but this was about as likely as Adolf Hitler had been for the Hindenburg, for the same reason.[6] The propaganda effects of another disaster would have been suboptimal. For Eckener, the name—sometimes the airship would incorrectly, if fondly, be called Graf Zeppelin II, even by its crew—was a way of honoring the Grand Old Man, of harking back to a more civilized time, and of reminding the world of the glories that had once attended the original Graf Zeppelin.

After the ceremony, Eckener, Dürr, and the old veterans Captains von Schiller and Sammt and First Officer Heinrich Bauer climbed aboard for the first test flight, which would last about ten hours. Two more short flights, visiting several cities, followed over the coming week.[7] The last of these, on September 22, would be Eckener's final voyage. It was a bittersweet one—more bitter than sweet.

Ostensibly, the trip was a farewell for Eckener to one of his favorite cities, Vienna, but as always with the airship in the Nazi era, the political was entwined with the personal. The flight took place at the apex of the Sudetenland crisis, in which Hitler had demanded the annexation of that Czech territory. Indeed, on the very day the Graf Zeppelin visited Vienna and was cheered on the rooftops, British prime minister Neville Chamberlain flew to Germany to discuss the issue with Hitler, presaging the Munich Agreement that would dismember Czechoslovakia.

On board the Graf that day were twenty Air Ministry radio technicians, belying the claim that the visit was merely for publicity. On the way back, the Graf Zeppelin was ordered to cruise along the Czech border to conduct radio surveillance. Just in case the Czechs got irritated by the airship's presence in a sensitive area, the *Graf Zeppelin* was escorted by four Messerschmitt fighters.

This was not the end Eckener wanted, but it had to be. The airship was no longer his and he had no say in its (mis)use. The Zeppelin was once vaunted as the bridger of commerce and connector of peoples, but all that was history now. As a conductor of spy missions against nations upon which Hitler had designs, the *Graf Zeppelin* had proven Ickes's perspicacity in seeing a potential "military importance" (and purpose, for that matter) to airships, whether helium- or hydrogen-filled.

That November, following the outrages and violence of the *Kristallnacht* pogroms against German Jews, Ickes derived great pleasure from publicly announcing that the original helium contract approved by the NMCB had officially expired and that there would be no chance of a sale in the future. By now, even Cordell Hull, his foe at the State Department, recognized that Ickes had been right and backed him.

The *Graf Zeppelin*, now commanded by Captain Sammt, made another twenty-seven flights. The vast majority of them were visits to various cities and flyovers of air shows and the like before Goebbels tired of propaganda tours and threw the problem over to Göring, who thought perhaps the airship could be used to spy on the British.

The former dining room was filled with banks of radio consoles and oscillographs, with the more top-secret equipment being kept in what had been the passenger cabins, whose berths and washbasins had been torn out. Dozens of Luftwaffe technicians and linguists, as well as several Gestapo agents to make sure the Zeppelin crew didn't get too inquisitive, were placed on board.

Since 1935–36, the British had been developing radar to locate and track incoming aircraft, but very little was known of their progress or how the technology worked. By mid-1939, the British had constructed seventeen large radar installations with tall antenna masts, known as the Chain Home network, along the southern and eastern coastline.

Flight 24, of August 2–4, 1939—a month before war was declared—was dispatched to elicit what the purpose and capabilities of these masts were. The *Graf Zeppelin* flew across the Channel and proceeded up the east coast to Aberdeen. There the crew cut the engines and let the airship drift at sea, listening for radio signals. At one point, several brand-new Spitfires darted toward them, forcing the airship to hightail it out to sea as the Luftwaffe personnel took photos of the unfamiliar airplanes.

Flight 24's biggest failure was reporting that Chain Home was non-operational because the technicians had not scanned a wide enough frequency spectrum. But the radar system had actually been active since April and had tracked LZ-130 for some time—indeed, the British could hardly have failed to notice such a giant, radio-wave-bouncing metallic object. That mistake would prove costly the following year, during the Battle of Britain, when the Luftwaffe would learn to its cost how effective radar really was.[8]

TRIPPE HAD BELIEVED that he could make hay from Eckener's travails. With the loss of the *Hindenburg* and the lack of helium, a dangerous competitor in the Atlantic had been removed. Alas, his own travails were only just beginning, just as Eckener had shrewdly expected.

The root of the problem was the delayed delivery of the six Boeing 314s he needed to open Atlantic service. The first of these was originally scheduled for December 1937, which had long since passed. Lindbergh had been right to warn Trippe against buying them: Boeing simply couldn't get the plane to work properly.

Initially neither able to float nor fly, the Boeing, it was soon discovered, could not land, either. Pilots coming in experienced what was known as "porpoising"—a bouncing and skipping off the water. Pan American told Boeing that the aircraft was unacceptable until the problem was solved. Trippe regretted not sticking loyally with Martin and Sikorsky.[9]

In the meantime, Imperial too was having a terrible time developing a suitable airplane for its reciprocal Atlantic flights. A stretched version of the S-23 Empire flying boat, the S-30, was decent but too short-ranged—it needed in-air refueling, a tough trick to pull off—and barely improved on its predecessor's already lackluster performance. A trial flight across the Atlantic was canceled at the end of 1938.

Politically, as well, the stars were aligning against Trippe. He had few friends in the Roosevelt administration. The influential Ickes, among others, considered him an "unscrupulous person who cajoles and buys his way. He has made quite an unsavory record in the South American countries."[10] Trippe's past was returning to haunt him.

In this instance, the cozy gentleman's agreement between Pan American and Imperial for exclusive landing rights on either side of the Atlantic was prompting dark, terrifying whispers that Trippe could be charged with violating the Sherman Antitrust Act for locking out potential competitors.

After several panicked phone calls from Trippe, Woods Humphery, loyal to the last, informed the Air Ministry that he could not allow Trippe to risk prosecution and released him from the longstanding exclusivity agreement. Then he dutifully resigned, after which Imperial Airways was reorganized under new management to form a single state airline, British Overseas Airways Corporation (BOAC), which had negligible interest in a route to New York.[11]

The threat of an antitrust suit proved a blessing in disguise. As of February 1939, Trippe was relieved, at last, from the reciprocity deal that had chained him to Imperial for so long. The path across the Atlantic was free, and finally, too, the first of the Atlantic Boeing 314s was christened as *Yankee Clipper* on March 3 by Eleanor Roosevelt at Anacostia Naval Air Station with a gold-rimmed bottle filled not with Champagne but, according to Trippe's PR man, William Van Dusen, with waters from the world's seven seas—a rather unlikely story. At the ceremony, a beaming Trippe announced to a national radio audience that the *Yankee Clipper*, the first of a fleet, would "carry the American flag across aviation's last frontier—the Atlantic Ocean—to link the old world with the new."[12]

He announced that May 20, 1939—the twelfth anniversary of the transatlantic Lindbergh flight—would be the day the *Yankee Clipper* would stride into history as herald of the eternal dominion of Pan American.

AT THE TIME, Trippe's airline served 54,072 route miles in forty-seven countries with 126 airliners, 145 ground radio stations, and over 5,000 employees. But rather less well advertised were its chaotic finances—not *as* dreadful as Zeppelin's, perhaps, but dreadful nonetheless.[13]

The bills were coming due for the huge Boeing 314 order, plus ones for other planes, and the vainglorious Pacific expansion remained an unplumbable sinkhole. In 1938 alone, the route to Asia lost $1.155 million, and as a consequence the year ended with Pan American squeaking net profits of just $46,672. Trippe's working capital (defined as current assets minus current liabilities)—basically, the funds needed to continue daily operations—had collapsed to negative $1.3 million. It was getting to the point where Pan American would find it difficult to pay wages. Worse, Trippe had loaded up the airline with debt: no less than $6 million.[14]

Trippe was sailing perilously close to the shoals of bankruptcy, but he seemed sanguine about the possibility of shipwreck. His view was that he was solidifying a strategic position that would pay off in the next year or two, so what did a few tactical defeats here and there matter?

His board did not see it the same way, but Trippe didn't care. As master of the house, he had long selected its members to make it a toothless entity that allowed him carte blanche. He rarely even bothered to inform the board of major decisions until after they'd been put in motion, and no one had complained because everyone was making money and Trippe was an acknowledged genius whose vision and judgment were not to be questioned.

Nevertheless, Trippe's board appointees always had a purpose. In the early days, its members were rich and well-connected friends of his with an interest in aviation; then came the former government officials who smoothed his path to the top; after that there were former and sometimes aggrieved competitors who were appeased with lucrative seats. Whenever Pan American needed loans or financing, bankers suddenly popped up in the boardroom, and if Trippe wanted to forge an industry alliance to face off against Congress, the heads of large transportation concerns (steamships, railroads) were invited to join. Once an individual had exhausted his utility, he was replaced by someone better suited to serving Trippe's ends.

The exception was Pan American's most useless director, Cornelius Vanderbilt "Sonny" Whitney, now forty years old, who had known Trippe since their Yale days and stuck with him since the very beginning. He had survived owing to his inoffensive, unthreatening impotence and his willingness to rubber-stamp any proposal Trippe presented to him.

The straitlaced Trippe considered Sonny, heir to not just one of America's great fortunes but three (Payne, Vanderbilt, and Whitney), to be nothing but an amusing, charming playboy. When he wasn't fending off paternity suits or carousing with the Crusaders (a young men's anti-temperance group), he financed movies, with *Gone with the Wind* to his credit—perhaps explaining the presence of so many tie-in copies of Margaret Mitchell's book in Pan American's onboard libraries—and collected art.

But he was always there when Trippe needed him: He'd put in $25,000 without a moment's thought when Trippe suggested they merge Eastern with Colonial Air Transport, put in double everyone else's stake when Trippe formed Aviation Corporation of America, and, while recovering from a blistering hangover with his head in his hands, murmured, "I'll take a million of that" when Trippe went around the table asking to raise $3 million for expansion in 1934.

But Trippe had underestimated Sonny, a sharper businessman than his fun-loving persona implied. Over time, Whitney had accumulated 154,432 shares (more than 10 percent of the company) in Pan American, while his cousin Jock held 56,400 more. Trippe, conversely, had begun with 25 percent of Pan American, but in recent years had sold off large portions to invest in real estate, leaving him with just 7,056 shares in his own name.[15]

Not even a man as wealthy as Sonny was immune to continued losses, and Pan American's falling stock price was costing him money. Trippe was getting too big for his boots, and Sonny wanted some accountability and more transparency. The Pacific had been a disaster, Trippe's Eckener fixation and his obsession with the Atlantic had wasted millions of dollars over the years, and for the first time, Pan American had not declared a dividend. It was time to kill the king.

On March 14, 1939, Trippe entered the boardroom on the fifty-eighth floor of the Chrysler Building, assuming that this would be yet another walkover meeting. The *Yankee Clipper* had arrived about ten days earlier and the Atlantic service was due to begin in a couple of months, so as far as Trippe was concerned, Pan American had turned a corner and, like a victorious Caesar returning from the wars in Gaul and Germania, he was due the accolade of a triumph and a crowning with laurels.

He received instead a knife in the back, wielded by his Brutus. Trippe, a man with few of them, realized too late that Sonny was not really his friend, more a business acquaintance who happened to be friendly. As Whitney had once said, Trippe and he "didn't live in the same community": Despite his every effort to blend in, to join the right clubs, and to camouflage his embarrassing background, Trippe was not really One of Us after all. It was now his turn to be dispensed with as ruthlessly as Trippe himself had done to others who became surplus to his requirements, like Sikorsky and Martin.

In the meeting, Trippe, much to his surprise, was faced with seven stony-faced men. The execution took just a few minutes. Sonny and Jock ordered a vote, secured "Ayes" from at least three of the others, and assumed control of Pan American. Trippe was removed from his position as chief executive, with Sonny taking his place, and demoted to president, a relatively powerless position in which he would have to report to Sonny as chairman. Major decisions would henceforth be taken by a stronger, more activist executive committee. Following Trippe's practice of keeping secrets, the coup occupied just a single line in the minutes and no record was taken of who voted for and against. With but 7,056 shares in his own company, Trippe had become a hired manager with a tiny salary of $17,500.

For the newspapers, Trippe claimed that he had resigned and voluntarily allowed Whitney to take over, and Sonny was gentlemanly enough not to humiliate him further with a public contradiction. In private, it was a different story. Trippe was obliged to vacate his vast, opulent office and move his famous three-foot globe and rolltop desk down the hall to what had once been a secretary's room. He kept his door open, though, so that he could watch which of his former loyalists made pilgrimages to their new master. And he took notes.

Trippe had no intention of ending his days as Whitney's lackey, as he suspected that Sonny would not last long, not with his variety of interests and calls on his time. Like it or not, Whitney needed him and his expertise; he had no hope of running an international airline by himself. It would just take a little time, though it would take longer than Trippe expected, for Whitney was serious about turning Pan American around. He even disbanded his polo team.

The executive committee, of which Trippe was a member (as were radio expert Leuteritz, Priester, Van Dusen, and George Rihl, the former dirty-tricks man in South America), was summoned to meet every Monday and Friday at 10:30 A.M. sharp to report to Whitney in his office, which had been redecorated with captain's chairs, a glass-topped table, and cheerful curtains in a pointed jab at its former occupant's secretiveness.

Until then, it had always been Trippe who spoke and others who nodded their agreement, but now he lived up to his Yale nickname, "Mummy," and rarely spoke a word. Whitney took to interrogating the executives on their activities, which didn't sit well with Trippe, of course, and he sullenly glared at anyone who confided too much.

Once, Leuteritz, proud of some recent success in Colombia, made the mistake of boasting about it to Whitney, only to be pulled aside later by Rihl, who asked him, "Why did you give Whitney all the details? Didn't you see Trippe's face?" Rihl himself changed sides with amoral alacrity, and when he was upbraided for his treachery by Trippe, he told him where to go. Trippe sputtered, "I'm your boss," at which Rihl sneered, "The hell you are."[16]

For the grand launch of the *Yankee Clipper* on May 20, however, Trippe was nothing but smiles. At 1:08 P.M., the *Yankee Clipper* took off on its first scheduled commercial (i.e., nonpassenger) flight to Europe. Captain Arthur LaPorte spoke on two-way radio with Trippe and a few other worthies, and listened to a congratulatory telegram from Roosevelt, before heading into the distance carrying 1,804 pounds of mail. He landed in the Azores for a few hours to refuel before successfully making it to Lisbon and thence on to France. On May 27, LaPorte returned home, completing the Atlantic round-trip.

On July 8, the *Yankee Clipper* took off at 8:30 A.M., this time bound for Southampton in England. Carrying twenty-two passengers, nearly all newspaper editors and publishers, it was an uneventful flight, the only remarkable thing being that, at last, Trippe had achieved his dream: regular passenger service across the Atlantic.[17] The term *clipper* would soon become redolent of Pan American glamour: Trippe's planes would grace the covers of *Time* (at least three times) and *Life* (twice) and were mentioned in *The Philadelphia Story*. Ilsa Lund and Victor Laszlo would even escape from Casablanca to Lisbon to take "the Clipper to America."[18]

Trippe would not be able to savor his Pyrrhic victory over Eckener for long. The outbreak of war less than three months later would force Pan American to halt the flights. But Trippe still had one more battle to fight.

FOR THE AIRSHIP, the end was pitiful. On the same day, September 1, as the Wehrmacht invaded Poland, the second *Graf Zeppelin* was deflated and hung up inside its hangar to join its retired predecessor, the conqueror of the globe, as a temporary, and depressing, museum piece. In the coming months, the airship would be stripped of its engines, electrics, control-car navigation, and gas cells, leaving but a carcass. There was talk of dismantling the hangar and even the *Graf Zeppelins* themselves.

On March 1, 1940, Göring himself came to Frankfurt to make the final decision about the two airships. Captain Pruss, formerly of the *Hindenburg* and now director of Zeppelin operations at Frankfurt airport, and Dr. Issel of the DZR board were on hand to greet the air minister and his entourage of thirteen generals. Göring and General Erhard Milch, his deputy, were taken on a tour of LZ-130. The air minister was not impressed.

When Issel mentioned that it was a pity that the *Reichsminister* had not yet taken an airship flight, Göring good-humoredly replied, "No, you won't get me into that, the thing burns." Issel then pointed out that only the *Hindenburg* had ever come to grief, but Göring dismissed the argument: "No, the thing is rubbish." The visit went downhill from there.

The party entered the skeletonized airship, with Göring making a joke about falling through the open girders. He asked Pruss for his views on airship travel to South America, and the captain proudly claimed that the *Hindenburg* had made the run in three days. Göring snorted and stated that after the war airplanes would be able to do it in eight hours.

Inspecting the crew cabins, what was left of the passenger rooms, and the public areas, Göring kept saying they were "rubbish." When he saw the fuel tanks, Göring asked about the risk of explosion. There was none at all with diesel fuel, Pruss assured him, but Göring wasn't having it: "A match would be enough, then the whole thing would burn."

In the dining room, he leaned outside a window to announce to the official photographer: "Just take a snapshot: the picture will become a [valuable] rarity."

And with that, the air minister drew Pruss and Issel aside. He'd once had some sympathy for the airship, he said, but now, to be frank, he considered it valueless. He addressed the scarred Pruss personally: "I well understand that you are fighting for your cause, as you have suffered for it," and shook his hand goodbye.

Then he descended to the hangar floor to chat with the workmen. Göring asked Rudolf Sauter, the chief engineer, whether he trusted the "thing." When the ever-loyal Sauter said yes, Göring amusedly observed, "Well, you have my admiration," before placing his hand tenderly on a parked Messerschmitt Bf 109 fighter, saying, "I prefer this."[19]

A week later, the demolition of the *Graf Zeppelins* began. It took no fewer than 376 men exactly one month, two weeks, and four days to erase Count von Zeppelin's gigantic inventions from existence. On May 6, the third anniversary of the death of the *Hindenburg*, the hangars at Frankfurt were unceremoniously flattened with dynamite.[20]

Göring wanted the deed done as quickly as possible. He had a war to run. Four days later, the invasion of France, Belgium, Luxembourg, and the Netherlands began.

· · · · ·

MEANWHILE, BACK IN New York, Chairman Whitney, as Trippe had predicted, found that running an airline was not as easy as he'd thought and was rapidly losing interest in the whole affair. He delegated the responsibility of overseeing the executive committee to Thomas Morgan, a former chairman of the Curtiss-Wright Corporation and an experienced aviation hand, but Morgan could make neither head nor tail of the corporate records: Trippe, he said, "had everything so snarled up nobody could ever untangle it." For years, Trippe had been writing policy with his right hand and executing it with his left, and nobody knew what either was doing. Anything important was locked inside Trippe's head, and he was becoming less and less communicative.

Trippe's position was strengthening by the day as Whitney's weakened. Whitney's fall was hastened by the staggering performance of Pan American in 1939, for which Sonny could take no credit. It was all down to the tough decisions Trippe had made during his years at the top that had finally paid off.

That year, Pan American's route network added nearly ten thousand miles and net profits exploded from 1938's $46,672 to nearly $2 million, making it the most successful year in the company's history by an order of magnitude. West of Honolulu remained a trouble area, partly owing to the ongoing Sino-Japanese War, but passenger numbers and mail from California to Hawaii were booming, cutting the loss of $1,150,000 in 1938 to a mere $200,000. Trippe had even managed to retire half the company debt while ordering another six Boeing 314s.[21]

On January 23, 1940, Trippe engineered a reverse coup by having the board vote on transferring the powers of chief executive officer from the chairman to him as president. Soon afterward, Whitney embarked on a long yachting vacation and never gave another thought to Pan American, having sold his stock at a huge profit.

The ax soon began swinging down upon the necks of the nobles on the executive committee who had so unwisely bent the knee to Whitney. Firings were never Trippe's style; his preferred method was far more sadistic: loss of access. William Van Dusen, who had served his new master a little too eagerly, was henceforth kept at arm's length, while George Rihl, who had signed his own death warrant by telling Trippe he didn't work for him, was exiled to Brazil and soon pushed into retirement.

The men who had soldiered with Trippe since the beginning were severed from power and turned into consultants; no longer would they be able to issue orders to managers in the field as they thought fit. Thus, Priester, who'd risen to the rank of vice president in Whitney's reign, kept his title but not Trippe's favor. Leuteritz remained as chief communications engineer but, like Priester, was barred from Trippe's office. He quit in disgust a few years later.

Lindbergh, had he thrown his weight behind Trippe, could probably have dissuaded Whitney from mounting his ouster during the palace coup. But he had remained neutral during the whole affair, confining himself to writing in his journal that "in many ways I am sorry to see [what happened], for I like Juan. I always felt he had great ability," and thus he polished perhaps the sharpest serpent's tooth.[22]

He and Trippe had once been the closest of friends, or rather, the closest that Trippe got to being friends, and without Lindbergh's dedication and knowledge, Pan American would never have succeeded, no matter how great Trippe's abilities. But they had grown apart in the later years, especially politically—Trippe thought Lindbergh's Nazi sympathies embarrassing. The aviator's unwillingness to take his side when it counted was treasonable, for Trippe's relationship with Lindbergh wasn't just "business," in which one sought alliances of convenience, but a personal one. They had fought in the trenches together, and to be left for dead (as Trippe saw it) by a comrade who could have saved him was unforgivable. Lindbergh would continue to be associated with Pan American for decades to come, but not as an intimate at the court of King Juan.

Trippe found new men and raised them from the dust to replace his fallen barons, but never again would he allow anyone to become close enough to threaten his crown.

The interregnum was over, and the king had returned from over the water. Trippe moved back into Whitney's vacated throne room, albeit without his famous globe.

That stage prop was taken by freight elevator to the basement and eventually wound up in the National Air and Space Museum. Trippe, who had become his own man, no longer had any use for such childish things.

He now, after all, ruled the world.

A VISITOR TO ECKENER in the summer of 1939 found him worn out and hopelessly pessimistic.[1] He knew a war was coming, he knew Göring was planning to scrap LZ-130, and he knew that the Zeppelin idea was finished. He collected his scores of honors—the Guggenheim Foundation Medal, the National Geographic Society Medal, the gold medal of the Royal Aeronautical Society, the medal of the Fédération International Aeronautique, the Leibniz Medal of the Academy of Science, a Lilienthal Medal, an Albert Einstein Medal, even the Golden Key for Atlantic City, among so many others—and threw them into a simple cardboard box with "Decorations" scrawled on the top. They had all been in vain.[2]

With the end of the airships, he told a friend, the Zeppelin Company "switched completely over to armament work," and its factories and workshops were converted to making Panzer gearboxes and aircraft engines.[3] By the fall of 1941, the company had signed contracts to produce rocket-propellant tanks and fuselage sections for Hitler's promised wonder weapon, the V-2.[4]

Eckener remained the honorary chairman with no decision-making role, despite having being appointed a *Wehrwirtschaftsführer* (leader of the war economy) in 1939—a title Eckener dismissed as meaning "practically nothing" and thought was ridiculous.

Becoming a "leader of the war economy" was a feel-good appointment for the four-hundred-odd executives of large corporations producing war materials, but before 1940 there was no political litmus test for recipients to prove their Party bona fides (which Eckener would have failed). After that, the title became much more closely identified with Nazi values.[5]

At some point, he picked up a temporary directorship of Deutsche Bank, where he did little, though the bank was a nexus of Nazi financing, and sometime in 1944 he became a short-term director of an Austrian company, Kontinentale Rohstoffe und Papierindustrie. It's possible that the firm was involved in producing gunpowder, but more likely it was a paper-making concern with interests in manufacturing film stock. He was also a deputy chairman of a part-Zeppelin-owned company called Aero Union, whose vague mandate was to promote "airship communication" between nations—essentially, nothing.[6]

What happened at Zeppelin was out of his hands. He was merely a convenient and rather pathetic figurehead. Why he stayed on when he could have claimed "illness" and retired—not that he did any work—was probably because he needed the money from his directorships and may have felt that it was his duty to help the country fight a war he loathed. It would prove an unfortunately compromising situation.[7]

He instead tried to repel the cult of Nazism from impinging on his daily life. Wagner he had always disliked, but he turned even against Beethoven, seeing him now as too nationalistic—or rather, used too often for nationalistic ends. He listened only to Mozart, "whose music, after all, is completely free" of politics.[8]

When he was obliged to travel to Peenemünde to watch a trial of Wernher von Braun's V-2 rocket in November 1942, he was much more interested in reading the works of Maxim Gorky, the Russian novelist, in order to try to understand the weirdness of his own Soviet experiences and feasting upon the contents of his Johanna-packed lunchbox ("rolls with roast veal, a roll with sausage and a fantastic egg . . . and a cheese roll") than he was in the liftoff and the rockets' military applications. All Eckener hoped for was to get it over with so he could go home.[9]

Throughout the war, as carefully expressed in his letters to his friends, which were read by the Gestapo, Eckener disguised his habitual sarcasm and irony with seemingly banal cheerleading and Hitler adoration to make it past the censors. "What good does it do to complain and to question?" he asked. "Now we just have to get through so we can end up in the realm of the promised thousand-year happiness and peace; otherwise, we'll be in a very bad way." After all, "a person thinks, but the Führer directs," so all would no doubt turn out well.[10]

If he had little to do at Zeppelin before, he had still less after the night of June 20–21, 1943, when the Royal Air Force bombed the Zeppelin factory producing the V-2 parts. (Eckener laconically commented that "it was a rather violent visit that brought a great deal of confusion.")[11] That marked the end of Zeppelin's contribution to the rocket program.

Neither did Friedrichshafen escape the attention of Bomber Command. The following year, in the lead-up to D-Day, when it was critical to hamper German tank production, a major raid by 322 Lancaster bombers occurred in late April. This time the damage was far worse, with some two-thirds of the town's industrial buildings destroyed.[12] The Zeppelin tank-gearbox factory was completely demolished.

Eckener's modest house sustained severe damage, with one bomb landing in the back garden and another near the front gate. Glass splinters and bomb fragments littered the interior. Eckener and Johanna had happened to be visiting their daughter Lotte in Constance at the time and so had not taken shelter in the cellar, which had caved in; they would surely both have perished.[13] From then on, Eckener and Johanna lived at Lotte's house.

On July 20, 1944, yet another raid, this time by the American 485th Bombardment Group, destroyed almost the entirety of whatever was left of the Zeppelin works. To Eckener, it was as if the whole history of the count's enterprise had been wiped from the record. Lotte noticed that her father for the first time looked "small and pale," though at least his gruffness soon returned. "He truly is a stoic, after all."[14]

The respite didn't last long. July 20 also happened to be the day on which Colonel Claus Schenk von Stauffenberg attempted to assassinate Hitler with a briefcase bomb at the Wolf's Lair headquarters. In the murderous aftermath, thousands of suspected anti-Nazis were rounded up and interrogated by the Gestapo, including Eckener, who was "interviewed" several times. He had known several of the plotters, though he had been uninvolved with the conspiracy.[15] Had Eckener been even tangentially privy to the plot, that would have been enough to prompt a show trial and a hanging—of seven thousand people arrested, nearly five hundred were executed—but he had cautiously avoided any and all politics since the 1930s.

After being released, Eckener waited for the war to end. In late April 1945, French troops entered Friedrichshafen. For a short time Eckener held out hope that the French occupying authorities would support the rebuilding of the Zeppelin facilities to help Germany's economic redevelopment. A Major Lasnier was willing to put Maybach back on its feet, but for the Zeppelin Company there was no prospect. The major said, in a remark harking back to the First World War, that "everyone in France regards the name Zeppelin as a symbol of Germany's warlike will."

Despite Eckener's decades-long insistence that the airship was a peaceful instrument, he had been done in again by the acts of others. Eckener himself appealed to the French that Zeppelin had not manufactured military equipment "for any length of time after the war began" and claimed, in a curious callback to Colsman's plan to make Zeppelin-branded aluminum pots and pans in 1918–19, that the company had focused instead on aluminum household goods.[16]

This was a desperate fib, and the French knew it. They ordered that anything left of the Zeppelin works was to be demolished, but the military government had many other things on its plate, and there was a delay.

Taking advantage of the unexpected reprieve, in February 1946 Eckener tried once again to save the company by admitting, belatedly, that the firm indeed had been a "subcontractor for arms manufacturers" but pleaded that Zeppelin was known throughout the world for bringing "understanding and cooperation among the nations." As a sop, he even suggested changing the name to the anodyne "Equipment and Vehicle Construction Company" if that would make the French feel better—but it was all to no good. The demolition order stayed.[17]

The following year, there was better news from America, where old friends like Meister, Litchfield, Arnstein, and Rosendahl were eager to see Eckener. Litchfield and Rosendahl arranged (as Eckener could hardly afford it) for him to travel to the United States for a six-month visit. The just-retired Rosendahl, whom even Eckener considered an airship obsessive— he would launch into long monologues on their virtues at the slightest opportunity—was working with Goodyear's Litchfield to resurrect the American airship program.[18]

To that end, Litchfield had recently published a book, rather plaintively titled *Why? Why Has America No Rigid Airships?*, and he and Rosendahl wanted Eckener to be their star attraction. Eckener was only too pleased to go, for it meant that he could send care packages back to his family from the Land of Plenty. He thought three or four months would be enough, as he and Johanna were "too old to want to live separated."

In early April 1947, Eckener, now aged almost eighty, began his trip to America, but before he departed he bade his farewell to Ludwig Dürr "in a deeply emotional way." His reason, unfortunately, is unclear, but the two may have had a political falling out, with Dürr refusing to see the reality that the Zeppelin Company would be, could be, no more.[19]

For the first time, Eckener was traveling by airplane to the United States, albeit by military transports rather than by Pan American. There was a layover in the Azores, which he had passed over so many times before, and he saw how far aviation had advanced since the 1930s. The Americans had at least twenty airplanes based there nowadays, as well as comfortable quarters and more food than Eckener had seen in years. After that, it was on to Newfoundland, where they again landed after a nine-hour flight, and then on to Connecticut, where a plane picked him up and took him to Lakehurst. It was quite a change from the old days.

Eckener took the time to visit New York to see Meister and Rosendahl and Jerome Hunsaker from PZT—and Karl von Wiegand, too, whom he'd known since the First World War. Wiegand had had a terrible Second one, having been interned with his *Graf Zeppelin*-adoring love, Lady Drummond-Hay, in a Japanese prisoner-of-war camp in the Philippines. When they were released in 1945, she was ill, and would die of coronary thrombosis several months later. Wiegand was still quite sprightly, traveling and churning out copy for Hearst (who remained friendly with Eckener), though he was almost totally blind.

Afterward, Eckener caught the plane to Akron—now just two hours away—to visit Litchfield and Karl Arnstein.[20] Eckener was happy again. He had plenty of food and good company, and it made a wonderful change to be free of his German troubles and to walk down the street and still be recognized by fans.

Then it was time for business. Back he went to New York for a jolly get-together with the old gang from AZT/IZT. Amid much chatter, there was excited talk about the Litchfield/Rosendahl plan and getting the financing together for a transatlantic airship line. While Eckener enjoyed himself, he doubted whether much, if anything, would come of it. He was right about that. A scheduled meeting in Washington with apparently interested politicians kept getting bumped, and the buzz, whatever there was of it, slowly died away.[21] AZT/IZT would be formally dissolved in June 1950, though for some reason its long-dormant Pacific counterpart, PZT—perhaps Litchfield and Hunsaker couldn't bear to let it go—lingered on until 1957.[22]

Eckener stayed a few months longer, "just to accumulate some more funds, which can be used to send packages out."[23] But he was getting homesick, and he knew he would never see his friends again.

Just before Christmas 1947, he came home to Germany, only to be confronted by one of the most horrible experiences of his life. On Christmas Eve, there was news that as part of the Allies' denazification process—intended to find, expose, "cleanse," and punish those closely identified with Nazism—he, Hugo Eckener, had been classified as a "beneficiary" of the regime owing to his directorships, his status as a *Wehrwirtschaftsführer* (leader of the war economy), and his chairmanship of Zeppelin.[24]

The punishment was a severe one. He was fined 100,000 reichsmarks, prohibited from all managerial activity for five years, and forbidden to vote or to hold elected office for the same period.[25]

To gain some idea of the absurd magnitude of the fine, which Eckener had nowhere near the means to ever pay—he had savings of 35,000 reichsmarks, and lived on 300 a month—consider that Ernst Heinkel, a fellow *Wehrwirtschaftsführer* but one who was an active member of the Nazi Party, the manufacturer of the Luftwaffe's eponymous bombers, an amasser of a colossal fortune, and an eager employer of slave labor, was fined 2,000 reichsmarks.[26]

Eckener was given no reason for the decision, and the mortal embarrassment and humiliation shocked him into silence for several weeks, but there was still hope. Denazification was often clumsily or randomly imposed or prompted by political reasons (Eckener suspected that Communists in the state government had put the French up to it), and a small fish could be hooked while many a big fish swam away. A decision could also be reversed with "positive evidence of anti-Nazi activity," and with that Eckener was handily endowed.[27]

Over the coming months, the newspapers were full of letters from readers indignantly pointing out Eckener's well-known opposition to Nazism. People, including tortured resistance leaders, sent notes to him saying the charge against him was laughable, and there was a letter, signed by a large number of eminent politicians, publishers, lawyers, scientists, and professors backing him to the hilt. Eckener himself wrote to the German state government and the French military authorities asking for clarification and citing all the episodes in the 1930s in which he had found himself in hot water with the Nazis.

On July 14, 1948, the judgment was completely rescinded by a U.S. tribunal, which found that Eckener had not been "incriminated" and had in fact "wanted nothing to do with the politics of National Socialism, but rather lived only for his business ventures and the old tradition of airship construction."[28] The German Federal Republic gave him the Order of Merit to make up for the affront.[29]

The news of his victory, which came a few weeks before Eckener's eightieth birthday, was a cause for necessarily modest celebration. He and Johanna were still living at his daughter's house, and no one had much money. Eckener had been reduced to begging Meister to send him spaghetti, coffee, shrimp, and cooked ham from the United States to help him through, and his birthday gifts consisted of flowers from the garden, some ancient cigars, and a little bowl or vase from the back of someone's cupboard. But he received three hundred letters and sixty telegrams from well-wishers. He was still remembered.[30]

Two years later, Eckener and Johanna were finally able to return to a home of sorts in Friedrichshafen, only for Johanna to suffer a stroke in October 1951. She continued to experience paralysis for many months afterward and had another stroke at the end of 1952, rendering her almost unable to speak and see, while Eckener despaired.[31]

On his eighty-fifth birthday in 1953, when former Zeppelin employees honored him with a traditional torchlit procession, Eckener surprised them by admitting publicly for the first time the obvious truth that he had lost the battle against Trippe and his airplanes. "You all must finally come to terms with the fact that the era of the airship is over for good," he told the consternated crowd. "If you expect that the airship will get a chance again, then it would be the same as if you wanted to stop the automobile traffic on the roads in order to reinstate the covered wagon." Even the safety of the airplane these days, he added, was the same as that of the airship, and as for speed and technological jumps, it wasn't even a competition.[32]

Eckener himself was weakening quickly. He gardened a little, as he had been doing when the count first visited him half a century before, listened to Mozart, read Goethe, and received the occasional guest at his (as The New York Times put it in a "Whatever Happened To . . . ?" type column) "shabby pavilion on the lake in Friedrichshafen." To these guests, he always said on their departure, "Look on this visit as your parting from me."[33]

A year later, on August 14, 1954, four days after his eighty-sixth birthday, Eckener died of heart failure (but was outlived, to many people's surprise, by Johanna, who'd staged a marvelous recovery). Described by The New

York Times as the "*Graf Zeppelin* Pilot"—which Eckener would have appreciated—he was interred next to six of the crewmen who had perished aboard his *Hindenburg*.[34]

WHEN HE HAD addressed the former Zeppelin workers on his eighty-fifth birthday, Eckener had already seen the future: jet-powered airplanes. But he died just fourteen months before Trippe ordered the first of what would become a 128-*strong* fleet of Boeing 707s—capable of 600 mph—for Pan American's Atlantic service. Carrying around 135 passengers, flying at an altitude of 32,000 feet, enjoying a 4,000-mile range, and with a maximum takeoff weight of some 168 tons, Trippe's Boeings were far beyond anything remotely imaginable in the golden years of the 1920s and 1930s.

Now they transported carefree tourists and busy executives across an ocean few remembered was once an unbridgeable moat—except by airship. By mid-1962, Pan American had completed no fewer than a hundred thousand nonstop transatlantic flights, and to mark his airline's global domination Trippe moved the company the following year to a new fifty-nine-story skyscraper towering above Grand Central Station with a gigantic "Pan Am" sign at the top. In 1964, Trippe retired as president and finally left Pan American four years later. He would die in 1981, just a decade before his airline did in bankruptcy court.

BY THEN, THAT there was once an age of the airship, a time when these silver giants had freely roamed the skies, had been almost completely forgotten. Hugo Eckener, too, for decades one of the most famous men in the world and counted as Lindbergh's equal, had vanished from the collective memory.

The omnipresence of the airplane was, and is, so overwhelming that it's easy to forget that it even had a competitor or that its rise was not necessarily inevitable. One could say the same, perhaps, about the internal combustion engine, which in the earliest days of the automobile was considered on a par with steam-driven engines and battery-powered motors. It required a certain concatenation of decisions, luck, and circumstances to emerge the victor, but nothing is permanent. Today the electric car is presenting a challenge to the internal combustion engine's once-unquestioned superiority.

Occasionally, there are similar attempts to revive the airship business.

Benefiting from new, advanced materials and improved design, a future airship would be, as Eckener's had been, extraordinarily energy-efficient, especially in terms of cargo-lifting. Some believe its modern incarnation would be ideal for bringing humanitarian aid to remote, war-ravaged locations.

After all these years, it would be nice to think of the airship as an emissary of peace and a bridger of national divides, as a young Count von Zeppelin and an old Hugo Eckener had hoped.

Acknowledgments

W̲RITING AN AIRSHIP-SIZED book like this one requires the contribution of many individuals. I would like to thank, in particular, Dr. John D. Anderson for his technical advice on aerodynamics, John Provan for an illuminating discussion of airship history, and Patrick Russell, who rendered invaluable assistance in correcting several points in my discussion of the last minutes of the *Hindenburg*. I must also include in that list Mark A. Heald, for sending me information regarding his father's witnessing of the disaster, as well as Alvaro Bellon, trustee of the Lighter-Than-Air Society of Akron, Ohio, and Professor Guillaume de Syon of Albright College, who provided copies of several hard-to-find articles. Any errors are mine, I should add.

Pamela Elbe, of the National Museum of American Jewish Military History, helped with some details about the Jewish War Veterans of the United States (JWV) organization, while Albrecht Graf von Brandenstein-Zeppelin clarified the count's religious views.

Of key importance were Karen Caruana and Karen Laakko, who heroically (and accurately and quickly) translated many of the longer or more complex German texts, as well as David Pfeiffer, who at the National Archives undertook for me the task of digitizing thousands of pages of testimony from the 1937 Board of Inquiry into the loss of the *Hindenburg*. And I should not forget Debbie Saracini and the staff of the San Diego Air & Space Museum for their assistance in procuring files from the Henry Cord Meyer Collection, and Patrizia Nava, curator of Aviation Archives at the Eugene McDermott Library, University of Texas at Dallas, who provided copies of important documents from the Charles Rosendahl, Hans von Schiller, and Douglas Robinson papers.

Molly Turpin of Random House shouldered the tremendous burden, in terms of both effort and time, of editing, weeding, and taming an overlong and digressive manuscript, for which hard labor and dedication she must be awarded highest honors, while Martin Schneider performed sterling service in copyediting the text, improving prose, and asking good questions. Because the making of a book is a team effort, recognition, too, is due to production editor Evan Camfield, production manager Jennifer Backe, designer Simon Sullivan, cover designer Pete Garceau, and art director Anna Bauer.

A thousand thanks should be extended to ace agent Eric Lupfer of Fletcher & Co., who instantly said yes to my offhand question ("Hey, what about a book about the *Hindenburg*?") and diligently piloted it from proposal to final draft. He remains a rock of sound advice, excellent judgment, and admirable dedication.

Family often comes last in the acknowledgments, though really they ought to be first. My late parents, Paul and Susan Rose, would have loved to see this book, and I dearly wish they could have. I'd like to thank my brother, Ari, and my sister, Zoë (and my brother-in-law, Craig), for their support over an occasionally trying couple of years, as well as my extended family (Liz and Chad, Ben and Jaime, Erna, David and Carolyn, and Freya). Special mentions, too, for Yvonne Taylor and William Liberman for all that they have done.

Pride of place, of course, goes to my beloved wife, Rebecca, the best partner one could imagine—she is my north star. My fine son, Edmund, though he's mostly a car guy, has also learned to appreciate the grandeur of the airship, and I'm beyond proud of him, and all that he shall accomplish.

Bibliography

Contemporary Newspapers and Magazines

Aircraft

Aircraft Journal

The Airship

American Economic Review

American Gas Engineering Journal

Annals of the American Academy of Political
and Social Science

Architectural Forum

Atlanta Constitution

Aviation

Aviation and Aeronautical Engineering

Aviation and Aircraft Journal

Baltimore American

Bemidji Daily Pioneer

Boston Daily Globe

The Century

Chicago Daily Tribune

Cincinnati Enquirer

Collier's

The Colonist

Crawford County Bulletin

Crockery & Glass Journal

Current Literature

Current Opinion

Daily Mail

Electrical Review

The Engineer

Everybody's Magazine

The Fatherland

Flight

Fortune

The Forum

Geographical Journal

Hampton's Magazine

Harper's Bazaar

Harper's Magazine

The Independent

The Inter-Ocean

Irish Times

Jewish Daily Bulletin

Journal of the American Society for Naval
Engineers

Journal of the Franklin Institute

Journal of the Royal Aeronautical Society

Journal of the Royal Society of Arts

The Journal of the Society of Arts

Journal of the Society of Automotive
Engineers

Kansas City Times

Life

Literary Digest

The Living Age

Los Angeles Times

Machinery

Manchester Guardian

McClure's Magazine

Milwaukee Daily Sentinel

Munsey's Magazine

Nashville American
National Geographic
Nature
The New Yorker
New York Herald Tribune
The New York Times
New York Tribune
North American Review
The Outlook
Popular Mechanics
Popular Science
Proceedings of the American Academy of Arts and Sciences
Proceedings of the American Philosophical Society
Railway Age
Reader's Digest

Royal United Services Institution Journal
St. Louis Post-Dispatch
Science
Science News-Letter
Scientific American
The Scientific Monthly
The Scots Journal
Sigma Xi Quarterly
Smart Set
Time
The Times (London)
U.S. Air Service
Washington Herald
The Washington Post
The World's Work
The Youth's Companion

Archives

Air Ministry Papers (Public Record Office, London)

Charles Rosendahl Papers (Eugene McDermott Library, University of Texas, Dallas)

Douglas Robinson Papers (Eugene McDermott Library, University of Texas, Dallas)

E. B. Ashmore Papers (Imperial War Museum, London)

Hans von Schiller Papers (Eugene McDermott Library, University of Texas, Dallas)

Henry Cord Meyer Papers (San Diego Air & Space Museum, California)

Record Group 197 (Records of the Civil Aeronautics Board), Bureau of Air Commerce (Records Relating to the Hindenburg Disaster, 1937), "Transcripts of Testimony" (A1/Entry 3), National Archives, College Park, Maryland

Sources

Ackroyd, J.A.D. "Sir George Cayley: The Invention of the Aeroplane near Scarborough at the Time of Trafalgar." Journal of Aeronautical History 1 (2011): 130–81.

Alexander, J. T. "Aeromania, 'Fire-Balloons,' and Catherine the Great's Ban of 1784." The Historian 58 (1996), no. 3: 497–516.

Allen, H. The Story of the Airship. Goodyear Tire & Rubber Co., 1931.

Anderson, J. D., Jr. The Airplane: A History of Its Technology. Reston, Va.: American Institute of Aeronautics and Astronautics, 2002.

Anon. America Builds: The Record of the Public Works Administration. Washington, D.C.: Division of Information, 1939.

Anon. "The Intrepid Professor Lowe." Civil War Monitor (Winter 2015): 20–21.

Archbold, R., and K. Marschall. Hindenburg: An Illustrated History. New York: Barnes & Noble (orig. Madison Press), 1997.

Armstrong, W. A. "William A. Moffett and the Development of Naval Aviation." In Aviation's Golden Age: Portraits from the 1920s and 1930s, edited by W. M. Leary, 60–73. Iowa City: University of Iowa Press, 1989.

Baldwin, M. *With Brass and Gas: An Illustrated and Embellished Chronicle of Ballooning in Mid-Nineteenth Century America*. Boston: Beacon Press, 1967.

Ballantyne, R. M. *Up in the Clouds, or Balloon Voyages*. London: James Nisbet & Co., 1880.

Ballendorf, D. A. "Secrets Without Substance: U.S. Intelligence in the Japanese Mandates, 1915–1935." *Journal of Pacific History* 19 (1984), no. 2: 83–99.

Ballendorf, D. A., and M. L. Bartlett. *Pete Ellis: An Amphibious Warfare Prophet, 1880–1923*. Annapolis, Md.: Naval Institute Press, 1997.

Barr, W., and E. B. Baldwin. "The First Tourist Cruise in the Soviet Arctic." *Arctic* 33 (1980), no. 4: 671–85.

Bartholomew, R. E. "Michigan and the Great Mass Hysteria Episode of 1897." *Michigan Historical Review* 24 (1998), no. 1: 133–41.

Bauer, M., and J. Duggan. *LZ-130 Graf Zeppelin and the End of Commercial Airship Travel*. Friedrichshafen, Germany: Zeppelin-Museum, 1996.

Beasley, D. *Who Really Invented the Automobile? Skulduggery at the Crossroads*. Simcoe, Ontario: Davus Publishing, 1997.

Beaty, D. *The Water Jump: The Story of Transatlantic Flight*. New York: Harper & Row, 1976.

Belafi, M. *The Zeppelin*. Barnsley, U.K.: Pen & Sword, 2015.

Bender, M., and S. Altschul. *The Chosen Instrument: Juan Trippe, Pan Am—the Rise and Fall of an American Entrepreneur*. New York: Simon and Schuster, 1982.

Bentele, E. *The Story of a Zeppelin Mechanic: My Flights, 1931–1938*. Translated by Simon Dixon. Friedrichshafen, Germany: Zeppelin-Museum, 1992.

Berson, M. J., K. T. Carano, and C. T. Sheffield. "Steam Man and Airships: Technology of the Future in the Past." *Social Education* 72 (2008), no. 3: 124.

Bilstein, R. E. *Flight Patterns: Trends of Aeronautical Development in the United States, 1918–1929*. Athens: University of Georgia Press, 1983.

———. "Edward Pearson Warner and the New Air Age," in *Aviation's Golden Age: Portraits from the 1920s and 1930s*, edited by W. M. Leary, 113–26. Iowa City: University of Iowa Press, 1989.

———. *Flight in America: From the Wrights to the Astronauts*. Baltimore: Johns Hopkins University Press, 2001.

Black, J. *Air Power: A Global History*. Lanham, Md.: Rowman & Littlefield, 2016.

Blanchard, J.-P. *The Principles, History, and Use of Air-Balloons*. New York: J. Fellows, 1796.

Bloor, D. *The Enigma of the Aerofoil: Rival Theories of Aerodynamics, 1909–1930*. Chicago: University of Chicago, 2011.

Botting, D. *Dr. Eckener's Dream Machine: The Great Zeppelin and the Dawn of Air Travel*. New York: Owl Books, 2001.

Bowdoin Van Riper, A. *Imagining Flight: Aviation and Popular Culture*. College Station, Tx.: Texas A&M University Press, 2004.

Bradshaw, P., Jr. "The Role of Technology in the Failure of the Rigid Airship as an Invention," Ph.D. diss., University of Florida, 1975.

Braun, H. "Das 'Wundergas' Helium, Die US-Amerikanische Innenpolitik Und Die Deutschen Zeppeline." *Vierteljahrshefte für Zeitgeschichte* 53 (2005), no. 4: 571–600.

Brittain, J. E. "The International Diffusion of Electrical Power Technology, 1870–1920." *Journal of Economic History* 34 (1974), no. 1: 108–121.

Brodherson, D. "'An Airport in Every City': The History of American Airport Design." In *Building for Air Travel: Architecture and Design for Commercial Aviation*, edited by J. Zukowsky, 67–95. Chicago/Munich: Art Institute of Chicago/Prestel-Verlag, 1996.

Brooks, P. W. *Zeppelin: Rigid Airships, 1893–1940*. Washington, D.C.: Smithsonian Institution Press, 1992.

Brown, W. E., Jr. "Pan Am: Miami's Wings to the World." *The Journal of Decorative and Propaganda Arts* 23 (1998): 144–61.

Bruce-Briggs, B. *The Shield of Faith: A Chronicle of Strategic Defense from Zeppelins to Star Wars.* New York: Simon & Schuster, 1988.

Bryant, L. "The Silent Otto." *Technology and Culture* 7, no. 2 (1966): 184–200.

Bureau of the Census. *Historical Statistics of the United States, 1789–1945.* Washington, D.C.: United States Department of Commerce, 1949.

Burleigh, M. *Death and Deliverance: Euthanasia in Germany, 1900–1945.* Cambridge, U.K.: Cambridge University Press, 1994.

Burns, R., ed. *Radar Development to 1945.* London: Institution of Engineering and Technology, 1988.

Cameron, J. D. "To Transform the Revolution into an Evolution: Underlying Assumptions of German Foreign Policy Toward Soviet Russia, 1919–27." *Journal of Contemporary History* 40 (2005), no. 1: 7–24.

Campbell, E. *Zeppelins: The Past and the Future.* St. Albans, U.K.: Campfield Press, 1918.

Castle, H. A. *History of St. Paul and Vicinity.* Chicago and New York: Lewis Publishing Co., 1912.

Chanute, O. *Aerial Navigation: A Lecture Delivered to the Students of Sibley College, Cornell University.* New York: Railroad and Engineering Journal, 1891.

Christopher, J. *Transatlantic Airships: An Illustrated History.* Ramsbury, U.K.: Crowood Press, 2010.

Clarke, B. *Atlantic Adventure: A Complete History of Transatlantic Flight.* London: Allan Wingate, 1958.

Clarke, D. "Scareships over Britain: The Airship Wave of 1909." *Fortean Studies* 6 (1999): 39–63.

Clarke, I. F. *Voices Prophesying War: Future Wars, 1763–3749.* 2nd ed. Oxford: Oxford University Press, 1992.

———. *The Great War with Germany, 1890–1914: Fictions and Fantasies of the War-to-Come.* Liverpool, U.K.: Liverpool University Press, 1997.

Cochrane, D., V. Hardesty, and R. Lee. *The Aviation Careers of Igor Sikorsky.* Washington, D.C.: National Air and Space Museum, 1989.

Colsman, A. *Luftschiff Voraus! Arbeit und Erleben am Werke Zeppelins.* Stuttgart: Deutsche Verlags-Anstalt, 1933.

Confino, A. *The Nation as a Local Metaphor: Württemberg, Imperial Germany, and National Memory, 1871–1918.* Chapel Hill, N.C./London: University of North Carolina Press, 1997.

Corn, J. J. *The Winged Gospel: America's Romance with Aviation.* Baltimore: Johns Hopkins University Press, 2002.

Courtwright, D. T. *Sky as Frontier: Adventure, Aviation, and Empire.* College Station, Tx.: Texas A&M University Press, 2005.

Crouch, T. D. *The Eagle Aloft: Two Centuries of the Balloon in America.* Washington, D.C.: Smithsonian Institution Press, 1983.

———. *A Dream of Wings: Americans and the Airplane, 1875–1905.* New York: W. W. Norton & Co., 2002.

———. *Lighter Than Air: An Illustrated History of Balloons and Airships.* Baltimore: Johns Hopkins University Press, 2009.

Cunningham, P., ed. *The Letters of Horace Walpole, Fourth Earl of Oxford.* London: Richard Bentley and Sons, 1891.

Daley, R. *An American Saga: Juan Trippe and His Pan Am Empire.* New York: Random House, 1980.

Davies, R.E.G. *Airlines of the United States Since 1914*. Rev. ed. Washington, D.C.: Smithsonian Institution Press, 1982.

———. *Pan Am: An Airline and Its Aircraft*. New York: Orion Books, 1987.

Davison, P. "The R.101 Story: A Review Based on Primary Source Material and First Hand Accounts." *Journal of Aeronautical History* 5 (2015): 43–167.

Denazification. Report of the Military Governor (1 April 1947–30 April 1948), No. 34. Office of Military Government for Germany (U.S.).

De Syon, G. "Faded Memories: The Wright Brothers and Germany, 1909–1913." Undated lecture.

———. "Searching for the German Hero: Biographies of Count Zeppelin, 1908–1938." In *Memory, History and Critique: European Identity at the Millennium* (Proceedings of the Fifth ISSEI Conference at the University for Humanist Studies, The Netherlands, August 1996), edited by F. Brinkhuis and S. Talmor. Cambridge, Mass.: MIT Press, 1998.

———. "Turning a National Icon into a Business Tool: The Zeppelin Airship, 1924–1933." *Proceedings of the Third International Airship Convention*, edited by G. Khoury. London: The Airship Association, 2000.

———. "Bangs and Whimpers: The German Public and Two Zeppelin Disasters, 1908–1937." In *Ordinary Reactions to Extraordinary Events*, edited by R. B. Browne and A. G. Neal, 172–81. Bowling Green, Oh.: Popular Press, 2001.

———. "Airplane or Airship? The Cultural Dimensions of a Transatlantic Debate." Paper, 4th International Airship Convention and Exhibition (2002).

———. *Zeppelin!: Germany and the Airship, 1900–1939*. Baltimore: Johns Hopkins University Press, 2002.

———. "What the Wrights Wrought: The Centennial of Flight in Recent Literature." *Technology and Culture* 45 (2004), no. 2: 350–57.

———. "Is it Really Better to Travel Than to Arrive?: Airline Food as a Reflection of Consumer Anxiety." In *Food for Thought: Essays on Eating and Culture*, edited by L. C. Rubin. Jefferson, N.C.: McFarland & Co., 2008.

———. "Toys or Tools? The Riddle of French Army Airships, 1884–1914." Paper, 8th International Airship Convention (2010).

Dewey, R. L. "The Merchant Marine Act of 1936." *American Economic Review* 27 (1937), no. 2: 240–52.

Dick, H. G., and D. H. Robinson. *The Golden Age of the Great Passenger Airships: Graf Zeppelin and Hindenburg*. Washington, D.C.: Smithsonian Institution Press, 1985.

Dienel, L., and M. Schiefelbusch. "German Commercial Air Transport Until 1945." *Revue Belge de Philologie et d'Histoire* 78 (2000), no. 3–4: 945–67.

Dittmann, F. "Innovations in the Electric Energy System." *Icon* 13 (2007): 62–85.

Dodd, W. E., Jr., and M. Dodd, eds. *Ambassador Dodd's Diary, 1933–1938*. London: Victor Gollancz, 1945.

Dooley, S. "The Development of Material-Adapted Structural Form." Ph.D. diss., École Polytechnique Fédérale de Lausanne, Switzerland (2004).

D'Orcy, L., ed. *D'Orcy's Airship Manual: An International Register of Airships with a Compendium of the Airship's Elementary Mechanics*. New York: Century Co., 1917.

Dornberger, Walter. *V-2: The Inside Story of Hitler's "Secret Weapon" and the Men and Events Behind It That Almost Changed the Course of History*. Translated by J. Cleugh and G. Halliday. The Viking Press, 1955.

Douglas, D. G. "Airports as Systems and Systems of Airports: Airports and Urban Development in America Before World War Two." In *From Airships to Airbus: The History of Civil and Commercial Aviation*, edited by W. M. Leary and W. F. Trimble, 55–84. Washington, D.C.: Smithsonian Institution Press, 1995.

Duggan, J. *Olympia-Fahrt 1936*. Ickenham, Middlesex, U.K.: Zeppelin Study Group, 2000.

———. *LZ-129 Hindenburg: The Complete Story*. Ickenham, Middlesex, U.K.: Zeppelin Study Group, 2002.

———. *Airships in the Arctic*. Ickenham, U.K.: Zeppelin Study Group, 2006.

Duggan, J., and J. Graue. *Commercial Zeppelin Flights to South America: The Commercial South America Flights and Airmails of the Zeppelin Airships*. Valleyford, Wash.: JL Diversified, 1995.

Duggan, J., and H. C. Meyer. *Airships in International Affairs, 1890–1940*. Basingstoke, U.K.; New York: Palgrave, 2001.

Dunn, M. B., R. R. Gilman, and F. von Zeppelin. "Zeppelin in Minnesota: The Count's Own Story." *Minnesota History* 40 (1967), no. 6: 265–78.

Dürr, L. *Twenty-five Years of Zeppelin Airship Construction*. Translated by A. Reid. Berlin, 1924; Lulu Publishing, 2013.

Dyakov, Y., and T. Bushuyeva. *The Red Army and the Wehrmacht: How the Soviets Militarized Germany, 1922–33, and Paved the Way for Fascism*. Amherst, New York: Prometheus Books, 1995.

Dyck, H. L. "German-Soviet Relations and the Anglo-Soviet Break, 1927." *Slavic Review* 25 (1966), no. 1: 67–83.

Eckener, H. "The First Airship Flight Around the World." *National Geographic* 57 (1930), no. 6: 653–88.

———. *Count Zeppelin: The Man and His Work*. Translated by L. Farnell. Berlin/London: Massie Publishing Company, 1938.

———. *Im Zeppelin über Länder und Meere*. Flensburg: Verlagshaus Christian Wolff, 1949.

———. *My Zeppelins*. Translated by D. Robinson. London: Putnam, 1958.

Elimination of German Resources for War. Hearings Before a Subcommittee of the Committee on Military Affairs, United States Senate, Part 5, Testimony of Treasury Department, July 2, 1945. Washington, D.C.: Government Printing Office, 1945.

Ellsworth, L., and E. H. Smith. "Report of the Preliminary Results of the Aeroarctic Expedition with *Graf Zeppelin*." *Geographical Review* 22 (1932), no. 1: 61–82.

Evaluation Report 38. "Interview with Dr. Eckener." May 25, 1945. Combined Intelligence Objectives Sub-Committee. Washington, D.C.: Office of the Publication Board, Department of Commerce, 1945.

Evans, R. J. *The Third Reich in Power*. New York: Penguin Books, 2006.

———. *The Third Reich in History and Memory*. Oxford: Oxford University Press, 2015.

Fitzpatrick, J., ed. *The Writings of George Washington: From the Original Manuscript Sources, 1745–1799*. Washington, D.C.: Government Printing Office, 1938.

Foss, C. "Russia's Romance with the Airship." *History Today*, December 1997, 10–16.

Freedman, A. "Zeppelin Fictions and the British Home Front." *Journal of Modern Literature* 27 (2004), no. 3: 47–62.

Freud, S. *Introductory Lectures on Psycho-Analysis*. London: Allen & Unwin, 1929.

Fritzsche, P. A *Nation of Fliers: German Aviation and the Popular Imagination*. Cambridge, Mass.: Harvard University Press, 1992.

Fromm, B. *Blood and Banquets: A Berlin Diary, 1930–1938*. New York: Simon and Schuster, 1992.

Fulton, G. "Helium Through World Wars I and II." *Naval Engineers Journal* 77 (1965), no. 5: 733–38.

Gandt, R. L. *China Clipper: The Age of the Great Flying Boats.* Annapolis, Md.: Naval Institute Press, 1991.

Ganz, C. R. "Aviation, Nationalism, and Progress." In C. R. Ganz, *The 1933 Chicago World's Fair: A Century of Progress,* 137–50. Urbana/Chicago, Illinois: University of Illinois Press, 2008.

Gardner, L. D. "Aboard the *Hindenburg.*" *Buoyant Flight* 9 (1962), no. 4: 2–5.

Gaudi, R. *African Kaiser: General Paul Von Lettow-Vorbeck and the Great War in Africa, 1914–1918.* New York: Caliber, 2017.

Gaulin, K. "The Flying Boats: Pioneering Days to South America." *The Journal of Decorative and Propaganda Arts* 15 (1990): 78–95.

Gibbs-Smith, C. H. "Sir George Cayley: 'Father of Aerial Navigation' (1773–1857)." *Notes and Records of the Royal Society of London* 17 (1962), no. 1: 35–56.

———. *The Invention of the Aeroplane, 1799–1909.* New York: Taplinger Publishing Co., 1966.

Gillespie, R. "Ballooning in France and Britain, 1783–1786: Aerostation and Adventurism." *Isis* 75 (1984), no. 2: 249–68.

Gilman, R. R. "Zeppelin in Minnesota: A Study in Fact and Fable." *Minnesota History* 39 (1965), no. 7: 278–85.

———. "Count Von Zeppelin and the American Atmosphere." *Smithsonian Journal of History* 3 (1968), no. 1: 29–40.

Gispen, K. "Engineers in Wilhelmian Germany: Professionalization, Deprofessionalization, and the Development of Nonacademic Technical Education." In *German Professions, 1800–1950,* edited by G. Cocks and K. H. Jarausch, 104–22. New York: Oxford University Press, 1990.

Glines, C. V., ed. *Lighter-Than-Air Flight.* New York: Franklin Watts, 1965.

Goldbeck, G. "Entwicklungsstufen des Verbrennungsmotors." *Motortechnische Zeitschrift* 23 (1962): 76–80.

Goldsmith, M. *Zeppelin: A Biography.* New York: William Morrow & Company, 1931.

Gollin, A. "England Is No Longer an Island: The Phantom Airship Scare of 1909." *Albion* 13 (1981), no. 1: 43–57.

———. *No Longer an Island: Britain and the Wright Brothers, 1902–1909.* Stanford, Calif.: Stanford University Press, 1984.

Gordon, A. *Naked Airport: A Cultural History of the World's Most Revolutionary Structure.* Chicago: University of Chicago Press, 2008.

Graham, M.B.W. "R&D and Competition in England and the United States: The Case of the Aluminum Dirigible." *The Business History Review* 62, no. 2 (1988): 261–85.

Gregor, N. *Daimler-Benz in the Third Reich.* New Haven, Conn.: Yale University Press, 1998.

Griffin, J. *Glass Houses and Modern War.* London: Chatto & Windus, 1938.

Grossman, D., C. Ganz, and P. Russell. *Zeppelin Hindenburg: An Illustrated History of LZ-129.* Stroud, Gloucestershire, U.K.: The History Press, 2017.

Gunga, H.-C.. *Nathan Luntz: His Life and Work in the Fields of High-Altitude Physiology and Aviation Medicine.* New York: Academic Press, 2008.

Hacker, G. *Die Männer von Manzell.* Frankfurt am Main: Societäts-Druckerei, 1936.

Haigh, R. H., D. S. Morris, and A. R. Peters. *German-Soviet Relations in the Weimar Era: Friendship from Necessity.* Totowa, N.J.: Barnes & Noble, 1985.

Hallion, R. P. "Daniel and Harry Guggenheim and the Philanthropy of Aviation." In *Aviation's Golden Age: Portraits from the 1920s and 1930s,* edited by W. M. Leary, 18–34. Iowa City: University of Iowa Press, 1989.

———. *Taking Flight: Inventing the Aerial Age, from Antiquity Through the First World War.* New York: Oxford University Press, 2003.

Hansen, J. R. *The Bird Is on the Wing: Aerodynamics and the Progress of the American Airplane*. College Station, Tx.: Texas A&M University Press, 2004.

Hansen, J. R., D. B. Taylor, J. Kinney, and J. L. Lee, eds. *The Wind and Beyond: A Documentary Journey into the History of Aerodynamics in America*. Washington, D.C.: National Aeronautics and Space Administration, 2003.

Hartcup, G. *The Achievement of the Airship: A History of the Development of Rigid, Semi-Rigid, and Non-Rigid Airships*. North Pomfret, Vt.: David & Charles, 1974.

Harvey, A. D. "'Against London': A Zeppelin Officer's Account." *Air Power History* (summer 2010): 14–21.

Hawley, E. "Three Facets of Hooverian Associationalism: Lumber, Aviation, and Movies, 1921–1930." In *Regulation in Perspective: Historical Essays*, edited by T. McCraw, 95–123. Cambridge, Mass.: Harvard University Press, 1981.

Hedin, R. *The Zeppelin Reader: Stories, Poems, and Songs from the Age of Airships*. Iowa City: University of Iowa Press, 1998.

Heppenheimer, T. A. *Turbulent Skies: The History of Commercial Aviation*. New York: John Wiley & Sons, 1995.

Herz, J. H. "The Fiasco of Denazification in Germany." *Political Science Quarterly* 63 (1948), no. 4: 569–94.

Hiam, C. M. *Dirigible Dreams: The Age of the Airship*. Lebanon, N.H.: ForeEdge, 2014.

Higham, R. *The British Rigid Airship, 1908–1931: A Study in Weapons Policy*. London: G. T. Foulis, 1961.

Hildebrandt, A. *Airships Past and Present*. Translated by H. W. Story. New York: D. Van Nostrand Co., 1908.

Holanda, R. *A History of Aviation Safety: Featuring the U.S. Airline System*. Bloomington, Ind.: AuthorHouse, 2009.

Holland, M. *Architects of Aviation*. New York: Duell, Sloan and Pearce, 1951.

Holman, B. "The Phantom Airship Panic of 1913: Imagining Aerial Warfare in Britain Before the Great War." *Journal of British Studies* 55 (2016): 99–119.

Holmes, R. *Falling Upwards: How We Took to the Air*. New York: Pantheon Books, 2013.

Hopkins, G. E. *Pan Am Pioneer: A Manager's Memoir from Seaplane Clippers to Jumbo Jets*. Lubbock, Tx.: Texas Tech University Press, 1995.

Hopkins, S. "Daredevil Balloonist John Steiner: America's Champion Aeronaut in the Civil War." *Military Images* 33 (2015), no. 4: 42–44.

Hühne, M. C. *Pan Am History, Design, and Identity*. Berlin: Callisto Publishers, 2016.

Ickes, H. L. *The Secret Diary of Harold L. Ickes*. New York: Simon and Schuster, 3 vols., 1954.

Italiaander, R. *Ferdinand Graf von Zeppelin: Reitergeneral, Diplomat, Luftschiffpionier: Bilder und Dokumente*. Konstanz, Germany: Verlag Friedr. Stadler, 1980.

——. *Ein Deutscher Namens Eckener: Luftfahrtpionier und Friedenspolitiker: Vom Kaiserreich bis in Die Bundesrepublik*. Konstanz, Germany: F. Stadler, 1981.

Jackson, D. D. *Flying the Mail*. Alexandria, Va.: Time-Life Books, 1982.

Jakab, P. L. *Visions of a Flying Machine: The Wright Brothers and the Process of Invention*. Washington, D.C.: Smithsonian Institution Press, 1990.

Jay, W., ed. *The Life of John Jay: With Selections from His Correspondence and Miscellaneous Papers*. 2 vols. New York: J.&J. Harper, 1833.

Jones, L. E. "Sammlung Oder Zersplitterung? Die Bestrebungen zur Bildung einer neuen Mittelpartei in der Endphase der Weimarer Republik 1930–1933." *Vierteljahrshefte für Zeitgeschichte* 25 (1977), no. 3: 265–304.

Josephson, M. *Empire of the Air: Juan Trippe and the Struggle for World Airways*. New York: Harcourt, Brace and Company, 1944.

Kaempffert, W. "How It Feels to Fly: From Châlons to Vincennes in a Biplane." *The World's Work* 16 (1910): 515–18.

Kalabash Wordsmith, R. (pseud.). "Notes on the International Zeppelin Transport Co. and Pacific Zeppelin Transport Co." *Buoyant Flight* 39 (1992), no. 4: 2, 4–5.

———. "Notes on the International Zeppelin Transport Co. and Pacific Zeppelin Transport Co." *Buoyant Flight* 39 (1992), no. 5: 4–5, 8.

Kauffman, S. B., and G. E. Hopkins. *Pan Am Pioneer: A Manager's Memoir from Seaplane Clippers to Jumbo Jets*. Lubbock, Tx.: Texas Tech University Press, 1995.

Keen, P. "The 'Balloonomania': Science and Spectacle in 1780s England." *Eighteenth-Century Studies* 39 (2006), no. 4: 507–35.

Kelly, C. J. *The Sky's the Limit: The History of the Airlines*. New York: Coward-McCann, 1963.

Kelly, M. *Steam in the Air: The Application of Steam Power in Aviation During the 19th and 20th Centuries*. Barnsley, U.K.: Pen & Sword, 2006.

Kershaw, I. *Hitler: 1899–1936: Hubris*. New York: W. W. Norton & Co., 2000.

Keynes, J. M. *A Revision of the Treaty: Being a Sequel to The Economic Consequences of the Peace*. London: Macmillan and Co., 1922.

Knäusel, H. G. *Zeppelin and the United States of America: An Important Episode in German-American Relations*. Friedrichshafen, Germany: Luftschiffbau Zeppelin, 1981.

Knight, R. W. *The Hindenburg Accident: A Comparative Digest of the Investigations and Findings, with the American and German Reports Included*. Bureau of Air Commerce Safety and Planning Division, Department of Commerce, Report No. 11 (August 1938). Riverdale, Conn.: 7 C's Press, 1976.

Kolm, S. L. "'Who Says It's a Man's World?': Women's Work and Travel in the First Decades of Flight." In *The Airplane in American Culture*, edited by D. A. Pisano, 147–64. Ann Arbor: University of Michigan Press, 2003.

Komons, N. A. "William P. MacCracken, Jr., and the Regulation of Civil Aviation." In *Aviation's Golden Age: Portraits from the 1920s and 1930s*, edited by W. M. Leary, 35–59. Iowa City: University of Iowa Press, 1989.

Köster, R. "Zeppelin: The Airship and the Need for Diversification After World War One (1918–1929)." Paper presented at the European Business History Association's 12th Annual Conference, Bergen, Norway, 2008.

Kramer, T. *Eduard Spelterini: Photographs of a Pioneer Balloonist*. Zurich: Scheidigger & Speiss, 2008.

Kravitz, S. A. "The Business of Selling the Soviet Union: Intourist and the Wooing of American Travelers, 1929–1939." M.A. diss., Concordia University, 2006.

Kulikowski, M. "Fantasy Flights: Technology, Politics and the Soviet Airship Programme, 1930–1938." *Icon* 21 (2015) : 66–80.

Lamm, M. "1932 Maybach Zeppelin Streamlined Limo." *Autoweek*, June 9, 1997.

Langins, J. "Hydrogen Production for Ballooning During the French Revolution: An Early Example of Chemical Process Development." *Annals of Science* 40 (1983), no. 6: 531–58.

Lardner, D. *Treatise on Hydrostatics and Pneumatics*. Boston: Stimpson and Clapp, 1832.

Lauritzen, F. "Propaganda Art in the Postage Stamps of the Third Reich." *Journal of Decorative and Propaganda Arts* 10 (1988): 62–79.

Leary, W. M. "At the Dawn of Commercial Aviation: Inglis M. Uppercu and Aeromarine Airways." *Business History Review* 53 (1979), no. 2: 180–93.

———. "Henry Ford and Aeronautics During the 1920s." In *Aviation's Golden Age: Portraits From the 1920s and 1930s*, edited by W. M. Leary, 1–17. Iowa City: University of Iowa Press, 1989.

———. "Safety in the Air: The Impact of Instrument Flying and Radio Navigation on U.S.

Commercial Air Operations Between the Wars." In *From Airships to Airbus: The History of Civil and Commercial Aviation*, edited by W. M. Leary and W. F. Trimble, 97–113. Washington, D.C.: Smithsonian Institution Press, 1995.

Lee, D. D. "Herbert Hoover and the Rise of Commercial Aviation, 1921–1926." *Business History Review* 58 (1984): 78–102.

———. "Herbert Hoover and the Golden Age of Aviation." In *Aviation's Golden Age: Portraits from the 1920s and 1930s*, edited by W. M. Leary, 127–48. Iowa City: University of Iowa Press, 1989.

Lehmann, E. A. *Zeppelin: The Story of Lighter-Than-Air Craft*. Edited by A. Sutton. Stroud, U.K.: Fonthill Media, 2015.

Levitt, M. L. "The Development and Politicization of the American Helium Industry, 1917–1940." *Historical Studies in the Physical and Biological Sciences* 30 (2000), no. 2: 333–47.

Lindbergh, C. A. *The Wartime Journals of Charles A. Lindbergh*. New York: Harcourt, Brace, Jovanovich, 1970.

———. *Autobiography of Values*. New York: Harcourt Brace Jovanovich, 1978.

Lloyd George, D. *War Memoirs of David Lloyd George*. London: Odhams Press, 2 vols., 1938.

Lochner, L. P. "Aboard the Airship *Hindenburg*: Louis P. Lochner's Diary of Its Maiden Flight to the United States." *The Wisconsin Magazine of History* 49 (1965), no. 2: 101–21.

Loftin, L. K., Jr. *Quest for Performance: The Evolution of Modern Aircraft*. Washington, D.C.: National Aeronautics and Space Administration, 1985.

Lorelli, J. A. *To Foreign Shores: U.S. Amphibious Operations in World War Two*. Annapolis, Md.: Naval Institute Press, 1995.

Lorenz, R. D. "Edmond Halley's Aeronautical Calculations on the Feasibility of Manned Flight in 1691." *Journal of Aeronautical History* 2 (2012): 28–31.

Lynn, M. R. *The Sublime Invention: Ballooning in Europe, 1783–1820*. London: Routledge, 2016.

Lyth, P. "The Empire's Airway: British Civil Aviation from 1919 to 1939." *Revue Belge de Philologie et d'Histoire* 78 (2000), no. 3: 865–87.

MacKay, R. A. "Foreign Governments and Politics: Newfoundland Reverts to the Status of a Colony." *American Political Science Review* 28 (1934), no. 5: 895–900.

Mackenzie, D. "Ireland, Canada and Atlantic Aviation, 1935–45: A Comparative Study." *The Canadian Journal of Irish Studies* 18 (1992), no. 2: 31–47.

Maitland, E. M. *The Log of H.M.A. 34: Journey to America and Back*. London: Hodder and Stoughton, 1920.

Martyn, T. *Hints of Important Uses, to Be Derived from Aerostatic Globes, with a Print of an Aerostatic Globe, and Its Appendages*. London: T. Martyn, 1784.

Maurer, H.-J. "A Dissident Nazi: Hans-Jörg Maurer's Würzburg Diary." *Wisconsin Magazine of History* 50 (1967), no. 4: 347–91.

Maxim, H. S. *Artificial and Natural Flight*. London: Whittaker and Co., 1908.

———. *My Life*. London: Methuen & Co., 1915.

Mayo, H. B. "Newfoundland's Entry into the Dominion." *The Canadian Journal of Economics and Political Science/Revue Canadienne d'Economique et de Science Politique* 15 (1949), no. 4: 505–22.

McCannon, J. "Sharing the Northern Skies: German-Soviet Scientific Cooperation and the 1931 Flight of the *Graf Zeppelin* to the Soviet Arctic." *Russian History* 30 (2003), no. 4: 403–31.

———. "Winged Prometheans: Arctic Aviation as Socialist Construction in Stalinist Russia, 1928–1939." *Scientia Canadensis: Canadian Journal of the History of Science, Technology and Medicine* 33 (2010), no. 2: 75–97.

McCullough, D. *The Wright Brothers*. New York: Simon & Schuster, 2015.

McDonough, F. *The Gestapo: The Myth and Reality of Hitler's Secret Police*. New York: Sky-horse Publishing, 2017.

McLaughlin, H. E. *Footsteps in the Sky: An Informal Review of U.S. Airlines Inflight Service, 1920–Present*. Denver: State of the Art, 1994.

Menne, B. *Blood and Steel: The Rise of the House of Krupp*. Translated by G. H. Smith. New York: Lee Furman, 1938.

Meyer, H. C. *Airshipmen, Businessmen, and Politics, 1890–1940*. Washington, D.C.: Smithsonian Institution Press, 1991.

——. *Count Zeppelin: A Psychological Portrait*. Auckland, New Zealand: Lighter-Than-Air Institute, 1998.

Miller, W. *I Found No Peace: The Journal of a Foreign Correspondent*. New York: Simon and Schuster, 1936.

Moedebeck, H.W.L. *Pocket-Book of Aeronautics*. Translated by W. M. Varley. London: Whittaker & Co., 1907.

Morgan, R. P. "The Political Significance of German-Soviet Trade Negotiations, 1922–5." *The Historical Journal* 6 (1963), no. 2: 253–71.

Nekrich, A. M. *Pariahs, Partners, Predators: German-Soviet Relations, 1922–1941*. Translated by G. L. Freeze. New York: Columbia University Press, 1997.

Nelson, S. B. "Airships in the Arctic." *Arctic* 46 (1993), no. 3: 278–83.

Nielsen, T. *The Zeppelin Story*. Translated by Peter Chambers. London: Allen Wingate, 1955.

Nitske, W. R. *The Zeppelin Story*. South Brunswick, N.J.: A. S. Barnes and Co., 1977.

Nixon, E. B., ed. *Franklin D. Roosevelt and Foreign Affairs*. Cambridge, Mass.: Harvard University Press, 3 vols., 1969.

Norway, N. S. *Slide Rule: The Autobiography of an Engineer*. 1954. Reprint, New York: Vintage, 2010.

Office of United States Chief of Counsel for Prosecution of Axis Criminality. *Nazi Conspiracy and Aggression*. Washington, D.C.: Government Printing Office, 1946.

O'Neill, R. A., and J. F. Hood. *A Dream of Eagles*. Boston: Houghton Mifflin, 1973.

Overton, J. "Economic Crisis and the End of Democracy: Politics in Newfoundland During the Great Depression." *Labour/Le Travail* 26 (1990): 85–124.

Overy, R. J. "Cars, Roads, and Economic Recovery in Germany, 1932–8." *Economic History Review* (1975) 28, no. 3: 466–83.

Pennoyer, P., and A. Walker. *The Architecture of Delano & Aldrich*. New York: W. W. Norton & Co., 2003.

Pfister, O. *The Psychoanalytic Method*. Translated by C. R. Payne. New York: Moffat, Yard, and Co., 1919.

Pickering, S. "Wake Island." *United States Naval Institute Proceedings* 48 (1922), no. 12: 2075–79.

Pisano, D. A. "The Greatest Show on Earth: The Confrontation Between Utility and Entertainment in Aviation." In *The Airplane in American Culture*, edited by D. A. Pisano, 39–74. Ann Arbor: University of Michigan Press, 2003.

Pomerantz, S. I. "George Washington and the Inception of Aeronautics in the Young Republic." *Proceedings of the American Philosophical Society* (1954) 98, no. 2: 131–38.

Ponce de Leon, C. "The Man Nobody Knows: Charles A. Lindbergh and the Culture of Celebrity." In *The Airplane in American Culture*, edited by D. A. Pisano, 75–101. Ann Arbor: University of Michigan Press, 2003.

Porter, R. *Aerial Navigation: The Practicability of Traveling Pleasantly and Safely from New-York to California in Three Days, Fully Demonstrated: With a Full Description of a Perfect Aerial Locomotive, with Estimates of Capacity, Speed, and Cost of Construction*. New York: H. Smith, 1849.

Powell, T. "Grand Classic." *Automobile Quarterly* 43 (2003), no. 1, 12–24.

Randall, S. J. "Colombia, the United States, and Interamerican Aviation Rivalry, 1927–1940." *Journal of Interamerican Studies and World Affairs* 14 (1972), no. 3: 297–324.

Rawlinson, A. *The Defence of London*. London: Andrew Melrose, 2nd ed., 1923.

Reagan, M. D. "The Helium Controversy." In *American Civil-Military Decisions: A Book of Case Studies*, edited by H. Stein, 45–57. Birmingham: University of Alabama Press, 1963.

Reinhart, C. M., and K. S. Rogoff. *This Time Is Different: Eight Centuries of Financial Folly*. Princeton, N.J.: Princeton University Press, 2009.

Rieger, B. "Modern Wonders: Technological Innovation and Public Ambivalence in Britain and Germany, 1890s to 1933." *History Workshop Journal* 55 (2003), no. 1: 152–76.

Robinson, D. H. *The LZ-129 Hindenburg*. New York: Arco, 1964.

———. *Giants in the Sky: A History of the Rigid Airship*. Seattle: University of Washington Press, 1973.

———. "David Schwarz: Tragödie Des Erfinders: Zur Geschichte Des Luftschiffes." *Buoyant Flight* 32 (1984), no. 1: 2–8.

Robinson, D. H., and C. H. Keller. *"Up Ship!": A History of the U.S. Navy's Rigid Airships, 1919–1935*. Annapolis, Md.: Naval Institute Press, 1982.

Röhl, J.C.G. *Kaiser Wilhelm II, 1859–1941: A Concise Life*. Translated by S. De Bellaigue. Cambridge, U.K.: Cambridge University Press, 2014.

Rose, A. "Radar Strategy: The Air Dilemma in British Politics, 1932–1937." Ph.D. diss., Cambridge University, 1996.

———. "Radar and Air Defence in the 1930s." *Twentieth-Century British History* 9 (1998), no. 2: 219–45.

Rotch, A. L. *The Conquest of the Air, or the Advent of Aerial Navigation*. New York: Moffat, Yard and Co., 1909.

Rust, D. L. *Flying Across America: The Airline Passenger Experience*. Norman: University of Oklahoma Press, 2009.

Sanford, G. "Illuminating Systems: Edison and Electrical Incandescence." *OAH Magazine of History* 4 (1989), no. 2: 16–19.

Schatzberg, E. "Ideology and Technical Choice: The Decline of the Wooden Airplane in the United States, 1920–1945." *Technology and Culture* 35 (1994), no. 1: 34–69.

Schurz, C. *The Reminiscences of Carl Schurz*. New York: The McClure Company, 1907.

Schwartz, R. *Flying Down to Rio: Hollywood, Tourists, and Yankee Clippers*. College Station, Tx.: Texas A&M University Press, 2004.

Seibel, C. W. *Helium: Child of the Sun*. Lawrence: University Press of Kansas, 1969.

Serling, R. J. *Legend & Legacy: The Story of Boeing and Its People*. New York: St. Martin's Press, 1991.

Siemens, D. *Stormtroopers: A New History of Hitler's Brownshirts*. New Haven, Conn.: Yale University Press, 2017.

Smith, R. K. "The Intercontinental Airliner and the Essence of Airplane Performance, 1929–1939." *Technology and Culture* 24 (1983), no. 3: 428–49.

Snider, J. D. "'Great Shadow in the Sky': The Airplane in the Tulsa Race Riot of 1921 and the Development of African-American Visions of Aviation, 1921–1926." In *The Airplane in American Culture*, edited by D. A. Pisano, 105–46. Ann Arbor: University of Michigan Press, 2003.

Squires, J. D. "Aeronautics in the Civil War." *American Historical Review* 42 (1937), no. 4: 652–69.

Stamp, J. "Aero-Static Warfare: A Brief Survey of Ballooning in Mid-Nineteenth-Century Siege Warfare." *Journal of Military History* 79 (2015), no. 3: 767–82.

Steiner, J. H. "Press Release, 1863." *Minnesota History* 40 (1967), no. 6: 279–80.

Stokes, P. "Power for Flight in the Victorian Era." *Aerospace* 11 (1984), no. 10: 10–18.

Swientek, H.-O. "Alfred Colsman (1873–1955): Ein Leben Für Die Deutsche Luftfahrt." *Tradition: Zeitschrift für Firmengeschichte und Unternehmerbiographie* 10 (1965), no. 3: 112–26.

Thebaud-Sorger, M. "Balloons in the Historiography of Aerial Mobility." *Mobility in History* 4 (2013): 83ff.

Thompson, S. P. *Dynamo-Electric Machinery: A Manual for Students of Electrotechnics.* 3rd ed. London: E. & F. N. Spon, 1888.

Tittel, L. *Graf Zeppelin: His Life and His Work.* Translated by Peter A. Schmidt. Friedrichshafen, Germany: Zeppelin-Museum, 1995.

Toland, J. *The Great Dirigibles: Their Triumphs and Disasters.* New York: Dover Publications, 1972.

Topping, D. *When Giants Roamed the Sky: Karl Arnstein and the Rise of Airships from Zeppelin to Goodyear.* Edited by E. Brothers. Akron, Oh.: University of Akron Press, 2000.

Trippe, B. S. *Pan Am's First Lady: The Diary of Betty Stettinius Trippe.* Edited by R.E.G. Davies. McLean, Va.: Paladwr Press, 1996.

Tuan, K. "Aviation Insurance in America." *Journal of Risk and Insurance* 32 (1965), no. 1: 1–22.

Turner, C. C. *Aerial Navigation of To-Day: A Popular Account of the Evolution of Aeronautics.* London: Seeley & Co., 1910.

Turnor, C. H. *Astra Castra: Experiments and Adventures in the Atmosphere.* London: Chapman and Hall, 1865.

Vaeth, J. G. *Graf Zeppelin: The Adventures of an Aerial Globetrotter.* New York: Harper & Brothers, 1958.

———. "Zeppelin Decor: The *Graf Zeppelin* and the *Hindenburg.*" *The Journal of Decorative and Propaganda Arts* 15 (1990): 48–59.

Van der Linden, F. R. *Airlines and Air Mail: The Post Office and the Birth of the Commercial Aviation Industry.* Lexington: University Press of Kentucky, 2002.

Vantoch, V. *The Jet Sex: Airline Stewardesses and the Making of an American Icon.* Philadelphia: University of Pennsylvania Press, 2013.

Van Vleck, J. *Empire of the Air: Aviation and the American Ascendancy.* Cambridge, Mass.: Harvard University Press, 2013.

Ventry, A., and E. M. Kolesnik. *Airship Saga: The History of Airships Seen Through the Eyes of the Men Who Designed, Built, and Flew Them.* Poole, Dorset, U.K.: Blandford Press, 1982.

Vissering, H. *Zeppelin: The Story of a Great Achievement.* Chicago: Henry Vissering, 1922.

Von Baumgarten, R. "Aviation Medicine: Its Past, Present, and Future." In *From Airships to Airbus: The History of Civil and Commercial Aviation,* edited by W. M. Leary and W. F. Trimble, 71–90. Washington, D.C.: Smithsonian Institution Press, 1995.

Von Gleich, A. "Germany and Latin America." RAND Corporation, Memorandum RM-5523-RC, 1968.

Von Schierbrand, W. *The Kaiser's Speeches: Forming a Character Portrait of Emperor William II.* New York/London: Harper & Brothers, 1903.

Von Tschudi, G. *Aus 34 Jahren Luftfahrt: Persönliche Erinnerungen.* Berlin: Reimar Hobbing, 1928.

Warner, G. "The Flying Boats of Foynes." *History Ireland* 9 (2001), no. 1: 40–44.

Warner, G. E., C. M. Foote, E. D. Neill, and J. F. Williams. *History of Ramsey County and the City of St. Paul.* Minneapolis: North Star Publishing Co., 1881.

Warntz, W. "Transatlantic Flights and Pressure Patterns." *Geographical Review* 51 (1961), no. 2: 187–212.

Watkins, T. H. *Righteous Pilgrim: The Life and Times of Harold L. Ickes, 1874–1952.* New York: Henry Holt and Co., 1990.

Wegerdt, A. "Germany and the Aerial Navigation Convention at Paris, October 13, 1919." *Journal of Air Law and Commerce* 1 (1930): 1–32.

Weirather, L. "Romancing the Clipper: America's Technological Coming of Age in Children's Literature." *Studies in Popular Culture* 13 (1990), no. 1: 27–45.

Wells, H. G. *Anticipations of the Reaction of Mechanical and Scientific Progress upon Human Life and Thought.* London: Chapman & Hall, 1902.

Whalen, S., and R. E. Bartholomew. "The Great New England Airship Hoax of 1909." *The New England Quarterly* 75 (2002), no. 3: 466–76.

Wheeler, G. E. "The United States Navy and the Japanese 'Enemy': 1919–1931." *Military Affairs* 21 (1957), no. 2: 61–74.

Wilkes, C. *Narrative of the United States Exploring Expedition During the Years 1838, 1839, 1840, 1841, 1842.* Philadelphia: 1849.

Wise, J. *Through the Air: A Narrative of Forty Years' Experience as an Aeronaut.* Philadelphia, New York, Boston: To-Day Printing and Publishing Company, 1873.

Wohl, R. *A Passion for Wings: Aviation and the Western Imagination, 1908–1918.* New Haven, Conn.: Yale University Press, 1994.

———. *The Spectacle of Flight: Aviation and the Western Imagination, 1920–1950.* New Haven, Conn.: Yale University Press, 2005.

Woolf, V. *The Question of Things Happening: Letters, 1912–1922.* Edited by N. Nicholson. London: Hogarth Press, 1976.

Zabecki, D. T. *The German 1918 Offensives: A Case Study in the Operational Level of War.* London/New York: Routledge, 2006.

Zahm, A. F. *Aerial Navigation: A Popular Treatise on the Growth of Air Craft and on Aeronautical Meteorology.* New York: D. Appleton and Co., 1911.

Notes

Prologue

1. This section is based on *"Hindenburg* at Lakehurst," October 8, "Dirigible to Visit Six States Today," October 9, and *"Hindenburg* Soars over Six States," October 10, 1936, *The New York Times*; H. C. Meyer, "F. W. (Willy) von Meister: Portrait of an Airship Businessman," in H. C. Meyer, *Airshipmen, Businessmen, and Politics, 1890–1940* (Washington, D.C.: Smithsonian Institute Press, 1991), pp. 209–10; "Zeppelin: Easy Rider in a Hurricane," *Literary Digest*, October 17, 1936, pp. 17–18 (describes the fountain-pen trick); J. Toland, *The Great Dirigibles: Their Triumphs and Disasters* (New York: Dover Publications, 1972), pp. 10–12. Of particular value is the full passenger list at airships.net/hindenburg/flight-schedule/millionaire-flight/. For numbers and statistics, see Series K 239–245, "Air Transport—Aircraft Production and Exports: 1913 to 1945," p. 224, Ser. K 246–256, "Air Transport—Scheduled Air Transportation, Domestic Only, 1926 to 1945," p. 224, and Ser. K 265–273, "Air Transport—Accidents: 1927 to 1945," p. 225, in Bureau of the Census, *Historical Statistics of the United States, 1789–1945* (Washington, D.C.: United States Department of Commerce, 1949); "Motor Vehicle Traffic Fatalities, 1900–2007," at fhwa .dot.gov/policyinformation/statistics/2007/pdf/fi200.pdf. On noise and other experiments, see P. R. Bassett, "The *Hindenburg*—Some Measurements," *U.S. Air Service* 21 (December 1936).

1. The Aeronaut

1. E. C. Belote, born in Ontario County, New York, in 1812, moved to Saint Paul in 1856 and managed the Merchants Hotel until 1861, when he took over the more upscale International (which would burn down in 1869). See G. E. Warner, C. M. Foote, E. D. Neill, and J. F. Williams, *History of Ramsey County and the City of St. Paul* (Minneapolis: North Star Publishing Co., 1881), p. 493. On the International Hotel, see H. A. Castle, *History of St. Paul and Vicinity* (Chicago and New York: Lewis Publishing Co., 1912), p. 417. The International had been built just a few years before—in 1856—and its five floors were crowned by some impressive crenellated battlements. It was one of the few brick-

work buildings in Saint Paul. General Schurz, in his *The Reminiscences of Carl Schurz* (New York: The McClure Company, 2 vols., 1907), volume 2, p. 340, notes that Zeppelin at the time "spoke English moderately well."

2. R. R. Gilman, "Zeppelin in Minnesota: A Study in Fact and Fable," *Minnesota History* 39 (1965), no. 7, p. 282.

3. R. R. Gilman, "Count Von Zeppelin and the American Atmosphere," *Smithsonian Journal of History* 3 (1968), no. 1, p. 35; M. B. Dunn, R. R. Gilman, and F. von Zeppelin, "Zeppelin in Minnesota: The Count's Own Story," *Minnesota History* 40 (1967), no. 6, pp. 273–75.

4. On Zeppelin's background, see H. C. Meyer, *Count Zeppelin: A Psychological Portrait* (Auckland, New Zealand: Lighter-Than-Air Institute, 1998), pp. 18–19, 21, 23; W. R. Nitske, *The Zeppelin Story* (South Brunswick, N.J.: A. S. Barnes and Company, 1977), pp. 15–17; H. Eckener (trans. L. Farnell), *Count Zeppelin: The Man and His Work* (Berlin and London: Massie Publishing Company, 1938), p. 6; L. Tittel (trans. P. A. Schmidt), *Graf Zeppelin: His Life and His Work* (Friedrichshafen, Germany: Zeppelin-Museum, 1995), p. 6. Zeppelin himself cracked the joke about his forebears gambling away all their lands; see E. A. Lehmann (ed. A. Sutton), *Zeppelin: The Story of Lighter-Than-Air Craft* (1937; repr. London: Fonthill Media, 2015), p. 145.

5. Quoted in Nitske, *The Zeppelin Story*, pp. 22–23.

6. Eckener, *Count Zeppelin*, pp. 37, 14–17, 41.

7. There are, as might be expected, conflicting versions of Zeppelin's meeting with Lincoln. Carl Schurz claims that it was he who introduced the count (whom he leaves unnamed) to the president. See his *Reminiscences*, volume 2, p. 340. Schurz also relates the comic anecdote concerning Zeppelin's bemusement at Lincoln's remark, though Nitske believes "there is serious doubt that the incident ever happened" as Zeppelin's "superb storytelling ability and high sense of humor" would have prompted him to tell the tale himself, rather than leaving it to Schurz. See Nitske, *Zeppelin Story*, p. 25. Notwithstanding Zeppelin's narrative skills, he took his family background very seriously and likely didn't catch the irony.

8. The most thorough descriptions of Zeppelin's travels are Dunn, Gilman, and Zeppelin, "Zeppelin in Minnesota: The Count's Own Story"; and Gilman, "Count von Zeppelin and the American Atmosphere," pp. 29–40.

9. On the high cost of balloons, see "Perilous Balloon Ascent in Lake Erie," *Daily News* (England), July 9, 1857; R. Holmes, *Falling Upwards: How We Took to the Air* (New York: Pantheon Books, 2013), p. 57.

10. T. D. Crouch, *The Eagle Aloft: Two Centuries of the Balloon in America* (Washington, D.C.: Smithsonian Institution Press), pp. 229–35. In the summer of 1853, Steiner experienced the terrible effects of flying into a thunderstorm when he took off from Camden, New Jersey. See his description in J. Steiner, "Press Release, 1863," *Minnesota History* 40 (1967), no. 6, pp. 279–80, which transcribes a lengthy interview originally published in the *St. Paul Pioneer* of August 9, 1863. The article was intended to advertise his upcoming tour in the city. For the name of his balloon on his current tour, see his advertisement in the *Milwaukee Daily Sentinel*, July 24, 1863, p. 4; for the number of balloons he'd lost, see "Perilous Balloon Ascent in Lake Erie."

11. S. Hopkins, "Daredevil Balloonist John Steiner: America's Champion Aeronaut in the Civil War," *Military Images* 33 (2015), no. 4, pp. 42–44. On Steiner's service at Island No. 10, *The Times* (London) of April 14, 1862, has a report, quoted in C. H. Turnor (ed.), *Astra Castra: Experiments and Adventures in the Atmosphere* (London: Chapman and Hall, 1865), p. 287.

12. Dunn, Gilman, and Zeppelin, "Zeppelin in Minnesota: The Count's Own Story," pp. 275–76.

13. R. von Baumgarten, "Aviation Medicine: Its Past, Present, and Future," in W. F. Trimble (ed.), *From Airships to Airbus: The History of Civil and Commercial Aviation* (Washington, D.C.: Smithsonian Institution Press, 2 vols., 1995), volume 2, pp. 75–76.

14. In Steiner's July 24, 1863, advertisement in the *Milwaukee Daily Sentinel*, he stated that "the weather should be perfectly calm in order to make the Army Ascensions."

15. Among the earliest popularizers of the term was Jean-Pierre Blanchard, one of the most famous of the early aeronauts. See his definition in *The Principles, History, and Use of Air-Balloons* (New York: J. Fellows, 1796), p. 3.

16. Letter, Zeppelin to his father, August 19 postscript, printed in Dunn, Gilman, and Zeppelin, "Zeppelin in Minnesota: The Count's Own Story," pp. 276–77.

17. Gilman, "Zeppelin in Minnesota: A Study in Fact and Fable," p. 279. On the number of passengers the *Hercules* could handle, see Steiner's July 24, 1863, advertisement in the *Milwaukee Daily Sentinel*. The interview is carried in full in *The Bemidji Daily Pioneer*, February 9 and 10, 1915.

18. Letter published in Dunn, Gilman, and Zeppelin, "Zeppelin in Minnesota: The Count's Own Story," pp. 276–77. The original letter to his father is dated August 18, but Zeppelin added a postscript dated August 19 after his ascent with Steiner.

19. Quote from Zeppelin's 1915 interview, in *The Bemidji Daily Pioneer*, February 10, 1915.

20. Gilman, "Zeppelin in Minnesota: A Study in Fact and Fable," p. 285.

21. Hopkins, "Daredevil Balloonist," pp. 43–44. The last mention of him in a newspaper was in 1875, when it was reported that he was trying to build a gigantic airship to cross the Atlantic. "Steiner's Balloon," *The Inter-Ocean*, May 27, 1875; "A Big Balloon," *Crawford County Bulletin*, June 17, 1875.

22. On his itinerary, see Dunn, Gilman, and Zeppelin, "Zeppelin in Minnesota: The Count's Own Story," pp. 277–78. In the count's own words, "While I was above St. Paul I had my first idea of aerial navigation strongly impressed upon me and it was there that the first idea of my Zeppelins came to me." Quoted in Gilman, "Zeppelin in Minnesota: A Study in Fact and Fable," p. 285. This recollection came much later, obviously: At this time, Zeppelin was introduced only to the problem of aerial navigation and some rudimentary design ideas.

23. Tittel, *Graf Zeppelin*, p. 19.

24. A. Confino, *The Nation as a Local Metaphor: Württemberg, Imperial Germany, and National Memory, 1871–1918* (Chapel Hill and London: University of North Carolina Press, 1997), pp. 17, 20.

25. Tittel, *Graf Zeppelin*, p. 73.

26. Confino, *Nation as a Local Metaphor*, pp. 17, 20–21.

27. Tittel, *Graf Zeppelin*, p. 73.

28. Eckener, *Count Zeppelin*, pp. 100, 106–7.

29. Letter of October 15, 1869, quoted in M. Belafi, *The Zeppelin* (Barnsley, U.K.: Pen & Sword, 2015), p. 9.

30. Recollection of Oskar Wilcke, printed in R. Italiaander, *Ferdinand Graf Von Zeppelin: Reitergeneral, Diplomat, Luftschiffpionier: Bilder und Dokumente* (Konstanz, Germany: Verlag Friedr. Stadler, 1980), pp. 140–41.

31. Belafi, *The Zeppelin*, pp. 9–10; Eckener, *Count Zeppelin*, pp. 109–17; "Zeppelin and His Troopers," *Cincinnati Daily Enquirer* (reprinting the *New York Evening Post*), September 5, 1870, has an exciting account of the raid. On the medal, see Nitske, *The Zeppelin Story*, p. 43.

32. Holmes, *Falling Upwards*, pp. 251–93, has a terrific narrative of the siege; J. Stamp, "Aero-Static Warfare: A Brief Survey of Ballooning in Mid-Nineteenth-Century Siege Warfare," *Journal of Military History* 79 (2015), no. 3, p. 780.

33. Eckener, *Count Zeppelin*, p. 125; Tittel, *Graf Zeppelin*, p. 75.

2. The Fever Dream

1. Belafi, *The Zeppelin*, p. i.

2. Meyer, *Count Zeppelin*, p. 39; Eckener, *Count Zeppelin*, pp. 155–56; Tittel, *Graf Zeppelin*, p. 25; D. Robinson, *Giants in the Sky: A History of the Rigid Airship* (Seattle: University of Washington Press, 1973), p. 13. Stephan's conclusion is quoted in Nitske, *The Zeppelin Story*, p. 43.

3. Eckener, *Count Zeppelin*, pp. 155–56; Robinson, *Giants in the Sky*, pp. 13–14; Belafi, *The Zeppelin*, p. 14.

4. See diary entry, printed in Eckener, *Count Zeppelin*, p. 157; Belafi, *The Zeppelin*, p. 15.

5. Quoted in Eckener, *Count Zeppelin*, p. 158.

6. Diary entry of June 12, 1876, quoted in Eckener, *Count Zeppelin*, pp. 130–31.

7. Eckener, *Count Zeppelin*, pp. 132–33, 159. The memorandum is printed in Robinson, *Giants in the Sky*, p. 14.

8. This section based on Meyer, *Count Zeppelin*, pp. 31–37; Eckener, *Count Zeppelin*, pp. 138, 142–51; Tittel, *Graf Zeppelin*, pp. 25, 77.

3. The Government of the Air

1. Quoted in Holmes, *Falling Upwards*, pp. 17–18.

2. H. Pückler-Muskau, "Frail Aerial Bark," in *Lighter-Than-Air Flight*, ed. C. V. Glines (New York: Franklin Watts, 1965), pp. 49–50.

3. J. T. Alexander, "Aeromania, 'Fire-Balloons,' and Catherine the Great's Ban of 1784," *The Historian* 58 (1996), no. 3, pp. 497–516.

4. P. Keen, "The 'Balloonomania': Science and Spectacle in 1780s England," *Eighteenth-Century Studies* 39 (2006), no. 4, pp. 510–11, 520–21.

5. "Aerostation," *The Scots Magazine* 46 (1784), pp. 560–61.

6. Letter, Washington to Louis Le Bèque du Portail, April 4, 1784, in *The Writings of George Washington: From the Original Manuscript Sources, 1745–1799*, ed. J. Fitzpatrick (Washington, D.C.: Government Printing Office, 1938), 27, p. 387; for a general overview, see S. I. Pomerantz, "George Washington and the Inception of Aeronautics in the Young Republic," *Proceedings of the American Philosophical Society* 98 (1954), no. 2, esp. pp. 131–32.

7. John Adams to Abigail Adams, September 7, 1783, at *Adams Family Papers: An Electronic Archive*, Massachusetts Historical Society, masshist.org/digitaladams/.

8. Letter, Jefferson to Francis Hopkinson, February 18, 1784, in *Founders Online*. National Archives, founders.archives.gov/documents/Jefferson/01-06-02-0402.

9. Crouch, *The Eagle Aloft*, pp. 60–66.

10. Quoted in Alexander, "Aeromania," p. 502.

11. Letter, Robert Morris to Jay, in a letter of November 27, 1783, printed in *The Life of John Jay: With Selections from His Correspondence and Miscellaneous Papers*, ed. W. Jay (New York: J. & J. Harper, 2 vols., 1833), volume 2, p. 138.

12. Quoted in Crouch, *The Eagle Aloft*, p. 26.

13. M. R. Lynn, *The Sublime Invention: Ballooning in Europe, 1783–1820* (London: Routledge, 2016), p. 45.

14. Holmes, *Falling Upwards*, pp. 30–31.

15. Alexander, "Aeromania," pp. 510–12.

16. A. M. Josephy, Jr. (ed.), *The American Heritage History of Flight* (New York, 1962), p. 44, cited in Alexander, "Aeromania," p. 516.

17. Crouch, *The Eagle Aloft*, pp. 97–102.

18. The phrase "aerial phrenzy" was coined in the anonymous work *London Unmask'd: Or the New Town Spy* (1785), quoted in Keen, "'Balloonomania,'" p. 510. Tiberius Cavallo, one of England's leading scientific advocates of ballooning, was particularly irritated by the cheap jokes. See his remarks, printed in Turnor (ed.), *Astra Castra*, p. 81.

19. Keen, "'Balloonomania,'" p. 517; letter, Walpole to Sir Horace Mann, December 2, 1783, printed in P. Cunningham (ed.), *The Letters of Horace Walpole, Fourth Earl of Oxford* (London: Richard Bentley and Sons, 9 vols., 1891), volume 8, p. 438.

20. Keen, "'Balloonomania,'" p. 517.

21. On Martyn, see his *Hints of Important Uses, to Be Derived from Aerostatic Globes, with a Print of an Aerostatic Globe, and Its Appendages* (London: T. Martyn, 1784), p. 15; R. Gillespie, "Ballooning in France and Britain, 1783–1786: Aerostation and Adventurism," *Isis* 75 (1984), no. 2, p. 262.

22. Letter to Banks, November 21, 1783, quoted in G. H. Bryan and E. H. Harper, "Modern Aeronautics," *Nature* 83 (1910), March 31, pp. 132–34.

23. Franklin reiterated the issue to other correspondents. As he wrote to his friend Richard Price, "All the conversation here at present turns on the Balloons . . . and the means of managing them so as to give Men the Advantage of Flying." Quoted in Crouch, *The Eagle Aloft*, p. 26.

24. Keen, "'Balloonomania,'" p. 515.

25. P. Bradshaw, Jr., "The Role of Technology in the Failure of the Rigid Airship as an Invention," Ph.D. diss., University of Florida (1975), pp. 32–35.

26. Keen, "'Balloonomania,'" p. 516.

27. The contemporary debates are discussed in Gillespie, "Ballooning in France and Britain, 1783–1786," pp. 257–58. See also the excellent summary in the *Westminster Review*, quoted at length in Turnor (ed.), *Astra Castra*, esp. p. 313.

28. Gillespie, "Ballooning in France and Britain, 1783–1786," pp. 257–58.

29. *The Colonist* (Australia), "Australian Steam Navigation," December 22, 1838, carried a report on the controversy.

30. For standard contemporary opinion on the problem, see D. Lardner, *Treatise on Hydrostatics and Pneumatics* (Boston, Mass.: Stimpson and Clapp, 1832), pp. 148–50. Lardner was a longtime enemy of Brunel's and lost consistently in their clashes.

31. For an account of the *Great Western*'s arrival in New York, see the lengthy reprint of the *Quarterly Review* article in Turnor (ed.), *Astra Castra*, pp. 300–301.

32. On the remarkable Porter, see J. Lipman, *Rufus Porter Rediscovered: Artist, Inventor, Journalist, 1792–1884* (New York: Clarkson W. Potter, 1980); Crouch, *The Eagle Aloft*, pp. 292–318; C. E. McCluer, "Rufus Porter and His 'Flying Ship,'" *Scientific American*, January 8, 1910, p. 30.

33. R. Porter, *Aerial Navigation: The Practicability of Traveling Pleasantly and Safely from New-York to California in Three Days, Fully Demonstrated: With a Full Description of a Perfect Aerial Locomotive, with Estimates of Capacity, Speed, and Cost of Construction* (New York: H. Smith, 1849).

4. Aerial Navigation

1. It must be borne in mind that figures and details regarding Giffard's dirigible are not just sparse but differ in several respects (for instance, his engine/boiler weight is written as either 350 lbs or 330 lbs, and the weight-to-horsepower ratio ranges from 110 to 117). The differences, however, do not vary so much as to materially affect conclusions as to general performance. A key near-contemporary resource on Giffard is O. Chanute, *Aerial Navigation: A Lecture Delivered to the Students of Sibley College, Cornell University* (New York: Railroad and Engineering Journal, 1891), pp. 7–9; see also R. P. Hallion, *Taking Flight: Inventing the Aerial Age from Antiquity Through the First World War* (New York: Oxford University Press, 2003), p. 83; M. Kelly, *Steam in the Air: The Application of Steam Power in Aviation During the 19th and 20th Centuries* (Barnsley, U.K.: Pen & Sword, 2006), pp. 27–30; Robinson, *Giants in the Sky*, p. 2; and P. Stokes, "Power for Flight in the Victorian Era," *Aerospace* 11 (1984), no. 10, p. 12. My back-of-the-envelope estimates as to Giffard's useful lift are based on a typical composition of coal gas consisting of hydrogen, methane, carbon monoxide, and ethylene, but figures vary depending on altitude, pressure, temperature, and humidity.

2. On Lowe's balloons, see M. Baldwin, *With Brass and Gas: An Illustrated and Embellished Chronicle of Ballooning in Mid-Nineteenth Century America* (Boston, Mass.: Beacon Press, 1967), pp. 211–38.

3. G. Glaser, "The Paris Electrical Exhibition," *Science* 2 (1881), no. 63, September 10, 1881, pp. 430–33; Anon., "Electric Railroads in Paris," *Science* 2 (1881), no. 71, November 5, 1881, pp. 526–27; F. Dittman, "Innovations in the Electric Energy System," *Icon* 13 (2007), p. 63; G. Sanford, "Illuminating Systems: Edison and Electrical Incandescence," *OAH Magazine of History* 4 (1989), no. 2, pp. 16–19.

4. J. E. Brittain, "The International Diffusion of Electrical Power Technology, 1870–1920," *Journal of Economic History* 34 (1974), no. 1, pp. 108–21; S. P. Thompson, *Dynamo-Electric Machinery: A Manual for Students of Electrotechnics*, 3rd ed. (London: E. & F. N. Spon, 1888), pp. 140–41; H.E.H. Clifford, "The Efficiency of Small Electro-Motors," *Proceedings of the American Academy of Arts and Sciences* 22 (1886), May–June, pp. 493–97.

5. This section is based on Hallion, *Taking Flight*, pp. 86–87; G. De Syon, "Toys or Tools? The Riddle of French Army Airships, 1884–1914," Paper 22, presented to the 8th International Airship Convention, Bedford, U.K., 2010, pp. 1–9; C. Renard, "The First Flight of *La France*," *The Airship* 2 (1936), no. 8, pp. 63–65; Belafi, *The Zeppelin*, p. 15; Robinson, *Giants in the Sky*, pp. 2–3. The quotes regarding average speed and the new standard are taken from the Chairman's Address, printed in *The Journal of the Society of Arts* 33 (1884), no. 1670, November 21, p. 16. The Eiffel Tower data come from Chanute, *Aerial Navigation*, p. 6.

5. The System

1. Letter, printed in Eckener, *Count Zeppelin*, p. 164.

2. Quoted in Meyer, *Count Zeppelin*, p. 48. A slightly different version can be found in Eckener, *Count Zeppelin*, pp. 199–200.

3. D. Clerk, "The Internal-Combustion Engine," *Journal of the Royal Society of Arts* 65 (1916), no. 3341, December 1, 1916, pp. 43–44.

4. The material regarding Daimler and Maybach's early years is based on a variety of online resources (gottliebdaimler.com, for instance), but see also L. Bryant, "The Silent Otto,"

Technology and Culture 7 (1966), no. 2, pp. 184–200; G. Goldbeck, "Entwicklungsstufen des Verbrennungsmotors," *Motortechnische Zeitschrift* 23 (1962), pp. 76–80; D. Beasley, *Who Really Invented the Automobile? Skulduggery at the Crossroads* (Simcoe, Ont.: Davus Publishing, 1997), pp. 113–14. On Woelfert, see Robinson, *Giants in the Sky*, p. 3.

5. Belafi, *The Zeppelin*, p. 16.

6. See the diagram in Belafi, *The Zeppelin*, p. 16.

7. Robinson, *Giants in the Sky*, pp. 17, 21.

8. Eckener, *Count Zeppelin*, p. 168.

9. See Robinson, *Giants in the Sky*, p. 21; Eckener, *Count Zeppelin*, p. 170. As late as 1908, Zeppelin still seemed to believe otherwise.

10. Letter to Schlieffen, printed in Eckener, *Count Zeppelin*, pp. 161–62; Robinson, *Giants in the Sky*, p. 17; Belafi, *The Zeppelin*, p. 16.

11. Belafi, *The Zeppelin*, pp. 16–17. On the choice of name, see J. Duggan and H. C. Meyer, *Airships in International Affairs, 1890–1940* (London: Palgrave, 2001), p. 23.

12. Quoted in Belafi, *The Zeppelin*, p. 19. Tschudi's warning about the PAB was recalled by Zeppelin in his diary entry of February 16, 1894; see Eckener, *Count Zeppelin*, pp. 188–89.

13. Eckener, *Count Zeppelin*, pp. 174–75.

14. Quoted in Eckener, *Count Zeppelin*, p. 175.

15. On the high ranking of the quality of Württemberg's scientific education, see "Science Notes," *Scientific American*, January 11, 1902, p. 19.

16. Belafi, *The Zeppelin*, p. 19.

17. On aluminum, see S. Dooley, "The Development of Material-Adapted Structural Form," Ph.D. diss., École Polytechnique Fédérale de Lausanne, Switzerland (2004), Appendix: "Aluminum, Plywood, and Rigid Airships," esp. pp. 151–80.

18. On Bach and Kober, see Belafi, *The Zeppelin*, p. 19; for Zeppelin's letter to Kober, see Meyer, *Count von Zeppelin*, pp. 43–44.

19. Quoted in Belafi, *The Zeppelin*, p. 19.

20. Letter from Zeppelin to General Bronsart von Schellendorf, January 30, 1894, printed in Eckener, *Count Zeppelin*, p. 187.

21. On car production, see Beasley, *Who Really Invented the Automobile?*, p. 139.

22. It was not likely Daimler himself, who was suffering from a serious heart condition at the time. He would be away for much of the following year. On Daimler's movements, see daimler.com/company/tradition/founders-pioneers/gottlieb-daimler.html.

23. See, for example, his 1893 description in Belafi, *The Zeppelin*, p. 20; Robinson, *Giants in the Sky*, p. 16.

24. Quoted in Robinson, *Giants in the Sky*, p. 17.

25. Belafi, *The Zeppelin*, p. 20; Robinson, *Giants in the Sky*, pp. 15–16.

26. Quoted in Eckener, *Count Zeppelin*, p. 184.

27. Letter of November 13, 1893, in Eckener, *Count Zeppelin*, pp. 185–86.

28. See, for instance, his letters to General Oskar von Lindequist, December 29, 1893, and Schellendorf, the minister of war, January 30, 1894, in Eckener, *Count Zeppelin*, pp. 186–87.

29. Diary, February 16, 1894, in Eckener, *Count Zeppelin*, pp. 188–89.

6. The Pivot

1. Much of the Schwarz material is based on Douglas Robinson's detailed and invaluable review of Cvi Rotem's limited-edition *David Schwarz: Tragödie des Erfinders: Zur Geschichte des Luftschiffes*, which appeared in *Buoyant Flight* 32 (1984), no. 1, pp. 2–8.

2. The names of the committee members are listed in Belafi, *The Zeppelin*, p. 20. Biographical information has been accumulated from various online sources, including the relevant Wikipedia entries.

3. H. Müller-Breslau, "Zur Geschichte des Zeppelin-Luftschiffes," in *Verhandlung zur Beförderung des Gewerbfleisses*, January 1914, p. 35, quoted in Robinson, *Giants in the Sky*, p. 19.

4. Zeppelin's letter, dated May 1894, is reproduced in Belafi, *The Zeppelin*, p. 21.

5. Belafi, *The Zeppelin*, p. 21.

6. Meyer, *Count Zeppelin*, pp. 45–46; marginalia quoted in Robinson, *Giants in the Sky*, p. 19; Nitske, *The Zeppelin Story*, p. 53.

7. Eckener, *Count Zeppelin*, p. 193.

8. Robinson, *Giants in the Sky*, pp. 20–21.

9. Belafi, *The Zeppelin*, p. 22.

10. Belafi, *The Zeppelin*, p. 22.

11. Eckener, *Count Zeppelin*, p. 195.

12. Printed in Eckener, *Count Zeppelin*, p. 197. On Gross, see B. Menne (trans. G. H. Smith), *Blood and Steel: The Rise of the House of Krupp* (New York: Lee Furman, 1938), pp. 83–84.

13. Letter to the king of Württemberg, no date, printed in Eckener, *Count Zeppelin*, pp. 198–99. By the end of August, Zeppelin was so certain of the merits of his sky train that he patented his concept of a "steerable air locomotive with several supporting bodies arranged in succession." See Patent No. 98580, August 31, 1895, printed in Belafi, *The Zeppelin*, p. 21.

14. Letter to the king of Württemberg, no date, printed in Eckener, *Count Zeppelin*, pp. 198–99.

15. Eckener, *Count Zeppelin*, p. 201; Belafi, *The Zeppelin*, pp. 22–23.

16. Belafi, *The Zeppelin*, p. 23.

17. Belafi, *The Zeppelin*, p. 24.

18. Eckener, *Count Zeppelin*, pp. 204–9; Belafi, *The Zeppelin*, p. 24. On the Association of German Engineers, see K. Gispen, "Engineers in Wilhelmian Germany: Professionalization, Deprofessionalization, and the Development of Nonacademic Technical Education," in *German Professions, 1800–1950*, ed. G. Cocks and K. H. Jarausch (New York; Oxford University Press, 1990), pp. 104–22.

19. H.-C. Gunga, *Nathan Luntz: His Life and Work in the Fields of High-Altitude Physiology and Aviation Medicine* (Academic Press; New York, 2008), pp. 162–63; Belafi, *The Zeppelin*, p. 24.

20. R. Italiaander, *Ein Deutscher Namens Eckener: Luftfahrtpionier und Friedenspolitiker: Vom Kaiserreich Bis in Die Bundesrepublik* (Konstanz, Germany: F. Stadler, 1981), p. 92. Gross's accusation seems to have been based on the count's recent change in plans from using hollow tubes—which were dangerously prone to bending under stress—for the girders to a lattice design, in which two horizontal beams are strengthened by means of thin, diagonal metal strips laced in an X pattern. It turned out that the latticed version had first been proposed by Berg's people. Eckener mentions the accusation, first raised by Melanie (Schwarz's redoubtable widow), very generally in *Count Zeppelin*, p. 211, while Robinson is more specific in his review of *David Schwarz*, pp. 7–8. On the design change, see Eckener, *Count Zeppelin*, p. 214; Dooley, "The Development of Material-Adapted Structural Form," p. A.195. On the duel, see Lehmann, *Zeppelin*, pp. 129–30; Belafi, *The Zeppelin*, pp. 25–26.

21. Figures vary depending on the source. See Belafi, *The Zeppelin*, pp. 25–26 (including a partial list of shareholders); Robinson, *Giants in the Sky*, p. 23.

22. Quoted in Belafi, *The Zeppelin*, p. 26.

23. Lehmann, *Zeppelin*, p. 119; Belafi, *The Zeppelin*, pp. 28–29.

24. Robinson, *Giants in the Sky*, p. 28; P. W. Brooks, *Zeppelin: Rigid Airships, 1893–1940* (Washington, D.C.: Smithsonian Institution Press, 1992), p. 32; Lehmann, *Zeppelin*, pp. 119–20. A detailed character description is provided in Italiaander, *Ferdinand Graf von Zeppelin*, pp. 74–76. On the missing finger, see G. W. Price, "How It Feels to Fall into a Forest," *Literary Digest*, August 6, 1910, p. 218.

25. E. L. Kuhn, "Zeppelin und die Folgen: Die Industrialisierung der Stadt Friedrichshafen," undated presentation, Henry Cord Meyer Papers, Box 3, Folder 2, pp. 2, 6–9.

26. Belafi, *The Zeppelin*, pp. 25–27; "Leona Dare's Trapeze" (June 9, 1879), "The Romance in a Trapeze Performer's Life" (November 26, 1880), and her obituary (May 25, 1922), *The New York Times*.

7. A Wonderfully Ingenious Toy

1. P. P. Wegener, *What Makes Airplanes Fly? History, Science, and Applications of Aerodynamics* (New York: Springer-Verlag, 1991), pp. 124–25; Hallion, *Taking Flight*, pp. 27–28; J. D. Anderson, Jr., *The Airplane: A History of Its Technology* (Reston, Va.: American Institute of Aeronautics and Astronautics, 2002), p. 10.

2. Quoted in A. F. Zahm, *Aerial Navigation: A Popular Treatise on the Growth of Air Craft and on Aeronautical Meteorology* (New York: D. Appleton and Co., 1911), pp. 17–18.

3. Anderson, *The Airplane*, pp. 4–5.

4. On Cayley's early experiments, see J.A.D. Ackroyd, "Sir George Cayley: The Invention of the Aeroplane near Scarborough at the Time of Trafalgar," *Journal of Aerospace History* 1 (2011), p. 149; C. H. Gibbs-Smith, *The Invention of the Aeroplane, 1799–1909* (New York: Taplinger Publishing Co., 1966), pp. 5–8; Hallion, *Taking Flight*, p. 110.

5. Stokes, "Power for Flight in the Victorian Era," p. 10; Ackroyd, "Sir George Cayley," p. 156.

6. See G. Cayley, "On Aerial Navigation," parts 1–3, 1809–1810, reprinted in *The Wind and Beyond: A Documentary Journey into the History of Aerodynamics in America*, ed. J. R. Hansen, D. B. Taylor, J. Kinney, and J. L. Lee (Washington, D.C.: National Aeronautics and Space Administration, 2003), Part I: *The Ascent of the Airplane*, pp. 34–56.

7. Gibbs-Smith, *The Invention of the Aeroplane*, pp. 8–9.

8. F. Wenham, "On Aerial Locomotion" (1866), "Report on the First Wind Tunnel" (1871), and "Minutes of Aeronautical Society" (1872), reprinted in Hansen et al. (eds.), *Wind and Beyond*, part 1, pp. 58–88.

9. Anderson, *The Airplane*, pp. 45–51. See also aerosociety.com/News/Insight-Blog/3949/The-Royal-Aeronautical-Society-Part-1-The-early-years.

10. H. S. Maxim, *Artificial and Natural Flight* (London: Whittaker and Co., 1908), p. 31.

11. H. S. Maxim, *My Life* (London: Methuen & Co., 1915), p. 294.

12. Anderson, *The Airplane*, pp. 52–57; Gibbs-Smith is particularly harsh. See Gibbs-Smith, *The Invention of the Aeroplane*, p. 21.

13. Letter of December 8, 1896, quoted in Hallion, *Taking Flight*, p. 167.

14. "Airship or Aeroplane—Which," *Scientific American*, December 9, 1899, p. 370.

15. Letter, W. Wright to the Smithsonian Institution, May 30, 1899, reprinted in Hansen et al. (eds.), *Wind and Beyond*, part 1, pp. 188–89.

8. The Folly

1. On the office, see Lehmann, *Zeppelin*, p. 119.

2. Belafi, *The Zeppelin*, p. 27.

3. Robinson, *Giants in the Sky*, p. 23; Belafi, *The Zeppelin*, p. 30.

4. For the list of names, see Belafi, *The Zeppelin*, p. 34.

5. E. Wolf, "The First Flight of Count Zeppelin's Air Ship," *McClure's Magazine*, November 1900, pp. 8–9.

6. On the hangar, see Brooks, *Zeppelin: Rigid Airships*, p. 30; Belafi, *The Zeppelin*, pp. 28–30. *Scientific American* had a detailed, early report on the hangar, though much of the information is inaccurate. See "Count von Zeppelin's Dirigible Airship," *Scientific American*, November 11, 1899.

7. Burr is usually cited in authoritative sources as being a member of the crew, including Lehmann, *Zeppelin*, p. 121. Robinson seems to be unique in omitting Burr and replacing him with a mechanic named Eisele. See Robinson, *Giants in the Sky*, p. 25.

8. For LZ-1's technical specifications and its flights, see Robinson, *Giants in the Sky*, pp. 23–26; Belafi, *The Zeppelin*, pp. 28–33; L. Dürr (trans. A. Reid), *25 Years of Zeppelin Airship Construction* (1924; repr. Lulu Publishing, 2013), pp. 25–27. Depending on which source one consults, there are different numbers given for the engine horsepower; they range from 14 to 16. H.W.L. Moedebeck, the most authoritative source, refers to "a H.P. Daimler motor giving 14.7 H.P. at 680 revolutions." H.W.L. Moedebeck (trans. W. M. Varley), *Pocket-Book of Aeronautics* (London: Whittaker & Co., 1907), p. 332. Eugen Wolf's detailed firsthand account is a crucial resource. Wolf, "The First Flight of Count Zeppelin's Air Ship," pp. 3–11.

9. On Kübler, see Lehmann, *Zeppelin*, pp. 117–18.

10. On Gemmingen, see D. Topping, *When Giants Roamed the Sky: Karl Arnstein and the Rise of Airships from Zeppelin to Goodyear*, ed. E. Brothers (Akron, Oh.: University of Akron Press, 2001), pp. 46–47.

11. On the volunteers, see Wolf, "The First Flight," p. 9.

12. Quoted in Belafi, *The Zeppelin*, pp. 35–37.

13. Wolf, "The First Flight," p. 10.

14. Robinson, *Giants in the Sky*, p. 27; Brooks, *Zeppelin: Rigid Airships*, p. 33.

15. Toland, *The Great Dirigibles*, p. 40; H.W.L. Moedebeck, "The Termination of the Trials of Count von Zeppelin's Airship," *Scientific American*, April 13, 1901, Supplement, pp. 21138–21140. On Krogh, see Belafi, *The Zeppelin*, p. 48; "Count von Zeppelin's Dirigible Airship," *Scientific American*, March 3, 1906, p. 195.

16. Brooks, *Zeppelin: Rigid Airships*, p. 33; Lehmann, *Zeppelin*, p. 124. On the banking crisis, see C. M. Reinhart and K. S. Rogoff, *This Time is Different: Eight Centuries of Financial Folly* (Princeton, N.J.: Princeton University Press, 2009), Appendix A.4, "Historical Summaries of Banking Crises," p. 365.

17. W. von Schierbrand (ed.), *The Kaiser's Speeches: Forming a Character Portrait of Emperor William II* (New York and London: Harper & Brothers, 1903), pp. 258–59.

18. G. de Syon notes German press reaction. See G. de Syon, *Zeppelin! Germany and the Airship, 1900–1939* (Baltimore: Johns Hopkins University Press, 2002), p. 23. The *Frankfurter Zeitung* is quoted in Nitske, *The Zeppelin Story*, p. 58.

19. "The Ascension of Count von Zeppelin's Airship," *Scientific American*, August 11, 1900, p. 88.

20. The report is quoted in Robinson, *Giants in the Sky*, pp. 26–27.

21. Italiaander, *Ein Deutscher Namens Eckener*, p. 76.

22. H. Eckener (trans. Robinson), *My Zeppelins* (London: Putnam, 1958), p. 11. For Ecken-

er's report from the second flight, see P. Fritzsche, *A Nation of Fliers: German Aviation and the Popular Imagination* (Cambridge, Mass.: Harvard University Press, 1992), pp. 7–8, and Italiaander, *Ein Deutscher Namens Eckener*, p. 94.

9. The Surprise

1. Recollection of Oskar Wilcke, printed in Italiaander, *Ferdinand Graf von Zeppelin*, pp. 140–41.
2. The letter of congratulation (dated January 7 and 15, 1901) is printed in Italiaander, *Ferdinand Graf von Zeppelin*, p. 86; and M. Goldsmith, *Zeppelin: A Biography* (New York: William Morrow and Co., 1931), pp. 120–21.
3. Belafi, *The Zeppelin*, p. 44.
4. As noted by Moedebeck in his "The Termination of the Trials of Count von Zeppelin's Airship," pp. 21138–21140.
5. Dürr, *25 Years of Zeppelin Airship Construction*, p. 39; Robinson, *Giants in the Sky*, p. 29.
6. Dürr, *25 Years of Zeppelin Airship Construction*, pp. 50–51; illustrations 51–54, p. 49; illustrations 55–56, p. 50.
7. Belafi, *The Zeppelin*, pp. 44–45; Brooks, *Zeppelin: Rigid Airships*, pp. 34–35. Fuel consumption fell by a third between LZ-1 and LZ-2. See Dürr, *25 Years of Zeppelin Airship Construction*, p. 83, Table 2, "Airship Engines."
8. Lehmann, *Zeppelin*, p. 125. On Hergesell's appearance, see his photograph in "How Zeppelin Plans to Try to Reach the North Pole by Airship," *The New York Times*, July 25, 1909.
9. Belafi, *The Zeppelin*, pp. 45–46, and associated photographs. It is unclear how movable the rudders were. Zeppelin was more concerned with controlling altitude than direction at this stage.
10. Letters of February 18 and February 16, 1903, quoted in Italiaander, *Ein Deutscher Namens Eckener*, p. 92.
11. Eckener, *Count Zeppelin*, p. 222.
12. Belafi, *The Zeppelin*, pp. 44–45; Nitske, *The Zeppelin Story*, p. 60. The *Die Woche* article proposed raising funds to transport a dismantled airship to New York and then sending it by train to St. Louis for the 1904 World's Fair, where it would be reassembled. On this outlandish scheme, soon quietly dropped, see Goldsmith, *Zeppelin*, p. 125. The aeronautical writer A. L. Rotch was the count's American contact for the St. Louis proposal, though Rotch was under the impression that the airship was to be built *in* St. Louis. See his *The Conquest of the Air, or the Advent of Aerial Navigation* (New York: Moffat, Yard and Co., 1909), p. 104. On Zeppelin's ham-handed appeal, see De Syon, *Zeppelin!*, p. 26. For the *Vaterland* anecdote, see Meyer, *Count Zeppelin*, p. 50.
13. Belafi, *The Zeppelin*, p. 45.
14. G. Hartcup, *The Achievement of the Airship: A History of the Development of Rigid, Semi-Rigid, and Non-Rigid Airships* (North Pomfret, Vt.: David & Charles, 1974), pp. 30–39. See "German-French Airship Rivalry," *Scientific American*, August 24, 1913; "Comparison of French and German Strength in Dirigible Airships," August 16, 1913, pp. 126, 133–34; De Syon, "Toys or Tools? The Riddle of French Army Airships, 1884–1914."
15. Regarding Zeppelin's finances, there are varying accounts available. The most authoritative seems to be Belafi, *The Zeppelin*, p. 45, but see also Nitske, *The Zeppelin Story*, pp. 60–61; Robinson, *Giants in the Sky*, p. 29; and Brooks, *Zeppelin: Rigid Airships*, pp. 34–35.
16. Lehmann, *Zeppelin*, pp. 118–19.

17. As related in Lehmann, *Zeppelin*, p. 125. For Krogh and Dürr's placement, see "Count von Zeppelin's Dirigible Airship," *Scientific American*, p. 196.

18. Italiaander, *Ferdinand Graf von Zeppelin*, p. 134.

19. Lehmann, *Zeppelin*, pp. 126–27.

10. The Equestrian

1. Lehmann, *Zeppelin*, p. 128.

2. Quoted in Belafi, *The Zeppelin*, p. 49.

3. Eckener (trans. Robinson), *My Zeppelins*, pp. 11–14; and the recollections in his *Count Zeppelin*, p. 225. Meyer cites the *der Luftikus* remark in *Count Zeppelin: A Psychological Portrait*, p. 50. Elphberg's article, "Wasted a Fortune on a Flying Machine," *Baltimore American*, December 30, 1906, appeared in several other papers, including the *Omaha World-Herald*. Zeppelin cited his lack of experience and the estimate of three hours of flying time in his article "Dirigible Airships," *Scientific American*, Supplement, December 22, 1906, p. 25887; see also Belafi, *The Zeppelin*, p. 50. Eckener's "vague" feeling about the dinner is recorded in Italiaander, *Ein Deutscher Namens Eckener*, p. 108. See also Eckener's much later recollection of the conversation, in T. Nielsen (trans. P. Chambers), *The Zeppelin Story* (London: Allen Wingate, 1955), pp. 35–39.

4. On the Wrights and their outlook on flying, see P. L. Jakab, *Visions of a Flying Machine: The Wright Brothers and the Process of Invention* (Washington, D.C.: Smithsonian Institution Press, 1990).

5. D. McCullough, *The Wright Brothers* (New York: Simon & Schuster, 2015), pp. 103–6.

6. H. G. Wells, *Anticipations of the Reaction of Mechanical and Scientific Progress upon Human Life and Thought* (London: Chapman & Hall, 1902), p. 191.

7. It's possible that Zeppelin was dimly aware of the Wrights this early. Moedebeck had published an article by Wilbur in the July 1901 issue of his journal, *Illustrated Aeronautical Reports*, "Die Wagerechte Lage Während des Gleitfluges" (The horizontal pilot-position in gliding flight). Gibbs-Smith, *Invention of the Aeroplane*, Appendix V, "European Aviation: 1901–1903," pp. 305–6.

8. J. Tobin, "The First Witness: Amos Root at Huffman Prairie" (September 28, 2001), *Following in the Footsteps of the Wright Brothers: Their Sites and Stories*, Paper 7. corescholar.libraries.wright.edu/following/symposium/program/7.

9. On the cost of *Flyer I*, see McCullough, *The Wright Brothers*, p. 108.

10. Only a month before the Wrights first took off, for instance, a German civil servant named Karl Jatho had made a 180-foot journey in a 9-horsepower-engine airplane equipped with rudimentary elevators and rudders. It turned out that the "flight" was merely a Maxim-style forced *Flugsprung* (leap into the air) by a craft unsustainably carried forward by its own momentum and no more work was ever done on it. Gibbs-Smith, *Invention of the Aeroplane*, p. 51; see also "Aeronautics Now Are Far Advanced," *The New York Times*, October 7, 1907.

11. Gibbs-Smith, *Invention of the Aeroplane*, p. 87.

12. Eckener (trans. Robinson), *My Zeppelins*, pp. 13–14; Nielsen (trans. Chambers), *The Zeppelin Story*, p. 43; Italiaander, *Ein Deutscher Namens Eckener*, p. 107.

13. Zeppelin, "The Truth About My Airship," dated mid-February 1906, and co-signed by Eckener, printed in Italiaander, *Ferdinand Graf von Zeppelin*, pp. 80–84.

14. Italiaander, *Ein Deutscher Namens Eckener*, pp. 14–20, 77.

15. Italiaander, *Ein Deutscher Namens Eckener*, pp. 52–57.

16. Letter to Ottomar Enking, printed in Italiaander, *Ein Deutscher Namens Eckener*, p. 54.

17. By his daughter. See Italiaander, *Ein Deutscher Namens Eckener*, p. 77.

18. Quoted in Italiaander, *Ein Deutscher Namens Eckener*, p. 78.

19. Quoted in Italiaander, *Ein Deutscher Namens Eckener*, p. 80.

20. Various letters, quoted in Italiaander, *Ein Deutscher Namens Eckener*, pp. 83, 62, 80.

21. Eckener, *Count Zeppelin*, p. 226.

22. Belafi, *The Zeppelin*, p. 54.

23. On Gross's comment, see Italiaander, *Ferdinand Graf von Zeppelin*, p. 87. The ill feeling would never quite dissipate. Years later, when Zeppelin was made aware that a Gross-Basenach would be present when one of his own airships made an appearance before the kaiser, he installed a small cannon in his gondola to fire a blank shot across its bow. It was only by the intervention of his friend, the aviation enthusiast Georg von Tschudi, that Zeppelin was dissuaded from using it. G. von Tschudi, *Aus 34 Jahren Luftfahrt: Persönliche Erinnerungen* (Berlin: Reimar Hobbing, 1928), pp. 94–95.

24. Hartcup, *Achievement of the Airship*, p. 69.

25. Meyer, *Count Zeppelin*, p. 58.

26. Belafi, *The Zeppelin*, p. 55.

27. Meyer, *Count Zeppelin*, p. 56. A memorandum to the minister of war from Zeppelin, marked "Secret," of February 10, 1906, exhibited the same confidence. Printed in Italiaander, *Ferdinand Graf von Zeppelin*, pp. 84–85. For Eckener's comment about Parseval, see his letter to Friedrich Maass, quoted in Italiaander, *Ein Deutscher Namens Eckener*, pp. 126–27.

28. Belafi, *The Zeppelin*, p. 51; Robinson, *Giants in the Sky*, p. 31.

29. Belafi, *The Zeppelin*, p. 50.

30. Hartcup, *Achievement of the Airship*, pp. 39–40.

31. H. Hergesell and Captain von Kehler, "The Flights of Zeppelin's Airship on October 9 and 10, 1906," *Scientific American*, Supplement, February 23, 1907, p. 26037.

32. Eckener, October 14, 1906, *Frankfurter Zeitung*, printed in Italiaander, *Ein Deutscher Namens Eckener*, pp. 101–3.

33. Parseval later sold his rights and patents to the army for 130,000 marks. Between 1909 and 1914, some twenty-three blimps would be built, but only six were bought by the army: Eight would be sold to foreign governments as far afield as Japan and the rest to companies to take tourists on brief sightseeing trips. Hartcup, *Achievement of the Airship*, p. 41.

34. Robinson, *Giants in the Sky*, pp. 32–33.

35. J.C.G. Röhl (trans. S. De Bellaigue), *Kaiser Wilhelm II, 1859–1941: A Concise Life* (Cambridge, U.K.: Cambridge University Press, 2014), pp. 103–8.

36. Belafi, *The Zeppelin*, pp. 53–54.

37. Robinson, *Giants in the Sky*, pp. 33, 35; Meyer, *Count Zeppelin*, p. 57; Italiaander, *Ferdinand Graf von Zeppelin*, pp. 89–91.

11. Up into the Empyrean

1. Belafi, *The Zeppelin*, p. 64; Brooks, *Zeppelin: Rigid Airships*, pp. 36–37.

2. "Remains in Air, 4h 17m," *Boston Daily Globe*, September 25, 1907.

3. Recollections of Oskar Wilcke, in Italiaander, *Ferdinand Graf von Zeppelin*, pp. 140–41. The first woman to fly a motorized dirigible, if only for a brief time, was Aida de Acosta, an American socialite who in June 1903 persuaded Alberto Santos-Dumont to allow her to pilot his one-man personal dirigible in Paris.

4. Robinson, *Giants in the Sky*, p. 34.

5. "Germany and Count Zeppelin's Airship," *Irish Times*, October 9, 1907.

6. Brooks, *Zeppelin: Rigid Airships*, p. 37; Belafi, *The Zeppelin*, p. 64; Nitske, *The Zeppelin Story*, p. 67; Meyer, *Count Zeppelin*, p. 57; Hartcup, *Achievement of the Airship*, p. 70.

7. A.M.B., "Count Zeppelin," *Daily Mail*, June 20, 1908.

8. "Count Zeppelin a Union Veteran," *The Washington Post*, July 23, 1908. I traced this genealogy by following the breadcrumbs in geni.com/people/Ferdinand-Graf-von-Zeppelin/6000000008469593415.

9. "Has Conquered the Air," *Cincinnati Enquirer*, November 25, 1907.

10. A. M. Krecker, "Seven Great Wonders of Science and Industry Perfected in 1907," *Chicago Daily Tribune*, December 15, 1907.

11. "Airship Club in Germany," *The New York Times*, December 22, 1907.

12. "What Will Society Do Next?" *St. Louis Post-Dispatch*, March 15, 1908.

13. "Aeroplane Crushed After Long Flight," *The New York Times*, May 15, 1908.

14. Zeppelin, "Dirigible Airships," *Scientific American*, Supplement, December 22, 1906, pp. 25886, 25888.

15. Belafi, *The Zeppelin*, p. 65; "Big Airship Flies," *The Washington Post*, June 21, 1908.

16. Robinson, *Giants in the Sky*, p. 35; Belafi, *The Zeppelin*, 65; Brooks, *Zeppelin: Rigid Airships*, p. 39.

17. Interestingly, Eckener was one of the very few who noted the September flight in an article for the *Hamburg Foreigners' Paper*, following it up with a look at the problem-prone progress of Parseval's balloon. Both pieces were, as might be expected, skeptical as to their merits, but one wonders, since these vehicles were secret, who the source of his information was: leaked by Zeppelin, no doubt. Eckener, "The Thirteen-Hour Flight of the Military Airship," September 15, 1908; and "An Expert About the Accident," September 17, 1908, in the *Hamburg Foreigners' Paper*. Italiaander, *Ein Deutscher Namens Eckener*, p. 104.

12. Conquerors of the Celestial Ocean

1. Nitske, *The Zeppelin Story*, p. 70.

2. Italiaander, *Ein Deutscher Namens Eckener*, pp. 103, 106.

3. Quoted in Italiaander, *Ein Deutscher Namens Eckener*, p. 108.

4. This section is based on the following accounts: Hergesell, "With Zeppelin in His Airship," *The Globe*, July 29, 1908; Hergesell's *Die Woche* narrative in Belafi, *The Zeppelin*, pp. 65–66; "Makes Flight of 12 Hours," *Boston Daily Globe*, July 2, 1908; E. Sandt, "A Trip in the Zeppelin Airship," *Scientific American*, Supplement, August 15, 1908.

5. A description and illustration of the Bassus device is in A. Hildebrandt (trans. H. W. Story), *Airships Past and Present* (New York: D. Van Nostrand Co., 1908), pp. 306–7. See also T. Kramer and H. Stadler (eds.), *Eduard Spelterini: Photographs of a Pioneer Balloonist* (Zurich: Scheidigger & Speiss, 2008).

6. Nitske, *The Zeppelin Story*, p. 70; Lehmann, *Zeppelin*, p. 133; "Kaiser Greets Zeppelin," *The New York Times*, July 3, 1908; "King and Queen Aeronauts," *Chicago Daily Tribune*, July 4; "A King in an Airship," *The Guardian and Observer*, July 5.

7. Nitske, *The Zeppelin Story*, pp. 70–71.

13. The Flames of Hell

1. Robinson, *Giants in the Sky*, p. 36.
2. "Zeppelin's Airship Disabled, Descends After 12-Hour Sail," *St. Louis Post-Dispatch*, August 4, 1908; "Air Conquered; Zeppelin Flies in His Machine," *Atlanta Constitution*, August 5, 1908.
3. "On Last Leg of Journey," *Courier-Journal*, August 5, 1908; "People Awed by Monster of Air," *Nashville American*, August 5, 1908.
4. "On Last Leg of Journey."
5. Fritzsche, *A Nation of Fliers*, p. 9.
6. "People Awed."
7. H. Eckener, *Im Zeppelin über Länder und Meere* (Flensburg, Germany: Verlagshaus Christian Wolff, 1949), p. 404. The Island of the Blessed of Greek mythology, sometimes known as Elysium, was an earthly paradise populated by the righteous and the heroic.
8. Robinson, *Giants in the Sky*, pp. 37–39; Belafi, *The Zeppelin*, pp. 70–71.
9. Quoted in Fritzsche, *A Nation of Fliers*, p. 13.
10. Fritzsche, *A Nation of Fliers*, p. 13.
11. Lehmann, *Zeppelin*, pp. 136–39.
12. "Sympathy with Count Zeppelin," *Manchester Guardian*, August 7, 1908. See also the photograph of Zeppelin's departure in Belafi, *The Zeppelin*, p. 73.
13. "Tragic End for Monarch of Air," *Boston Daily Globe*, August 6, 1908; Lehmann, *Zeppelin*, pp. 139–40.
14. "Germany Rushes to Zeppelin's Aid," *The New York Times*, August 7, 1908.
15. D. Lloyd George, *War Memoirs of David Lloyd George* (London: Odhams Press, 2 vols., 1938 ed.), volume 1, p. 19.
16. "Sympathy with Count Zeppelin," *Manchester Guardian*.

14. The Miracle

1. Lehmann, *Zeppelin*, pp. 139–40.
2. Belafi, *The Zeppelin*, pp. 71–72.
3. Fritzsche, *A Nation of Fliers*, pp. 15–17; Nitske, *The Zeppelin Story*, p. 75; Lehmann, *Zeppelin*, pp. 140–42; Robinson, *Giants in the Sky*, p. 41; Georg Hacker, *Die Männer von Manzell* (Frankfurt am Main: Societäts-Druckerei, 1936), p. 104. On currency conversion, see measuringworth.com and history.ucsb.edu/faculty/marcuse/projects/currency .htm. I have used an approximate exchange rate of 4 marks to the pre-1914 dollar. An article in *Machinery*, June 1, 1909, states that Zeppelin received $1,300,000.
4. On LZ-3, see H. Eckener, "The Remodeled 'Zeppelin III' Airship," *Scientific American*, Supplement, December 12, 1908, pp. 380–81.
5. Quoted in Lehmann, *Zeppelin*, p. 130.
6. For Wilhelm's speech, see Italiaander, *Ferdinand Graf von Zeppelin*, p. 107.
7. Robinson, *Giants in the Sky*, p. 43; Meyer, *Count Zeppelin*, p. 67; photos in Italiaander, *Ferdinand Graf von Zeppelin*, p. 145, and Belafi, *The Zeppelin*, p. 153.
8. As referenced in "Monument to Zeppelin," *The New York Times*, October 18, 1908.
9. Goldsmith, *Zeppelin*, pp. 143–44; Fritzsche, *A Nation of Fliers*, pp. 17–18.
10. Advertisement (Thermos) in *Life*, August 18, 1910, p. 257; Fritzsche, *A Nation of Fliers*, pp. 36–37 (games), 20 (large Zeppelin); M. Cutler, "New Christmas Toys," *Harper's Ba-*

zaar, December 1909, p. 1223 (see also "Airship Toys," *Crockery and Glass Journal*, July 22, 1909, p. 10); *The Youth's Companion*, October 20, 1910, p. 575 (mini-Zeppelins).

11. De Syon, *Zeppelin!*, pp. 45, 63.

12. See letters reproduced in H. G. Knäusel, *Zeppelin and the United States of America: An Important Episode in German-American Relations* (Friedrichshafen, Germany: Luftschiffbau Zeppelin, 1981), pp. 26, 170.

13. Freud's comment appears in his *Introductory Lectures on Psycho-Analysis* (London: Allen & Unwin, 1929), p. 155. A sixteen-year-old, diagnosed by Oskar Pfister with obsessional neurosis, dreamt that "I saw a Zeppelin airship and went after it. It landed in H. on a meadow. Then there was something with maps in the car or somewhere else. Then I went off and was finally in C. near the station there. I asked for directions how to get home and was led to a house. There were various dried fish and thick green seaweed, out of which, a white worm came. Then, I finally came home. Everything was full of laundry in great disorder. Then the Zeppelin flew directly over our house and made a kind of salt hail. Then someone said to me, that is a trial of a method by which it could destroy all crops in case of war." O. Pfister (trans. C. R. Payne), *The Psychoanalytic Method* (New York: Moffat, Yard, and Company, 1919), p. 283.

14. Fritzsche, *A Nation of Fliers*, pp. 17, 20.

15. *The New York Times* ridiculed the pseudonymous "Rudolph von Elphberg," author of that December 1906 article devastatingly lampooning Zeppelin as a bankrupt fool, in "Zeppelin's Strenuous Career," July 25, 1909.

16. H. Kaufman, "Learn to Believe in Yourself," *Chicago Daily Tribune*, August 30, 1908. See, similarly, "Compensation," *The Youth's Companion*, September 3, 1908. For a summary of changing views of Zeppelin, see G. de Syon, "Searching for the German Hero: Biographies of Count Zeppelin, 1908–1938," in *Memory, History, and Critique: European Identity at the Millennium*, ed. F. Brinkhuis and S. Talmor, Proceedings of the Fifth ISSEI Conference at the University for Humanist Studies, Utrecht, The Netherlands, August 1996 (Cambridge, Mass.: MIT Press, 1998).

15. Kings of the Sky

1. "Zeppelin Tells How His Airship Was Destroyed," *St. Louis Post-Dispatch*, August 6, 1908.

2. Robinson, *Giants in the Sky*, p. 66.

3. Fritzsche, *A Nation of Fliers*, p. 25. For a fuller excerpt, see Italiaander, *Ferdinand Graf von Zeppelin*, p. 96.

4. G. De Syon, "Bangs and Whimpers: The German Public and Two Zeppelin Disasters," in R. B. Browne and A. G. Neal (eds.), *Ordinary Reactions to Extraordinary Events* (Bowling Green, Oh.: Popular Press, 2001), pp. 173–74.

5. "The Zeppelin Airship Disaster," *Scientific American*, August 15, 1908; "The Wreck of the Zeppelin IV," *Scientific American*, Supplement, September 5, 1908, p. 149.

6. As early as 1906, Eckener had proposed building a network of safe "airship harbors" throughout Germany. Eckener's article, dated October 14, 1906, is quoted in Belafi, *The Zeppelin*, p. 52.

7. "Tragic End for Monarch of Air," *Boston Daily Globe*, August 6, 1908.

8. McCullough, *The Wright Brothers*, pp. 152–53, 160.

9. McCullough, *The Wright Brothers*, pp. 155–73, 210; *The Observer*, August 16, 1908; "The Wright Aeroplane Tests," *Scientific American*, August 29, 1908. For Wilbur as man of the

year and century, see George Prade's article in *Les Sports*, January 1, 1909, cited in R. Wohl, *A Passion for Wings: Aviation and the Western Imagination, 1908–1918* (New Haven, Conn.: Yale University Press, 1994), p. 33.

10. A. Gollin, *No Longer an Island: Britain and the Wright Brothers, 1902–1909* (Stanford, Calif.: Stanford University Press, 1984), p. 291.

11. McCullough, *The Wright Brothers*, pp. 183–92; Gollin, *No Longer an Island*, pp. 371–72.

12. McCullough, *The Wright Brothers*, p. 174.

13. Advertisements in *Current Literature*, July 1909, p. 115; *Smart Set*, November 1, 1909.

14. *The Youth's Companion*, October 20, 1910, p. 575; M. Cutler, "New Christmas Toys," *Harper's Bazaar*, December 1909, p. 1223.

15. "Zeppelin in Berlin; Welcomed by Kaiser," *The New York Times*, August 30, 1909. See also the report in the *Tägliche Rundschau*, printed in Italiaander, *Ferdinand Graf von Zeppelin*, pp. 110–11.

16. Quoted in G. De Syon, "Airplane or Airship? The Cultural Dimensions of a Transatlantic Debate," paper presented at the 4th International Airship Convention and Exhibition, Cambridge, England, July 31, 2002, p. 2.

17. "Count Zeppelin Praises Wright," *Atlanta Constitution*, August 14, 1908.

18. "Wright Sees No Hope in Zeppelin Airship," *The New York Times*, September 4, 1909.

19. "Wright Makes Balloon Trip," *The New York Times*, September 16, 1909.

20. "Wright 1,600 Feet Up in Berlin Flight," *The New York Times*, October 3, 1909; "The Record Flights of Orville and Wilbur Wright," *Scientific American*, October 16, 1909, p. 274.

21. For Wilbur's views, see Gollin, *No Longer an Island*, pp. 299–300.

16. Zeppelin City

1. For a compilation of aircraft developers in 1909, see Gibbs-Smith, *Invention of the Aeroplane*, Appendix 1, pp. 268–87. For the later numbers, see "Aviation as a Business," *The Washington Post*, August 27, 1911.

2. McCullough, *The Wright Brothers*, pp. 252–58.

3. *Machinery*, June 1, 1909, p. 795.

4. For these vignettes, see Lehmann, *Zeppelin*, pp. 117–18.

5. On Colsman, see H.-O. Swientek, "Alfred Colsman (1873–1955): Ein Leben Für Die Deutsche Luftfahrt," *Tradition: Zeitschrift für Firmengeschichte und Unternehmerbiographie* 10 (1965), no. 3, pp. 112–26. Much of the following information is drawn from this key source, as well as from Italiaander, *Ferdinand Graf von Zeppelin*, pp. 97–98. On his physical appearance, see Price, "How It Feels to Fall into a Forest," *Literary Digest*, p. 218.

6. Robinson, *Giants in the Sky*, p. 41; "School of Aeronauts," *The Washington Post*, July 2, 1908, reporting on a speech by Zeppelin to the Assocation of German Engineers.

7. "Zeppelin Building Factory," *The New York Times*, September 7, 1908; Robinson, *Giants in the Sky*, p. 42.

8. Swientek, "Alfred Colsman," p. 122.

9. Lehmann, *Zeppelin*, p. 142.

10. For Eckener's comment, see Swientek, "Alfred Colsman," p. 117.

11. Belafi, *The Zeppelin*, pp. 73, 113; E. L. Kuhn, "Zeppelin und die Folgen: Die Industrialisierung der Stadt Friedrichshafen," undated transcript of a seminar talk, in the Henry Cord Meyer Papers, Box 3, Folder 2, p. 23. On women employees, see De Syon, *Zeppelin!*, p. 65.

12. Belafi, *The Zeppelin*, pp. 113–16; H. Vissering, *Zeppelin: The Story of a Great Achievement* (Chicago: Henry Vissering, 1922), pp. 41–42; Meyer, "Building Rigid Airships: Three Communities and Their Changing Fortunes," in *Airshipmen, Businessmen, and Politics*, pp. 99–100; Kuhn, "Zeppelin und die Folgen," pp. 19–20 (there had been threats of a strike in 1911). Such model villages as Zeppelin's were not uncommon in Germany, though by this time they were usually run by municipalities rather than by private firms. See, for instance, F. C. Howe, "Düsseldorf: A City of Tomorrow," *Hampton's Magazine*, December 1, 1910. Krupp, the industrial concern, originated the concept of a company town serving as a state within a state in the early 1860s. Zeppelin's opinion on the matter is encapsulated in a diary entry of June 1896: "I too see in the advanced state of discontent among agricultural and industrial wage-earners and of many of the petty bourgeoisie and small farmers a great and imminent danger to the existing political and social order. I share the view that the authors and disseminators of this discontent must be the more energetically opposed, the less real ground there is for discontent or where it exists, as in a general trade depression—the less the government and employers are able to remedy it. I do not base my opinion, like others, upon the right of self-preservation—no one has a moral right to preservation at the cost of his betters—but upon the unconditional duty of defending the existing weal of the whole community against forces which merely destroy without being able to create anything valuable." Quoted in Eckener, *Count Zeppelin*, pp. 202–3.

13. Quoted in Meyer, "In Search of the Real Count Zeppelin," in *Airshipmen, Businessmen, and Politics, 1890–1940*, p. 41.

17. The Wonder Weapons

1. See I. F. Clarke, *The Great War with Germany, 1890–1914: Fictions and Fantasies of the War-to-Come* (Liverpool, U.K.: Liverpool University Press, 1997), pp. 233–47; R. Wohl, *A Passion for Wings: Aviation and the Western Imagination, 1908–1918* (New Haven, Conn.: Yale University Press, 1994), pp. 76–79.

2. Quoted in Fritzsche, *A Nation of Fliers*, p. 40.

3. Quoted in Meyer, *Count Zeppelin*, p. 55.

4. Quoted in De Syon, *Zeppelin!*, p. 76.

5. Meyer, "In Search of the Real Count Zeppelin," in Meyer, *Airshipmen, Businessmen, and Politics*, p. 42.

6. Meyer, "In Search of the Real Count Zeppelin," pp. 40, 45.

7. In an interview with the *Daily Mail*, July 11 and October 9, 1908; A. M. Gollin, "England Is No Longer an Island: The Phantom Airship Scare of 1909," *Albion* 13 (1981), no. 1, pp. 46, 48; Wohl, *A Passion for Wings*, pp. 80, 296, n. 37. Martin may have flattered himself, or liked to boast, that he had the kaiser's ear, but he was in fact dismissed from his Statistical Office position by a disciplinary hearing around the time of the interview.

8. D. Clarke, "Scareships over Britain: The Airship Wave of 1909," *Fortean Studies* 6 (1999), pp. 39–63.

9. For an early mention of the concept, see "Lines of Airships," *Los Angeles Times*, July 9, 1908.

10. De Syon, *Zeppelin!*, p. 78. Moltke only kept those three, Colsman claimed, because he trembled at the thought of the public backlash that would ensue if the army yet again "ignored" St. Zeppelin. Colsman, cited in Robinson, *Giants in the Sky*, p. 62.

11. "Zeppelin Airships No Longer in Favor," *The New York Times*, May 23, 1909.

18. The Lucky Ship

1. O. C. von Verschuer, "Airship Anchorages," *Scientific American*, June 17, 1911, p. 384.

2. Belafi, *The Zeppelin*, pp. 91–92.

3. On the *Reichshalle*, see Robinson, *Giants in the Sky*, p. 52.

4. "The Aeronautic Exposition in Frankfort," *Scientific American*, Supplement, no. 1755, August 21, 1909, pp. 121–22.

5. Robinson, *Giants in the Sky*, p. 52; Italiaander, *Ein Deutscher Namens Eckener*, p. 112.

6. "Riding on the Air Flyer," *Kansas City Times*, June 23, 1910.

7. Belafi, *The Zeppelin*, p. 93; Robinson, *Giants in the Sky*, p. 55. See also "The First Aerial Liner," *Literary Digest*, July 2, 1910, pp. 3–4.

8. G. Ward Price, "The Psychology of Air Travel," *Daily Mail*, July 5, 1910.

9. C. Dienstbach, "The Wreck of the *Deutschland*," *Scientific American*, July 9, 1910, p. 26.

10. "The *Deutschland*'s Fall," *Literary Digest*, July 23, 1910, p. 129.

11. "*Deutschland*'s Fall," p. 129.

12. Price, "The Psychology of Air Travel." On the crash, see Robinson, *Giants in the Sky*, p. 55.

13. Quoted in Belafi, *The Zeppelin*, p. 93.

14. For the contract between Eckener and the DELAG, dated August 1911, see Italiaander, *Ein Deutscher Namens Eckener*, p. 114.

15. Robinson, *Giants in the Sky*, p. 56; Brooks, *Zeppelin: Rigid Airships*, pp. 61–63; Italiaander, *Ein Deutscher Namens Eckener*, p. 115.

16. On the average wage, see Duggan and Meyer, *Airships in International Affairs*, p. 38. On promotional materials, see Robinson, *Giants in the Sky*, pp. 56–58. On weekly fatalities between 1910 and 1914, see B. Williams, "The War Aeroplane Here and Abroad," *Scientific American*, November 4, 1916, p. 412.

17. Robinson, *Giants in the Sky*, p. 58.

18. Robinson, *Giants in the Sky*, pp. 58–59. Eckener mentioned the 150-trip rule in a 1913 speech, printed in Italiaander, *Ein Deutscher Namens Eckener*, p. 117.

19. "Wind Statistics for Aeronauts," *Scientific American*, April 30, 1910.

20. Zeppelin had previously cited Hergesell's work in an article, probably drafted with Eckener's help. See "Dirigible Airships," *Scientific American*, Supplement, December 22, 1906, pp. 25886–25888.

21. Lehmann, *Zeppelin*, p. 45.

22. T. R. MacMechen and C. Dienstbach, "Over Sea by Airship," *The Century*, May 1910, pp. 120–22.

23. Charles C. Turner, "Flying in the Future," *Scientific American*, Supplement, no. 1783, March 5, 1910, pp. 154–55; MacMechen and Dienstbach, "Over Sea by Airship," pp. 120–21; "How to Find Your Way in the Air," *Scientific American*, July 8, 1911, pp. 24–25. Eckener discussed the creation of the weather-information network in his memoirs, quoted in Italiaander, *Ein Deutscher Namens Eckener*, p. 115.

24. On Dürr, Arnstein, Jaray, and the relationship between design and analysis, see Topping, *When Giants Roamed the Sky*, pp. 34–43.

25. De Syon, *Zeppelin!*, pp. 84–85.

26. C. Dientsbach, "A Journey in a Passenger-Carrying Zeppelin Airship," *Scientific American*, June 29, 1912, pp. 580, 591–92. Parts of this section are based on his follow-up article, "A Journey in a Zeppelin: Impressions of a Trip in the Airship *Viktoria Luise*," *Scientific American*, May 17, 1913, pp. 449–50, but the circumstances were identical. See also W. Kaempffert, "How It Feels to Fly: From Châlons to Vincennes in a Biplane," *The World's*

Work 16 (1910), October, pp. 515–18; and T. Baldwin, "How It Feels to Fly," *The World's Work* 15 (1910), January, pp. 127–30. The poet Hermann Hesse described similar experiences, printed as "A Jaunt in the Air," printed in Italiaander, *Ferdinand Graf von Zeppelin*, pp. 120–22. For menus and beverages, see D. Robinson, "Passenger Flying in the Big Zeppelins," unpublished essay, in the Robinson Papers, Box 10, Folder 4, pp. 2, 4; G. de Syon, "Is it Really Better to Travel Than to Arrive? Airline Food as a Reflection of Consumer Anxiety," in *Food for Thought: Essays on Eating and Culture*, ed. L. C. Rubin (Jefferson, N.C.: McFarland & Co., 2008), p. 200; on stamps and postcards, see Robinson, "Passenger Flying in the Big Zeppelins," in Robinson Papers, Box 10, Folder 4, p. 4; on Heinrich Kubis, see facesofthehindenburg.blogspot.com/2008/12/heinrich-kubis.html.

19. The High Priests

1. Undated 1912 letter to Colsman, printed in Italiaander, *Ein Deutscher Namens Eckener*, p. 121.

2. Robinson, *Giants in the Sky*, p. 356.

3. Undated 1912 letter to Colsman, printed in Italiaander, *Ein Deutscher Namens Eckener*, p. 121. On the movie, see De Syon, *Zeppelin!*, p. 63.

4. T. R. MacMechen and C. Dienstbach, "The Greyhounds of the Air," *Everybody's Magazine*, September 1912, p. 295.

5. Robinson, *Giants in the Sky*, p. 61.

6. Robinson, *Giants in the Sky*, p. 67, n. 18; Belafi, *The Zeppelin*, pp. 103–4.

7. MacMechen and Dienstbach, "Greyhounds of the Air," pp. 300–301.

8. C. Dienstbach, "A Study of the Giant Airship of the Future," *Scientific American*, August 26, 1911, pp. 185–86.

9. MacMechen and Dienstbach, "Over Sea by Airship," pp. 127, 129.

10. MacMechen and Dienstbach, "Over Sea by Airship," pp. 115–16; E. Alt, "By Air Across the Atlantic Ocean," *Scientific American*, Supplement, no. 1800, pp. 13–14; W. Warntz, "Transatlantic Flights and Pressure Patterns," *Geographical Review* 51 (1961), no. 2, pp. 192–96.

11. Anon., "Proposed Transatlantic Airship Flight," *Scientific American*, October 1, 1910, pp. 259, 264–66; "Wellman Airship Is Safely Launched," *The New York Times*, September 25, 1910; "Wellman Tells of His Record Airship Voyage," *The New York Times*, October 20, 1910; "Sailor to Pilot Wellman Airship," *The New York Times*, September 6, 1910; J. Christopher, *Transatlantic Airships: An Illustrated History* (Ramsbury, U.K.: Crowood Press, 2010), pp. 20–27.

12. Duggan and Meyer, *Airships in International Affairs*, pp. 38–40; De Syon, *Zeppelin!*, p. 80.

13. Brooks, *Zeppelin: Rigid Airships*, pp. 68–69; Robinson, *Giants in the Sky*, p. 80; Belafi, *The Zeppelin*, p. 122.

14. C. Dienstbach, "Lessons of the Disasters of the L-II," *Scientific American*, November 1, 1913, p. 340; "Expert Report on the Last Zeppelin Disaster," *Scientific American*, November 22, 1913, p. 395; E. Forlanini, "Causes of Accidents with Airships," *Scientific American*, Supplement, no. 1998, April 18, 1914, pp. 255–56; Belafi, *The Zeppelin*, pp. 123–24. The composition of the crew is given variously, but Belafi's seems the most authoritative.

15. Brooks, *Zeppelin: Rigid Airships*, pp. 68–71; Robinson, *Giants in the Sky*, p. 81.

16. R. Holanda, *A History of Aviation Safety: Featuring the U.S. Airline System* (Blooming-ton, Ind.: AuthorHouse, 2009), p. 25.

17. Dienstbach, "Lessons of the Disasters of the L-II," *Scientific American*, p. 340.

18. See the 1915 edition of *Yahrbuch der Schiffbautechrische Gesselschaft*, published as "Zep-pelin Airships: An Address, Historical and Descriptive, by Their Designer," *Scientific American*, Supplement, no. 2091, January 29, 1916, pp. 77–79.

19. Brooks, *Zeppelin: Rigid Airships*, Appendix 3, "DELAG Operations in Germany, 1910–1914," p. 198.

20. Memorandum, April 2, 1914, printed in Italiaander, *Ein Deutscher Namens Eckener*, pp. 182–83.

21. On DELAG's four-year revenue and expenditure, see Brooks, *Zeppelin: Rigid Airships*, Appendix 2, "Commercial Operations with Rigid Airships," p. 194. Between 1910 and 1914, operating costs amounted to 4,260,000 marks.

22. MacMechen and Dienstbach, "Over Sea by Airship," pp. 126–27.

23. See Brooks, *Zeppelin: Rigid Airships*, tables in Appendix 3, "DELAG Operations in Ger-many, 1910–1914," p. 198.

24. Eckener, *Count Zeppelin*, p. 256.

20. Z-Ships

1. Lehmann, *Zeppelin*, p. 43.

2. Meyer, "In Search of the Real Count Zeppelin," p. 45. On Zeppelin's militarism on the eve of war, see also Duggan and Meyer, *Airships in International Affairs*, p. 43.

3. Quoted in Meyer, *Count Zeppelin*, p. 103.

4. Kuhn, "Zeppelin und die Folgen," Henry Cord Meyer Papers, Box 3, Folder 2, p. 27.

5. Nitske, *The Zeppelin Story*, p. 87.

6. Lehmann, *Zeppelin*, p. 50.

7. Goldsmith, *Zeppelin*, p. 124.

8. Lehmann, *Zeppelin*, p. 45; Nitske, *The Zeppelin Story*, p. 110; Italiaander, *Ein Deutscher Namens Eckener*, p. 124; for his reasons for volunteering, see letter to Maass, August 24, 1914, p. 12; Nielsen (trans. Chambers), *The Zeppelin Story*, p. 121.

9. Lehmann, *Zeppelin*, p. 52.

10. Nitske, *The Zeppelin Story*, p. 87.

11. Lehmann, *Zeppelin*, p. 89; Robinson, *Giants in the Sky*, p. 82.

12. Hartcup, *Achievement of the Airship*, p. 93.

13. General Friedrich von Bernhardi, in his 1911 book *Germany and the Next War*, was typi-cal in cautiously judging that air vehicles *might* be a "practical means of reconnoitering the enemy" but considered even that possibility an "open question." Quoted in J. W. McConaughy, "Aeroplanes in War," *Munsey's Magazine*, August 1915, p. 465.

14. W. Kaempffert, "Aircraft and the Future," *The Outlook*, June 28, 1913, p. 456.

15. Anon., "Practicability of the Contemplated Raid of Zeppelin Airships," *Current Opinion*, December 1914, pp. 410–11; Anon., "Comparison of French and German Strength in Dirigible Airships," *Scientific American*, August 16, 1913, pp. 126, 133; Anon., "Aircraft in War: How Aircraft Will Affect Strategy; the Air Strength of Europe," *Scientific American*, September 5, 1914, p. 172.

16. Abstract of article by Nicolas Flamel, *La Nature* (July 31, 1914), "The Use of Aluminium in War," published in *Scientific American*, Supplement, no. 2073, September 25, 1915.

17. C. Dienstbach, "The Prospects of Aerial Fighting in the Present War," *Scientific American*, August 22, 1914, p. 131; and "Has the Fighting Dirigible Arrived?," *Scientific American*, May 16, 1914, p. 412.

18. Anon., "Aircraft in War," p. 172; Kaempffert, "Aircraft and the Future," p. 456; Dienstbach, "Has the Fighting Dirigible Arrived?" p. 412.

19. Goldsmith, *Zeppelin*, p. 217; Brooks, *Zeppelin: Rigid Airships*, p. 73. Numbers vary slightly.

20. Lehmann, *Zeppelin*, p. 46.

21. Lehmann, *Zeppelin*, pp. 48, 55.

22. Nitske, *The Zeppelin Story*, p. 88; Robinson, *Giants in the Sky*, pp. 85–86; Lehmann, *Zeppelin*, p. 49.

23. Lehmann, *Zeppelin*, pp. 53–54; Anon., "The Effect of the Zeppelin Bombardment of Antwerp," *Scientific American*, September 26, 1914, p. 248. On the difficulties of aiming bombs, see A. Buttner, "Throwing Bombs from Airships," *Scientific American*, Supplement no. 2092, February 5, 1916, p. 85. For an eyewitness account of the raid from the ground, see Anon., "An Airship in the Field: A Personal Narrative from a German Observer," *Scientific American*, March 27, 1915, p. 200.

24. Quoted in Hartcup, *Achievement of the Airship*, p. 93.

25. On Eckener's views of Strasser, see his letters to Johanna, December 9 and 12, 1917, and February 12, 1918, printed in Italiaander, *Ein Deutscher Namens Eckener*, pp. 155–56. On Strasser's appearance and habits, see Lehmann, *Zeppelin*, p. 189; on nicknames, see H. von Schiller, "A Million Miles in a Zeppelin," Schiller Papers, p. 60. This lengthy memorandum seems to have been written sometime between 1937 and 1939.

26. Letter, Eckener to Maass, August 24, 1914, printed in Italiaander, *Ein Deutscher Namens Eckener*, p. 127.

27. Robinson, *Giants in the Sky*, p. 83; Lehmann, *Zeppelin*, p. 91; Nitske, *The Zeppelin Story*, p. 97.

28. Letter, Eckener to Johanna, March 20, 1915, printed in Italiaander, *Ein Deutscher Namens Eckener*, pp. 134–35.

29. Letter, Eckener to Colsman, April 1, 1915, printed in Italiaander, *Ein Deutscher Namens Eckener*, pp. 135–37. When one reporter later visited the fleet, he "heard expressions of enthusiasm over the latest developments in Zeppelins. I personally know that not many navy people were enthusiastic over them before the beginning of the war. I find now that Zeppelins are held in high estimation by many officers as auxiliaries for the navy." K. Von Wiegand, "Zeppelins Have Won Confidence of German Navy," *St. Louis Post-Dispatch*, October 26, 1915.

30. L. d'Orcy, "Cost of the War in Airships," *Scientific American*, October 2, 1915, pp. 294, 307–8; Lehmann, *Zeppelin*, pp. 61, 134; Nitske, *The Zeppelin Story*, p. 90.

31. Topping, *When Giants Roamed the Sky*, pp. 36–40; letter, Eckener to Zeppelin, March 26, 1915, printed in Italiaander, *Ein Deutscher Namens Eckener*, p. 128; L. D'Orcy, "Evolution of the Rigid Airship Design," *Scientific American*, December 23, 1916, pp. 576, 579, 581; C. Dienstbach, "The War-Zeppelin," *Scientific American*, June 10, 1916, p. 619.

32. Brooks, *Zeppelin: Rigid Airships*, p. 78; Dürr, *25 Years of Zeppelin Airship Production*, pp. 40–41; on wartime production issues, see Topping, *When Giants Roamed the Sky*, p. 44. For a general survey, see W. B. Stout, "Duraluminum," *Scientific American*, March 1922, p. 196.

33. Robinson, *Giants in the Sky*, pp. 101, 137–38; Brooks, *Zeppelin: Rigid Airships*, p. 94.

21. Pirates of the Air

1. See "The Disappointing Zeppelin," *Literary Digest*, January 23, 1915, p. 144.

2. Quoted in A. Freedman, "Zeppelin Fictions and the British Home Front," *Journal of Modern Literature* 27 (2004), no. 3, pp. 50–51. See also Anon., "Practicability of the Contemplated Raid of Zeppelin Airships," *Current Opinion*, December 1914, pp. 401–11.

3. P. Scott, "As a British Admiral Saw It," *North American Review*, July 1919, p. 65; on the inadequacies of guns at this stage, see C. Dienstbach, "Aircraft Artillery and Bomb-Dropping," *Scientific American*, February 6, 1915, pp. 126–27.

4. Anon., "The Zeppelin Raid into England," *Literary Digest*, February 6, 1915, p. 235.

5. Robinson, *Giants in the Sky*, p. 92; M. Dudley, "The War Told in Medals," *Munsey's Magazine*, April 1918, pp. 605–15; Anon., "The German Mind as Betrayed in War-Medals," *Literary Digest*, February 9, 1918, pp. 27–28.

6. Quoted in Anon., "The Zeppelin Raid into England," *Literary Digest*, February 6, 1915, p. 235.

7. The General Staff and Admiralty in Berlin argued their case in a statement widely published abroad: "There is nothing in international law, or in any international agreements, against [bombing]. The standpoint taken is that London is a defended city, that its bombardment by Zeppelins would constitute no violation of the laws of war . . . any more than if London were under the guns of the German Army or Navy, and that the docks, shipyards, arsenals, barracks, railway stations, Government buildings, military establishments, buildings where aerial guns are mounted, etc., are all, by laws of war, proper targets for Zeppelins." See "A Zeppelin Raid on London," *Irish Times*, April 2, 1915.

8. Interview by K. von Wiegand, "Zeppelins Have Come to Stay, Says Inventor," *The Bemidji Daily Pioneer*, February 9 and 10, 1915.

9. For instance, E. Campbell, *Zeppelins: The Past and the Future* (St. Albans, U.K.: Campfield Press, 1918), p. 13.

10. This section is based on A. D. Harvey, "Against London: A Zeppelin Officer's Account," [memoir of *Oberleutnant zur See* Hans Gebauer], *Air Power History*, Summer 2010, pp. 16–18; Robinson, *Giants in the Sky*, pp. 95–102, 130; Nitske, *The Zeppelin Story*, pp. 99–101, 107; Lehmann, *Zeppelin*, pp. 64, 67–68, 148–50, 156; Goldsmith, *Zeppelin*, pp. 218–19; interview by Karl von Wiegand of Lieutenant Commander Mathy, "Not a Zeppelin Lost in Raids," *Boston Daily Globe*, September 22, 1915, and a reprint, "How I Bombed London," *Peking Gazette* (from the *Daily Mail* and the New York *World*), October 25, 1915; Anon., "The Thrills of Air-Raiding," *Aircraft*, July 1, 1915; A. Seim (trans. C. W. Sykes), "The Rigger Tells a Tale," in R. Hedin (ed.), *The Zeppelin Reader: Stories, Poems, and Songs From the Age of Airships* (Iowa City: University of Iowa Press, 1998), pp. 77–80.

11. L. R. Freeman, "Sharks of the Air over London," *Current Opinion*, June 1916; F. L. Mayhew, "Report from Royal Engineers," in Hedin (ed.), *The Zeppelin Reader*, p. 90; "Zeppelin Bombs," *Scientific American*, Supplement no. 2084, December 11, 1915, p. 375; P. Maxwell, "The Death-Ship in the Sky," *Forum*, July 1916.

12. One option, mysteriously not acted upon, called for airplane pilots to martyr themselves by ramming airships. See J. A. Steinmetz and C. Dienstbach, "The Air Bomb: A New Method of Mining the Air and of Thwarting an Attack by Flying Machine or Dirigible: The Inventor's Explanation and a Critic's Objection," *Scientific American*, August 15, 1915, pp. 113–14. More promising was a French scheme to use a network of huge megaphones to pick up and locate the sound of approaching airships and then use triangulation to calculate their position and altitude. Anon., "Aerial Range-Finding with Electrical

'Ears,'" *Scientific American,* October 30, 1915, p. 377. The British, on similar lines, developed "sound locators" employing blind men, whose fineness of hearing was considered acute, to listen to large conical horns to detect direction and range, but they did not work as well as had been hoped. On this matter, see diary entries, October 14, November 2, 3, 7, 1917, Ashmore Papers 66/75/1, Imperial War Museum, London; A. Rawlinson, *The Defence of London,* 2nd ed. (London: Andrew Melrose, 1923), pp. 108, 112. On average, the ear's range is 4.49 miles; sound locators were only able to increase this by 1.29 miles. Table 7.1, in R. Burns, "Air Defence, Some Problems," in *Radar Development to 1945,* ed. Burns (London: Institution of Engineering and Technology, 1988), p. 116.

13. Report by J. T. Babington, approx. September 27, 1915; letter, Rawlinson to Percy Scott, September 25, 1915, AIR 1/2316/223/19/24, Public Record Office, London. See P. Scott, "The Defence of London Against Zeppelins, 1915 to 1916," *The Strand Magazine,* May 1919, pp. 349–54. Much of the material in this section is based on my "Radar Strategy: The Air Dilemma in British Politics, 1932–1937," Ph.D. diss., Cambridge University (1996), ch. 1.

14. L. R. Freeman, "The Passing of a Zeppelin," *The Living Age,* January 20, 1917.

15. On parachutes, see Nitske, *The Zeppelin Story,* p. 104.

16. O. Mieth, "Shot Down by the British: A Zeppelin Officer's Story," *The Living Age,* April 17, 1926, pp. 143–47; Lehmann, *Zeppelin,* pp. 164–66. Lehmann's and Mieth's accounts differ in some minor ways.

17. Interview by Karl von Wiegand of Lieutenant Commander Mathy, "Not a Zeppelin Lost in Raids," *Boston Daily Globe,* September 22, 1915.

18. Mathy, quoted at wwi.lib.byu.edu/index.php?title=Mathy&redirect=no.

19. W. Forest, "London's War Slogan 'Business as Usual' Disrupted by First Zeppelin Raid," *New York Herald Tribune,* May 22, 1932.

20. M. MacDonagh, "In London During the Great War, October 1, 1916," in *The Zeppelin Reader,* ed. Hedin, pp. 98–100. Those assigned to guard the wreckage often made the biggest profits from selling souvenirs. When some men of the Royal Buckinghamshire Hussars were given the unpleasant task of "gathering up the crew," who had fallen into marshy ground, "there were indentations in the soft soil of the shape of their bodies, arms, legs, everything—a mould of the bodies really." But the good news was that when their officers weren't looking, they flogged sticks of "wicker chairs, loaves of German bread and bits of burnt silk and pieces of aluminum" to collectors and sightseers. "It kept us in beer for months!" said one. C. Williams, "Report of the Royal Buckinghamshire Hussars, September 25, 1916," in *The Zeppelin Reader,* ed. Hedin, pp. 94–95.

21. Robinson, *Giants in the Sky,* p. 123.

22. Campbell, *Zeppelins: The Past and the Future,* pp. 32–33; "Friedrichshafen in Mourning," *The New York Times,* October 24, 1917.

23. See Flight Lieutenant Faulkner's comment, quoted in Anon., "The Answer to the Zeppelin," *Literary Digest,* February 10, 1917; Robinson, *Giants in the Sky,* p. 125; Nitske, *The Zeppelin Story,* p. 97; "Flying by Night," *Literary Digest,* November 4, 1916, pp. 1170–71.

24. Robinson, *Giants in the Sky,* pp. 127, 130; Nitske, *The Zeppelin Story,* p. 107; Hartcup, *Achievement of the Airship,* p. 100. On hypoxia and the 12,000-foot principle, see Baumgarten, "Aviation Medicine: Its Past, Present, and Future," in *From Airships to Airbus,* ed. Trimble, volume 2, p. 80; see also D. T. Courtwright, *Sky as Frontier: Adventure, Aviation, and Empire* (College Station: Texas A&M University Press, 2005), p. 41.

25. Nitske, *The Zeppelin Story,* p. 110; Robinson, *Giants in the Sky,* p. 138.

26. V. Woolf, *The Question of Things Happening: Letters, 1912–1922,* ed. N. Nicolson (London: Hogarth Press, 1976), p. 214.

27. Lehmann ghostwrote, or advised the ghostwriter of, an article by Paul Litchfield, "Lighter-Than-Air Craft," *Annals of the American Academy of Political and Social Science* 131 (1927), May, p. 81. See also Lehmann, *Zeppelin*, p. 198.

28. For a typical claim that Zeppelins forced Britain to devote significant resources to home defense, see Nitske, *The Zeppelin Story*, p. 106. On British counter-bomber defenses, see W. Raleigh and H. A. Jones, *The War in the Air* (London, 6 vols.), cited in J. Griffin, *Glass Houses and Modern War* (London: Chatto & Windus, 1938), p. 56.

29. The British official historian notes that in 51 airship raids about 196 tons of bombs were dropped, killing 557 people, injuring 1,358, and inflicting "material damage estimated at £1,527,585." Quoted in Hartcup, *Achievement of the Airship*, p. 104. The figure for Zeppelin costs comes from Campbell, *Zeppelins: The Past and the Future*, pp. 20–21.

30. Figures vary slightly, though the effect is the same. See Robinson, *Giants in the Sky*, Appendix A, "The 161 Rigid Airships Built and Flown, 1897–1940," pp. 330–39; L. d'Orcy, "German Airship Construction During the War," Table II, "German Airship Losses, 1914–1918," *Scientific American*, May 1921, p. 458. Vissering claims there were 88; see Vissering, *Zeppelin*, p. 24.

31. Cited in Anon., "Germany's Dominion of the Air," *Literary Digest*, September 16, 1916, p. 660; British figures are in Anon., "Do Zeppelins Pay?" *Literary Digest*, October 7, 1916, pp. 880–81. Of the British tally, a French military expert named Gustave Hervé chirped, "Why, a single machine gun, properly placed on the day of an assault, has disposed of as many infantrymen in half an hour!" Quoted in Anon., "More and More Zeppelins," *Literary Digest*, December 2, 1916, p. 1458.

32. When the American William Cooper Stevenson was traveling through German-occupied France he happened to run into an old friend, a Captain Schroeder, who introduced him to a lieutenant assigned to LZ-74. Stevenson asked the lieutenant, who had invited him to come aboard for the next raid, why he bothered with the bombing when it accomplished so little, only to be told that the British censor had been blacking out reports of real damage. W. C. Stevenson, "Raiding England from the Sky," *Outlook*, October 25, 1916.

33. "Eight Million Cubic Feet of Manufactured Gas Burned in a Few Seconds, Lighting Up All of London," *American Gas Engineering Journal*, June 28, 1919. In another instance, the acquisition from an intelligence source of the transcript of a House of Lords debate in March 1916 was music to German ears when it reached Berlin. During the proceedings, Lord Montagu of Beaulieu asserted that two major factories, one manufacturing munitions and the other airplanes, had missed destruction by "only a few yards." Report found among Eckener's papers, printed in Italiaander, *Ein Deutscher Namens Eckener*, pp. 131–32.

34. On the tragic story of L-19, see "The Last Messages of the Doomed Zeppelin L-19 Crew," *The Fatherland*, October 18, 1916; Lehmann, *Zeppelin*, p. 93; Anon., "A Zeppelin Tragedy," *Literary Digest*, March 11, 1916, pp. 634–35. *The Fatherland* was a pro-German American newspaper.

35. Quoted in Campbell, *Zeppelins: The Past and the Future*, p. 27.

36. The chief of the naval staff informed the kaiser on April 11, 1916, of the results of a series of raids on England between March 21 and April 1: "At Grimsby . . . a battleship . . . was heavily damaged by a bomb, and had to be beached. At Kensington an aeroplane hangar was wrecked, near Tower Bridge [in London] a transport ship damaged, in Great Tower Street a factory wrecked. . . . It was reported that a big fire had broken out at West India Docks, and that at Tillbury Docks a munition boat exploded (400 killed)." Again, an impressive achievement—but not a word of it was true. Quoted in Robinson, *Giants in the Sky*, pp. 115–16.

37. On the maps, see Schiller's letter to Wilbur Cross, November 14, 1961, p. 3, attached to Schiller, "A Million Miles in a Zeppelin," Schiller Papers.

38. Harvey, "Against London: A Zeppelin Officer's Account," p. 16.

39. Lehmann, *Zeppelin*, p. 65.

40. Robinson, *Giants in the Sky*, pp. 104–5.

41. Letter, Eckener to Colsman, February 3, 1916, printed in Italiaander, *Ein Deutscher Namens Eckener*, pp. 137–38.

42. Letter, Eckener to Colsman, February 3, 1916, printed in Italiaander, *Ein Deutscher Namens Eckener*, pp. 137–38; Eckener, *My Zeppelins*, pp. 17–18.

43. Letter, Eckener to Johanna, December 8, 1917, printed in Italiaander, *Ein Deutscher Namens Eckener*, p. 155.

44. Letters, Eckener to Johanna, June 25, 1917, and December 8, 1917, in Italiaander, *Ein Deutscher Namens Eckener*, pp. 147, 155.

45. Letter, Eckener to Johanna, May 25, 1917, printed in Italiaander, *Ein Deutscher Namens Eckener*, pp. 140–41.

46. Robinson, *Giants in the Sky*, pp. 136–39; letter, Eckener to Johanna, August 6, 1918, printed in Italiaander, *Ein Deutscher Namens Eckener*, p. 159.

47. Robinson, *Giants in the Sky*, pp. 139–40; Lehmann, *Zeppelin*, pp. 194–95.

22. China Show

1. Goldsmith, *Zeppelin*, pp. 227–28, 231–32; Robinson, *Giants of the Sky*, pp. 117–20.

2. Goldsmith, *Zeppelin*, p. 229.

3. Goldsmith, *Zeppelin*, p. 235.

4. Nitske, *The Zeppelin Story*, pp. 117–18; Goldsmith, *Zeppelin*, p. 238; Topping, *When Giants Roamed the Sky*, pp. 45–46.

5. W. Kaempffert, "What Zeppelin Really Achieved as a Pioneer Aeronaut," *The New York Times*, quoted in *Current Opinion*, May 1917; Anon., "Zeppelin in Success and Failure," *Literary Digest*, March 24, 1917.

6. Anon., "Count Ferdinand von Zeppelin," *Aviation and Aeronautical Engineering*, March 15, 1917; Anon., "Count Zeppelin," *Outlook*, March 21, 1917.

7. Letter, Eckener to Johanna, February 12, 1918, printed in Italiaander, *Ein Deutscher Namens Eckener*, p. 156.

8. In a letter to Johanna dated October 26, 1917, Eckener wrote, "After the war? Yes, dear wife, the DELAG will probably come back in some form or another, as Colsman says. . . . I'll be there, of course, but I'll leave the traveling to others." Printed in Italiaander, *Ein Deutscher Namens Eckener*, p. 152. Eckener would actually do *more* traveling than others.

9. D. H. Robinson and C. L. Keller, *"Up Ship!": A History of the U.S. Navy's Rigid Airships, 1919–1935* (Annapolis, Md.: Naval Institute Press, 1982), p. 1.

10. Instruction printed in Robinson, *Giants in the Sky*, p. 110.

11. Robinson, *Giants in the Sky*, p. 143.

12. P. W. Baker, "Why the Airship Failed," p. 17, paper delivered to the Historical Group of the Royal Aeronautical Society, November 26, 1974, in Henry Cord Meyer Papers, Box 2, File 19.

13. On backgrounds, see Italiaander, *Ein Deutscher Namens Eckener*, pp. 193–95; "Lehmann Trained Years for Career," *The New York Times*, May 9, 1936; J. A. Sinclair, "Capt. Ernst August Lehmann," *The Airship* 4 (1937), no. 14, p. 21.

14. For Lehmann's claims, see Goldsmith, *Zeppelin*, p. 226, and Robinson, *Giants in the Sky*,

p. 139n. Lehmann's bragging was occasioned by the construction of the last of the war-time Zeppelins, L-72, though Robinson points out that L-72 was actually intended for long-range reconnaissance in the North Sea.

15. Lehmann, *Zeppelin*, pp. 174–77; Nitske, *The Zeppelin Story*, p. 107; Robinson, *Giants in the Sky*, pp. 250–51.

16. Details of the L-59 story in this section are based on Robinson, *Giants in the Sky*, pp. 252–55; Lehmann, *Zeppelin*, pp. 182–88; Nitske, *The Zeppelin Story*, pp. 107–9; letters, Eckener to Johanna, October 5 and October 19, 1917, printed in Italiaander, *Ein Deutscher Namens Eckener*, pp. 148–49, 150–51; R. Gaudi, *African Kaiser: General Paul von Lettow-Vorbeck and the Great War in Africa, 1914–1918* (New York: Caliber, 2017), pp. 8–14, 360–73.

23. The Beginning or the End?

1. R. J. Evans, "The Defeat of 1918," in Evans, *The Third Reich in History and Memory* (Oxford: Oxford University Press, 2015), p. 29.

2. Topping, *When Giants Roamed the Sky*, p. 88. The official is quoted in Kuhn, "Zeppelin und die Folgen," Henry Cord Meyer Papers, Box 3, Folder 2, p. 30.

3. Topping (ed. Brothers), *When Giants Roamed the Sky*, p. 59. The report is quoted in Kuhn, "Zeppelin und die Folgen," p. 31; on the events in Friedrichshafen, see pp. 33–36. Troops were eventually dispatched to clear Friedrichshafen of Socialists and Communists. See "Rioters in Halle Kill 55 Persons and Wound 170," *New York Tribune*, March 18, 1919.

4. Meyer, "Building Rigid Airships," p. 88.

5. J. C. Segrue, "Barbed Wire and Bored Dog Only Guards at Zep Plant," *New York Tribune*, January 16, 1919.

6. Italiaander, *Ein Deutscher Namens Eckener*, p. 182.

7. Letter, Eckener to his brother, Alex, October 27, 1918, in Italiaander, *Ein Deutscher Namens Eckener*, pp. 177–78.

8. Meyer, "In the Shadow of the Titan: Thoughts on the Life and Work of Naval Engineer Johann Schütte," in Meyer, *Airshipmen, Businessmen, and Politics, 1890–1940*, p. 68.

9. Swientek, "Alfred Colsman," p. 117; Italiaander, *Ein Deutscher Namens Eckener*, p. 130.

10. Meyer, "Building Rigid Airships," p. 101; R. Köster, "Zeppelin: The Airship and the Need for Diversification After World War One (1918–1929)," paper presented at the 12th Annual Conference of the European Business History Association, August 2008, pp. 3–7; Meyer, "Eckener's Struggle to Save the Airship for Germany, 1919–1929," in Meyer, *Airshipmen, Businessmen, and Politics*, pp. 135–38.

11. Köster, "Zeppelin: The Airship and the Need for Diversification After World War One (1918–1929)," pp. 2, 9, emphasis added.

12. Köster, "Zeppelin: The Airship and the Need for Diversification After World War One (1918–1929)," pp. 2–8; Swientek, "Alfred Colsman," pp. 118–19; De Syon, *Zeppelin!*, p. 112; Duggan and Meyer, *Airships in International Affairs*, p. 52; Meyer, "Eckener's Struggle to Save the Airship for Germany, 1919–1929," pp. 135–38. For Eckener's views of Colsman as "impulsive and rash [and] downright dangerous," see his letter to Friedrich Maass, February 18, 1921, in Italiaander, *Ein Deutscher Namens Eckener*, p. 200–201.

13. Duggan and Meyer, *Airships in International Affairs*, p. 53.

14. Treaty of Versailles, June 28, 1919, Clauses 198–210.

15. Quoted in Robinson, *Giants in the Sky*, p. 256.

16. Lehmann, *Zeppelin*, p. 198; Nielsen (trans. Chambers), *The Zeppelin Story*, pp. 132–33.

17. Lehmann, *Zeppelin*, p. 198. On the bombing rumor about Lehmann, see "Building Zeppelin for the United States," *The New York Times*, August 26, 1923.

18. Nielsen (trans. Chambers), *The Zeppelin Story*, pp. 132–33.

19. Lehmann, *Zeppelin*, p. 199.

20. J. Hunsaker, "The Present Status of Airships in Europe," *Journal of the Franklin Institute* 177 (1914), no. 6, pp. 633–35; Robinson, *Giants in the Sky*, pp. 182–84.

21. Quoted in R. Maksel, "Airplane Versus Zeppelin in 1917," *Air and Space*, March 3, 2016, at airspacemag.com/daily-planet/l-49-180958236/.

22. Anon., "The New Zeppelins," *Scientific American*, December 29, 1917, p. 411; Anon., "Some Super-Zeppelin Secrets," *Scientific American*, November 24, 1917, pp. 386, 392, 394; L. d'Orcy, "The Modern Zeppelin Airship," *Aviation and Aeronautical Engineering*, December 15, 1917; W. Mitchell, "Memoirs of World War One," in *The Zeppelin Reader*, ed. Hedin, pp. 101–5; Topping, *When Giants Roamed the Sky*, p. 54.

23. Robinson and Keller, *"Up Ship!"* pp. 10–11, 13–14; "Globe-Circling Zeppelin Plan of U.S. Navy," *New York Tribune*, June 16, 1919.

24. Robinson and Keller, *"Up Ship!"* p. 117; Duggan and Meyer, *Airships in International Affairs*, p. 57.

25. Robinson, *Giants in the Sky*, p. 256; Memorandum, "Hafenanlagen für die Weltluftfahrt," February 1919, printed in Italiaander, *Ein Deutscher Namens Eckener*, pp. 228–29.

26. Lehmann, *Zeppelin*, p. 199.

27. Robinson and Keller, *"Up Ship!"* pp. 119–20; Duggan and Meyer, *Airships in International Affairs*, p. 59; Italiaander, *Ein Deutscher Namens Eckener*, p. 153.

28. Duggan and Meyer, *Airships in International Affairs*, p. 57; Robinson and Keller, *"Up Ship!"* p. 118.

24. Bringing Back the Dead

1. W. Hensley, "America and Air Transportation," *Aircraft Journal*, January 17, 1920, p. 11.

2. "How It Looks and Feels to Fly in an Air-Pullman," *Literary Digest*, March 13, 1920. On passenger numbers, see Robinson, "Passenger Flying in the Big Zeppelins," in Robinson Papers, Box 10, Folder 4, p. 5.

3. Hensley, "America and Air Transportation," p. 11.

4. Italiaander, *Ein Deutscher Namens Eckener*, p. 186; on Johanna never flying, see p. 120. See also "Germany's Passenger Service Ceases," *Railway Age*, October 24, 1919.

5. Italiaander, *Ein Deutscher Namens Eckener*, p. 185–86; E. Klein, "The Log of a Zeppelin Journey: Lake Constance to Berlin in Five Hours," *The Living Age*, April 10, 1920, p. 82, originally published in *Neue Freie Presse*, January 1920.

6. Italiaander, *Ein Deutscher Namens Eckener*, p. 86.

7. Italiaander, *Ein Deutscher Namens Eckener*, pp. 85–86.

8. Letter to Johanna, October 2, 1919, in Italiaander, *Ein Deutscher Namens Eckener*, p. 187.

9. Meyer, "In the Shadow of the Titan," p. 65.

10. De Syon, *Zeppelin!*, p. 114.

11. Klein, "The Log of a Zeppelin Journey," p. 84.

12. "A Europe–South America Airship Scheme," *Aviation and Aircraft Journal*, December 5, 1921.

13. Brooks, *Zeppelin: Rigid Airships*, Appendix 3, "Operating Statistics of Rigid Airships," p. 198.

14. Klein, "The Log of a Zeppelin Journey," p. 85.

15. Lehmann, *Zeppelin*, p. 161; Robinson, *Giants of the Sky*, pp. 120, 162–63, 174–75; "The Super-Zeppelin," *Railway Age Gazette*, November 10, 1916; "Gigantic Airships to Link Continent," *Los Angeles Times*, January 23, 1919. See also the puff piece by E. M. Maitland, a leading British airshipman, "The Airship for Commercial Purposes," *Aviation and Aeronautical Engineering*, February 1, 1919; "Foresees Airships in Overseas Trade," *The New York Times*, January 23, 1919; and "The Commercial Future of Airships," *Journal of the Royal Society of Arts* 68 (1920), no. 3524, pp. 461–75. Also interesting is A. H. Ashbolt, "An Imperial Airship Service," *Journal of the Royal Society of Arts* 70 (1921), no. 3605, pp. 102–21.

16. See "Preparations for the Transatlantic Flight," *Aviation and Aeronautical Engineering*, April 1, 1919; "Flight Delayed Across Continent," *Los Angeles Times*, April 4, 1919; "Americans Off First?" *Los Angeles Times*, May 4, 1919; "Navy Fliers Ready for Transatlantic Start Tomorrow," *The New York Times*, May 5, 1919; "All Three Navy Planes Nearing the Azores; Halfway Point Passed in Seven Hours," *New York Tribune*, May 17, 1919; "30-Minute Lead Beat Fog to Finish," *The New York Times*, May 18, 1919; "Final Jump to Plymouth Expected Today," *New York Tribune*, May 28, 1919; E. Magnani, "The NC-4, First Across the Atlantic," *Aviation History*, November 2002, at historynet.com/aviation-history-three-us-flying-boats-were-the-first-to-fly-across-the-atlantic-in-1919.htm.

17. E. M. Maitland, *The Log of H.M.A. R34: Journey to America and Back* (London: Hodder & Stoughton, 1920), p. 24; Robinson, *Giants in the Sky*, pp. 164–66; "The Trans-Atlantic Dirigible," *Scientific American*, July 19, 1919, pp. 58–59, 66–68, 70, 72. For the official Royal Air Force report, see "Trans-Atlantic Flight of R.34," undated, in Henry Cord Meyer Papers, Box 2, Folder 19.

18. *The New York Times*, July 7, 1919; Duggan and Meyer, *Airships in International Affairs*, p. 64.

19. De Syon, "Turning a National Icon into a Business Tool: The Zeppelin Airship, 1924–1933," in *Third International Airship Convention*, ed. G. Khoury (London: The Airship Association, 2000), p. 3.

20. "Two Notable Airship Voyages," *Aviation and Aeronautical Engineering*, March 1, 1919; S. Johnson, "Britain's Conquest of the Skies," *The Forum*, March 1919, p. 333. See also, among many others, L. d'Orcy, "The Case for the Airship," *Journal of the Society of Automotive Engineers* 4 (April 1919), p. 304; A. C. Lescarboura, "Hitch Your Wagon to a Gas Bag," *The Independent*, June 7, 1919, p. 359.

21. Anon., "Count Zeppelin," *Aircraft*, May 1, 1917.

22. A. Euler, "The Internationality of Aviation," March 15, 1919, printed in Italiaander, *Ein Deutscher Namens Eckener*, pp. 183–84.

23. W. J. Abbott, "From New York to Chicago by the Air Mail," *Munsey's Magazine*, February 1918, pp. 47–55.

25. The Visionary

1. R. E. Bilstein, *Flight in America: From the Wrights to the Astronauts* (Baltimore: Johns Hopkins University Press, 2001), pp. 32–37, 42, 74–75; Courtwright, *Sky as Frontier*, p. 46; Hallion, *Taking Flight*, pp. 375–77, 388–91. See also B. Williams, "The War Aeroplane Here and Abroad," *Scientific American*, November 4, 1916, p. 412; on Curtiss production, see F. Eppelsheimer, "Where Wings Are Made for Fighting Men: How the War Has Stimulated the Aeroplane Industry in this Country," *Scientific American*, September 4, 1915, pp. 204–5.

2. Bilstein, *Flight in America*, pp. 74–75.

3. Bilstein, *Flight in America*, pp. 60–61; Courtwright, *Sky as Frontier*, pp. 46–48; J. Van Vleck, *Empire of the Air: Aviation and the American Ascendancy* (Cambridge, Mass.: Harvard University Press, 2013), p. 42.

4. D. A. Pisano, "The Greatest Show on Earth: The Confrontation Between Utility and Entertainment in Aviation," in *The Airplane in American Culture*, ed. D. A. Pisano (Ann Arbor: University of Michigan Press, 2003), p. 51.

5. Courtwright, *Sky as Frontier*, pp. 49–50.

6. J. D. Snider, "'Great Shadow in the Sky': The Airplane in the Tulsa Race Riot of 1921 and the Development of African-American Visions of Aviation, 1921–1926," in *The Airplane in American Culture*, ed. Pisano, pp. 134–36.

7. Courtwright, *Sky as Frontier*, pp. 48, 52; Pisano, "The Greatest Show on Earth," p. 52.

8. Pisano, "The Greatest Show on Earth," p. 57; Courtwright, *Sky as Frontier*, pp. 52–53. On Jack Chapman, see J. J. Corn, *The Winged Gospel: America's Romance with Aviation* (Baltimore: Johns Hopkins University Press, 2002), photo insert.

9. R.E.G. Davies, *Airlines of the United States Since 1914* (Washington, D.C.: Smithsonian Institution Press, 1982), pp. 17–19, 25; D. L. Rust, *Flying Across America: The Airline Passenger Experience* (Norman: University of Oklahoma Press, 2009), p. 23; F. R. Van der Linden, *Airlines and Air Mail: The Post Office and the Birth of the Commercial Aviation Industry* (Lexington: University Press of Kentucky, 2002), pp. 1–8.

10. Courtwright, *Sky as Frontier*, p. 64; "How the Airplane Will Speed Up Business," *Literary Digest*, March 30, 1929, p. 18; Van Vleck, *Empire of the Air*, p. 41.

11. Bilstein, *Flight in America*, p. 53; Davies, *Airlines of the United States Since 1914*, p. 21.

12. Bilstein, *Flight in America*, p. 53; Courtwright, *Sky as Frontier*, p. 56.

13. M. R. Stearns, "All Aboard by Air," *World's Work*, April 1929, pp. 34–41, 144, 146, 148, 150.

14. Bilstein, *Flight in America*, p. 53. On Smith's telegram, see W. B. Courtney, "Mail from the Moon," *Collier's*, November 24, 1928, p. 8.

15. Courtwright, *Sky as Frontier*, pp. 58–59.

16. Davies, *Airlines of the United States Since 1914*, p. 28.

17. Davies, *Airlines of the United States Since 1914*, pp. 1–15; W. M. Leary, "At the Dawn of Commercial Aviation: Inglis M. Uppercu and Aeromarine Airways," *Business History Review* 53 (1979), no. 2, pp. 180–93.

18. On Trippe's background, see M. Bender and S. Altschul, *The Chosen Instrument: Juan Trippe, Pan Am—The Rise and Fall of an American Entrepreneur* (New York: Simon & Schuster, 1982), pp. 19–29, 36, 57; M. Josephson, *Empire of the Air: Juan Trippe and the Struggle for World Airways* (New York: Harcourt, Brace & Co., 1944), pp. 22–23.

19. R. Daley, *An American Saga: Juan Trippe and His Pan Am Empire* (New York: Random House, 1980), p. 7; Josephson, *Empire of the Air*, pp. 22–23.

20. On Trippe's personal habits, see Bender and Altschul, *Chosen Instrument*, p. 104; Josephson, *Empire of the Air*, pp. 78–79; B. S. Trippe (R.E.G. Davies, ed.), *Pan Am's First Lady: The Diary of Betty Stettinius Trippe* (McLean, Va.: Paladwr Press, 1996), p. 2.

21. Quoted in Bender and Altschul, *Chosen Instrument*, pp. 13–15.

22. Bender and Altschul, *Chosen Instrument*, p. 31; Daley, *American Saga*, pp. 6–7; Trippe, *Pan Am's First Lady*, p. 6.

23. Josephson, *Empire of the Air*, pp. 24–25.

24. Daley, *American Saga*, pp. 10–11.

25. Daley, *American Saga*, pp. 7–8; Josephson, *Empire of the Air*, p. 24.

26. Bender and Altschul, *Chosen Instrument*, p. 61.

27. S. Johnson, "Britain's Conquest of the Skies," *The Forum*, March 1919, pp 328–34; G. C.

Westervelt and H. B. Sanford, "Possibilities of a Trans-Pacific Flight," *Aviation and Aeronautical Engineering*, September 1, 1920; L. d'Orcy, "Airship Versus Airplane," *Scientific American*, February 1, 1919, pp. 98–99, 104.

28. "The Rigid Airship Is Not a Failure," *Current Opinion*, March 1919, p. 173.

29. L. d'Orcy, "The Case for the Airship," p. 306. D'Orcy had been discussing the subject since at least 1916: see L. d'Orcy,"Possibilities and Conditions of Crossing the Atlantic by Airship," *Scientific American*, August 26, 1916, pp. 188, 196.

30. A. Klemin, "With the Men Who Fly," *Scientific American*, January 1924, p. 72.

31. Josephson, *Empire of the Air*, p. 24. Bender and Altschul, *Chosen Instrument*, pp. 44–45, has it as "gigantic game."

32. On Lawson, see Corn, *Winged Gospel*, pp. 40–41.

33. M. Simon, "Fantastically Wrong: The Inventor of the Airliner Also Invented This Hilariously Absurd 'Science,'" *Wired*, October 8, 2014, at wired.com/2014/10/fantastically-wrong-lawsonomy/.

34. Giovanni Caproni, the Italian builder of larger bombers during the war, exerted a similar influence on young Trippe. In April 1919, a month before Trippe's article appeared, Caproni had claimed that in a matter of years "you will be able to climb aboard an airplane in New York, have the porter take your luggage to your berth, stretch out for a comfortable siesta, spend a few pleasant hours with fellow-passengers in the lounge, dine in regular Ritz fashion, and be in London the following evening in time to enjoy your supper at the Savoy." His prediction was widely regarded as fantastical—and, indeed, proved to be so. In early 1921, he finally unveiled his creation: a *nine*-winged (in three sets, placed forward, midway, and stern, of three), eight-engine, 77-foot-long Goliath of a flying boat that could carry 100 passengers sitting in pairs on wooden benches. Wide panoramic windows allowed passengers to view the most marvelous scenery. Such a complex machine was predictably plagued by technical and mechanical problems, and on a test flight it crashed into a lake a few moments after taking off. It was too expensive to even consider building another one. Caproni quoted in W. Hart Smith, "Via Air-Line, Now: The New Skyways of the World," *Forum*, April 1919; J. Binns, "Marvels of Aerial Progress to Be Seen This Year; Monster Aircraft Are Going In for Transatlantic Service," *New York Tribune*, April 25, 1920.

35. Quoted in T. A. Heppenheimer, *Turbulent Skies: The History of Commercial Aviation* (New York: John Wiley & Sons, 1995), p. 17.

36. Hallion, *Taking Flight*, pp. 316–17.

37. Topping, *When Giants Roamed the Sky*, p. 57; Hartcup, *Achievement of the Airship*, p. 110; Belafi, *The Zeppelin*, p. 131; D'Orcy, "The Case for the Airship," p. 303; Brooks, *Zeppelin: Rigid Airships*, Appendix 5, pp. 202–3. Some of the figures given differ between the various accounts. On the disproportionate increases by the airship, see "Is the Dirigible Outstripping the Airplane?" *Scientific American*, April 12, 1919, pp. 366–67; R. B. Price, "The Development of Commercial Dirigibles," *Scientific American*, August 18, 1917, p. 103.

38. Paul Jaray at Zeppelin contributed several dense and learned articles to specialist journals on this point. See, for instance, "The Development of Aircraft with Special Reference to the Zeppelin Airships," *Aviation and Aircraft Journal*, April 4, 1921 (Part 1); April 11, 1921 (Part 2).

39. The debate over the mathematics of airplane versus airship is covered in D'Orcy, "The Case for the Airship," p. 304; W. Lockwood Marsh, "The Case for the Airship," *Aviation and Aeronautical Engineering*, January 1, 1919; Anon., "The Position of Airships in the Future of Aeronautics," *Scientific American*, December 28, 1918, p. 407; G. Whale, "To-

morrow's Airships: A Survey of What Has Been Done in Commercial Aviation and Its Bearing on the Future," *Scientific American*, July 30, 1921, pp. 80, 88; "The Airship of the Near Future," *Scientific American*, July 19, 1919; "British Airship Development and Operations," *Aviation and Aeronautical Engineering*, January 15, 1919; "Are We 'Making Good' in Aviation?" *Literary Digest*, March 9, 1918, p. 13.

26. The Stolen Horse

1. J. A. Lorelli, *To Foreign Shores: U.S. Amphibious Operations in World War Two* (Annapolis, Md.: Naval Institute Press, 1995), p. 10; G. E. Wheeler, "The United States Navy and the Japanese 'Enemy': 1919–1931," *Military Affairs* 21 (1957), no. 2, pp. 61–74; Robinson, *Giants in the Sky*, p. 182; Duggan and Meyer, *Airships in International Affairs*, pp. 72–74.

2. "America's New Super-Zeppelin," *Literary Digest*, August 27, 1921, p. 19. See also G. H. Dacy, "Our ZR-2 Airship and Its Shed," *Scientific American*, September 3, 1921, pp. 160, 171; R. Higham, *The British Rigid Airship, 1908–1931: A Study in Weapons Policy* (London: G. T. Foulis, 1961), pp. 203–29; Robinson and Keller, "*Up Ship!*" pp. 17–19, 21, 29–49. On the terms of the giveaway offer of the airships, see "Disposal of Airship: Terms of Government Offer," *Flight*, June 2, 1921, p. 374.

3. The *New York Tribune* gushed that at 700 feet long ZR-2, "aristocrat of the sky," was so large that if it were stood upright next to the tallest skyscraper in the world, the Woolworth Building in New York, one would only be able to see 92 feet of stonework peeking up at the top. Quoted in "America's New Super-Zeppelin," p. 19.

4. L. d'Orcy, "The Final Solution of the Airship Problem," *Scientific American*, January 25, 1919, pp. 73, 84–86; Squier quoted in "Fire-Proof Balloons," *Literary Digest*, February 22, 1919, pp. 27–28 (see also "Aeronautics in the United States," *Electrical Review*, January 25, 1919); "Helium Was to Aid in Bombing Germans," *The New York Times*, March 17, 1919; Anon., "If Germany Had Had Our Supply of Helium Gas," *Munsey's Magazine*, August 1919, pp. 524–25; Anon., "The Use of Helium for Airships," *The Scientific Monthly* 8 (1919), no. 4, pp. 383–84; S. G. Roberts, "The Industrial Production of Helium," *Scientific American*, May 1922, pp. 308–9; "Navy Had New Aero Terror for Germany When War Ended," *New York Tribune*, March 17, 1919; C. W. Seibel, *Helium: Child of the Sun* (Lawrence: University Press of Kansas, 1969), pp. 1, 20, 41–49; M. L. Levitt, "The Development and Politicization of the American Helium Industry, 1917–1940," *Historical Studies in the Physical and Biological Sciences* 30 (2000), no. 2, pp. 335–39.

5. Robinson and Keller, "*Up Ship!*" p. 120.

6. Robinson and Keller, "*Up Ship!*" pp. 120–21; Belafi, *The Zeppelin*, p. 161. On the critical distinction between a "compensation" and a "reparations" airship, see Robinson and Keller, "*Up Ship!*" p. 226, n. 16. On the Rathenau deal, see De Syon, *Zeppelin!* pp. 115–16; De Syon, "Turning a National Icon into a Business Tool," p. 2, n. 5.

7. Robinson, *Giants in the Sky*, pp. 171–73; on Wicks's message, see "The Downfall of a Dirigible," *The Independent*, September 17, 1921.

8. "What Crumpled Up the ZR-2?" *Literary Digest*, September 24, 1921, pp. 20–21; "The Loss of the R-38," *Aviation and Aircraft Journal*, September 5, 1921; L. d'Orcy, "The Lesson of the ZR-2 Disaster," *Scientific American*, September 17, 1921, p. 200.

9. "Is Airship Travel Profitable?" *Scientific American*, September 24, 1921. For similar views, see "The Loss of the R-38," *Aviation and Aircraft Journal*, September 5, 1921.

10. Robinson and Keller, "*Up Ship!*" pp. 120–21.

11. Knäusel, *Zeppelin and the United States of America*, p. 35.

12. Robinson and Keller, *"Up Ship!"* p. 124. On Jaray, see Topping, *When Giants Roamed the Sky*, p. 89.

13. Appendix 7.1, "Excerpts from Contracts to Build ZR-3 (LZ-126) from June 26, 1922, between the *Luftschiffbau Zeppelin*, Friedrichshafen, and the United States Department of the Navy," printed in Knäusel, *Zeppelin and the United States of America*, pp. 106–12.

14. Fulton's notes on personalities is in Robinson and Keller, *"Up Ship!"* p. 125; Weyerbacher's comment on Eckener, p. 119; for Eckener's remark about Lehmann, see Nielsen (trans. Chambers), *The Zeppelin Story*, p. 132. Lehmann's views, which accorded with Dürr's, are printed in Topping, *When Giants Roamed the Sky*, p. 75.

15. Robinson and Keller, *"Up Ship!"* p. 126.

16. Letter, Captain Frank Upham to Moffett, March 1, 1923, quoted in Robinson and Keller, *"Up Ship!"* p. 130.

17. Robinson and Keller, *"Up Ship!"* pp. 125–28.

18. Letter, Moffett to Henry Vissering, February 24, 1923, in Robinson and Keller, *"Up Ship!"* pp. 127–28.

19. Köster, "Zeppelin: The Airship and the Need for Diversification After World War One (1918–1929)," p. 14. On the corporate structure, see Vissering, *Zeppelin*, pp. 10, 55; Topping, *When Giants Roamed the Sky*, p. 80.

20. Letters, Fulton to Moffett, March 5 and March 6, 1923; Moffett to Fulton, April 16, in Robinson and Keller, *"Up Ship!"* pp. 128–29.

21. Robinson and Keller, *"Up Ship!"* p. 131; H. Mingos, "The Airship Programme," *The Outlook*, September 12, 1923; "Coming—One New Zeppelin, Our Only 'Spoils of War,'" *Literary Digest*, April 21, 1923, pp. 58–60.

22. "Building Zeppelin for the United States," August 26, 1923; "Zeppelins to Keep Plant in Germany," November 30, 1923; "Hundreds of Workmen to Lose Jobs on ZR-3," *The New York Times*, May 12, 1924.

23. Quoted in Robinson and Keller, *"Up Ship!"* p. 135.

27. One Card

1. Meyer, "In the Shadow of the Titan," p. 71; Duggan and Meyer, *Airships in International Affairs*, p. 78; Topping, *When Giants Roamed the Sky*, pp. 90–93; "Goodyear Firm Buys Zeppelin Rights," *The New York Times*, November 2, 1923; "Goodyear to Build Zeppelins Here," *The New York Times*, November 8, 1923; "The American Zeppelin," *The Living Age*, November 17, 1923. On Eckener and his bicycle, see "Remarks by Commander C. E. Rosendahl upon Occasion of Acceptance of Guggenheim Award for Dr. Hugo Eckener," December 17, 1937, in Rosendahl Papers, Box 66, Folder 8, p. 3.

2. Robinson and Keller, *"Up Ship!"* pp. 131, 134–36. In mid-March, the American press in Berlin was speculating, too optimistically, that Eckener would be taking the "giant airship on its trial trips in next few days." T. R. Ybarra, "Prepare Zeppelin to Fly to America," *The New York Times*, March 17, 1924.

3. Topping, *When Giants Roamed the Sky*, pp. 80–81, 83; "America's New Zeppelin," *The Living Age*, October 18, 1924, p. 148.

4. "America's New Zeppelin," *The Living Age*, p. 145.

5. On the Organization Consul, see "Secret Orders and Murder in Germany," *Literary Digest*, May 5, 1923, pp. 52, 54–56; De Syon, *Zeppelin!*, pp. 119, 116; Topping, *When Giants*

Roamed the Sky, p. 96; "The Last Zeppelin—Ours," *Literary Digest*, October 24, 1924, p. 12. On Arnstein, "German Press Discusses Jewish Origin of Arnstein, Builder of the ZR-3," *Jewish Daily Bulletin*, October 16, 1924.

6. De Syon, "Turning a National Icon into a Business Tool," p. 3; De Syon, *Zeppelin!*, pp. 120–21; Meyer, "France Perceives the Zeppelins, 1924–1937," in *Airshipmen, Businessmen, and Politics*, pp. 169, 171.

7. Appendix 7.1, "Excerpts from Contracts to Build ZR-3 (LZ-126)," in Knäusel, *Zeppelin and the United States of America*, p. 109.

8. Italiaander, *Ein Deutscher Namens Eckener*, p. 217.

28. Queen of the Air

1. On the captains, Belafi, *The Zeppelin*, p. 164. On Eckener's all-or-nothing view that "it was clear that not only the lives of the participants, but also the fate of the Zeppelin idea, would depend on our success," see Eckener (trans. Robinson), *My Zeppelins*, p. 21.

2. "Hundreds of Germans to Lose Jobs on ZR-3," *The New York Times*, May 12, 1924.

3. Robinson and Keller, *"Up Ship!"* p. 130; Italiaander, *Ein Deutscher Namens Eckener*, p. 205.

4. Lehmann, *Zeppelin*, pp. 212–14.

5. Lehmann, *Zeppelin*, p. 214.

6. Eckener (trans. Robinson), *My Zeppelins*, p. 23; Lehmann, *Zeppelin*, p. 215.

7. Eckener (trans. Robinson), *My Zeppelins*, pp. 24–27; Lehmann, *Zeppelin*, pp. 217–20. Eckener's telegrams to Arnstein, Dürr, and Maybach are quoted in Topping, *When Giants Roamed the Sky*, p. 96. On meteorology in general on the trip, see W. Scherz, "In Three Days to America," *The Living Age*, August 15, 1925 (orig. *Uhu* [Berlin], July); and Anon., "The Transatlantic Voyage of ZR-III," *Scientific American*, August 1924, p. 115.

8. "The Last Zeppelin—Ours," p. 12; on the receipt, see Meyer, "Zeppelin Intermezzos in Detroit, 1920 and 1924," in *Airshipmen, Businessmen, and Politics*, p. 140.

9. Italiaander, *Ein Deutscher Namens Eckener*, p. 126.

10. H. Mingos, "ZR-3—The New Leviathan of the Skies," *The Outlook*, September 17, 1924, pp. 92–94.

11. Lehmann, *Zeppelin*, pp. 219–20.

12. Belafi, *The Zeppelin*, p. 167.

13. Italiaander, *Ein Deutscher Namens Eckener*, pp. 201–2; Fritzsche, *A Nation of Fliers*, p. 139; Belafi, *The Zeppelin*, pp. 161, 166. On Eckener's views of Beethoven and Wagner, see letter to his daughter, February 17, 1928, printed in Italiaander, p. 250; for his comment about changing musical tastes, see J. G. Vaeth, *Graf Zeppelin: The Adventures of an Aerial Globetrotter* (New York: Harper & Brothers, 1958), p. 128.

14. Letter to Johanna, October 19, 1924, printed in Italiaander, *Ein Deutscher Namens Eckener*, pp. 206–8; for Mayor Hylan's proclamation, p. 216. Curtis Wilbur is quoted in Topping, *When Giants Roamed the Sky*, p. 99.

15. Letter, Eckener to Johanna, undated but end of October 1924, in Italiaander, *Ein Deutscher Namens Eckener*, p. 209–10; for Ford's comment to Eckener, see "Eckener Visit to Ford Arranged 4 Years Ago," *The New York Times*, October 13, 1928.

16. Quoted in Meyer, "Zeppelin Intermezzos in Detroit, 1920 and 1924," p. 147.

17. Meyer, "Building Rigid Airships," pp. 107–8; Meyer, "Zeppelin Intermezzos in Detroit, 1920 and 1924," p. 147.

18. Fritzsche, *A Nation of Fliers*, p. 138.

19. Topping, *When Giants Roamed the Sky*, pp. 96, 98. Letter, Johanna to Eckener, October 22, 1924, in Italiaander, *Ein Deutscher Namens Eckener*, p. 211.

20. Quoted in Fritzsche, *A Nation of Fliers*, pp. 139–40.

21. Letter, Johanna to Eckener, October 22, 1924, in Italiaander, *Ein Deutscher Namens Eckener*, p. 211.

22. On Mussolini, see R. Wohl, *The Spectacle of Flight: Aviation and the Western Imagination, 1920–1950* (New Haven, Conn.: Yale University Press, 2005), pp. 62–66.

23. Göring, quoted in Bender and Altschul, *Chosen Instrument*, p. 109.

24. On the airship in general, see Anon., "Our First Rigid Airship, the *Shenandoah*," *Scientific American*, February 1924, pp. 82–83.

25. Robinson, *Giants in the Sky*, pp. 198–200; Robinson and Keller, *"Up Ship!"* pp. 91–95.

26. On the price of helium, Robinson, *Giants in the Sky*, pp. 194, 201.

27. G. Fulton, "Fuel Gas and Helium for Airships," *The Royal United Services Institution Journal* 74 (1929), no. 495, pp. 553–56; Robinson and Keller, *"Up Ship!"* p. 96.

28. Anon., "Can Huge Dirigibles Supplant Our Expresses and Liners?" *Literary Digest*, January 10, 1925, pp. 52–56, quoting Eckener's interview with *Motor Life*.

29. Bilstein, *Flight in America*, p. 42; Davies, *Airlines of the United States Since 1914*, p. 35.

30. On the Round-the-World flight, see Van Vleck, *Empire of the Air*, pp. 30–35; on the number of stops, see H. G. Dick and D. H. Robinson, *The Golden Age of the Great Passenger Airships: Graf Zeppelin and Hindenburg* (Washington, D.C.: Smithsonian Institution Press, 1985), p. 39.

31. Italiaander, *Ein Deutscher Namens Eckener*, p. 246.

29. Annus Horribilis

1. Italiaander, *Ein Deutscher Namens Eckener*, p. 239.

2. Meyer, "Eckener's Struggle to Save the Airship for Germany, 1919–1929," p. 130; letter from Eckener, July 6, 1925, printed in Meyer, "The Political Origins of the Airship *Graf Zeppelin*, 1924–1928," pp. 161–62.

3. Meyer, "Eckener's Struggle to Save the Airship for Germany, 1919–1929," p. 133; Meyer, "In the Shadow of the Titan," pp. 75–76; Meyer, "The Political Origins of the Airship *Graf Zeppelin*, 1924–1928," in *Airshipmen, Businessmen, and Politics*, p. 162.

4. Italiaander, *Ein Deutscher Namens Eckener*, pp. 234–41, 141–42; Eckener (trans. Robinson), *My Zeppelins*, pp. 29–31. On the hectic schedule, see Schiller, "A Million Miles in a Zeppelin," Schiller Papers, p. 68.

5. On Colsman's estimate, see Köster, "Zeppelin: The Airship and the Need for Diversification After World War One (1918–1929)," p. 22, n. 110.

6. On the disaster, see the detailed accounts in Robinson and Keller, *"Up Ship!"* pp. 105–15 and Appendix D, "The *Shenandoah*'s Midwest Flight Itinerary," p. 204; Anon., "Technical Aspects of the Loss of the U.S.S. *Shenandoah*," *Journal of the American Society of Naval Engineers* 38 (1926), no. 3, pp. 487–580; Robinson, *Giants in the Sky*, pp. 203–6. Arnstein's comments are in Topping, *When Giants Roamed the Sky*, pp. 116–17.

7. On the differences between German and American techniques, see Topping, *When Giants Roamed the Sky*, pp. 116–17.

8. For Eckener's views on traversing the United States, now disappointed, see "The *Los Angeles* and Commerce," *The Outlook*, October 29, 1924, p. 310.

9. "No Turning Back in America's Conquest of the Air," *Literary Digest*, September 19, 1925, pp. 5–8.

30. The Fox

1. Rust, *Flying Across America*, p. 24.
2. Van der Linden, *Airlines and Air Mail*, pp. 10–11.
3. N. A. Komons, "William A. MacCracken, Jr. and the Regulation of Civil Aviation," in Leary (ed.), *Aviation's Golden Age*, p. 38; K. Tuan, "Aviation Insurance in America," *Journal of Risk and Insurance* 32 (1965), no. 1, pp. 2–4.
4. Hoover quoted in Komons, "William A. MacCracken," p. 38. D. D. Lee, "Herbert Hoover and the Golden Age of Aviation," in Leary (ed.), *Aviation's Golden Age*, pp. 127–28; D. D. Lee, "Herbert Hoover and the Rise of Commercial Aviation, 1921–1926," *Business History Review* 58 (1984), pp. 78–102; E. Hawley, "Three Facets of Hooverian Associationalism: Lumber, Aviation, and Movies, 1921–1930," in T. McCraw (ed.), *Regulation in Perspective: Historical Essays* (Cambridge, Mass.: Harvard University Press, 1981), pp. 95–123.
5. Davies, *Airlines of the United States Since 1914*, pp. 33–35; Komons, "William A. MacCracken," pp. 35–59.
6. Komons, "William A. MacCracken," p. 46. On insurance rates, see A. Klemin, "American Passenger Air Transport," *Scientific American*, October 1929, p. 361.
7. Davies, *Airlines of the United States Since 1914*, p. 165.
8. Pisano, "The Greatest Show on Earth," pp. 56–57, 59.
9. Komons, "William A. MacCracken," pp. 45, 38; Bilstein, *Flight in America*, p. 72.
10. Letter, Daniel Guggenheim to Hoover, January 16, 1926; and memorandum, California Institute of Technology, "Development of Aeronautics," August 1926, printed in J. R. Hansen et al. (eds.), *Wind and Beyond*, Documents 2–23 (a–b), pp. 604–10; R. Schwartz, *Flying Down to Rio: Hollywood, Tourists, and Yankee Clippers* (College Station: Texas A&M University Press, 2004), p. 218; R. Hallion, "Daniel and Harry Guggenheim and the Philanthropy of Aviation," in Leary (ed.), *Aviation's Golden Age*, pp. 18–34; Bilstein, *Flight in America*, p. 73.
11. Davies, *Airlines of the United States Since 1914*, pp. 38–39. See also D. Bloor, *The Enigma of the Aerofoil: Rival Theories in Aerodynamics, 1909–1930* (Chicago: University of Chicago Press, 2011).

31. The Trap

1. Bender and Altschul, *Chosen Instrument*, pp. 66–67, 70; Daley, *American Saga*, pp. 15–16.
2. Bender and Altschul, *Chosen Instrument*, pp. 63–64, 103; Trippe, *Pan Am's First Lady*, pp. 1–2, 9–10.
3. On the development of aviation law, see Komons, "William A. MacCracken," pp. 39–40; W. E. Berchtold, "Trade Routes in the Sky," *The New Outlook*, October 1934, p. 16; Josephson, *Empire of the Air*, p. 199; on Trippe's lock-out methods, see F. W. Wile, "Pan American Airways," *Fortune* 13 (1936), April, p. 159.
4. Daley, *American Saga*, pp. 17–18; Bender and Altschul, *Chosen Instrument*, pp. 70–71; on "old fogies," see Josephson, *Empire of the Air*, p. 28.
5. Bilstein, *Flight in America*, pp. 56–57.
6. Rust, *Flying Across America*, p. 36.
7. Bender and Altschul, *Chosen Instrument*, pp. 73–74; Davies, *Airlines of the United States Since 1914*, pp. 49–50; Courtwright, *Sky as Frontier*, p. 67.
8. These shenanigans are covered in Bender and Altschul, *Chosen Instrument*, pp. 74–77,

81; Josephson, *Empire of the Air*, pp. 20–25, 27; Davies, *Airlines of the United States Since 1914*, pp. 61, 195.

32. The Boom

1. Corn, *Winged Gospel*, pp. 18–24; Van Vleck, *Empire of the Air*, pp. 43–44; Rust, *Flying Across America*, p. 7. On Trippe and Lindbergh, Daley, see *American Saga*, pp. 60–63.

2. Fritsche, *A Nation of Fliers*, pp. 146–47.

3. Courtwright, *Sky as Frontier*, p. 76.

4. Snider, "'Great Shadow in the Sky,'" pp. 105–8; Van Vleck, *Empire of the Air*, pp. 44–46, 49–51.

5. W. M. Leary, "Safety in the Air: The Impact of Instrument Flying and Radio Navigation on U.S. Commercial Air Operations Between the Wars," in Leary and Trimble (eds.), *From Airships to Airbus*, volume 1, pp. 106–7; Courtwright, *Sky as Frontier*, p. 62; Davies, *Airlines of the United States Since 1914*, p. 28. The "prowling cat" quote is in Bender and Altschul, *Chosen Instrument*, p. 188.

6. Corn, *Winged Gospel*, pp. 113–22; Van Vleck, *Empire of the Air*, p. 52; "Air Enthusiasts Advised on Jobs," *The New York Times*, October 7, 1928; R. M. Cleveland, "What Is the Future of Aviation?" *Scientific American*, October 1929, p. 313.

7. Douglas, "Airports as Systems and Systems of Airports," in Leary and Trimble (eds.), *From Airships to Airbus*, volume 1, p. 72; A. Klemin, "American Passenger Air Transport," *Scientific American*, October 1929, p. 326.

8. Van der Linden, *Airlines and Air Mail*, pp. 47–48; Rust, *Flying Across America*, pp. 32–36; Davies, *Airlines of the United States Since 1914*, pp. 58–62, 67. C. S. Clancy's lengthy interview with Rogers mentions his time in Munich and Zurich, though Rogers seems to have been so uninterested in airships he did not cross Lake Constance to Friedrichshafen. See Clancy, "Aviation's Patron Saint," *Scientific American*, October 1929, pp. 283–86.

9. A. Gordon, *Naked Airport: A Cultural History of the World's Most Revolutionary Structure*, 2nd ed. (Chicago: University of Chicago Press, 2008), p. 32. For typical examples, see J. F. O'Ryan (Trippe's nemesis), "For Safety, the Air," *North American Review*, November 1928, pp. 616–20; and W. P. MacCracken, Jr., and W. B. Courtney, "Flight Savers," *Collier's*, March 29, 1930, pp. 14–15, 46, 48–50.

10. For numbers, see Series L 265–273, "Air Transport—Accidents: 1927 to 1945," in *Historical Statistics of the United States, 1789–1945*, p. 225; Leary, "Safety in the Air," in Leary and Trimble (eds.), *From Airships to Airbus*, volume 1, p. 106; "Motor Vehicle Traffic Fatalities, 1900–2007," at www.fhwa.dot.gov/policyinformation/statistics/2007/pdf/fi200.pdf; "31,000 More Killed in Our Motor Massacre," *Literary Digest*, June 14, 1930, p. 11. Admittedly, unlicensed or undertrained pilots flying (or stunt-flying) private planes (sometimes homemade) added another 2,875 accidents and 965 deaths to the 1927–29 air total, but even then, one could hardly compare the lethality of the two modes of transport. For a harsh view of amateur fliers, see T.J.C. Martyn, "'Fool Flying' Takes a Grim Toll," *The New York Times*, October 7, 1928; a typical example is "8 Killed, 11 Injured in 6 Plane Crashes," *The New York Times*, October 8, 1928. Trains were, statistically speaking, by far the safest way to travel; see "A New Era for the Railroads," *Railway Age*, February 21, 1931, p. 397.

11. W. B. Courtney, "Up and Going," *Collier's*, March 23, 1929, p. 9.

12. Courtwright, *Sky as Frontier*, p. 102; Courtney, "Up and Going," p. 9.

13. The most precise figures for passengers are given in Series K 246–256, "Air Transport—

Scheduled Air Transportation, Domestic Only: 1926–1945," in *Historical Statistics of the United States*, p. 224.

14. Courtwright, *Sky as Frontier*, p. 10; Rust, *Flying Across America*, pp. 59, 51; Kolm, "'Who Says It's a Man's World?'" p. 153.

15. Rust, *Flying Across America*, p. 78.

16. Series K 239–245, "Air Transport—Aircraft Production and Exports: 1913–1945," in *Historical Statistics of the United States*, p. 224.

17. Bilstein, *Flight in America*, pp. 62–69; Anon., "How the Airplane Will Speed Up Business," *Literary Digest*, March 30, 1929, pp. 18–19; Cleveland, "What Is the Future of Aviation?" p. 314.

18. Anon., "All Aboard the Lindbergh Limited," *Literary Digest*, March 2, 1929, p. 54; Anon, "The Fokker 'F-10' Monoplane," *Flight*, December 13, 1928, pp. 1048–49; M. R. Stearns, "All Aboard by Air," *World's Work*, April 1929, p. 144.

19. Van Vleck, *Empire of the Air*, p. 51.

20. Douglas, "Airports as Systems and Systems of Airports," in Leary and Trimble (eds.), *From Airships to Airbus*, volume 1, pp. 55–58, 60.

21. Anon., "Airports," *World's Work*, November 1929, p. 37; W. E. Arthur, "How Shall We Design Our Airports?" *Scientific American*, October 1929, pp. 298–301; Gordon, *Naked Airport*, pp. 25, 28. See also Series K 257–264, "Air Transport—Airports, Aircraft, Pilots, and Miles Flown: 1926–1945," *Historical Statistics of the United States*, p. 225; W. B. Courtney, "Up and Going," *Collier's*, March 23, 1929, pp. 54, 56.

22. Rust, *Flying Across America*, pp. 48, 68–69; A. Phillips, "Over the Scenic Southwest," *Air Travel News*, November 1929, p. 11; W. B. Courtney, "High-Flying Ladies," *Collier's*, August 20, 1932, p. 45.

23. Darling, "Across the Continent in Forty-Eight Hours," p. 55; H. E. McLaughlin, *Footsteps in the Sky: An Informal Review of U.S. Airlines Inflight Service, 1920–Present* (Denver, Colo.: State of the Art, 1994), p. 18; Rust, *Flying Across America*, pp. 52–54; Phillips, "Over the Scenic Southwest," p. 11.

24. V. G. Darling, "Across the Continent in Forty-Eight Hours," *World's Work*, September 1929, p. 54.

25. Rust, *Flying Across America*, pp. 52–54; Darling, "Across the Continent in Forty-Eight Hours," p. 55.

26. McLaughlin, *Footsteps in the Sky*, p. 5.

27. McLaughlin, *Footsteps in the Sky*, pp. 10–14; V. Vantoch, *The Jet Sex: Airline Stewardesses and the Making of an American Icon* (Philadelphia: University of Pennsylvania Press, 2013), pp. 15–20.

28. Corn, *Winged Gospel*, pp. 72, 75, 86; Rust, *Flying Across America*, p. 55; B. Gould, "Milady Takes the Air," *North American Review*, December 1929, pp. 691–97. No woman was employed by a commercial carrier until 1934, when Central Airlines (later part of United) hired Helen Richey as co-pilot on the Washington, D.C.–Detroit route. She was forced out by the all-male pilots' union, and it wouldn't be until 1973 that another woman (Emily Warner) was taken on by an American airline. Interestingly, there was reverse sexism, as well: Male stewards were forbidden in January 1942, after the United States entered the war, but could be designated as "pursers." It was not until 1972 that men could work as stewards, or "flight attendants," as they and stewardesses would come to be called. McLaughlin, *Footsteps in the Sky*, p. 5.

29. McLaughlin, *Footsteps in the Sky*, pp. 11–12; Vantoch, *The Jet Sex*, p. 22.

30. Hasle, interviewed in Courtney, "High-Flying Ladies," p. 45; S. L. Kolm, "'Who Says It's

a Man's World?' Women's Work and Travel in the First Decades of Flight," in *The Airplane in American Culture*, ed. Pisano, pp. 151–52.

31. Courtwright, *Sky as Frontier*, p. 106; Rust, *Flying Across America*, pp. 70–71.

32. Courtney, "High-Flying Ladies," p. 45.

33. "Movie Entertains Air Travelers Flying from St. Paul to Chicago," *The New York Times*, February 18, 1929.

34. Rust, *Flying Across America*, p. 64.

35. Darling, "Across the Continent in 48 Hours," pp. 52–56. On the interiors, see also Phillips, "Over the Scenic Southwest," p. 11.

36. Courtney, "Up and Going," pp. 54, 56; McLaughlin, *Footsteps in the Sky*, pp. 4, 16–17; Vantoch, *The Jet Sex*, p. 18; Courtwright, *Sky as Frontier*, pp. 103–4; Rust, *Flying Across America*, pp. 62–63.

37. Bender and Altschul, *Chosen Instrument*, p. 80; Van der Linden, *Airlines and Air Mail*, pp. 47, 50; Davies, *Airlines of the United States Since 1914*, pp. 162–63; Anon., "Aviation," *World's Work*, July 1929, p. 34; Cleveland, "What Is the Future of Aviation?" p. 313. Estimates of Rentschler's wealth vary a little, depending on sources. Rentschler himself stated to a Senate committee in 1934 the figure given here, of which he realized $9,500,000 after selling some shares. See G. Robinson, "The Rentschlers Fly the Dollar," New York *Daily News*, January 19, 1934, p. 33.

33. Terra Incognita

1. Daley, *American Saga*, pp. 36–39, 69–71, 88; Josephson, *Empire of the Air*, p. 38.

2. Daley, *American Saga*, p. 37. On Leuteritz, pp. 45–47.

3. Josephson, *Empire of the Air*, pp. 32, 38–39, 40–41; O. E. Dunlap, Jr., "Why Fly Without Radio?" *Scientific American*, September 1927, pp. 226–28; A. Perry, "Radio Guides the Airway Traveler," *Scientific American*, March 1929, pp. 238–39.

4. Trippe, *Pan Am's First Lady*, p, 15; Daley, *American Saga*, pp. 43–55.

5. Daley, *American Saga*, pp. 60–63.

6. On Hoyt, see "Builders of the Aviation Industry," *Scientific American*, March 1929, p. 231.

7. Davies, *Airlines of the United States Since 1914*, pp. 212–14; Bender and Altschul, *Chosen Instrument*, 82–87; Daley, *American Saga*, pp. 27–32, 36, 58; W. E. Brown, Jr., "Pan Am: Miami's Wings to the World," *Journal of Decorative and Propaganda Arts* 23 (1998), p. 148.

34. El Dorado

1. Daley, *American Saga*, pp. 40–41. On "the eagle," see Anon., "Lindbergh Unites the Americas," *Literary Digest*, January 21, 1928, p. 6.

2. Josephson, *Empire of the Air*, pp. 33–35. Capone appears to have been a fan of Pan American. Betty Trippe recalled that when she was at a restaurant in Nassau with some friends (Trippe was away in Washington) the wine waiter came over to say, "Mr. Capone at the next table wants to know if you'd like some Champagne?" Betty said no after looking over and seeing a party of eight men who "looked like real thugs in their flashy tropical suits with padded shoulders." When Betty was about to leave, the waiter came over again: "Mr. Capone would like to know if the ladies would like to go with him to the Jungle Club [a Nassau nightclub]?" Betty and Co. shook their heads and locked their bedroom doors.

The next day the newspapers reported the St. Valentine's Day Massacre. Trippe, *Pan Am's First Lady*, p. 18.

3. Anon., "Lindbergh Unites the Americas," *Literary Digest*, January 21, 1928, p. 5; Van Vleck, *Empire of the Air*, p. 70.

4. Daley, *American Saga*, pp. 39, 56–65; Josephson, *Empire of the Air*, pp. 61–62; Trippe, *Pan Am's First Lady*, p. 12; Bender and Altschul, *Chosen Instrument*, p. 103. On aircraft acquisitions, see R.E.G. Davies, *Pan Am: An Airline and Its Aircraft* (New York: Orion Books, 1987), pp. 12–13, 27. On 36th Street and airport design, see Brown, Jr., "Pan Am: Miami's Wings to the World," pp. 148–49; P. Pennoyer and A. Walker, *The Architecture of Delano and Aldrich* (New York: W. W. Norton & Co., 2003), p. 69; Gordon, *Naked Airport*, pp. 44–63; on stewards, see McLaughlin, *Footsteps in the Sky*, pp. 3–8, and Vantoch, *The Jet Sex*, p. 25; on the ground staff, see M. C. Hühne, *Pan Am: History, Design, and Identity* (Berlin: Callisto Publishers, 2016), p. 29.

5. Quoted in Josephson, *Empire of the Air*, p. 44.

6. Daley, *American Saga*, p. 68; Josephson, *Empire of the Air*, pp. 66–69; Bender and Altschul, *Chosen Instrument*, pp. 117–20, 147.

7. S. J. Randall, "Colombia, the United States, and Interamerican Aviation Rivalry, 1927–1940," *Journal of Interamerican Studies and World Affairs* 14 (1972), no. 3, pp. 297–324; Josephson, *Empire of the Air*, pp. 63–69; Bender and Altschul, *Chosen Instrument*, pp. 140–46; Daley, *American Saga*, p. 66.

8. Daley, *American Saga*, pp. 69–86; Josephson, *Empire of the Air*, pp. 71–76; R. A. O'Neill and J. F. Hood, *A Dream of Eagles* (Boston, Mass.: Houghton Mifflin, 1973), pp. 17, 201–3, 229, 283, 307–8; Bender and Altschul, *Chosen Instrument*, pp. 166–75; Davies, *Pan Am*, pp. 20–21. After the legal documents were signed to wind up NYRBA, O'Neill hastened to leave the Union League Club in New York but was followed, irritatingly, by Trippe as he headed out. Trippe hailed a cab and asked whether O'Neill would like to come with him. O'Neill reluctantly agreed and Trippe spent the ride alternating "between puffs on a big Havana and attempts at witticisms." He clearly wanted to ask something, and finally, as the cab pulled over, Trippe leaned in: "Ralph, listen. I'll make you a very attractive offer to manage our Latin American east-coast division." "Never," snapped O'Neill. "There's nothing more to say, except that you're the last man on earth I would work for." He got out of the taxi and the two never saw each other again. O'Neill ended up running a gold-dredging operation in deepest Bolivia, where he nursed his grievances and forever cursed the name of Trippe. C. J. Kelly mentions the author bumping into O'Neill in Bolivia in 1960. C. J. Kelly, *The Sky's the Limit: The History of the Airlines* (New York: Coward-McCann, 1963), p. 129n.

9. Bender and Altschul, *Chosen Instrument*, pp. 111–12; Daley, *American Saga*, p. 66.

10. Josephson, *Empire of the Air*, pp. 53–54.

11. Bender and Alschul, *Chosen Instrument*, pp. 127–28, 130.

12. Van Vleck, *Empire of the Air*, p. 75; Bender and Altschul, *Chosen Instrument*, p. 101; Daley, *American Saga*, pp. 95–96.

13. Bender and Altschul, *Chosen Instrument*, pp. 72, 76; Trippe, *Pan Am's First Lady*, pp. 26–46. On Van Dusen, see Daley, *American Saga*, p. 52.

14. Bilstein, *Flight in America*, p. 78; "Pan American Airways System," *Aviation*, May 3, 1930, p. 70; Bender and Altschul, *Chosen Instrument*, p. 176; Josephson, *Empire of the Air*, p. 59; Van Vleck, *Empire of the Air*, p. 68.

15. Bender and Altschul, *Chosen Instrument*, pp. 121–22.

16. Josephson, *Empire of the Air*, p. 58; S. B. Kauffman and G. E. Hopkins, *Pan Am Pioneer: A Manager's Memoir from Seaplane Clippers to Jumbo Jets* (Lubbock: Texas Tech Univer-

sity Press, 1995), p. 20. A surprisingly large number of Trippe's managers in the most remote airfields were Yale graduates, for some reason.

17. Anon., "Wiping Out Thousands of Miles Between the Two Americas," *Literary Digest*,
May 18, 1929, pp. 77–79.

18. Bender and Altschul, *Chosen Instrument*, pp. 104–5.

35. The Emissary

1. Bender and Altschul, *Chosen Instrument*, pp. 303, 151; Josephson, *Empire of the Air*, pp.
196, 78–79; Trippe's "religion" is quoted in Anon., "America's Aerial Bridge to the Orient," *Literary Digest*, November 23, 1935, pp. 28–29. On Eckener, see Vaeth, *Graf Zeppelin*, pp. 124–37.

2. A. Wegerdt, "Germany and the Aerial Navigation Convention at Paris, October 13, 1919,"
Journal of Air Law and Commerce 1 (1930), pp. 27–29; Duggan and Meyer, *Airships in
International Affairs*, pp. 123–24.

3. Memorandum, April 24, 1926, in Box 3, Folder 25, Henry Cord Meyer Papers; Italiaander,
Ein Deutscher Namens Eckener, pp. 243–45.

4. Eckener (trans. Robinson), *My Zeppelins*, pp. 29–31. Regarding the subsidiaries, Julius
Oesterle, a Zeppelin director, believed Eckener had been itching to do this for some
time: "He had only the one goal in mind—to lead the airship in Germany to its objective," and keeping Colsman's subsidiaries afloat had long "been a thorn in his side." Oesterle's notes are printed in Italiaander, *Ein Deutscher Namens Eckener*, p. 273.

5. Duggan and Meyer, *Airships in International Affairs*, pp. 126–27.

6. Robinson, *Giants in the Sky*, pp. 261–64; C. E. Rosendahl, "Inside the *Graf Zeppelin*,"
Scientific American, March 1929, pp. 201–5; Dick and Robinson, *Golden Age*, pp. 32–37
(technical figures usually vary slightly between sources), 76 (comparison of Blau-gas performance); menu card for October 11, 1932, reproduced in J. G. Vaeth, "Zeppelin Decor:
The *Graf Zeppelin* and the *Hindenburg*," *Journal of Decorative and Propaganda Arts* 15
(1990), p. 52; on the *Los Angeles* figures, see C. E. Rosendahl, "Bigger and Better Blimps,"
World's Work, July 1929, p. 69; "Blue Gas and Hydrogen," *Time*, October 15, 1928; "Zeppelin Fuel a Familiar Gas," *The New York Times*, October 7, 1928. On the power differences in heat units, see E. A. Lehmann, "Lehmann Describes the *Graf Zeppelin*," *The
New York Times*, October 13, 1928. American newspapers habitually referred to Blau gas
as "Blue gas"—*blau* is German for "blue"—though it wasn't that color.

7. Eckener (trans. Robinson), *My Zeppelins*, p. 31.

8. "Zeppelin Carries 82 on Long Flight," *The New York Times*, September 21, 1928.

9. "Last Zeppelin Test Will Be Made Today," *The New York Times*, October 2, 1928.

10. Robinson, *Giants in the Sky*, p. 265; "New Zeppelin Home After 35-Hour Trip," *The New
York Times*, October 4, 1928.

11. Meyer, "France Perceives the Zeppelins, 1924–1937," pp. 172–73.

12. "Reich Nationalists Ask a Referendum on Ending Republic," *The New York Times*, September 25, 1928.

13. De Syon, *Zeppelin!*, pp. 128–31.

14. Litchfield, "Lighter-Than-Air Craft," *Annals of the American Academy*, p. 81, citing these
invented numbers based on Lehmann's estimates.

15. De Syon, *Zeppelin!*, pp. 142–43.

16. "Airship Publicity Urged by Eckener," *The New York Times*, October 21, 1928; "Eckener
Proposes a 5-Airship Line," *The New York Times*, October 13, 1928.

17. Obituary of Rosendahl, *The New York Times*, May 15, 1977.

18. Robinson, *Giants in the Sky*, p. 220; Robinson and Keller, *"Up Ship!"* p. 178.

19. Topping, *When Giants Roamed the Sky*, pp. 142–44. See also "Uncle Sam's 9-Acre Wonder Hangar on Wheels," *Literary Digest*, January 4, 1930, pp. 50–51, 54–55; W. P. MacCracken, Jr., and W. B. Courtney, "Shrinking the Seas," *Collier's*, May 10, 1930, p. 70.

20. On Arnstein and Hughes, see Topping, *When Giants Roamed the Sky*, p. 137.

36. Survival of the Fittest

1. For this sweet story, see "Awaits Fiancé on Zeppelin," *The New York Times*, October 13, 1928. For the captain's descriptions, see C. M. Hiam, *Dirigible Dreams: The Age of the Airship* (Lebanon, N.H.: ForeEdge, 2014), p. 208.

2. Another was Robert Reiner, an importer of sewing equipment and old friend of Eckener's. He had booked his fare, not caring what it cost, years earlier. "Zeppelin Soars over Mediterranean," *The New York Times*, October 12, 1928.

3. "Zeppelin Soars over Mediterranean," *The New York Times*, October 12, 1928; C. E. Rosendahl, "Inside the Zeppelin," *Scientific American*, March 1929, p. 204; Hiam, *Dirigible Dreams*, p. 206. On Drummond-Hay's changes of clothing, confidentiality agreements, and the canary, see "Story of the Flight of the Big Dirigible," *The New York Times*, October 16, 1928.

4. "Braving Ocean Winds in the Graf Zeppelin," *Literary Digest*, November 3, 1928, pp. 37–38, 43.

5. "Zeppelin Travelers Insure for Millions," *The New York Times*, October 8, 1928.

6. "11½ Hour Trip Set a Duration Record," *The New York Times*, October 16, 1928; "Rosendahl Tells of Zeppelin Voyage," *The New York Times*, October 18, 1928; Anon., "Braving Ocean Winds in the *Graf Zeppelin*," *Literary Digest*, November 3, 1928, pp. 37–38; Hiam, *Dirigible Dreams*, p. 209.

7. Lehmann, *Zeppelin*, p. 245; "Eckener's Own Son Risks Life for Ship," October 14; "Story of the Flight of the Big Dirigible," October 16; "Dr. Eckener Lauds Courage of Crew," October 16; "Rosendahl Tells of Zeppelin Voyage," October 18, 1928, *The New York Times*; "Braving Ocean Winds in the *Graf Zeppelin*," pp. 37–38.

8. "Braving Ocean Winds in the *Graf Zeppelin*," p. 43; "Zeppelin to Tour Cities of America," *The New York Times*, September 25, 1928; Hiam, *Dirigible Dreams*, p. 210. On the food, see "Zeppelin Travelers Insure for Millions," *The New York Times*, October 8, 1928; for the menu card, see Vaeth, "Zeppelin Decor," p. 52.

9. "Story of the Flight of the Big Dirigible," *The New York Times*, October 16, 1928.

10. "Coolidges Thrilled by Zeppelin's Visit," *The New York Times*, October 16, 1928.

11. "Airship Hypnotizes City Gazing Upward," *The New York Times*, October 16, 1928.

12. "Airship Undergoes Customs Inspection," *The New York Times*, October 16, 1928.

13. "Dr. Eckener Lauds Courage of Crew," October 16, "Passengers Happy to Be on Land Again," October 16, "Germans Hear Police Offended Fliers Here," October 17, "Questions Jersey Police," October 18, 1928, *The New York Times*; "Braving Ocean Winds in the *Graf Zeppelin*," *Literary Digest*, November 3, 1928, pp. 44, 48; "A Million Miles in a Zeppelin," p. 74, Schiller Papers.

14. "20,000 View Zeppelin in Lakehurst Hangar," October 18; "150,000 Sightseers See Zeppelin in Day," October 21; "Odd Doings Amuse Visitors at Hangar," October 16, 1928, *The New York Times*.

15. "250 Reporters Held Lakehurst Sector," *The New York Times*, October 17, 1928.

16. "Germany Rejoices in Airship Victory," *The New York Times*, October 16, 1928.

17. Schiller, "A Million Miles in a Zeppelin," p. 75, Schiller Papers. "Dr. Eckener to Push Ocean Airship Line," October 18, "Throng Acclaims Guests," October 17, "Chicago Welcomes Zeppelin Party," October 22, 1928, *The New York Times*.

18. "City Hails Zeppelin Fliers in Parade Up Broadway," *The New York Times*, October 17, 1928.

19. "Lehmann Describes the *Graf Zeppelin*," *The New York Times*, October 13, 1928; photo accompanying "Frauleins Bear Off Terhune in Triumph," *The New York Times*, November 2, 1928.

20. "Braving Ocean Winds in the *Graf Zeppelin*," *Literary Digest*, November 3, 1928, pp. 43–44; "Zeppelin Passenger Sails," *The New York Times*, October 18, 1928.

21. On different opinions of the *Graf Zeppelin*'s flight, see newspaper summaries in "Germany's 'Merchantman of the Skies,'" *Literary Digest*, October 27, 1928, pp. 13–15. Though all financial estimates should be taken with a large spoonful of salt, see "Trip Much Shorter Than to America," *The New York Times*, November 1, 1928; "Zeppelin Ends Test; Ready for Trip Here," *The New York Times*, October 9, 1928.

22. A. R. Blessing, "Airships Versus Airplanes," *North American Review* 226 (1928), no. 1, pp. 53–63; "The *Graf Zeppelin*'s Second Visit," *Literary Digest*, August 17, 1929, p. 11; C. E. Rosendahl, "Lighter-Than-Air Machines," *Proceedings of the American Philosophical Society* 67 (1928), no. 4, pp. 320–21.

23. On the conference, see "Germany Studies Zeppelin's Flight," *The New York Times*, October 21, 1928, and "Airplane Liners Shown at Berlin," *The New York Times*, November 4, 1928. On Trippe's presence, see Bender and Altschul, *Chosen Instrument*, p. 183.

24. The exact flight time varies with the source. I've used Lehmann's, given in *Zeppelin*, p. 252. See also "Zeppelin Starts Flight to Germany with 63 on Board," October 29, "Bells, Bands, Guns Welcome Zeppelin," November 2, "Trip Much Shorter Than to America," November 1, "Germans Exultant at Zeppelin Return," October 30, "Zeppelin Far Out to Sea, Racing 100 Miles an Hour with a Gale at Her Heels," October 30, 1928, *The New York Times*.

25. "Disaster in Storm Menaced Zeppelin over Newfoundland," *The New York Times*, November 2, 1928.

37. Around the World

1. "Zeppelin's Start Likely on Thursday," *The New York Times*, August 13, 1929.

2. On costs and income, see Eckener (trans. Robinson), *My Zeppelins*, pp. 67–69; on the Hearst conditions, pp. 70–71. Eckener's claim that he earned a profit is debatable, as it's possible he was disappointed at the revenue: In early August, he was forecasting income of $500,000, which he would have fallen short of, even with Hearst money behind him. "Zeppelin's Cruise to Bring In $500,000 from Her Payload," *The New York Times*, August 6, 1929. On Japanese rights, see "Life's Ups and Downs on a Globe-Girdling Zep," *Literary Digest*, September 14, 1929, p. 41.

3. "Zeppelin Home in 55½ Hrs., Hailed by Joyous Crowds; Off to Tokyo Wednesday," August 11, "Animals on *Graf Zeppelin*," August 2, "Thieves Steal Safe as Zeppelin Sails," August 9, "Zeppelin to Start World Trip Tonight as Weather Favors," August 7, 1929, *The New York Times*.

4. Eckener (trans. Robinson), *My Zeppelins*, pp. 71–72; "Eckener Marks Birthday in Air," August 11; "Dr. Eckener Tells of Battling Winds," August 5, 1929, *The New York Times*.

5. D. Botting, *Dr. Eckener's Dream Machine: The Great Zeppelin and the Dawn of Air Travel* (New York: Owl Books, 2001), p. 151.

6. For these paragraphs, see A. M. Nekrich (trans. G. L. Freeze), *Pariahs, Partners, Predators: German-Soviet Relations, 1922–1941* (New York: Columbia University Press, 1997), pp. 39–40, 50–61; R. H. Haigh, D. S. Morris, and A. R. Peters, *German-Soviet Relations in the Weimar Era: Friendship from Necessity* (Totowa, N.J.: Barnes and Noble, 1985), pp. 131–36; C. Foss, "Russia's Romance with the Airship," *History Today*, December 1997, pp. 10–16; M. Kulikowski, "Fantasy Flights: Technology, Politics, and the Soviet Airship Program, 1930–1938," *Icon* 21 (2015), pp. 66–80; J. D. Cameron, "To Transform the Revolution into an Evolution: Underlying Assumptions of German Foreign Policy Toward Soviet Russia, 1919–27," *Journal of Contemporary History* 40 (2005), no. 1, pp. 7–24; J. McCannon, "Winged Prometheans: Arctic Aviation as Socialist Construction in Stalinist Russia, 1928–1939," *Comparative Issues in the History of Circumpolar Science and Technology* 33 (2010), no. 2, pp. 75–97; R. P. Morgan, "The Political Significance of German-Soviet Trade Negotiations, 1922–5," *Historical Journal* 6 (1963), no. 2, pp. 253–71; H. L. Dyck, "German-Soviet Relations and the Anglo-Soviet Break, 1927," *Slavic Review* 25 (1966), no. 1, pp. 67–83. On Moscow's official interest in airship development, see also letter to Eckener, "About Airship Building and About Aerial Trips on Airships of the Zeppelin Type over Russia and Siberia," [September] 1930, and "Draft Resolution of Politburo . . . on Airship Building," December 12, 1930, printed in Y. Dyakov and T. Bushuyeva, *The Red Army and the Wehrmacht: How the Soviets Militarized Germany, 1922–33* (Amherst, N.Y.: Prometheus Books, 1994), pp. 97–99.

 A fine example of warped Soviet thinking was provided by a high-level discussion as to whether the infallible Lenin, the Bolshevik pope, would have backed an airship program. Stalin argued that Lenin had not alluded to the matter and thus there was no ideological justification for airships; the money would be better spent on airplanes.

 But airship advocates eked out a minor victory when one of Lenin's old cronies, Fyodor Ilin, recollected that in 1914 the exile had seen a DELAG airship and said, "One day we will have those too." That was as good as Scripture, then, and it meant that airships had to be built by hook or by crook, but Stalin struck back by downplaying the revelation in *Pravda*, the state newspaper. In the USSR, the hierarchy of news was reflected in the location of articles. Major Party decisions always appeared on page one, with other news following in a strictly regulated order of importance. Revealingly, Ilin's recollection was printed at the very bottom of page 4, the last page. (Ilin later fatally defenestrated himself under mysterious circumstances.) I'm indebted to Professor Stephen Kotkin, Princeton University, for information regarding Ilin.

7. "Zeppelin in Russia Skirts Bad Weather," *The New York Times*, August 16, 1929; Eckener, *My Zeppelins*, p. 73; Botting, *Dream Machine*, p. 153.

8. M. Kozloff, "Why *Graf Zeppelin* Flew Around Moscow," *Pravda*, August 22, 1929, quoted in Italiaander, *Ein Deutscher Namens Eckener*, pp. 255–59. See also "Reds Annoyed at Zeppelin," *The New York Times*, August 23, 1929.

9. Botting, *Dream Machine*, p. 155.

10. "*Graf Zeppelin* Ends 21-Day World Trip; City's Tribute Today," *The New York Times*, August 30, 1929.

11. Eckener, *My Zeppelins*, pp. 82–83; Botting, *Dream Machine*, pp. 169–70.

12. H. Byas, "Thousands at Field All Night," *The New York Times*, August 22, 1929; Botting, *Dream Machine*, pp. 172–74.

13. Memorandum, Department of Far Eastern Affairs, "Reception of Airship *Graf Zeppelin*

in Japan Enroute from Friedrichshafen to Los Angeles," August 27, 1929, in Henry Cord Meyer Papers, Box 2, File 21.

14. "*Graf Zeppelin* Reaches Pacific Coast; Passes San Francisco, Nearing Goal; Thousands Wait at Los Angeles Field," *The New York Times*, August 26, 1929.

15. Eckener (trans. Robinson), *My Zeppelins*, p. 87.

16. Eckener (trans. Robinson), *My Zeppelins*, pp. 89–94; "Zeppelin in Texas on Lakehurst Trip; Dodges Storm Area," *The New York Times*, August 28, 1929.

17. "Chicago Welcome Heard over Radio," *The New York Times*, August 29, 1929; Eckener (trans. Robinson), *My Zeppelins*, p. 93.

18. H. Eckener, "The First Airship Flight Around the World," *National Geographic Magazine* 57 (1930), no. 6 (June), pp. 653–88.

19. See "The *Graf Zeppelin*'s Second Visit," *Literary Digest*, August 17, 1929, p. 11.

38. The Monster

1. "The Story of the Summer," *Literary Digest*, September 14, 1929, p. 16.

2. "American Ovations Move Dr. Eckener," August 30, "Zeppelin Air Line Reported in Prospect," August 15, "Zeppelin Lawyer Coming for Parley," August 21, 1929, *The New York Times*.

3. "Airships to Ply Between the Coast and Hawaii; New Company Plans 36-Hour Zeppelin Service," October 23, "Agree on Program for Pacific Airway," October 29, 1929, *The New York Times*; MacCracken, Jr., and Courtney, "Shrinking the Seas," pp. 70–71; "Goodyear Plans Several Projects," *Aviation*, November 16, 1929, p. 992. On Hunsaker's interest in Zeppelins, see his article, "The Present Status of Airships in Europe," *Journal of the Franklin Institute* 177 (1914), no. 6, pp. 597–639.

4. On Meister, see Meyer, "F. W. (Willy) von Meister," pp. 194–98.

5. R. M. Cleveland, "The Truth About Aviation Stocks," *Scientific American*, January 1930, pp. 30–31.

6. The most detailed analysis of IZT and PZT operations is R. Kalabash Wordsmith (pseud.), "Notes on the International Zeppelin Transport Co. and Pacific Zeppelin Transport Co.," *Buoyant Flight* 39 (1992), no. 4, Part 1, pp. 2, 4–5, and *Buoyant Flight* 39 (1992), no. 5, Part 2, pp. 4–5, 8. See also "NY Groups Plan Airship Lines," *Aviation*, October 26, 1929, p. 864; "Ocean Zeppelin Line Backed by Millions; Speeds Plans Here," March 25, 1930, "Air Lines Merger Being Negotiated," March 27, 1930, "Dirigible Service by 1934 Predicted," January 31, 1931, *The New York Times*; "More Interests Back Trans-Atlantic Company," *Aviation*, April 5, 1930, p. 734; J. C. Hunsaker, "The Day of the Dirigible," *North American Review* 229 (1930), no. 4, pp. 434–35; "Zepping to Europe on Regular Schedule at Express Speed," April 12, pp. 50, 52, "Germany Plans Bigger and Better Zeps," July 19, p. 33, "American Fuel for German Dirigibles," August 9, 1930, p. 27, *Literary Digest*. On South American immigration, see A. von Gleich, "Germany and Latin America," RAND Corporation, Memorandum RM-5523-RC, June 1968, pp. 6–7, and Table 1, "German Overseas Migration, 1871–1957"; on Condor's activities, see Josephson, *Empire of the Air*, pp. 81–84; "Scrambling for South America's $8,000,000 Air Trade," *Literary Digest*, May 2, 1931, p. 42. Schiller, in the Schiller Papers, confirms that Condor was to be the airships' connection in South America. See Schiller, "A Million Miles in a Zeppelin," p. 104. The McNary bill is covered in "Federal Bill Aids Ocean Zeppelins," *The New York Times*, April 26, 1930. On the Empire State Building (non-)idea, see Litchfield's suggestion in

MacCracken and Courtney, "Shrinking the Seas," p. 71; C. Gray, "Not Just a Perch for King Kong," *The New York Times*, September 23, 2010; I. Clavan, "Empire State Building: The Mooring Mast," *Architectural Forum* 54 (1931), February, pp. 229–34.

7. "Germany Expects Aviation Expansion," *The New York Times*, January 2, 1930.

8. L. D. Lyman, "Bridging the Ocean by Super-Airships," *The New York Times*, March 30, 1930.

9. Eckener (trans. Robinson), *My Zeppelins*, pp. 97–98.

10. Eckener (trans. Robinson), *My Zeppelins*, pp. 99–114. The storm story is told from the perspective of an anonymous "transatlantic traveller," in "The Graf Zeppelin Weathers a Storm," *The Airship* (1935), no. 6 (Summer), pp. 27–28. On Zeppelin plans, "Bankers Go by Train to See Dr. Eckener," *The New York Times*, May 31, 1930.

39. Engage the Enemy More Closely

1. See "Predicts Big Role for New Airships," *The New York Times*, December 8, 1929.

2. C. A. Lindbergh, "Lessons of the *Graf Zeppelin*'s Voyage," *The New York Times*, October 21, 1928.

3. D. Cochrane, V. Hardesty, and R. Lee, *The Aviation Careers of Igor Sikorsky* (Washington, D.C.: National Air and Space Museum, 1989), pp. 62–97; Daley, *American Saga*, p. 93; Bender and Altschul, *Chosen Instrument*, pp. 183–85.

4. Davies, *Pan Am*, pp. 12–13.

5. Daley, *American Saga*, pp. 92–95; D. Beaty, *The Water Jump: The Story of Transatlantic Flight* (New York: Harper & Row, 1963), p. 102; C. A. Lindbergh, *Autobiography of Values* (New York: Harcourt Brace Jovanovich, 1978), pp. 110–12, 115–16; Bender and Altschul, *Chosen Instrument*, pp. 182–86, 213; M. Holland, *Architects of Aviation* (New York: Duell, Sloan, & Pearce, 1951), p. 188. On Priester's adherence to the strictest standards, see "International Transport," *Aviation*, September 1932, pp. 374–77.

6. This section is based on J. R. Hansen, *The Bird Is on the Wing: Aerodynamics and the Progress of the American Airplane* (College Station: Texas A&M University Press, 2004), pp. 41–62; Anderson, Jr., *The Airplane*, pp. 183–282; L. K. Loftin, Jr., *Quest for Performance: The Evolution of Modern Aircraft* (Washington, D.C.: National Aeronautics and Space Administration, 1985), chs. 3 and 4, pp. 67–101, 186, and Appendix A, "Physical and Performance Data," Tables I and II, pp. 480–83; see also L. Breguet, "Aerodynamical Efficiency and the Reduction of Air Transport Costs," *Aeronautical Journal*, August 1922, and B. Melvill Jones, "The Streamline Airplane," both reprinted in Hansen et al. (eds.), *Wind and Beyond*, volume 2, Document 3–4, pp. 102–9, and Document 3–7, pp. 130–62. The airshipmens' flawed assumptions can be seen in C. E. Rosendahl, "Bigger and Better Blimps," *World's Work*, July 1929, p. 67; Hunsaker, "Day of the Dirigible," pp. 433–34; Blessing, "Airplanes versus Airships," pp. 55–56; and Eckener (as late as 1934!), "The Rigid Airship and Its Possibilities," *The Airship* 2 (1935), no. 5 (Spring), pp. 6, 8. For comments pointing out that their assumptions were wrong, see "Dornier Analyzes Routes," *The New York Times*, September 14, 1930. T.J.C. Martyn's "annihilation" remark is in "Do-X Is Compared with Huge R-101," *The New York Times*, November 3, 1929.

7. On LZ-128 and the Americans' disappointment, see J. Duggan, *LZ-129 Hindenburg: The Complete Story* (Ickenham, U.K.: Zeppelin Study Group, 2002), pp. 3–4, 9–10.

40. The Duelists

1. "His Majesty's Ship R-100," *Flight*, December 6, 1929, pp. 1275–80; "The Loss of H.M. Airship R-101," *Flight*, October 10, 1930, pp. 1107–1114, 1126; "How the R-100 Is Realizing Britain's Dreams," *Literary Digest*, August 16, 1930, pp. 29–30; "Britain's Dirigible Horror," *Literary Digest*, October 18, 1930, pp. 10–11; F.L.M. Boothby, "The Loss of R-101," *The Airship* 1 (1934), no. 2, pp. 24–27. For the fullest narrative, see P. Davison, "The R-101 Story: A Review Based on Primary Source Material and First-hand Accounts," *Journal of Aeronautical History* 5 (2015), pp. 43–167.

2. Hartcup, *Achievement of the Airship*, pp. 189–90; Robinson, *Giants in the Sky*, p. 297. Nevil Norway, of the R-100 team, describes the noxious atmosphere of the competition. Norway, *Slide Rule: An Autobiography of an Engineer* (New York: Vintage, 2010), chs. 2 and 3. See also Meyer, "Politics, Personality, and Technology: Airships in the Manipulations of Dr. Hugo Eckener and Lord Thomson, 1919–1930," in *Airshipmen, Businessmen, and Politics*, pp. 186–87.

3. Robinson, *Giants in the Sky*, pp. 303–4; Brooks, *Zeppelin: Rigid Airships*, p. 203, n. 6; Hiam, *Dirigible Dreams*, p. 187; Hartcup, *Achievement of the Airship*, pp. 180, 197–98. Beaty notes that R-100 cost £411,113 and R-101 £717,165. Beaty, *Water Jump*, p. 104.

4. "Offer Bill to Back Merchant Airships," *The New York Times*, December 5, 1930; Bender and Altschul, *Chosen Instrument*, p. 197.

5. When airship enthusiasm was at its height in 1929, United's net income had been $8,966,032, but in 1930 that figure had fallen to $3,302,206. That was bad, but the coming year was forecast to be still worse — as indeed it would be, when United's 1931 numbers plummeted to $2,712,570. "Annual Reports," *Aviation*, May 1931, p. 266; "Financial Reports," *Aviation*, April 1932, p. 191; "Aeronautical Finance," *Aviation*, June 21, 1930, p. 191.

6. Bender and Altschul, *Chosen Instrument*, p. 197; Duggan and Meyer, *Airships in International Affairs*, p. 262, n. 22, on his maneuverings and Hunsaker's report on the affair, pp. 193–94. See also letters from J. R. Parnell, secretary of the PZT, to Trippe, May 19, 1931 (appointment to board) and April 12, 1932 (reappointment), as well as the very friendly letter from Trippe to Hunsaker, December 10, 1931, inviting him to lunch and thanking him for sending a copy of Charles Rosendahl's new book on airships. That Trippe was in receipt of confidential information is beyond doubt: On July 1, 1931, he was sent the PZT Balance Sheet and Report of the Chairman of the Board of Directors (which detailed Litchfield's plans for the coming years in Congress). He also used a variety of underhanded methods to discover more. George Rihl, his dirty-tricks specialist in South America, was detailed to elicit intelligence on Zeppelin operations in Brazil. On September 13, 1933, he sent Trippe a memorandum, "Zeppelin Contract," in which he said that "I could only see this contract for a short time and as yet cannot secure a copy," but he managed to jot down the salient points, including Eckener's payment schedules, expenses, timetable for building an airport, and Zeppelin payments to the Brazilian government. Rihl also saw a letter from the German ambassador to the minister of foreign affairs in Brazil saying that Berlin was subsidizing Zeppelin with 900,000 marks. All of these documents are in Henry Cord Meyer Papers, Box 2, File 3. Hunsaker's, Trippe's, and Litchfield's statements are printed in *Hearing Before the Committee on Commerce, United States Senate*, S.5078, February 6, 1931, pp. 32–38, copy in Henry Cord Meyer Papers, Box 3, File 25. Pan American's financials are in "Profits and Losses," *Aviation*, May 1932, p. 233.

7. P. Lyth, "The Empire's Airway: British Civil Aviation from 1919 to 1939," *Revue Belge de Philologie et d'Histoire* 78 (2000), no. 3, pp. 865–72.

8. Beaty, *Water Jump*, p. 81; Bender and Altschul, *Chosen Instrument*, p. 199.

9. Josephson, *Empire of the Air*, p. 93.

10. On the complexities of the Bermuda/Azores negotiations, see Josephson, *Empire of the Air*, p. 94; Daley, *American Saga*, p. 107; Italiaander, *Ein Deutscher Namens Eckener*, p. 479.

11. For instance, "All-Red Cape-Cairo Complete," *Aviation*, March 1932, p. 144; "Trans-Atlantic Air Mail in Sight," *Aviation*, October 1932, pp. 420–21.

41. The Great Circle

1. "The 'Zep' Did Not Go 'Around the World,'" *Literary Digest*, September 28, 1929, p. 28.

2. J. Duggan, *Airships in the Arctic* (Ickenham, U.K.: Zeppelin Study Group, 2006), pp. 186–92.

3. Vaeth, *Graf Zeppelin*, pp. 113, 118–20; Duggan, *Airships in the Arctic*, pp. 165–67.

4. A full narrative of the flight can be found in Duggan, *Airships and the Arctic*, pp. 212–30; and J. McCannon, "Sharing the Northern Skies: German-Soviet Scientific Cooperation and the 1931 Flight of the *Graf Zeppelin* to the Soviet Arctic," *Russian History* 30 (2003), no. 4, pp. 403–31, which also investigates the Soviet purposes behind the expedition.

5. Eckener (trans. Robinson), *My Zeppelins*, p. 138.

6. Eckener (trans. Robinson), *My Zeppelins*, pp. 132–33, 139. L. Ellsworth and E. H. Smith note that "we were fortunate in being able to take advantage of favorable meteorological conditions" only because Eckener had delayed departure at one point for a few hours. See Ellsworth and Smith, "Report of the Preliminary Results of the Aeroarctic Expedition with *Graf Zeppelin*, 1931," *Geographical Review* 22 (1932), no. 1, pp. 61–82.

7. Before the Arctic flight, Eckener was reported as "being very optimistic over the possibility of establishing an inter-continental air route via the northern regions, and looked to this flight to establish whether safe navigation was possible during the unfavorable Arctic summer, which was characterized by thick fog." Quoted in Duggan, *Airships in the Arctic*, p. 198. Afterward, it was later reported, Eckener said he'd "given up the idea of trying to establish a northern air route to America by way of Greenland because of the fog disturbance." See "Eckener Drops New Route Idea," *The New York Times*, September 2, 1934.

8. Meyer, "Building Rigid Airships," pp. 95–96; Topping, *When Giants Roamed the Sky*, p. 166.

9. Statistics for these flights can be found in J. Duggan and J. Graue, *Commercial Zeppelin Flights to South America: The Commercial South America Flights and Airmails of the Zeppelin Airships* (Valleyford, Wash.: JL Diversified, 1995), pp. 17, 161–62.

10. Davies, *Airlines of the United States Since 1914*, p. 241.

11. Davies, *Airlines of the United States Since 1914*, p. 124; B. Clarke, *Atlantic Adventure: A Complete History of Transatlantic Flight* (London: Allan Wingate, 1958), p. 91.

12. On Newfoundland affairs at this time, see R. A. MacKay, "Foreign Governments and Politics: Newfoundland Reverts to the Status of a Colony," *American Political Science Review* 28 (1934), no. 5, pp. 895–900; J. Overton, "Economic Crisis and the End of Democracy: Politics in Newfoundland During the Great Depression," *Labour/Le Travail* 26 (1990), pp. 85–124; H. B. Mayo, "Newfoundland's Entry into the Dominion," *Canadian Journal of Economics and Political Science/Revue Canadienne d'Economique et de Science Politique* 15 (1949), no. 4, pp. 505–22; D. Mackenzie, "Ireland, Canada, and Atlantic Aviation, 1935–45: A Comparative Study," *Canadian Journal of Irish Studies* 18 (1992), no. 2, pp. 31–47. Trippe's agent's cable is quoted in Daley, *American Saga*, pp. 134–35.

13. Beaty, *Water Jump*, pp. 88–89, 92; Bender and Altschul, *Chosen Instrument*, pp. 211–12; Josephson, *Empire of the Air*, pp. 95–97.

14. Lindbergh, *Autobiography of Values*, pp. 108–9, 113, 115. "Merchant Aerial," *Time*, July 31, 1933; for Lindbergh's report to Trippe, see Daley, *American Saga*, p. 129. Anne Morrow Lindbergh describes the flight in detail in her article "Flying Across the North Atlantic," *National Geographic* 66 (1934), no. 3 (September), pp. 261–337, which is accompanied by spectacular photographs.

42. Master of Ocean Aircraft

1. "The Sikorsky S-40 Amphibian," *Aviation*, October 1931, pp. 594–98; "Putting Luxury in the Air," *Literary Digest*, January 9, 1932, p. 34; "Hailing the *American Clipper*, Our Biggest Plane," *Literary Digest*, October 24, 1931, pp. 36–37; Bender and Altschul, *Chosen Instrument*, pp. 187–99; Holland, *Architects of Aviation*, p. 188; Daley, *American Saga*, p. 95.

2. Davies, *Airlines of the United States Since 1914*, p. 243; Daley, *American Saga*, p. 109; Josephson, *Empire of the Air*, p. 95 (inaccurately dates the letter to August 15).

3. Daley, *American Saga*, pp. 101–4; Josephson, *Empire of the Air*, p. 81. W. B. Courtney, "Europe—On the Nose!" *Collier's*, December 1, 1934, pp. 10–11, 40–43, has a picturesque narrative of his flight on the route aboard the *Caribbean Clipper*, which includes his "lunch in South America" line and the description of the run as "Prep School." On Sikorsky's flying skills, see Bender and Altschul, *Chosen Instrument*, p. 187.

4. On cost comparisons, see Davies, *Pan Am*, p. 39.

5. On the sale to United, see Cochrane, Hardesty, and Lee, *Aviation Careers of Igor Sikorsky*, p. 99.

6. R. Burlingame, "From Barnstorming to Bombers," *Popular Science*, September 1941, pp. 51–58; "Giants and Near-Giants," *Literary Digest*, October 19, 1935, pp. 28–29; obituary, "Glenn L. Martin Dies of Stroke," *The New York Times*, December 5, 1955; "Tomorrow's Airplane," *Fortune*, July 1938, p. 88.

7. Beaty, *Water Jump*, pp. 113–14; Mayo's report is quoted in R. K. Smith, "The Intercontinental Airliner and the Essence of Airplane Performance, 1929–1939," *Technology and Culture* 24 (1983), no. 3, p. 437.

8. Statistics for these flights are catalogued in Duggan and Graue, *Commercial Zeppelin Flights to South America*, pp. 24–25.

9. Stamp income is discussed by Henry Cord Meyer, "How Philatelists Kept the Zeppelin Flying," printed in Italiaander, *Ein Deutscher Namens Eckener*, pp. 302–5. See also K. B. Stiles, "*Hindenburg* on Stamps," May 16, 1937, *The New York Times*, which covers the history of issues.

10. Wordsmith (pseud.), "Notes on the International Zeppelin Transport Co. and Pacific Zeppelin Transport Co.," part 1, p. 5. Eckener was particularly cagey when he addressed the Federal Aviation Commission on the subject of his financials in October 1934. Of the South American flights, he claimed that "the *Graf Zeppelin*, when fully booked, already now nearly covers its operating cost." Note that he did not use the phrase "makes a profit" and omitted to clarify that he was referring only to the latest year's flights; also worth noting is that the *Graf Zeppelin* was never "fully booked." See Eckener, "The Rigid Airship and Its Possibilities," *The Airship* 2 (1935), no. 5, p. 8 (for the "safety and regularity" argument, p. 5).

11. Wordsmith (pseud.), "Notes on the International Zeppelin Transport Co. and Pacific Zeppelin Transport Co.," part 1, p. 5; Topping, *When Giants Roamed the Sky*, pp. 179–80.

12. Robinson and Keller, *"Up Ship!"* pp. 185–86; W. A. Armstrong, "William A. Moffett and the Development of Naval Aviation," in Leary (ed.), *Aviation's Golden Age*, pp. 60–73. Topping, *When Giants Roamed the Sky*, pp. 181–85, examines the question of human error in the *Akron's* destruction.

43. The Hooked Cross

1. This section based on I. Kershaw, *Hitler: 1899–1936: Hubris* (New York: W. W. Norton & Co., 2000), pp. 493–678; D. Siemens, *Stormtroopers: A New History of Hitler's Brownshirts* (New Haven, Conn.: Yale University Press, 2017); R. J. Evans, "Men He Could Trust," *London Review of Books*, February 22, 2018, pp. 37–39. Eckener's January 1932 radio address is covered in "Bruening Attacks Hitler's Motives," *The New York Times*, January 24, 1932, and reprinted in full (with associated correspondence) in Italiaander, *Ein Deutscher Namens Eckener*, pp. 282–88; for other relevant material (Paul Löbe's quote on fusileered Marxists and heads rolling in sand, for instance), "Germans Uniting Behind Hindenburg," *The New York Times*, January 28, 1932; for Eckener's activities in 1932–33, see Duggan and Meyer, *Airships in International Affairs*, pp. 180–82; German middle-class politics and the German National Association are analyzed in L. E. Jones, "Sammlung Oder Zersplitterung? Die Bestrebungen zur Bildung Einer Neuen Mittelpartei in der Endphase der Weimarer Republik 1930–1933," *Vierteljahrshefte für Zeitgeschichte* 25 (1977), no. 3, pp. 265–304, as well as Italiaander, pp. 289–91. Eckener's bizarre horoscope in the *Neues Deutschland* is reprinted in Italiaander, pp. 295–98. De Syon, *Zeppelin!*, p. 261, n. 11, notes Eckener's absence for much of January.

2. De Syon, *Zeppelin!*, pp. 174–77.

3. For Wittemann and Pruss's Party membership, see De Syon, *Zeppelin!*, pp. 182, 265, n. 66. Owing to numerous shorter flights, Lehmann had actually captained the *Graf Zeppelin* more often than Eckener. Between September 18, 1928, and December 10, 1935: Lehmann had flown 272 times; Eckener, 133 times; Flemming, 34 times; Schiller, 27 times; Wittemann, 23 times; and Pruss, 16 times. Lehmann, *Zeppelin*, p. 311.

4. Vaeth, *Graf Zeppelin*, p. 164; Italiaander, *Ein Deutscher Namens Eckener*, p. 294. Diels, slippery as an eel, later submitted two affidavits *for the prosecution* at Nuremberg, one undated and the other October 31, 1945, printed as Documents 2472-PS and 2544-PS in Office of United States Chief of Counsel for Prosecution of Axis Criminality (ed.), *Nazi Conspiracy and Aggression* (Washington, D.C.: Government Printing Office, 1946), volume 5, pp. 224–26, 288–90. His self-exculpations should be treated with caution. For the development of the Gestapo and Diels, see F. McDonough, *The Gestapo: The Myth and Reality of Hitler's Secret Police* (New York: Skyhorse Publishing, 2017), pp. 14, 20–25.

5. On the Diels meeting and Eckener's incautious remarks, see Vaeth, *Graf Zeppelin*, pp. 158–59, which is based on later conversations with Eckener. For Eckener's phrase about "evil qualities," see note by Rosendahl, May 17, 1945, in Rosendahl Papers, Box 107, Folder 36. Nielsen (trans. Chambers), *The Zeppelin Story*, pp. 208–9, contains some additional details of the meeting but is by no means an authoritative source. Diels's behavior, menacing yet willing to give Eckener a chance to make amends, was certainly curious, but it becomes more explicable when one considers that he had a habit of doing so. At around the same time, for instance, he summoned to his office the Hearst journalist Karl von Wiegand, Eckener's friend and a longtime Zeppelin-watcher, for a little warning chat about a scoop he'd had concerning Hindenburg's will. See W. E. Dodd, Jr., and M. Dodd (eds.), *Ambassador Dodd's Diary, 1933–1938* (London: Victor Gollancz,

1945), entry of August 29, 1933, p. 43. It is said that Dodd's daughter, Martha, was one of Diels's paramours.

6. On the *Graf Zeppelin*'s trips that year, see Duggan, *Commercial Zeppelin Flights to South America*, pp. 31, 35.

7. Quoted in Italiaander, *Ein Deutscher Namens Eckener*, p. 301; Duggan, *LZ-129 Hindenburg*, p. 127.

8. Quoted in W. Dornberger (trans. J. Cleugh and G. Halliday), *V-2: The Inside Story of Hitler's "Secret Weapon" and the Men and Events Behind It That Almost Changed the Course of History* (New York: Viking Press, 1954), p. 66. Once, when Eckener took Dr. Fritz Todt, Hitler's highway builder, for an aerial survey by Zeppelin of the new road system, he asked Todt whether the Führer would ever consider doing the same; Todt replied that he'd already discussed the matter with him and been told he "didn't want to have anything to do" with airships. Eckener (trans. Robinson), *My Zeppelins*, p. 149.

9. Quoted in De Syon, *Zeppelin!*, p. 179.

10. Quoted in Vaeth, *Graf Zeppelin*, p. 163.

11. The most thorough discussion of the Chicago visit is C. Ganz's chapter "Aviation, Nationalism, and Progress," in her *The 1933 Chicago World's Fair: A Century of Progress* (Urbana/Chicago: University of Illinois Press, 2008), pp. 137–50; see also Meister's recollections in Meyer, "F. W. (Willy) von Meister," pp. 202–3. Luther's letter to Hull warning of Communist threats, October 13, 1933, is in the Henry Cord Meyer Papers, Box 2, File 21, as are various related correspondence and the State Department Division of Protocol's "Memorandum Concerning the Visit of the *Graf Zeppelin* and the German Ambassador to Chicago," October 26, 1933.

44. The Ledgers

1. Figures are based on the data in Duggan, *Commercial Zeppelin Flights to South America*, pp. 31–37, 46–50. I have omitted the ninth flight of the *Graf Zeppelin* in 1933, which was a special extended tour for the Chicago World's Fair. On the government subsidy, Meyer, "Problems of Helium and Spy Flights: The Brief Career of LZ-130," in Meyer, *Airshipmen, Businessmen, and Politics*, p. 211.

2. On Arnstein's visit, Topping, *When Giants Roamed the Sky*, pp. 188–89.

3. Memorandum, "The United States of America, Airship Construction, and Airship Service," March 30, 1934, quoted in Italiaander, *Ein Deutscher Namens Eckener*, pp. 319–22. De Syon, *Zeppelin!*, pp. 266–67 and 267, n. 95, calls this document a "historiographic puzzle." Who composed it is unknown. It may have been the work of the SS or the Gestapo, though the Air Ministry might have been the source. Its scarcity (five copies) implies it was heavily restricted. For the recipients, see Henry Cord Meyer's notes, in Meyer Papers, Box 6, File 20, p. 7 (of partial manuscript), n. 20.

4. Eckener (trans. Robinson), *My Zeppelins*, pp. 147–48.

5. See Italiaander, *Ein Deutscher Namens Eckener*, p. 309.

6. R. J. Evans, *The Third Reich in Power* (New York: Penguin Books, 2006), pp. 82–85.

7. Eckener's own thinking on the speech is covered in Italiaander, *Ein Deutscher Namens Eckener*, pp. 307–8, and p. 440, where in a note dated February 15, 1947, Eckener clarified that the speech was "extorted" from him and that the Propaganda people had subtly altered the concluding sentences.

8. Litchfield's lighthearted farewell note is printed in Dick and Robinson, *Golden Age*, p. 24.

9. G. Robinson, "The Rentschlers Fly the Dollar," p. 33, Anon., "Are the Rentschler Boys Good?" p. 31, editorial, "Capitalism at Its Damnedest," p. 29, January 19, 1934, New York Daily News.

10. The Air Mail scandal and Trippe's financial arrangements are covered in Bender and Alschul, Chosen Instrument, pp. 239–41.

11. Cochrane, Hardesty, and Lee, Aviation Careers of Igor Sikorsky, pp. 101–2; Davies, Pan Am, p. 36; "Three Spectacular Events Herald Aviation Progress," Literary Digest, August 11, 1934, p. 39.

12. R. Schwartz, Flying Down to Rio: Hollywood, Tourists, and Yankee Clippers (College Station: Texas A&M University Press, 2004), pp. 293–301, 314; Josephson, Empire of the Air, pp. 88–89; Brown Jr., "Pan Am: Miami's Wings to the World," pp. 151, 153; "Getting Acquainted with Winter Playlands," Literary Digest, December 19, 1931, pp. 26–47; "Rolling Down to Rio—in Five Days," Literary Digest, December 14, 1934, p. 28.

13. I. I. Sikorsky, "The Development and Characteristics of a Long-Range Flying Boat (the S-42)," Journal of the Royal Aeronautical Society 39 (1935), no. 292, pp. 263–81; Smith, "Intercontinental Airliner," pp. 441–42, and Table 2, "Sikorsky S-42B Versus Short S.23."

14. Daley, American Saga, pp. 108, 489n.

15. C. Wilkes, Narrative of the United States Exploring Expedition During the Years 1838, 1839, 1840, 1841, 1842 (Philadelphia: 5 vols., 1849), volume 5, pp. 267–68.

16. Quoted in Daley, American Saga, p. 136.

17. S. Pickering, "Wake Island," United States Naval Institute Proceedings 48 (1922), no. 12, pp. 2075–79.

18. Bender and Alschul, Chosen Instrument, p. 230.

19. Daley, American Saga, p. 139; Bender and Altschul, Chosen Instrument, pp. 231–33.

20. "Transpacific Air Service to Start Soon with Brazilian Clipper Heading Fleet," The New York Times, October 15, 1934.

21. Dick and Robinson, Golden Age, pp. 97–100.

22. M. Lamm, "1932 Maybach Zeppelin Streamlined Limo," Autoweek, June 9, 1997, p. 23; T. Powell, "Grand Classic," Automobile Quarterly 43 (2003), no. 1, pp. 12–24.

23. R. J. Overy, "Cars, Roads, and Economic Recovery in Germany, 1932–8," Economic History Review 28 (1975), no. 3, pp. 466–83.

24. N. Gregor, Daimler-Benz in the Third Reich (New Haven, Conn.: Yale University Press, 1998).

25. D. H. Robinson, The LZ-129 "Hindenburg" (New York: Arco Publishing Co., 1964), "Airship Diesels," no page number.

26. On Eckener's building of LZ-129 with helium rather than hydrogen in mind, see Dick and Robinson, Golden Age, pp. 100–101; Duggan, LZ-129 Hindenburg, pp. 37–38; Meyer, "Problems of Helium and Spy Flights: The Brief Career of LZ-130," pp. 211–12; M. Bauer and J. Duggan, LZ-130 Graf Zeppelin and the End of Commercial Airship Travel (Friedrichshafen, Germany: Zeppelin-Museum, 1996), p. 79.

27. A. Colsman, Luftschiff Voraus! Arbeit und Erleben am Werke Zeppelin (Stuttgart, Germany: Deutsche Verlags-Anstalt, 1933), p. 135.

28. Quoted in Meyer, "F. W. (Willy) von Meister," p. 202.

29. H. Braun, "Das 'Wundergas' Helium: Die US-Amerikanische Innenpolitik und Die Deutschen Zeppeline," Vierteljahrshefte für Zeitgeschichte 53 (2005), no. 4, p. 575, n. 20.

30. "Eckener on Way to Parleys Here," The New York Times, October 12, 1934.

45. The Medusa

1. See correspondence between Henry Cord Meyer, who was friendly with Schiller and Meister, and Chester Lewis (director of the archives of *The New York Times*), dated June 3, June 24, July 8, September 5, 1974, in the Henry Cord Meyer Papers, Box 5, Folder 21. The quotation is Meyer's paraphrasing of Schiller's remarks. Meister said the same thing in an interview with Meyer, reprinted in Meyer, "F. W. (Willy) von Meister," pp. 201–2. Eckener may have first met Ochs in 1929 during negotiations to sell the press rights (which eventually went to Hearst) for the *Graf Zeppelin*'s Round-the-World flight. On the occasion of the 1934 visit, the *Times* gave an enthusiastic review to a little book of photographs and poems extolling Nature by Eckener's daughter, Lotte. It was a rather obscure volume to which to devote so much space, but it kept Eckener in the public eye and gave the Nazis pause with respect to any crackdowns they may have been mulling. For the review, see "New German Poetry and Prose," *The New York Times*, October 7, 1934.

2. "Federal Plan Is Seen to Aid Ocean Air Mail," *The New York Times*, December 9, 1934; "Sea Dirigible Lines Win Federal Help," *The New York Times*, October 30, 1934. Eckener's complete testimony is published in Eckener, "The Rigid Dirigible and Its Possibilities," *The Airship* 2 (1935), no. 5, pp. 1–9, 23. A week earlier, rather clumsily, Eckener and Litchfield had contradicted each other. Eckener, trying to avoid any entanglement in the Air Mail scandal revelations, was quoted as saying that "he depended in no way upon government mail subsidies for the success of his ambitious aim—the beginning of a North Atlantic transport of passengers and mail." His proposition was a "purely business one." Once he received the helium, he would pay for a new hangar in America, start constructing two new 100-passenger airships (LZ-130 and LZ-131) in Friedrichshafen, and have LZ-129 performing demonstration flights to New York by the summer of 1935. The very next day, Litchfield clarified that in fact building two airships in Akron and two in Friedrichshafen for a transatlantic airline would be "feasible" only if the government approved a loan to Goodyear and guaranteed an air-mail contract. "Eckener Seeking Air Terminal Here," *The New York Times*, October 20, 1934; L. D. Lyman, "Ocean Lines to Fly Soon," *The New York Times*, October 21, 1934.

3. "Congress to Get Airship Program," *The New York Times*, November 11, 1934.

4. Figures based on the report *America Builds: The Record of the Public Works Administration* (Washington, D.C.: Division of Information, 1939), pp. 186–87, 191, and Table 13, "Summary of PWA Transportation Projects, Federal and Non-Federal Programs, March 1, 1939," p. 281.

5. On helium, see G. Fulton, "Helium Through World Wars I and II," *Naval Engineers Journal* 77 (1965), no. 5, pp. 733–38; M. D. Reagan, "The Helium Controversy," in *American Civil-Military Decisions: A Book of Case Studies*, ed. H. Stein (Birmingham: University of Alabama Press, 1963), pp. 45–47; Braun, "Das 'Wundergas' Helium," p. 575; Bauer and Duggan, *LZ-130 Graf Zeppelin*, p. 79. Eckener's entire statement is printed in "The Rigid Airship and Its Possibilities," pp. 1–8, 23. See also the Helium Control Act (H.R. 15344), P.L. 69-758, March 3, 1927.

6. Reagan, "The Helium Controversy," p. 46; Braun, "Das 'Wundergas' Helium," pp. 573–74.

7. "Sea Dirigible Lines Win Federal Help," October 30, "Eckener Gets Landing Rights," October 31, "Eckener Praises Air Field," November 4, 1934, *The New York Times*. See also Eckener's letter to Roosevelt, October 29, 1934, printed in Italiaander, *Ein Deutscher Namens Eckener*, pp. 312–13. On Eckener's not applying formally for helium, see A. Krock, "In Washington: A Star Witness on Our Helium Export Policy," May 12, " 'Smart-

Aleck' Talk on Helium Scored," May 13, "German *Hindenburg* Committee Which Arrived Last Night," May 14, 1937, *The New York Times*.

8. H. B. Miller, "The Violent Death of America's Last Dirigible," reprinted in Hedin (ed.), *The Zeppelin Reader*, pp. 230–41, originally published in *True Magazine*, August 1963; "Dirigibles in Disrepute After *Macon*'s Loss," *Literary Digest*, February 23, 1935, p. 8; "Launching a Campaign for New Dirigibles," *Literary Digest*, April 20, 1934, p. 18; "The Future of the Airship Revived," *Literary Digest*, April 7, 1934, p. 50; "Airship Building Will Come to a Halt," *The New York Times*, February 14, 1935. See also Series L 265–73, "Air Transport—Accidents: 1927 to 1945," in *Historical Statistics of the United States, 1789–1945*, p. 225. On German reactions, see "Berlin Impressed by *Macon* Disaster" and "Eckener to Carry Out Flight," *The New York Times*, February 14, 1935. In the summer of 1936, Rosendahl provided updated figures for the *Graf Zeppelin* (up to December 1935): "More than fifty round trips to South America, 111 ocean crossings and 505 flights covering 847,420 miles. She had carried 78,600 pounds of mail, 111,500 pounds of freight and 32,962 persons." See "Rosendahl Urges Dirigibles for U.S.," *The New York Times*, May 9, 1936. On the cost of the U.S. airships, see Robinson and Keller, "*Up Ship!*" Appendix E, "Airships' Performance Data and Chronology," pp. 204–7.

9. "Will Shift Air Terminal," December 28, 1934, "Plan Zeppelin Line Circling the Globe,"January 6, 1935, *The New York Times*.

10. Duggan, *LZ-129 Hindenburg*, pp. 5–8; Eckener (trans. Robinson), *My Zeppelins*, pp. 148–49; Duggan and Meyer, *Airships in International Affairs*, p. 204; De Syon, *Zeppelin!*, pp. 187–88; Vaeth, *Graf Zeppelin*, p. 167. For the photograph, see Bauer and Duggan, *LZ-130 Graf Zeppelin*, p. 18. Eckener's "lackadaisical" saluting is quoted in Italiaander, *Ein Deutscher Namens Eckener*, p. 327. On the Luftwaffe, see "Reich's Air Force Becomes Official," *The New York Times*, March 12, 1935.

11. De Syon, *Zeppelin!*, pp. 179–80; Eckener (trans. Robinson), *My Zeppelins*, p. 158. On rumors of *Adolf Hitler*, see H.-J. Maurer, "A Dissident Nazi: Hans-Jörg Maurer's Würzburg Diary," *Wisconsin Magazine of History* 50 (1967), no. 4, entry of April 16, 1936, p. 367. As early as June 1934, there were rumors the name would be *Hindenburg*; see O. D. Tolischus, "Big Airship to Carry 50," *The New York Times*, June 10, 1934. By the spring of 1935, it was confirmed when *The Airship*, a British periodical, referred to it as such in "News and Notes," *The Airship* 2 (1935), no. 5, p. 22. On Eckener's relief, see B. Fromm, *Blood and Banquets: A Berlin Diary, 1930–1938* (New York: Simon & Schuster, 1992), entry of March 26, 1936, p. 219.

12. Eckener (trans. Robinson), *My Zeppelins*, p. 156.

13. Duggan, *LZ-129 Hindenburg*, pp. 22–30, 47; Dick and Robinson, *Golden Age*, pp. 102–4.

14. There are sparse details available on Arpke, but see de.wikipedia.org/wiki/Otto_Arpke. On Breuhaus, see fritz-august-breuhaus.com/breuhaus-biography.html; W. Lambrecht, "A Flying Hotel," *Scientific American*, June 1934, p. 288; W. R. Storey, "Furniture That Tells Its Own Story," March 23, 1930, and Storey, "Harmony and Color in Table Settings," August 23, 1931, *The New York Times Magazine*. The interiors are covered in Vaeth, "Zeppelin Decor," p. 56, and especially airships.net/hindenburg/interiors/. On the lightness of the chairs, see "Tour of Dirigible an Airy Adventure," *The New York Times*, May 12, 1936; the china is described in Robinson, *LZ-129 Hindenburg*.

15. Duggan, *LZ-129 Hindenburg*, pp. 60–63; Dick and Robinson, *Golden Age*, pp. 105–8.

16. Wagner's process is detailed in his recollections of the event at typografie.info/3/topic/30940-der-hindenburg-schriftzug/. Bismarck, stout old nationalist that he was, had refused to read any document not printed in Fraktur, though a rival font, Antiqua, descended from medieval Latin scripts, was easier to read and was preferred, for political

reasons, by the more cosmopolitan Goethe and Nietzsche. The Nazi government, for its own political reasons, insisted on Fraktur, though in one of Hitler's more bizarre decisions, he would ban Fraktur in 1941 as being a sneaky "Jewish" font, as opposed to Antiqua, which was now seen as representing the true European heritage as defined by the Wehrmacht. In an embarrassing mix-up, the resulting edict, printed in Antiqua, would be circulated on official notepaper still letterheaded in Fraktur.

46. The Labyrinth

1. Dick and Robinson, *Golden Age*, pp. 109–10; De Syon, *Zeppelin!*, p. 181, Duggan, *LZ-129 Hindenburg*, pp. 66–77; Italiaander, *Ein Deutscher Namens Eckener*, p. 309, Eckener's memorandum to the seven captains, pp. 332–33; Eckener (trans. Robinson), *My Zeppelins*, pp. 155–56; Vaeth, *Graf Zeppelin*, p. 169.

2. This section based on Eckener (trans. Robinson), *My Zeppelins*, pp. 151–54, 158–61; Italiaander, *Ein Deutscher Namens Eckener*, pp. 342–43, the identification of Undersecretary Berndt as Goebbels's informant, p. 446; Vaeth, *Graf Zeppelin*, p. 170; Dick and Robinson, *Golden Age*, p. 117; Duggan, *Commercial Zeppelin Flights to South America*, pp. 71–72; Duggan and Meyer, *Airships in International Affairs*, pp. 211, 213; "Nazis Put Eckener in Coventry," *Literary Digest*, April 11, p. 15, "Praise for Eckener Ceases," April 1, "Eckener Refused Election Plea for Hitler; Name Barred from the Press as a Result," April 3, "Is It Envy?", April 4, "Orders Barring Eckener's Name," April 4, "Eckener Read About Ban," April 11, "Eckener Seeks End of Clash with Nazis," April 14, "Eckener Command Unaffected in Row," April 16, "Eckener and Officials Confer," April 19, "Eckener Will Lose Honor of a Street Name," April 23, "Eckener Is Strengthened by Roosevelt Invitation," April 25, 1936, *The New York Times*. For Eckener's ploy to get Roosevelt's attention, see Dodd, Jr. and Dodd (eds.), *Ambassador Dodd's Diary*, entry of April 16, 1936, p. 339; and letter, Dodd to R. Walton Moore, April 18, 1936, in *Franklin D. Roosevelt and Foreign Affairs*, ed. E. B. Nixon (Cambridge, Mass.: Harvard University Press, 3 vols., 1969), volume 3, pp. 290–91. For rumors that Eckener had been executed or had committed suicide, see Lady Drummond-Hay, "Mass Will Be Said in the *Hindenburg*," *The New York Times*, May 6, 1936. In his memoirs, Eckener is curiously forgetful on how many times he'd met FDR, saying of his February 1936 meeting, "I was not yet acquainted with President Roosevelt." See also Eckener (trans. Robinson), *My Zeppelins*, p. 159; Meyer, "F. W. (Willy) von Meister," p. 203. For Eckener's description of Goebbels as Mephistopheles, see Braun, "Das 'Wundergas' Helium," p. 581.

47. Here Be Dragons

1. "Pan American Airways," *Fortune*, April 1936, p. 167; "Aerial 'Causeway' Across the Pacific," *Literary Digest*, March 23, 1935, p. 15; Bender and Altschul, *Chosen Instrument*, pp. 142–43; Holland, *Architects of Aviation*, pp. 185–86; Daley, *American Saga*, pp. 234–35.

2. "China Clipper," *Time*, December 2, 1935, pp. 46–51.

3. Daley, *American Saga*, pp. 157–64; R. Gandt, *China Clipper: The Age of the Great Flying Boats* (Annapolis, Md.: Naval Institute Press), p. 77.

4. Daley, *American Saga*, pp. 165–73; Bender and Altschul, *Chosen Instrument*, pp. 247–51; "Tomorrow's Airplane," *Fortune*, July 1938, p. 64 (Martin's loss of $850,000). On Van Dusen's *Cosmopolitan* article, see quote in "America's Aerial Bridge to the Orient," p. 28.

5. Daley, *American Saga*, pp. 176–80; Beaty, *Water Jump*, pp. 116–17; Bender and Altschul, *Chosen Instrument*, pp. 251–53; "Pan American Airways," *Fortune*, pp. 79–80, 171–74. On delays, see "Clipped Clippers," *Time*, February 24, 1936; for prices and orders, see Davies, *Pan Am*, pp. 42–43; on the 314's toilets, see R. J. Serling, *Legend and Legacy: The Story of Boeing and Its People* (New York: St. Martin's Press, 1991), pp. 38–41. On the pro-airship arguments, see "Is the *Hindenburg* to Be the Final Answer?" *The Airship* 3 (1937), no. 12, p. 50. The M-130 interiors are covered in Hühne, *Pan Am History, Design, and Identity*, pp. 104–7. On the "blue-ribbon" route, see W. B. Courtney, "You Fly the Atlantic," *Collier's*, January 9, 1937, p. 8.

48. Master and Commander

1. Courtney, "You Fly the Atlantic," pp. 8–9.
2. The preceding section is based on Wordsmith (pseud.), "Notes on the International Zeppelin Transport Co. and Pacific Zeppelin Transport Co.," Part 2, p. 4; Duggan and Meyer, *Airships in International Affairs*, p. 211; W. Miller, *I Found No Peace: The Journal of a Foreign Correspondent* (New York: Simon & Schuster, 1936), pp. 306–16; Duggan, *LZ-129 Hindenburg*, pp. 120–21, 125; Robinson, *LZ-129 Hindenburg*, no page numbers, but an excellent resource; M. Burleigh, *Death and Deliverance: Euthanasia in Germany, 1900–1945* (Cambridge: Cambridge University Press, 1994), p. 47; Daley, *American Saga*, p. 183 (*Gone with the Wind*); Fromm, *Blood and Banquets*, p. 204, entry of July 28, 1935; Nixon (ed.), *Franklin D. Roosevelt and Foreign Affairs*, volume 3, p. 291n; on crew routine and food, see E. Bentele (trans. Dixon), *The Story of a Zeppelin Mechanic: My Flights, 1931–1938* (Friedrichshafen, Germany: Zeppelin-Museum, 1992), pp. 61–64; on the Bund, see Dick and Robinson, *Golden Age*, p. 135; on Frankfurt airport, see J. Breithaupt, "Possibilities of Universal Airship Traffic," *The Airship* 3 (1936), no. 11 (Fall), p. 33; L. P. Lochner, "Aboard the Airship *Hindenburg*: Louis P. Locher's Diary of Its Maiden Voyage to the United States," *Wisconsin Magazine of History* 49 (1965–66), pp. 101–21; L. Gardner, "Aboard the *Hindenburg*," *Buoyant Flight* 9 (1962), no. 4, pp. 2–5, has details of his later trip, including a description of the staterooms' lack of drawers and wardrobe space as well as comments on the lack of vibration, noise, and spillage; W. B. Courtney, "Sky Cruise," *Collier's*, May 8, 1937, pp. 12–13, 45–48, mentions uninspired food and the library being all in German; for letters between Eckener and Lehmann, see Italiaander, *Ein Deutscher Namens Eckener*, pp. 338–40. The following articles all appeared in *The New York Times* in 1936: "Airship Schedule Set," May 3; L. D. Lyman, "Rival Wings over the Ocean," May 3; Drummond-Hay, "Zeppelin to Avoid Racing New Liner," May 5; Drummond-Hay, "Mass Will Be Said in the *Hindenburg*," May 6; Drummond-Hay, "*Hindenburg* Begins First U.S. Flight," May 7; "Planes Must Keep from Airship," May 7; "Several Thousand See Take-Off," May 7; "Lakehurst Ready to Dock Zeppelin," May 8; Drummond-Hay, "*Hindenburg* Covers Half Ocean Flight; Is Due Tomorrow," May 8; "Eckener Is Gleeful," May 8; "Broadcast from Zeppelin," May 8; "Radio Greetings Sent from Airship," May 9; "Lakehurst Ground Crews Ready to Moor the *Hindenburg* Quickly," May 9; "Airship Largest, Fastest of Kind," May 9; "Eckener a Symbol of Success in Air," May 9; "Big Decrease in Fare Since *Graf Zeppelin* Trip," May 9; "Lehmann Trained Years for Career," May 9; "Thrilled by Mass in Air," May 10; "Wellman Survivor Hails Air Conquest," May 10; F. R. Daniell, "Crowds View Dirigible; Record Pleases Eckener; Off Tomorrow Night," May 10; "Throngs at Dawn Greet Big Airship," May 10; "8,000 at Field Get Close View of Ship," May 10; "Eckener Belittles His Rift with Nazis," May 10; "Eckener Not

Mentioned in German Flight News," May 10; "Lakehurst Is Glum as Bonanza Fails," May 11; "Zeppelin's Home Ready," May 11; "Luther Stresses Our Aid to Air Line," May 11; "Zeppelin Flies over City, Then Heads for Germany; Racing for New Record," May 12; "Statue of Liberty Blinks Her Salute to Airship," May 12; "Woman, 87, Seeks Thrill on Zeppelin," May 12; "Tour of Dirigible an Airy Adventure," May 12; "Turkeys and Spinach Put Aboard Zeppelin," May 12; "Nazis to Welcome Eckener," May 13; "300,000 Words Wired on Zeppelin," May 14; "Frankfort Honors Eckener with Cup," May 15; O. E. Dunlap, Jr., "Broadcasting from the Sky," May 17; "No Fun Found in Airship," May 23; L. D. Lyman, "Ocean Air Routes Sought by Powers," May 24; "Eckener to Command Flight to U.S. Today," June 19.

The relevant webpages at airships.net/hindenburg/flight-schedule/maiden-voyage/#maiden and airships.net/blog/hindenburg-piano/ are invaluable for reconstructing the passenger manifest and entertainment on board. On Bund activities at Lakehurst, the New York Police Department was already on alert and had sent officers of its "Radical Squad" to keep watch. According to Special Agent in Charge Burr, the Radical Squad "jumped into the fray at an opportune moment and quickly brought order out of the pending chaos." Letter, Burr to Chief Special Agent Bannerman of the Department of State, May 16, 1936, in Henry Cord Meyer Papers, Box 2, File 21.

Regarding Waldeck, sometime later she divorced Ullstein and married a Hungarian count, who divorced her for being Jewish. During World War II, now converted to Catholicism, she would spend much time in Hungary, where American intelligence kept an eye on her for associating with known Nazis, but she also attempted to contact the OSS for recruitment as a spy. The Gestapo was just as watchful, for she was friends with senior officers involved in the abortive July 1944 assassination attempt on Hitler. In that sense, she was a good match for Ritter, one being as opaque and as ambiguous as the other, though she ended up working as an American journalist after the war while he was sentenced to four years in prison at Nuremberg. Memorandum, "(Countess) Rosie Waldek [sic]," at cia.gov/library/readingroom/docs/DOC_0000389121.pdf.

Lochner's editor believes that Eckener's nickname of *Die Giftkröte* referred not to Goebbels but to Hitler, but this is highly unlikely: Eckener had very few dealings with Hitler, and even *he* would not have dared to say such a thing. Further, it was Goebbels who had put him on the no-name list, and the diminutive propaganda minister was commonly believed to be both venomous in character and toadlike in appearance.

3. J. Duggan, *Olympia-Fahrt, 1936* (Ickenham, U.K.: Zeppelin Study Group, 2000), pp. 3–6, 21–23; F. T. Birchall, "100,000 Hail Hitler; U.S. Athletes Avoid Nazi Salute to Him," *The New York Times*, August 2, 1936.

49. Mr. and Mrs. Brown

1. J. A. Sinclair, "Zeppelin News," *The Airship* 3 (1937), no. 12; and Editorial Notes—see table, "Travelling Time Europe–U.S.A. by Boat and Airship" in *The Airship* 4 (1937), no. 15, p. 35.
2. D. Grossman, C. Ganz, and P. Russell, *Zeppelin Hindenburg: An Illustrated History of LZ-129* (Stroud, U.K.: The History Press, 2017), pp. 90–91, 95.
3. Cleveland, "Hindenburg Ends Season," *The New York Times*, October 18, 1936.
4. Topping, *When Giants Roamed the Sky*, pp. 204, 206.
5. "Plans Federal Aid to Build Airships," *The New York Times*, October 11, 1936; Dick and Robinson, *Golden Age*, pp. 140–41.

6. Bauer and Duggan, *LZ-130 Graf Zeppelin*, pp. 28–29, 47, 51–56.

7. "Four Zeppelins to Ply Atlantic Routes by 1940," *The New York Times*, February 27, 1937.

8. "Eckener Offers Airship Aid Here," *The New York Times*, January 10, 1937.

9. Daley, *American Saga*, pp. 181–90, has a full account of the Trippes' around-the-world voyage (for Anne Lindbergh's impression of Trippe, see p. 495n); see also Betty's diary in Trippe (ed. Davies), *Pan Am's First Lady*, pp. 85–121. On the Empires' amazing reclining seats, see Courtney, "You Fly the Atlantic," p. 42; on the sights, see Bentele (trans. Dixon), *The Story of a Zeppelin Mechanic*, pp. 30–32, 51. Trippe's employment of Wronsky in New York seems to have been a favor to his father; see letter, J. C. Leslie to Henry Cord Meyer, November 13, 1975, in Henry Cord Meyer Papers, Box 2, File 3. On Trippe's fear of the airship, see transcript of interview by John C. Leslie with William Van Dusen, October 24, 1972, in Henry Cord Meyer Papers, Box 2, File 3. Trippe also kept tabs on Goodyear. He heard from Woods Humphery that Litchfield had visited London but was assured that Goodyear had no chance of receiving a subsidy or mail payments from the British government if he started an airship line there. "Memorandum of Telephone Conversation Held July 24th, 1934, at 11:30am," in Henry Cord Meyer Papers, Box 2, File 3. Imperial's travails in negotiating with local governments in the Middle East and India, which contributed heavily to its delays, are covered in Berchtold, "Trade Routes in the Sky," p. 17.

10. Memoranda by Eckener, "Zeppelin Plans in North America" and "Zeppelin Enthusiasm in the United States," in the company newsletter, quoted in Italiaander, *Ein Deutscher Namens Eckener*, pp. 342–43.

11. "Airship Group Forming," *The New York Times*, October 23, 1936; Wordsmith (pseud.), "Notes on the International Zeppelin Transport Co. (IZT) and Pacific Zeppelin Transport Co. (PZT)," Part 2, p. 4.

12. Duggan and Meyer, *Airships in International Affairs*, p. 216.

13. On the JWV, email to me from Pamela Elbe of the National Museum of American Jewish Military History, March 1, 2018; on Weinstein's day job (assuming it's the same Weinstein), see "$500,000 Housing Center for Asbury Park Site," *The New York Times*, September 13, 1947. See also "Jewish War Veterans Fight Reich Loans," September 3, 1934, F. T. Birchall, "Berlin Riots Mar Olympic Planning," July 26, 1935, "Dickstein in Clash in House on Reich," July 26, "Reich Curbs Drive on 'Foes' of State to Halt Criticism," July 10, "Jewish Veterans Ask Olympic Ban" September 1, "Even Split in A.A.U. on Olympic Issue," December 7, "Brundage Expects Complete U.S. Representation in Berlin Olympic Events," January 25, "Nazi Code Dwarfs Individual Rights," November 6, "Ship Act Benefits Urged for Airships," December 3, 1936, *The New York Times*. For the text of Roper's letter to Weinstein (and Meister's reply, made in vain), see "Treaty with Reich on Airline Denied," *The New York Times*, November 6, 1936. For Frank Bohn's views, see "Hitler Is Assailed by Non-Nazis Here," *The New York Times*, December 16, 1935. On Goodyear, see Topping, *When Giants Roamed the Sky*, pp. 216–18.

14. Dick and Robinson, *Golden Age*, p. 138.

15. To that end, Meister wrote to R. Walton Moore at the State Department on December 21, 1936, to inform him that AZT was asking permission from the navy to use Lakehurst, and on February 8 DZR formally applied for permission to enter U.S. territory on the basis of that old standby, "a limited number of experimental demonstration flights." Henry Cord Meyer Papers, Box 2, File 21. In spite of the application openly declaring that "passenger capacity" would be 72, DZR could not admit that the *Hindenburg*'s primary purpose was to carry passengers, as that would naturally imply that its primary purpose was to carry passengers, thereby making it *un*-experimental and risking being denied permission. In-

stead, DZR claimed the flights had been "proposed in order to further explore meteorological conditions prevailing [in] the North Atlantic during the Spring, Summer, and Fall seasons." It was precisely from this absurd situation that Eckener had sought to extricate himself. Knäusel, *Zeppelin and the United States of America*, Appendix 7.4, "Application for Permission to Fly the Airship Hindenburg from Germany into United States Territory in 1937," February 8, 1937, pp. 131–37.

16. Duggan, *LZ-129 Hindenburg*, pp. 139–52. On Gardner, see "Aboard the *Hindenburg*," p. 2.

17. "18 Trips This Year for the *Hindenburg*," *The New York Times*, January 13, 1937.

18. G. Warner, "The Flying Boats of Foynes," *History Ireland* 9 (2001), no. 1, pp. 40–44; Daley, *American Saga*, pp. 211–13; Bender and Altschul, *Chosen Instrument*, p. 263; Mackenzie, "Ireland, Canada, and Atlantic Aviation, 1935–45," pp. 31–37.

50. Es Ist Das Ende

1. The literature on the *Hindenburg* fire is enormous. Unless otherwise noted in the text, the following two chapters are based on the sources below.

Of key importance for the movements, transmissions, and cause of destruction of the airship is Acting Chief of the Air Transport Section R. W. Knight (ed.), *The Hindenburg Accident: A Comparative Digest of the Investigations and Findings, with the American and Translated German Reports Included*, Bureau of Air Commerce Safety and Planning Division, Department of Commerce, Report No. 11 (August 1938), reprinted by 7 C's Press; Riverdale, Connecticut, 1976. On Morrison, Nehlsen, WLS, and radio recordings in general, see wlshistory.com/WLS30/; and the intensive research of Michael Biel of Morehead State University, archived at web.archive.org/web/20070814144201/http://members .aol.com/jeff1070/hindenburg.html. Biographies of the passengers and crew may be found in Patrick Russell's extraordinarily comprehensive website, facesofthehindenburg .blogspot.com, and the many relevant pages at Dan Grossman's website, airships.net, are of great interest. Dr. Mark M. Heald's recollections of the "blue flame" were originally quoted in Robinson, *LZ-129*; and his son, former professor of physics Mark. A. Heald, kindly sent me additional correspondence. Patrick Russell also provided invaluable advice on the final minutes of the *Hindenburg*, especially vis-à-vis the subject of electrostatic discharge and the possibility of St. Elmo's fire. On lightning and the *Hindenburg*, see "Zeppelin: Easy Rider in a Hurricane," *Literary Digest*, October 17, 1936, p. 18, and on Pruss's expertise at landings, see Dick and Robinson, *Golden Age*, p. 137. On Lehmann's family, see J. A. Sinclair, "Capt. Ernst August Lehmann," *The Airship* 4 (1937), no. 14, p. 21. Margaret Mather's recollections, "I Was on the *Hindenburg*," *Harper's Magazine*, November 1937, and L. Adelt, "The Last Trip of the *Hindenburg*," *Reader's Digest*, November 1937, pp. 69–72, are extremely useful. On Eckener's movements before and after the fire, see letter, Eckener to Johanna, May 12, 1937, in Italiaander, *Ein Deutscher Namens Eckener*, p. 357; on his visit to Ambrosi and additional details (including his "impaired political health"), see Eckener (trans. Robinson), *My Zeppelins*, pp. 166–68. Arnstein's recollection (dated May 28, 1958) of his talk with Eckener is printed in Italiaander, pp. 349–50. Quotations and opinions from dozens of witnesses and experts (Eckener, for example) are taken from the voluminous Record Group 197 (Records of the Civil Aeronautics Board), Bureau of Air Commerce (Records Relating to the *Hindenburg* Disaster, 1937), "Transcripts of Testimony" (A1/Entry 3), at the National Archives, College Park, Maryland. The FBI case files can be found at "The *Hindenburg*," File No. 62-48190, available at vault.fbi.gov/Hindenburg%20. For reports on the Cuxhaven funeral, see Report, U.S. Naval Attache to Germany, "Funeral Rites at Cuxhaven for German Dead of the *Hindenburg* Accident," May 22, 1937, in Henry Cord Meyer Papers, Box 2, File 21.

The following all appeared in *The New York Times* in 1937:"Disaster Ascribed to Gas by Experts," May 7; "Airship Like a Giant Torch on Darkening Jersey Field," May 7; R. B. Porter, "Ships Falls Ablaze," May 7; "Eckener Grieves over Loss of Ship," May 7; H. O'Laughlin, "Passenger Tells of Escape Jump," May 7; "65 Engaged Passage for Return Voyage," May 7; L. A. Kieran, "Dirigible Touches Ground in Landing," May 7; "Lehmann a Veteran in Zeppelin Field," May 7; "Cause Is a Mystery," May 7; "Experts Lay *Hindenburg* Tragedy to Electrostatic or Backfire," May 8; "Burns Are Fatal to Capt. Lehmann," May 8; "Relief Fund Is Set," May 8; "Eckener Lauds Lehmann," May 8; "Pruss Gets Blood to Combat Burns," May 8; "Germans Found Helium Unfeasible," May 8; "Ship Official Died Trying to Save Wife," May 8; "Eckener Flies to Berlin," May 8; "Lehmann Wrote Views on Airships," May 9; R. B. Porter, "*Hindenburg* Board Starts Hunt Today for Cause of Blaze," May 10; "Rosendahl Is Mystified by the *Hindenburg* Fire; Advised Captain to Land," May 11; "Boy's Miraculous Escape from *Hindenburg* Is Told," May 11; "Funeral Services Today," May 11; "3 Countries Honor *Hindenburg* Dead," May 12; "Capt. Pruss Out of Danger," May 14; "*Hindenburg* Rigger Tells of a Blast," May 18; "Ocean Travelers," May 21; "Germany Pays Tribute to *Hindenburg* Dead; 25 Bodies Received with Military Honors," May 22; "Wife of Captain Pruss Here," May 22; "Eckener Lays Airship Fire to Static and Leaking Gas," May 23; "*Hindenburg* Theory of Eckener Studied," May 24; "Pruss Out of Danger," May 25; "New Static Theory for Dirigible Fire," May 25; "Rosendahl Returns to Sabotage Idea," May 27; "Inquiry on Airship Is Shifted to City," May 28; "Sabotage Rumor on Airship Bared," May 29; W. Kaempffert, "This Week in Science: Defining Static Electricity," May 30; "Germans See 8 Causes for Airship Disaster," June 11; "Sister Ship of *Hindenburg* Under Construction, May Be Somewhat Larger," July 11.

52. Resurrection

1. W. B. Courtney, *Collier's* aviation correspondent, was presumably among the most shocked. He had, with exquisite bad timing, published an article, "Sky Cruise," for the May 8, 1937, issue stating that "it is the firm conviction of this skeptical reporter . . . that only a stroke of war or an unfathomable act of God will ever mar this German dirigible['s] passenger safety record" (p. 46).

2. See Biel's careful reconstruction of the schedule at web.archive.org/web/20070814144201/ http://members.aol.com/jeff1070/hindenburg.html. "Behind the Scenes," *The New York Times*, May 16, 1937, outlines the ban on recordings. On calls to the switchboard, see "Flood of Phone Inquiries Answered by *The Times*," *The New York Times*, May 7, 1937; see also "Films of Disaster Viewed by Millions," *The New York Times*, May 8, 1937.

3. F. S. Nugent, "Newsreels in the News," *The New York Times*, May 23, 1937.

4. Eckener (trans. Robinson), *My Zeppelins*, p. 170; letter, Eckener to Johanna, May 12, 1937, in Italiaander, *Ein Deutscher Namens Eckener*, pp. 357–58.

5. "Sky Horror," *Literary Digest*, May 15, 1937, p. 11; "Germany Shocked by the Tragedy," May 7, "Men Continue Work on the New Zeppelin," May 8 (on Dürr), "New Airships Due," May 8, "Reich Halts Trips of *Graf Zeppelin*," May 10, "Hitler Prohibits Hydrogen Flights," May 12, 1937, *The New York Times*; De Syon, "Bangs and Whimpers: The German Public and Two Zeppelin Disasters, 1908–1937," pp. 176–78. Very interesting is an analysis of German reactions written by George Makinson, the American consul general

in Frankfurt, which was sent to Ambassador William Dodd and the State Department. "Plans of the Zeppelin Company," June 23, 1937, in Henry Cord Meyer Papers, Box 4, File 18.

6. Letter, Eckener to Johanna, May 12, 1937, printed in Italiaander, *Ein Deutscher Namens Eckener*, p. 357; Eckener, *My Zeppelins* (trans. Robinson), p. 168, and for Göring's belief in sabotage, p. 173. Eckener later gave the impression that he had not been on the radio, but the statement was carried on NBC-WJZ and was reported at the time. See "Eckener to Sail for Inquiry Here," *The New York Times*, May 8, 1937.

7. "Sabotage Reports Denied by Embassy," May 8, 1937; Kuhn quoted in "3 Countries Honor *Hindenburg* Dead," May 12, 1937, *The New York Times*.

8. See letter, Eckener to Johanna, May 12, 1937, printed in Italiaander, *Ein Deutscher Namens Eckener*, p. 357.

9. "Goering Grateful for U.S. Rescue Work" and "Hitler Thanks Roosevelt for Sympathy Message," *The New York Times*, May 8, 1937; letter, Göring to Mrs. Rosendahl, June 1937, in Rosendahl Papers, Box 18, Folder 14.

10. Memorandum, "American Commercial Airships," August 1, 1937, in the Henry Cord Meyer Papers, Box 3, Folder 25, pp. 11–12.

11. See "Sky Horror," pp. 12, 10; Series L 265–273, "Air Transport—Accidents: 1927 to 1945," in *Historical Statistics of the United States, 1789–1945*, p. 225; Holanda, *A History of Aviation Safety*, pp. 81–85.

12. "Roper Predicts Sale of Helium by U.S. for 'Commercial' Use on Foreign Airships," *The New York Times*, May 13, 1937; on editorials, see Meister memorandum on "American Commercial Airships," p. 12. See also, for a representative example, "'Useless' Gas," *Literary Digest*, May 29, 1937, pp. 6–7, which is pro-helium by dint of concluding, "Helium means greater airship safety and that means airship travel" and raising no objections to the potential sale to Germany.

13. "Washington Found Sympathetic to the Sale of Helium to Eckener," *The New York Times*, May 20, 1937. Eckener declared, "I am very happy over the noble step taken by President Roosevelt [in setting up the committee]. I knew he was interested in lighter-than-air transport. Now I learn that he is seriously interested in making lighter-than-air craft as safe as possible." Quoted in R. B. Porter, "*Hindenburg* Freed Hydrogen 4 Times," *The New York Times*, May 20, 1937.

14. Quoted in T. H. Watkins, *Righteous Pilgrim: The Life and Times of Harold L. Ickes, 1874–1952* (New York: Henry Holt and Co., 1990), p. 668.

15. "Favors Helium Export," *The New York Times*, May 14, 1937.

16. "3 in Airship Crew Fix Point of Blast," May 19, "Eckener Denies War Aim," May 27, "Bureau to Broaden Air Crash Inquiries; Eckener Asks Congress to Lift Helium Ban," May 28, 1937, *The New York Times*.

17. Eckener (trans. Robinson), *My Zeppelins*, pp. 176–77. Diary entry of June 17, 1937, in H. L. Ickes, *The Secret Diary of Harold L. Ickes* (New York: Simon & Schuster, 3 vols., 1954), volume 2, pp. 145–46.

18. Braun, "Das 'Wundergas' Helium," p. 589.

19. Eckener (trans. Robinson), *My Zeppelins*, p. 177.

20. "Atlantic Air Service to Start with Trial Flights Tomorrow," *The New York Times*, July 2, 1937; R. Owen, "New Era in Travel," *The New York Times*, July 11, 1937; Daley, *American Saga*, pp. 216–17.

21. "An Act Authorizing the Conservation, Production, Exploitation, and Sale of Helium Gas," Section 4, September 1, 1937, P.L. 75-411, S.1567. On Ickes's view, see entry of July 23, 1938, Ickes, *Secret Diary*, volume 2, p. 428. See also Memorandum by L. H. Price,

Office of Arms and Munitions Control, "In Re: Helium Act of September 1, 1937, and Regulations Issued Hereunder," April 8, 1938, in Henry Cord Meyer Papers, Box 2, File 13.

22. On these points, see especially Ickes's notes on a particularly contentious cabinet meeting. Diary entry of May 15, 1938, in Ickes, *Secret Diary*, volume 2, pp. 396–98.

23. The fullest narrative of the helium process is Bauer and Duggan, *LZ-130 Graf Zeppelin*, pp. 81–97; see also memorandum by L. H. Price, Office of Arms and Munitions Control, April 27, 1938, and confidential memorandum, "Interchange of Telegrams on the Subject of Helium," April 22, 1938, in Henry Cord Meyer Papers, Box 2, File 13.

24. Entry, March 19, 1938, in Ickes, *Secret Diary*, volume 2, p. 344, and memorandum by Price, "Re: Helium Act," April 8, 1938.

25. Memorandum by Price, "Re: Helium Act," April 8, 1938.

26. Diary entry, July 23, 1938, in Ickes, *Secret Diary*, volume 2, p. 428. Ickes professed that he was "really glad of a good excuse for disapproving this shipment."

27. Diary entry, March 19, 1938, in Ickes, *Secret Diary*, volume 2, p. 344.

28. Entries of April 17 and April 21, 1938, in Ickes, *Secret Diary*, volume 2, pp. 368–69, 372–73. Ickes's memorandum for the press, "The Military Importance of Helium," July 2, 1938, attached to a memorandum by Joseph Green of the Office of Arms and Munitions Control (who says Ickes was giving a "very false impression"), July 11, 1938, in Henry Cord Meyer Papers, Box 2, File 13. Dr. Issel, a DZR board member, who accompanied Eckener to America as his custodian but was not present at meetings, believed that Roosevelt was playing for votes. See his confidential report, printed in Italiaander, *Ein Deutscher Namens Eckener*, pp. 363–66. On congressional opinion, see letter, Rep. J. G. Polk (Democrat, Ohio) to Cordell Hull, March 28, 1938, registering his "very strong and urgent protest" against the sale, in Henry Cord Meyer Papers, Box 2, File 13. Senator Vandenberg (Republican, Michigan) also introduced a bill preventing helium exports. J. O'Donnell, "Eckener Sent Here to Force Helium Issue," *Washington Herald*, April 24, 1938. On hostility to Ickes, see the editorial accompanying P. H. Wilkinson, "Helium—Hope of the Airship," *Scientific American*, August 1938, p. 72.

29. In Friedrichshafen, Harold Dick informed Litchfield that "there is only one major topic of discussion here and that is helium. As each day rolls around the uncertainty and accompanying unrest grows tremendously." Letter, Dick to Litchfield, April 8, 1938, printed in Dick and Robinson, *Golden Age*, p. 163.

30. Braun, "Das 'Wundergas' Helium," p. 598; M. L. Levitt, "The Development and Politicization of the American Helium Industry, 1917–1940," *Historical Studies in the Physical and Biological Sciences* 30 (2000), no. 2, p. 346; Reagan, "The Helium Controversy," pp. 51–52; Bauer and Duggan, *LZ-130 Graf Zeppelin*, p. 94.

31. Diary entry, May 1, 1938, in Ickes, *Secret Diary*, volume 2, p. 385.

32. Quoted in Reagan, "The Helium Controversy," pp. 52–53.

33. Diary entry, May 12, 1938, in Ickes, *Secret Diary*, volume 2, pp. 391–93. Jackson, quoted in Watkins, *Righteous Pilgrim*, p. 669. Jackson's comment about "legal rubbish" is quoted by Joseph Green of the Office of Arms and Munitions Control, in a temperamental memorandum of May 13, 1938, in Henry Cord Meyer Papers, Box 2, File 13. Hull informed the U.S. Embassy in Berlin of the results of the conference the next day, concluding that "it would appear probable that the proposed sale will not repeat not take place." Telegram, Hull to Embassy, May 12, 1938, in Meyer Papers, Box 2, File 13.

34. Still recovering from an operation, Eckener had even been forced to miss the awarding to him of a Guggenheim Medal for achievement in the aeronautical sciences in December

and had asked Rosendahl to give an acceptance speech in his place. Rosendahl had delivered a cracker of a speech, filled with optimistic predictions as to Zeppelin's glorious future in the sky. Echoing Eckener's words from so many years earlier, Rosendahl proclaimed that the airship and its role in international commerce and transportation would "finally triumph over all that which now separates peoples and nations in an atmosphere of chauvinism." Eckener himself was not quite so sanguine, telling Rosendahl that he could not have done better himself but that he had one criticism: Rosendahl had painted the airship's chances in "somewhat too bright a shade." Letter, Eckener to Rosendahl, January 3, 1938, and "Translation of Dr. Eckener's Speech of Acceptance of the Guggenheim Medal, to Be Delivered Through the Courtesy of Commander Chas. E. Rosendahl," December 17, 1937, in Rosendahl Papers, Box 107, Folder 36. On Eckener's earlier bouts with a poor stomach, see letters to Johanna, May 12 and June 6, 1937, printed in Italiaander, *Ein Deutscher Namens Eckener*, pp. 358, 360. His article was published abroad as "The Zeppelin Crisis," Parts 1 and 2, *The Airship* 5 (1938), nos. 17 and 18.

35. On the Ickes-Eckener meeting, see entry, May 15, 1938, in Ickes, *Secret Diary*, volume 2, p. 399. Meyer, "Problems of Helium and Spy Flights: The Brief Career of LZ-130," p. 217, reprints a confidential report from Thomas Knowles (a Goodyear-AZT official) to Joseph Green, dated May 17, 1938, on the conversation, though Knowles had not been present. Eckener related the story of his damning admission to Louis Lochner some time later, with Lochner adding it as a postscript to his original diary in January 1966; see Lochner, "Aboard the Airship *Hindenburg*," p. 121. Eckener later met Roosevelt for the last time in a five-minute meeting, during which the president repeated Ickes's points. See Issel's report in Italiaander, *Ein Deutscher Namens Eckener*, p. 366. In his memoirs, Eckener gives a distorted account of this meeting. Much of it agrees with Knowles's report regarding their initial head-butting over *could* versus *would*, but Eckener writes that at that point he "stood up, saying, 'Mr. Secretary, I believe there would be no purpose in continuing the conversation,' and I left." This is most unlikely, and it curiously replicates his claim that he did the same in his 1934 conversation with Diels, the head of the Gestapo. If it did happen, then Eckener seems to have been in the habit of stomping off in the middle of important meetings with influential or dangerous men, but there are several accounts (Knowles, Issel, Ickes, even Eckener's own words to Lochman) that belie this version of events. Ickes, for one, never would have missed mentioning a detail like the most famous airshipman in the world cutting him off early and walking out in a huff. Eckener thus stayed until the end of the meeting and it ended disappointingly, but not angrily. Most revealingly, Eckener says on a later page of his memoirs that "privately I had to agree with Secretary Ickes, when he had declared [in the meeting] so emphatically, 'Your Hitler is going to make war!'" It was in this context that he said he could not guarantee that helium would never be used for military purposes, even if on May 8, just a week before his meeting with Ickes, Eckener had given an interview for the Movietone and Paramount newsreels in which he stated, once more, that "I am firmly convinced that under European conditions any military use of . . . helium inflated aircraft of any type is absolutely impossible, and that therefore, certain fears that have been expressed in the United States are fully unfounded." *My Zeppelins*, pp. 179–81. A mimeographed transcript is in the Rosendahl Papers, Box 107, Folder 36.

36. Eckener removed the flag during his Austrian lecture tour in May 1937; see Eckener (trans. Robinson), *My Zeppelins*, p. 166. In the case of Baldur von Schirach of the Hitler Youth, Eckener had dealt with the touchy matter with a wonderfully dry wit that insulted the odious Schirach without openly doing so. The publisher, the Central Office for the

Struggle for Freedom, had sent him a sample copy, inscribed with the blurb, "This marvelous book won't be missing in any company library," only for Eckener to ironically respond that he'd "like to express our considerable doubt that [Schirach] is aware of this way of recommending the book." And then, "P.S. The sample volume is going back to you at the same time as this letter." See Eckener's reply, printed in Italiaander, *Ein Deutscher Namens Eckener*, p. 360.

37. Minutes of meeting, printed in Italiaander, *Ein Deutscher Namens Eckener*, pp. 361–62.
38. De Syon, *Zeppelin!*, p. 270, n. 156. The poll was taken in September.
39. "Tomorrow's Airplane," *Fortune*, July 1938, pp. 53–54, 63–65, 84–88, 91.
40. Italiaander, *Ein Deutscher Namens Eckener*, p. 367.

53. King Across the Water

1. Memorandum, "American Commercial Airships," Schedule I, "Profit and Loss Estimates," p. 15; Schedule VII, "Estimated Cost of Terminal Equipment to Be Owned by Landing-Field Owners," p. 22; Schedule II, "Summary of Estimated Operating Costs," p. 17. For the Boeing 314 operating costs, see "Tomorrow's Airplane," *Fortune*, Table, "Planes of Yesterday, Today, and Tomorrow—A Comparison," p. 86.
2. "Zeppelin Insured for £500,000; Ship Got Low Rate as Good Risk," *The New York Times*, May 8, 1937.
3. On DZR budget estimates, see Bauer and Duggan, *LZ-130 Graf Zeppelin*, p. 175.
4. Eckener, "The Zeppelin Crisis," *The Airship* 5 (1938), nos. 18–20.
5. Bauer and Duggan, *LZ-130 Graf Zeppelin*, pp. 101–2.
6. Bauer and Duggan, *LZ-130 Graf Zeppelin*, p. 101.
7. "Trials of the LZ-130 *Graf Zeppelin*," *The Airship* 5 (1938), no. 19, p. 89.
8. Meyer, "Problems of Helium and Spy Flights," pp. 220–27; Bauer and Duggan, *LZ-130 Graf Zeppelin*, pp. 144–56; memoir by Eduard Boetius, printed in Italiaander, *Ein Deutscher Namens Eckener*, pp. 369–71, who confirms that Eckener was on none of the spy flights, "whose purpose was also completely antithetical to his views." On radar, see my article, "Radar and Air Defence in the 1930s," *Twentieth Century British History* 9 (1998), no. 2, pp. 219–45; and also Rose, "Radar Strategy: The Air Dilemma in British Politics, 1932–1937."
9. Daley, *American Saga*, pp. 225, 228.
10. Quoted in Bender and Altschul, *Chosen Instrument*, pp. 347–48.
11. Daley, *American Saga*, p. 229; Bender and Altschul, *Chosen Instrument*, pp. 289–91.
12. "Plane Christened by Mrs. Roosevelt," *The New York Times*, March 4, 1939; Daley, *American Saga*, pp. 229–30.
13. Daley, *American Saga*, p. 231.
14. "Consolidated Source and Application of Funds Statement, 1930–1939," printed in Daley, *American Saga*, p. 453, but I've used the more precise "Pan American Airways Reported $1,984,438 Profit Peak in 1939," *The New York Times*, May 2, 1940, for 1938's net profit figure. On the Pacific costs, see Benders and Altschul, *Chosen Instrument*, p. 284; on debt, p. 296.
15. Bender and Altschul, *Chosen Instrument*, p. 297; Daley, *American Saga*, pp. 233–37; on Trippe's stock holdings, pp. 500–501n.
16. Daley, *American Saga*, pp. 237–40; Bender and Altschul, *Chosen Instrument*, p. 297–99.
17. "First Passenger Flight Today on Northern Route to England," *The New York Times*, July 8, 1939.

18. Van Vleck, *Empire of the Air*, pp. 101–2; see also L. Weirather, "Romancing the Clipper: America's Technological Coming of Age in Children's Literature," *Studies in Popular Culture* 13 (1990), no. 1, pp. 27–45.

19. M. Pruss, "Visit of Göring on 1st March 1940 to Airship Port, Rhein-Main," printed in Bauer and Duggan, *LZ-130 Graf Zeppelin*, pp. 176–79.

20. Bauer and Duggan, *LZ-130 Graf Zeppelin*, pp. 181–88. Göring's order for scrapping, February 29, 1940, printed in Italiaander, *Ein Deutscher Namens Eckener*, pp. 374–75.

21. "Pan American Airways Reported $1,984,438 Profit Peak in 1939," *The New York Times*, May 2, 1940; Davies, *Pan Am*, p. 43; Bender and Altschul, *Chosen Instrument*, p. 301; on Morgan, see Daley, *American Saga*, p. 253.

22. Daley, *American Saga*, pp. 254–55, 242, 503n; C. A. Lindbergh, *The Wartime Journals of Charles A. Lindbergh* (New York: Harcourt, Brace, Jovanovich, 1970), entry of May 23, 1939.

Epilogue

1. Italiaander, *Ein Deutscher Namens Eckener*, p. 376.

2. Letter, Eckener to Italiaander, March 12, 1940, printed in Italiaander, *Ein Deutscher Namens Eckener*, p. 378; on the box, p. 530.

3. Letter, Eckener to Italiaander, January 25, 1940, printed in Italiaander, *Ein Deutscher Namens Eckener*, p. 377.

4. Eckener was briefly interviewed by RAF Squadron Leader B.J.M. Robinson and Captain T. Chapell of U.S. Ordnance Department regarding Zeppelin's participation in the V-2 program. The interviewee was Hugo, though the document refers to "Dr. Kurt Eckener." See Evaluation Report 38, May 25, 1945, "Interview with Dr. Eckener," Combined Intelligence Objectives Sub-Committee (Washington, D.C.: Office of the Publication Board, Department of Commerce, 1945).

5. On Eckener's views, see letter to Italiaander, December 19, 1941, in Italiaander, *Ein Deutscher Namens Eckener*, p. 379. On the political and legal distinction between pre- and post-1940 appointments as a Leader of the War Economy, see "Decision to Stop Proceedings," July 14, 1948, printed in Italiaander, *Ein Deutscher Namens Eckener*, p. 478.

6. Eckener's affiliations are noted in *Elimination of German Resources for War*, Hearings Before a Subcommittee of the Committee on Military Affairs, United States Senate, Part 5, Testimony of Treasury Department, July 2, 1945 (Washington, D.C.: Government Printing Office, 1945), pp. 766, 771. Oddly, the Austrian directorship is mentioned once, on the earlier page, but not listed in the main entry for Eckener.

7. On November 12, 1945, Eckener addressed accusations that he was a "war profiteer" by pointing out that he had not received "increased income or profits" as the head of Zeppelin; quite the opposite, he had suffered declines owing to his withdrawal "from management of its business dealings." Memorandum by Eckener, printed in Italiaander, *Ein Deutscher Namens Eckener*, p. 420.

8. Letter, Eckener to Italiaander, February 9, 1940, printed in Italiaander, *Ein Deutscher Namens Eckener*, p. 378.

9. Letters to Johanna, November 5 and November 6, 1942, printed in Italiaander, *Ein Deutscher Namens Eckener*, pp. 381–83.

10. Letters, Eckener to Dr. Louis Kiep, July 7 and August 8, 1943, printed in Italiaander, *Ein Deutscher Namens Eckener*, pp. 383–84.

11. Letter, Eckener to Kiep, July 7, 1943, printed in Italiaander, *Ein Deutscher Namens Eckener*, p. 383.

12. Italiaander, *Ein Deutscher Namens Eckener*, p. 407.

13. Letters, Lotte to Paul Simon (husband), April 26, 1944, and Eckener to the (French) Military Government, May 14, 1945, printed in Italiaander, *Ein Deutscher Namens Eckener*, pp. 401, 408.

14. Letters, Lotte to Simon, July 20 and 21, 1944, printed in Italiaander, *Ein Deutscher Namens Eckener*, pp. 401–2.

15. Italiaander, *Ein Deutscher Namens Eckener*, pp. 384, 401.

16. Memorandum by Dr. Karl Schmid, "Notes About a Meeting with Major Lasnier from the *Section Technique*," June 23, 1945, printed in Italiaander, *Ein Deutscher Namens Eckener*, pp. 411–12; Memorandum by Eckener, July 3, 1945, in Italiaander, p. 413.

17. Letter, Eckener to French Military Government, February 2, 1946, printed in Italiaander, *Ein Deutscher Namens Eckener*, pp. 427–29.

18. Obituary, Charles Rosendahl, *The New York Times*, May 15, 1977; letter, Meister to Eckener, January 8, 1947, printed in Italiaander, *Ein Deutscher Namens Eckener*, p. 439. On Rosendahl as a "fanatical believer," see letter, Eckener to Johanna, May 8, 1947, printed in Italiaander, p. 451.

19. "Goodyear, Eckener Plan Airship Work," *The New York Times*, April 25, 1947; letter, Eckener to Johanna, April 2, 1947, printed in Italiaander, *Ein Deutscher Namens Eckener*, p. 447.

20. Letters, Eckener to Johanna, May 1, May 8, and May 12, 1947; on Wiegand and Hearst, letters of October 10 and 11 to Johanna, printed in Italiaander, *Ein Deutscher Namens Eckener*, pp 449–53, 461. "German Expert Visiting Naval Airship Chief," *The New York Times*, May 2, 1947.

21. Letters, Eckener to Johanna, July 12 and 13, and October 10, 1947, printed in Italiaander, *Ein Deutscher Namens Eckener*, pp. 456–57, 461.

22. Wordsmith (pseud.), "Notes on the International Zeppelin Transport Co. and Pacific Zeppelin Transport Co.," Part 2, p. 8.

23. Letter, Eckener to Johanna, September 7, 1947, printed in Italiaander, *Ein Deutscher Namens Eckener*, p. 460.

24. On the classifications, see U.S. Forces, European Theater, "Removal of Nazis and Militarists," July 7, 1945, in *Denazification*, Report of the Military Governor (April 1, 1947–April 30, 1948), No. 34 (Office of Military Government for Germany [U.S.], Annex C, pp. 31, 34.

25. Italiaander, *Ein Deutscher Namens Eckener*, p. 466.

26. J. H. Herz, "The Fiasco of Denazification in Germany," *Political Science Quarterly* 63 (1948), no. 4, p. 584. On Eckener's savings, see letter to Meister, June 24, 1948, in Italiaander, p. 488.

27. U.S. Forces, European Theater, "Removal of Nazis and Militarists," p. 36.

28. "Decision to Stop Proceedings," July 14, 1948, printed in Italiaander, *Ein Deutscher Namens Eckener*, pp. 478–79. On the process in general, letters of support, etc., Italiaander, pp. 467–78. For Eckener's suspicions of Communists, see letter to Wiegand, March 1948, pp. 472–73.

29. D. Middleton, "Then and Now," *The New York Times*, March 22, 1953.

30. Italiaander, *Ein Deutscher Namens Eckener*, pp. 480, 483. On Eckener's request for food, see letter to Meister, undated, in Italiaander, p. 516.

31. See letters, Eckener to Meister, December 13, 1951, January 13 and April 30, 1952, in Italiaander, *Ein Deutscher Namens Eckener*, pp. 500–503.

32. Quoted in Italiaander, *Ein Deutscher Namens Eckener*, p. 514.

33. Recollection of Lotte Eckener, in Italiaander, *Ein Deutscher Namens Eckener*, p. 523; Middleton, "Then and Now," March 22, 1953.

34. Obituary, "Hugo Eckener Dies; *Graf Zeppelin* Pilot," *The New York Times*, August 15, 1954.

Index

ALEXANDER ROSE's previous books include *Men of War, American Rifle, Kings in the North,* and *Washington's Spies,* adapted into AMC's drama *Turn: Washington's Spies,* for which he served as a writer and producer.